# Contemporary Climatology

## Second Edition

Peter J Robinson & Ann Henderson-Sellers

Longman

**Pearson Education Limited**
Edinburgh Gate, Harlow
Essex CM20 2JE
England

*and Associated Companies throughout the world*

*Visit us on the world wide web at:*
*www.pearsoned-ema.com*



First published 1986
Reprinted 1987, 1989, 1991 (twice), 1992, 1994, 1996 (twice)
Second edition published 1999
Second impression 1999

ISBN 0 582 27631 4

**British Library Cataloguing-in-Publication Data**
A catalogue record for this book is available from the British Library

**Library of Congress Cataloging-in-Publication Data**
A catalog record for this book is available from the Library of Congress.

Typeset by 35 in 9½/11pt Times
Produced by Pearson Education Asia Pte Ltd.
Printed in Singapore (KKP)

Contemporary Climatology

# Contents

# Preface to the second edition

There have been great advances in the study of the climate of the Earth in the decade or more which has passed since *Contemporary Climatology* first appeared. No longer are we amazed by the range of information provided by satellites; we take it for granted and use it routinely. No longer do we regard computer models of the climate system as the specialised research tool of a small group. Such models are routinely used in climatic analyses. The most tangible result of these advances is our ability to make specific forecasts for climatic conditions several months in advance. In parallel with the scientific breakthrough that this signifies have been advances in our understanding of the potential impacts of climate change and variability on human activities and of methods which can be used to apply the new information for the solution of practical problems. In all of these areas our increasing knowledge is itself generating new questions and challenges. Thus we have advanced, but the objectives of the present work must remain unchanged from those of the first edition: we set out to write a book for undergraduates about the Earth's climate which reflects the various disciplines involved, provides the basic factual information, suggests ways in which this information can be used and indicates where the challenges, and excitement, lie. We have tried to emphasise the importance of climatic information without making its acquisition and understanding seem too daunting. The fundamental physics of climate has not changed, and therefore much of the early part of the book remains unchanged from the earlier edition. However, once we start to build on that foundation, there is a vast amount of new information. We have had to be highly selective.

The Bibliography provides not only the complete citations for all figures which are not original to this book, but also a select list of references designed to provide an introduction to the literature. The list is an eclectic mix of classic papers, lucid analyses, interesting sidelines and current research. We have also provided, as before, a glossary of common meteorological and climatological terms, and an appendix listing measurement units in the Système Internationale (SI). A second appendix gives pointers to a few selected sites on the Internet which contain climatological data or information. As an aid to clarity, we have also separated consideration of both the application of climate information and the measurement of climatic parameters into a series of side-bars.

The selection of material has been guided by discussions with numerous colleagues. Their comments, especially about what really constitutes the contemporary outlook on climatology, have been stimulating and thought provoking. Special thanks are due to colleagues and students at the University of North Carolina at Chapel Hill who, wittingly and unwittingly, helped to test ideas and concepts through our lecture notes and questions. Colleagues in the Climate Prediction Center of the US National Oceanic and Atmospheric Administration answered many questions about long-range forecasting, while the staff at the National Climatic Data Center, Asheville, NC, and at the National Weather Service Forecast Office in Raleigh, NC, continue to answer any and all questions with courtesy and wit. We thank the students at Macquarie University, past and present, and especially Dr Kendal McGuffie of the University of Technology, Sydney. We are also grateful to friends and colleagues at the Bureau of Meteorology Research Centre in Melbourne for re-reading text and answering questions. We reiterate our thanks to Mr L. Dent, Principal Meteorological Officer, Manchester Airport, and Mr Kiff at the Meteorological Office Training College, Shinfield Park, for allowing us to photograph their equipment. We are grateful to all who have granted permission to reproduce copyright material, which is duly acknowledged in the appropriate location. As before, this book would not have been completed without the constant encouragement, comments and help of our spouses. We thank Shirley and Brian for their many efforts.

Peter J. Robinson
Department of Geography,
University of North Carolina
and
Ann Henderson-Sellers
Australian Nuclear Science & Technology Organisation
14 September 1998

# Preface to the first edition

The study of climate has always been challenging because it draws upon many disciplines. We set out to write a book for undergraduates about the Earth's climate which reflects the various disciplines involved, provides the basic factual information, suggests ways in which this information can be used and indicates where the challenges, and excitement, lie. We have tried to emphasise the importance of climatic information without making its acquisition and understanding seem too daunting. To assist those who are less well acquainted with meteorological and climatological terminology there is a glossary at the end of the book. Information about Système Internationale (SI) units is given in an appendix.

People who study or need to be able to understand the climate come from a wide range of backgrounds with a large variety of motives. Indeed, one of the major difficulties which has beset the study of climate has been the naming of the people who do it! Speaking of climatology in 1978 in an address reported in the *Bulletin of the American Meteorological Society* (**60**, 1171–1174), Professor Kenneth Hare said, '. . . you hardly heard the word professionally in the 1940s. It was a layman's word. Climatologists were the halt and the lame . . . in the British service you actually had to be medically disabled in order to get into the climatological division . . . It was clearly not the age of climate. Now it is. It's the respectable thing to do . . . This is obviously the decade in which climate is coming into its own.' We hope that this text will encourage our readers in their quest for climatological excellence.

Ann Henderson-Sellers and Peter J. Robinson
Departments of Geography
Universities of Liverpool
and North Carolina
8 November 1984

# Acknowledgements

We are grateful to the following to reproduce copyright material:

The American Association for the Advancement of Science for figures 12.5, 12.12c and 13.13; the American Geophysical Union for figure 12.9; the American Meteorological Society for plate 4 and figures 1.3, 2.8, 3.8, 3.A.2, 4.A.1, 7.4, 7.7, 10.5a, 10.5b, 10.9, 11.10, 13.5, 13.11, 13.12, 13.A.1 and 14.I.1; the Association of American Geographers for figure 5.10; Cambridge University Press for Plates 1 and 3, and for figure 14.13; Cambridge University Press and NASA Langley Research Center for plate 2 and figure 2.11; the Crop Science Society of America for figure 2.9; William Dawson & Sons for figures 11.3a, b and c, and 11.4; WH Freeman & Co for figure 4.11; the Geological Society of America Inc. for figure 12.A.1; Harcourt, Brace & Co Ltd (Academic Press) for figures 5.1, 5.4, 5.8, 6.7, 6.9, 6.11, 7.5, 7.6, 8.14, 8.20, 9.4, 10.2, 10.15a, 10.A.1 and 10.A.3; the NASA Langley Research Center for Plate 1 and figure 2.11; Harvard University Press for figure 10.13; Wilmont Hess for figure 1.8 and 10.6, Hodder Headline (Edward Arnold) for figures 2.7 and 4.3; Malcolm Hughes and Keith Briffa for figure 12.2; the Intergovernmental Panel on Climate Change for plates 6 and 7, and for figures 13.10, 14.1, 14.8, 14.9, 14.10 and 14.12; Kluwer Academic Publishers for figures 12.13 and 14.7; David Robinson for figure 9.23; Macmillan Journals for figure 14.14; Enslow Publishing for figure 12.10; McGraw-Hill for figures 5.2, 5.5 and 6.10; TML Wigley and the National Center for Atmospheric Research for figure 12.6; the National Climatic Data Center for figure 9.I.1; the New York Academy of Sciences for figure 12.12a; the National Water & Climate Center for figure 3.A.3; the National Gallery for figure 12.7; Oxford University Press for figure 9.3; Prentice-Hall Inc, Upper Saddle River, NJ for figures 3.4, 8.4 and 8.6; HMSO for figure 5.I.2; NASA for plates 2 and 3; the Royal Meteorological Society for figures 2.I.1, 8.16, 9.16, 9.A.1, 12.1, 12.I.2, 14.6 and 14.15; the University of Chicago Press for figures 3.2, 3.3, 5.11, 7.2, 10.1, 10.10, 11.8 and 11.9; the University of Dundee for figures 9.7 and 9.11; the University of North Carolina Press for figures 3.A.1 and 3.A.3; Fredrich Vieweg & Sohn for figure 10.13; John Wiley & Sons Inc. for figures 1.8, 7.8, 7.9 and 10.6; and the World Meteorological Organization for figure 1.5.

Although every effort has been made to trace owners of copyright material, in a few cases this has proved impossible and we take this opportunity to offer our apologies to any copyright owners whose rights we may have infringed.

# List of symbols

All symbols used to represent constants and variables are defined at their first occurrence in the text. A limited number are used in other textual locations separate from this definition and these are collected here for easy reference.

*Roman*

|  |  |
|---|---|
| $A$ | albedo |
| $c_p$ | specific heat at constant pressure |
| $C$ | specific heat |
| $C^*$ | conductive capacity |
| $E$ | energy |
| $E^*$ | radiant energy |
| $g$ | acceleration |
| $G$ | heat flux into the ground |
| $H$ | sensible heat flux |
| $K^*$ | thermal diffusivity |
| $K$ | thermal conductivity |
| *or* $K$ | solar radiation ($K\downarrow$ = downward, $K\uparrow$ = upward) |
| $L$ | long-wave (terrestrial) radiation ($L\downarrow$ = downward, $L\uparrow$ = upward) |
| *or* $L$ | latent heat of vaporisation of water |
| $LE$ | latent heat flux (so defined because it equals the product of $L$ and rate of evaporation) |
| $p$ | pressure |
| $P$ | precipitation |
| PET | potential evapotranspiration |
| PWV | precipitable water vapour |
| $Q^*$ | net radiative flux at the surface |
| $R\downarrow$ & $R\uparrow$ | net incoming and outgoing planetary radiation |
| $S_F$ | solar (flux) constant (= 1370 W $m^{-2}$) |
| $S$ | instantaneous top-of-the-atmosphere solar flux (= $S_F/4$) |
| $t$ | time |
| $T$ | temperature |
| $T_d$ | dew point temperature |
| $V_g$ | geostrophic wind |
| $z$ | height (in the atmosphere) |
| $Z$ | solar zenith angle |

*Greek*

| | |
|---|---|
| $\gamma$ | environmental lapse rate |
| $\Gamma_d$ | dry adiabatic lapse rate (DALR) |
| $\Gamma_s$ | saturated adiabatic lapse rate (SALR) |
| $\Delta$ | indicates a small change in the associated variable (e.g. $\Delta T$ = small change in temperature) |
| $\varepsilon$ | emissivity |
| $\phi$ | potential temperature |
| $\lambda$ | wavelength (when a subscript indicates occurrence at a specific wavelength) |
| $\rho$ | density |
| $\sigma$ | Stefan–Boltzmann constant (= $5.67 \times 10^{-8}$ W $m^{-2}$ $K^{-4}$) |
| *or* $\sigma$ | standard deviation |
| $\tau$ | optical thickness (of atmosphere or cloud) |
| $\theta$ | latitude |
| $\Omega$ | angular rotation rate of the Earth |

CHAPTER 1

# The scope and controls of the climate

The envelope of air that surrounds the Earth affects us in many ways as we go about our day-to-day activities. Sometimes we respond to it almost unconsciously, as when we choose the type of clothes we will wear. At other times a conscious decision is needed: do we carry an umbrella today? On a longer time scale, our houses reflect the influence of climate, since, if winters are likely to be cold, we install a heating system. To alleviate hot summers, industrial societies install air conditioning, while non-industrial societies select building sites and designs that allow natural cooling. We predict future conditions when we decide what to plant in our gardens or fields, or when we schedule the time and place of our vacation. Institutions, as well as individuals, are influenced by the atmosphere. An electricity generating company must ensure that it has enough capacity to meet the demand on the coldest, or hottest, day. A water supply authority must plan so that it has enough storage to supply the needs during a long drought. A construction firm must determine the strongest winds likely to be encountered in order to build structures which can resist them.

Such problems, local and practical, and almost as old as civilisation itself, are climatological problems. More recently, climatological problems that are less visible and obvious, but more widespread, have arisen. These are associated with the human activities which are altering the composition of the Earth's atmosphere and the nature of its surface, and which may be leading to changes in the climate of the whole globe. Solutions to these problems, whether long-standing and local, or newer and global, require predictions of future conditions. The provision of such predictions, through increased understanding of atmospheric processes, fuller descriptions of the planet's climate, and better specification of the links between human activities and climate, is the aim of contemporary climatology.

As the range of climatological questions increases, so does the number of observations and analysis tools available to try to answer them. For many years the prime function of climatology was to synthesise the many observations of the 'elements' that constitute climate to provide a description of the varied climates of the Earth's surface, and to analyse the results to gain insight into the processes controlling climate. Although the observations were mainly of surface conditions, and therefore could lead to only partial explanations of climate, much of the information we commonly associate with climate – monthly average temperature and monthly total precipitation, for example – derives from these observations. The scope of climatology over the last few decades has increased immensely. Satellite observations, providing three-dimensional and global coverage, have literally and figuratively transformed our viewpoint. Almost at the other extreme, the recent development of enhanced radar systems is now allowing analysis of clouds, wind and rain in very fine detail. These observational advances have forced an explicit realisation that there is a *climate system* in which the climate of a particular place is constantly changing and dependent not only on the climate of all other places on Earth, but also on the changes that are taking place in the oceans, within the Earth's snow and ice cover, and on the land itself. This realisation has also stimulated advances in our understanding of the climate system. These advances are reflected in the development of climatic models. These models, usually couched in terms of mathematical equations expressing the physical laws governing atmospheric behaviour, are beginning to allow us to understand the causes of climate distributions, variations and changes, and to predict the future course of climate.

The recent rapid increase in climatological knowledge has been accompanied by an increase in public awareness of and concern with climate. Most of this stems from 'global warming' and the possible consequences, usually regarded as detrimental to society, associated with climate change. The concern extends from this global, long-term phenomenon to many shorter time and smaller area events, whether droughts in the Sahel or Australia, floods in India or the American Midwest, high winds in Britain, an increase in hurricane activity in the eastern United States, or lack of snow-pack in the polar regions. Climatologists are being called upon to address all of these concerns, and increasingly are being expected to provide the appropriate climate predictions.

A major aim of contemporary climatology is clearly to 'predict' future climatic conditions. These predictions may involve conditions a few months or years ahead in a specific locality, or those for a time far in the future and covering a major portion of the globe. The present state of development of climatology is such that we are far from providing definitive answers in either of these conditions. Nevertheless, it is clear that significant progress is being made. It is the aim of this book to indicate the present state of our knowledge, point out where further work is possible and suggest areas where further basic research is needed.

## 1.1 The science of climatology

The science of climatology is founded upon the observation, description and explanation of the physical properties of the Earth's atmosphere. In general, these physical properties are the familiar climatic elements such as temperature, precipitation, wind, humidity or clouds. The aim of the science is to explain, and eventually predict, these properties. As our understanding and explanation of the observations become more realistic we find that much information and many techniques from other scientific disciplines must be added to the original emphasis on physics (Figure 1.1). This information comes from the traditional allied sciences of chemistry and biology, from the other Earth sciences of geography, geology and oceanography, and, as the linkage between humans and climate becomes more apparent, from the social sciences and in many cases from disciplines such as history, normally regarded as part of the humanities. Nevertheless, throughout this book we shall place emphasis on the climate as a physical system, invoking other disciplines as necessary.

The atmosphere is a body of matter that is constantly in motion. The scales of the motion can range from the molecular, creating heat, to the global,

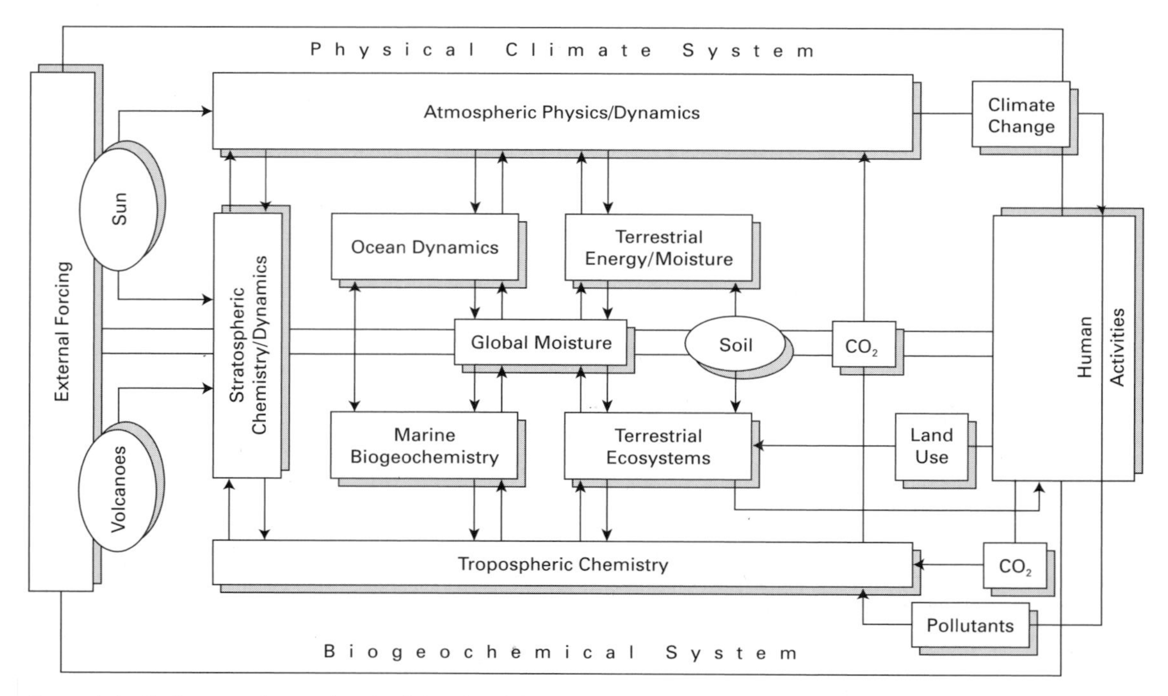


*Figure 1.1* Factors creating and maintaining the global climate system. Understanding of many of the activities within the various symbolic rectangles must involve many scientific disciplines.

creating the wind systems of the Earth. These motions, on all scales, themselves lead to modifications in the structure and composition of the atmosphere, most notably in the cycling of water and water vapour, which leads to cloud formation and precipitation. All of these motions and their effects are part of climatology and will be considered in detail subsequently. However, as an organising framework for the whole of climatology it is advantageous to use the concept of the energetics of the atmosphere.

The source of energy for all atmospheric motions is the Sun. Energy from the Sun passes through the atmosphere to the Earth's surface. During its passage a little energy is absorbed and leads to atmospheric heating, but most of the energy is absorbed at the surface. This surface warms and in turn heats the overlying air, so that the Earth's surface becomes the main source of heating for the atmosphere. The amount of heating depends greatly on the type of surface as well as the time of day and year. Thus it varies spatially and temporally. The unequal distribution of heat leads directly to the horizontal motions we know as winds, and to the vertical motions which create clouds and precipitation. Eventually the energy that has been received from the Sun and has taken part in the various activities within the atmosphere is returned to space. Hence the climate as we know it can be viewed as a series of energy transformations and exchanges within and between the atmosphere and the underlying surface. These exchanges and transformations act in such a way as to distribute energy over the globe and to maintain an energy balance by returning as much energy to space as is received from the Sun.

The solar energy which drives the climate system changes on a variety of time scales. Consequently the climate also changes. The day-to-day variability, the seasonal shifts, and the differences between one year and the next are all obvious parts of climate, and may have consequences for human activity. Although not as readily detected by human senses, but clearly apparent from various types of historical records, longer-term changes are equally common, and may be equally important from the human perspective. The long-term global rise in temperature over the last 100 years (Figure 1.2), for

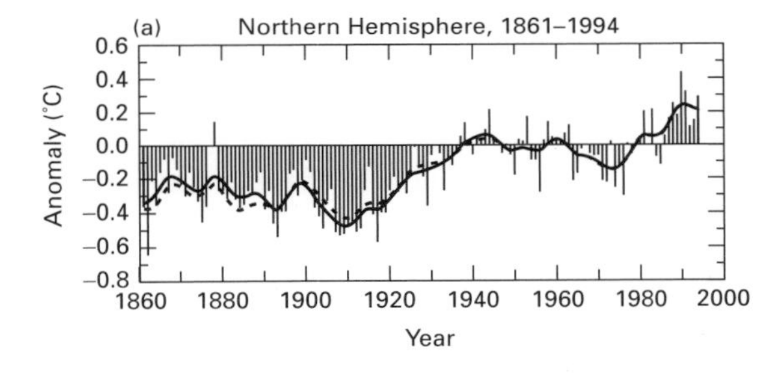


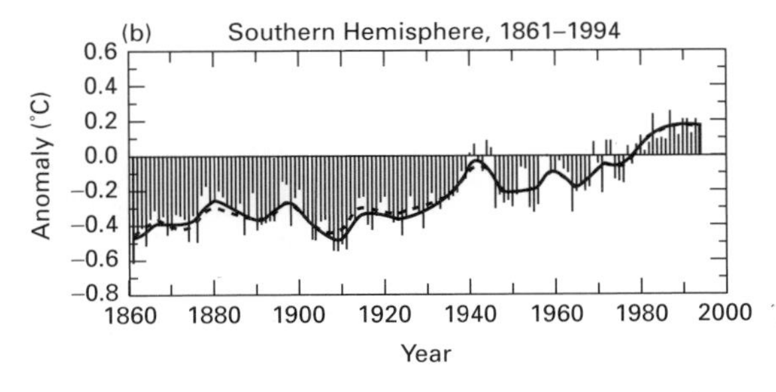


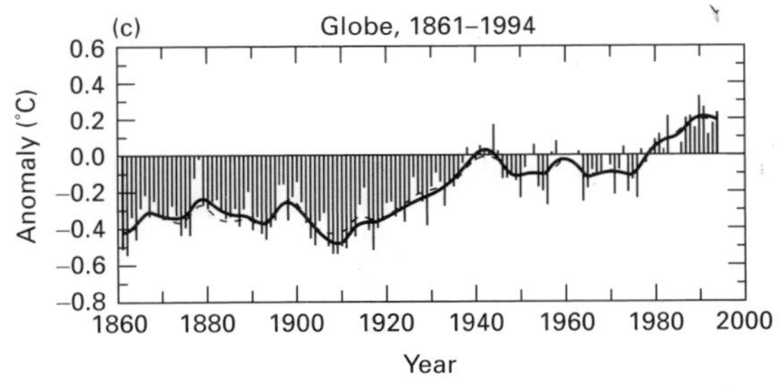


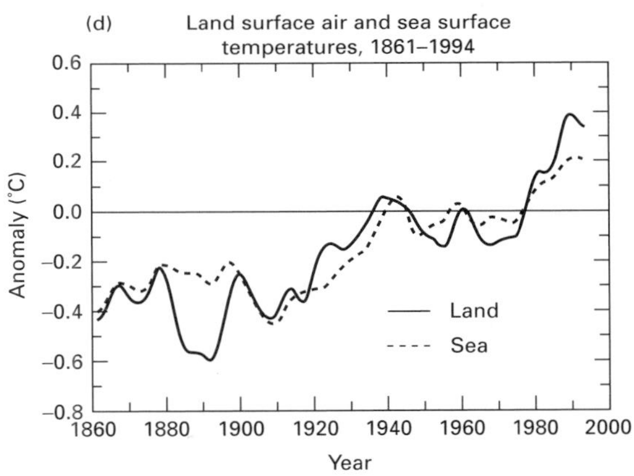


*Figure 1.2* Temperature trends, 1861–1994, expressed as deviations (anomalies) from the 1961–1990 average ('normal'). These are shown for (a) Northern and (b) Southern Hemispheres separately, and for (c) the whole globe. The same overall trend in both hemispheres, but the different detail, can be seen. In (d) the global anomalies for land and sea are separated. This indicates the slow response of the ocean to atmospheric changes. In (a)–(c) the bars represent individual years, the line a smoothed average of five sucessive years. The bars indicate the great annual variability of climate, even on this global scale, and the smoothed line emphasises the overall warming trend, with the various short-term cooling periods incorporated. (From Nicholls *et al.*, 1996.)

*Table 1.1* The composition of the atmosphere

| Constituent | Chemical formula | Abundance by volume |
|---|---|---|
| Nitrogen | $N_2$ | 78.08% |
| Oxygen | $O_2$ | 20.95% |
| Argon | Ar | 0.93% |
| Water vapour | $H_2O$ | variable (0–4%) |
| Carbon dioxide[b] | $CO_2$ | 340 ppmv[a] |
| Neon | Ne | 18 ppmv |
| Helium | He | 5 ppmv |
| Krypton | Kr | 1 ppmv |
| Xenon | Xe | 0.08 ppmv |
| Methane[b] | $CH_4$ | 2 ppmv |
| Hydrogen | $H_2$ | 0.5 ppmv |
| Nitrous oxide[b] | $N_2O$ | 0.3 ppmv |
| Carbon monoxide[b] | CO | 0.05–0.2 ppmv |
| Ozone[b] | $O_3$ | varies (0.02–10 ppmv) |
| Ammonia | $NH_3$ | 4 ppbv[a] |
| Nitrogen dioxide | $NO_2$ | 1 ppbv |
| Sulphur dioxide | $SO_2$ | 1 ppbv |
| Hydrogen sulphide | $H_2S$ | 0.05 ppbv |
| Halocarbons[b] | CFC, $CCl_4$, etc. | trace |

[a] ppmv and ppbv are parts per million and parts per billion by volume.
[b] Gases involved in the greenhouse effect.

example, is creating concerns about global warming arising from increasing atmospheric concentrations of greenhouse gases (Table 1.1). Although we think of the atmospheric composition prior to human activity as being unchanging, there has been a slow evolution of the atmosphere during the life of the Earth which has influenced climate.

There are also spatial variations in the distribution of incoming solar energy, with perpetual change in the distribution around the globe as a consequence of diurnal and seasonal variations. Nevertheless, equatorial regions generally receive the most, and polar regions the least. Again, this has clear consequences for the climate as we know it.

All of the processes associated with these energy flows obey the laws of physics, which themselves are commonly couched in mathematical terms. Indeed the development of mathematical models of climate is one of the major ways in which we are using both our increasing knowledge of climate processes and our improving observations of that climate. These models enhance our understanding of the science of climatology and increase our ability to predict climate.

## 1.2 The development of climatology

The rise of climatology as a science is closely linked with the increase in our ability to observe the atmosphere. As with many sciences, new observations often provide the basic information needed to give us new insights into processes acting, while often new theories about such processes demand that we obtain new measurements to test them. Certainly any understanding of how the world's climate system works, how it varies from time to time and place to place, and any use that can be made of resources provided by climate, depends on observing the climate at many places over a long period of time.

### 1.2.1 Descriptive climatology

The earliest climatological observations were simply visual or otherwise 'sensed' observations about nature made without instruments or sophisticated techniques. Descriptions of the march of the seasons and of the annual flooding of the Nile in ancient Egypt are examples. These observations at specific places had become sufficiently organised by the time of the ancient Greek civilisation for the Greeks to be able to divide the world into three zones: torrid, temperate and frigid. Although explanations for the observed phenomena abounded, there was little of what we would regard as scientific enquiry into their nature and causes.

The emphasis on description, rather than explanation, of the atmosphere continued for centuries. The development of the barometer and thermometer, together with the maintenance of records of wind direction and rainfall amount, added a quantitative dimension to our knowledge. By the late nineteenth century it was possible to describe the climate of much of the Earth's land surface, and some ocean areas, in reasonable detail. These descriptions relied heavily on the available observations, mainly of precipitation and temperature. Probably it is no coincidence that these were the main elements. Not only were instruments available to measure them, but also they were extremely useful, particularly in agriculture.

As the number of observations grew, one great problem was encountered by people trying to describe the climate: there were so many numbers that a simple tabulation was impractical and some form of summation was needed. Monthly values provided a convenient method. However, since monthly values vary from year to year, further averaging over several years became necessary. The result was the development of the concept of the climatic **normal**: an average over at least 30 years. It was felt that this long period was sufficient to smooth out the small-scale, year-to-year fluctuations and thus provide a true measure of the climate. Today, to most people, the climate is described by the monthly normals of average temperature and total rainfall.

The introduction of the climatic normal provided a means of summarising data for a single station. Summaries for areas were provided by the introduction of the concept of the climatic region. It was found that stations could be grouped together because they had similar normal values, or a similar monthly pattern of normals. Several schemes for climatic regions were proposed. Although defined almost entirely by analysis and comparison of available data, these climate classifications were often developed with a view to their eventual use. Most divided the climate in ways which had some significance for plant growth, and so the climatic regions reflected vegetation regions. Consequently they could be used, for example, in assessing the suitability of a plant, currently growing in a particular region with a particular climate, for introduction into a different region. Although the division into regions in this way paid no account to the causes of spatial variation, it was possible to infer causes in some cases. Indeed, this kind of descriptive approach can still yield useful information.

### 1.2.2 Meteorological advances

At the same time that these descriptions of climate were being generated, an entirely different approach was being taken by workers in the emerging field of meteorology. With the advent of rapid telegraphic communications it became possible to collect observations from a variety of places together quickly at a single point and analyse them with a view to forecasting future weather. The prime interest initially was in the short-range forecasting of storm tracks. Nevertheless, the questions needing to be answered to solve this problem led to a search for understanding of the physical laws governing the atmosphere. As understanding increased, so did the demand for observations. Pressure, wind speed and direction, visibility, cloud type and amount, and hourly temperatures were all needed and therefore observed. Later, as aviation meteorology gained in importance, the need for upper air observations grew. In response to this, the radiosonde was developed. Still later, after the Second World War, methods of radar observations of clouds and precipitation were developed. All these new observations and observational techniques served to advance our theoretical understanding of atmospheric processes and to improve our ability to forecast the weather.

During this time of great advance in meteorology, our understanding of climate was progressing slowly. The traditional observations continued to be made and some explanations of global climate attempted. The climatological data themselves often proved to be of great benefit. For example, during the Second World War operations in unfamiliar territory with equipment which could be sensitive to atmospheric conditions demanded both weather forecasts and longer-term estimates of the probability of occurrence of particular conditions. This wartime experience fostered several advances in applied climatology, notably the use of the climatic water balance in agricultural and water supply planning activities.

### 1.2.3 Assessing the variable nature of climate

Until the middle of the 1950s many developments in both meteorology and climatology were based on the assumption that the climate was constant. The weather varied with time and was responsible for the minor variations in climate from year to year, but in the long term the normals did not change. Although this was disputed by many climatologists, who could easily cite the ice ages several thousands of years ago and the Little Ice Age a few hundred years ago as evidence of change, the concept of a static climate was a useful concept to retain while a basic theory of climatic processes was being developed.

It is now clear that climate is never constant. Indeed, it is the departures from supposedly 'normal' conditions that often provide great insight into climatic processes, as well as having the greatest human impact. Three examples will suffice. The 30-year dry period in the USA at the beginning of this century culminated in the 'Dust Bowl' disaster, which reduced the output of the five main corn-growing states by more than 15% and demonstrated human vulnerability to climate fluctuations. The 1968–72 severe drought in the Sahel, when precipitation was around 50% of the 1931–60 average, demonstrated not only the human suffering caused by climatic fluctuations, but also the international nature of climatic events. Finally, investigations of the global climate during the periods of anomalously heavy rainfall in Peru in the 1972, 1982 and 1997 El Niño events led to insights into the operation of the phenomenon now known as the El Niño-Southern Oscillation (ENSO).

These highly visible examples of climatic fluctuations ensured that by the 1970s the concept of a fluctuating climate was well established and that climatic fluctuations, through their impact on humans, were of concern to more than climatologists.

### 1.2.4 Advances in measurement and analysis

Along with the increasing realisation that climate was not static came an increasing ability to monitor the atmosphere to determine that variability. At the same time, there was an increase in the ability to analyse this observational information. This allowed not only a better description of climate and its fluctuations, but also fostered new understanding of climatic processes, leading to opportunities for the prediction of future climates.

The key event of what amounted to a revolution in climatology was the coming of information from satellites. They were first used in the 1960s. Prior to this all our information was either surface based or from balloon-borne radiosondes, both restricted to a fixed location in space and a particular instant in time (Figure 1.3). These point data could be synthesised to give areal values and some notion of the three-dimensional structure of the atmosphere. Satellites give a global, three-dimensional picture, with observations available virtually constantly, so that variability on many time and space scales can be considered (Figure 1.4). Satellites actually observe the energy fluxes entering and leaving the atmosphere, which can be used to investigate processes associated with atmospheric energy flows and to deduce more familiar climatic elements such as temperature and winds. Hence the view of climatology has changed from one restricted to point-specific surface-based studies, with global implications available only indirectly, to one incorporating a global view and encouraging investigations of climate processes.

Further advances, but at the opposite end of the time and space scale, are being fostered by the increasing availability of advanced weather radars. These allow detailed analyses of atmospheric processes for small areas and short times. As yet, only small amounts of data are available, and these have been used primarily for the analysis of individual weather events, particularly those associated with cloud processes. They may eventually lead to better, more useful climatologies of clouds, precipitation and severe weather, but are not considered explicitly in this book.

Climatology at present has three sources of data: (a) the traditional surface-based observations, point specific but with a long period of record; (b) upper air data, again approximately point specific, and with a reasonably long record length; and (c) satellite data, with wide areal coverage of the globe but a comparatively short record. Climatology must use all these types in order to fulfil its mission. Much of our basic knowledge, and much of the material in subsequent chapters, has come from surface and upper air observations. Satellite data are allowing us to refine, and in some cases revolutionise, our understanding of the workings of the atmosphere on global and local scales.

A tremendous amount of data is generated by these observational systems, and it is a challenge to assimilate and analyse it in order to create useful information. Fortunately, advances in computer technology over the last few decades make data handling manageable, while sophisticated data compilations and complex analyses virtually unimaginable a few years ago can now be accomplished routinely. Indeed, the availability of climatological and statistical analysis packages, along with Internet access to data, allows almost anyone to become directly involved. It is difficult to give up-to-date guidance in this rapidly changing field, but some resources are indicated in Appendix B.

For the climatologist the rise in public concern for the environment in general, and climatic change in particular, has added to the excitement generated by the new data sources and analysis techniques. Concern about global warming first emerged in the early 1970s. Since then our concern and knowledge has grown, but it is still far from clear, when, where and how much warming will occur. Certainly one purpose of this text is to provide the basic information which indicates why it is not a simple matter to go from the well-established facts of an increase in anthropogenic gases in the atmosphere and a continuing trend of temperature increase to a forecast of future climate. Indeed, the complexity of the issues involved and the potential for a major climatic change, led the United Nations to create the Intergovernmental Panel on Climate Change (IPCC). The scientists and policy-makers associated with IPCC have synthesised a wealth of information and provided a set of benchmark publications helping national governments and international organisations understand and respond to the potential impact of climate change.

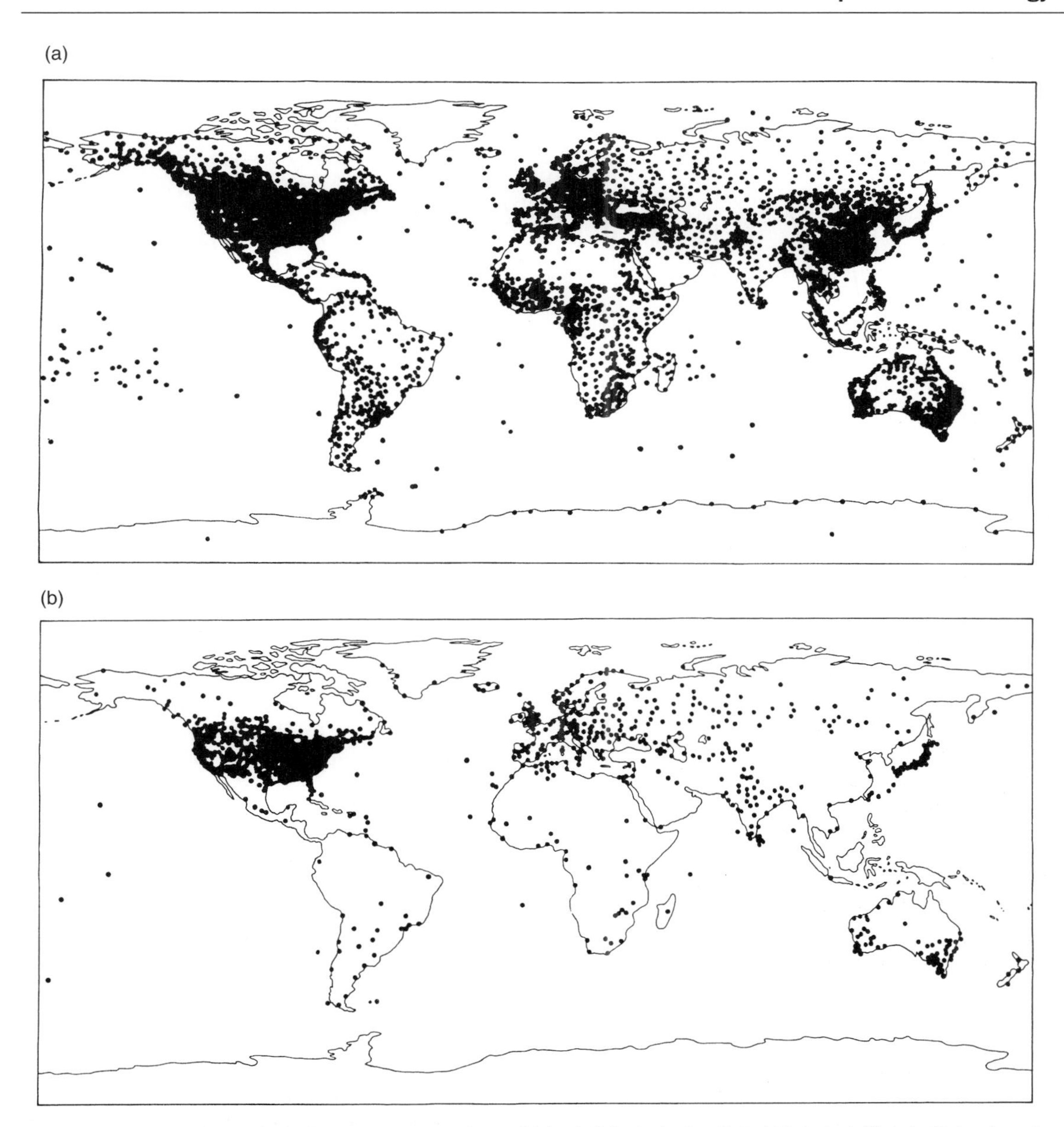


*Figure 1.3* Distribution of (a) all temperature stations which participate in the Global Historical Climate Network and (b) those participating stations with records commencing before 1900. The concentration of stations over the mid-latitude land areas, especially in (b), is clear. Most of the ocean stations are located on islands. This network is the prime input, supplemented by some shipboard and buoy observations, to the estimate of global average temperatures (Figure 1.2). (From Peterson and Vose, 1997.)

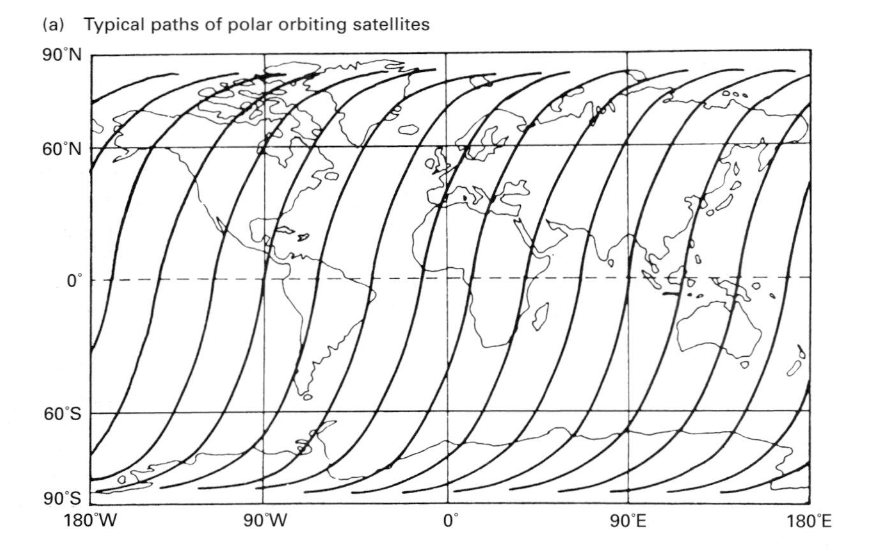


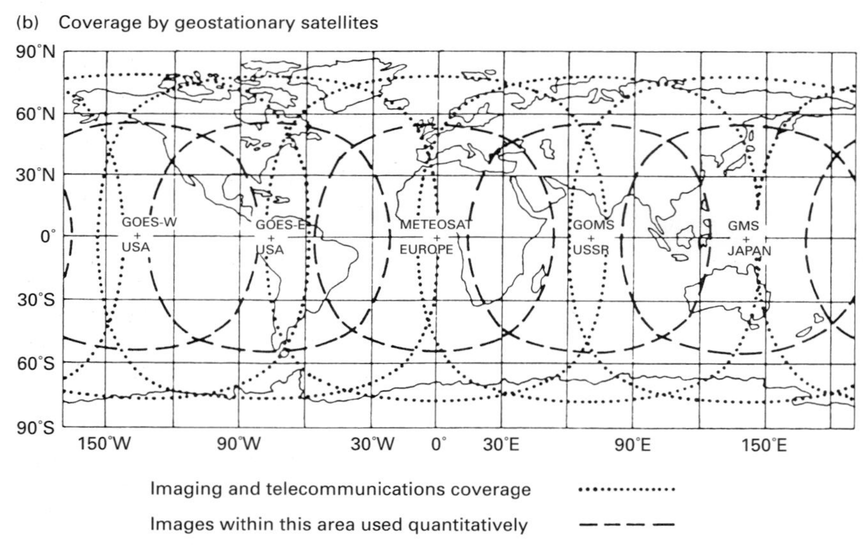


*Figure 1.4* Areal coverage of satellite surveillance for (a) polar orbiting satellites and (b) geostationary satellites. The upper diagram shows typical swaths; the sensor scans are made perpendicular to the centreline of the swath. Note that each of the swath line curves is a half-orbit path, but the projection distorts these great circles into S shapes. (Courtesy of NASA.)

### 1.2.5 Forecasting climate

The rise of public concern, the increased range and number of observations, and the advances in statistical analysis and climate modelling have all combined to foster efforts to forecast the climate. Although it can be suggested that climatology has always been primarily concerned with estimating future conditions, there is now an opportunity to change the nature of the forecasts.

Until recently, climatological forecasts were restricted to extrapolations of past conditions into the future. Although some climatological understanding might be incorporated, the forecast had to be given in terms of the probability of an event occurring within some specific time period and broad region. Thus it was possible to forecast the number of winter snow storms likely, or the probability of receiving three such storms, around any given observation station. This would be determined statistically from the past observational record, and the forecast would be the same from year to year. This might give useful baseline information to someone involved in snow removal planning, but it would say nothing about the chance that the next winter would be exceptionally snowy.

The new combination of models, analysis and observations allows us to begin making forecasts for specific times and places. To use our snow example, it would provide a forecast indicating that this particular coming winter is predicted to have nine snow storms, whereas our prediction based only on past records indicates that there are five storms on average. This greater detail commonly provides much more useful information. Indeed, it begins to provide information akin to that given by the familiar daily weather forecast. We are still a long way from the precision or detail of a daily forecast. Nevertheless advance statements about a particular winter at a particular location, even if currently restricted to estimates of the probability of above or below normal temperatures or precipitation, represent a major achievement.

## 1.3 Climatic elements

A variety of physical elements make up what we know as climate. For many people, temperature and precipitation epitomise climate. Indeed, for many years climatology focused on these two, but they are by no means the only climatic elements. Wind speed and direction, cloud type and amount, sunshine duration, atmospheric humidity, air pressure and visibility are obvious additions to the list of elements that we notice every day. Other elements may be equally, or more, important in particular situations. For example, soil moisture, soil temperatures and evaporation are vital in agriculture, pollutant concentration and the acidity of precipitation are of concern for human health, while radiant energy fluxes are of great interest to the climatologist seeking to understand atmospheric processes.

Climate therefore consists of a mixture of these elements, and as climatologists, we must be concerned with all of the elements. Not all will be equally important in all contexts. Furthermore, we know relatively little about some of them. This is particularly the case for those elements for which we have few observations. Some, such as the global extent of lying snow, can be observed only by satellite so that we have a relatively short period of record. Others, such as atmospheric turbulence, require delicate instruments, and measurement is restricted to specialised research sites. Still others, the acidity of precipitation for example, have not been recognised as important until recently, so that few long-term observations exist. On the other hand, elements such as temperature and precipitation are regularly measured at tens of thousands of sites world-wide. In general, not surprisingly, we know and understand more about the elements that are commonly measured than about those for which we have only a small number of observations.

Details of the observing system for various climatic elements are given in a series of separate boxes, one associated with each chapter. However, here we can give a broad overview of them. They can conveniently be divided into three types: measured elements, derived elements, and proxy elements.

### 1.3.1 Measured elements

The most common and familiar type of climatic elements are those that are measured directly. The instrument used may give either **contact measurements**, where the instrument is in direct contact with the entity being measured, or **remote sensing** measurements, where the instrument measures radiation from the entity, which is then converted into an observation of the required element. With minor exceptions, observations from the surface and in the upper air are contact measurements, while those from satellites and radar are remote measurements.

Contact measurements are point and time specific. Over land surfaces observations of the common elements of temperature and precipitation have been taken for many years at many stations and provide the most familiar information source for both temporal and spatial climate variability. As several of the boxes in later chapters discuss, accurate observation and analysis is not always straightforward, especially when long periods and diverse terrain are involved. Over the oceans the sites may be more homogeneous but, not surprisingly, observations are more difficult and less common. Some countries maintain fixed-location weather ships, which take standard observations. Increasingly, ocean buoys are being used to supplement our knowledge of ocean climate conditions.

The standard contact method for obtaining upper air data is by using a **radiosonde**. This is a package suspended below a balloon, which senses temperature, humidity and pressure as it ascends. The pressure observation is used to determine altitude. Some stations use a more sophisticated version, called a **rawinsonde**, which additionally measures wind speed and direction. Specially instrumented aircraft also measure conditions in the free atmosphere, but this is not done on a routine basis. Their use is restricted to experimental investigations of particular atmospheric features. A similar role, usually for the higher reaches of the atmosphere, is played by instrumented rockets.

A satellite is a platform for remote sensing instruments. The instrument, which may be designed to measure surface or atmospheric temperatures, or water vapour amounts in various atmospheric layers, or simply to take a cloud picture, actually senses the radiation emitted or reflected by the object under investigation. This object is generally at least a few square kilometres in area. Thus satellite observations are fundamentally different from surface observations. In many cases the areal integration is a great advantage. Further, satellites provide access to remote areas with no alternative data source and give continuous global coverage. The other major remote sensing instruments are associated with radar, and are not treated in detail here.

### 1.3.2 Derived elements

In addition to the elements that are actually measured, there are elements which are derived from them. Several of these will be discussed in greater detail later in the book. Examples include heating or growing season degree days, the summation for a given time period of temperatures above a certain threshold, which is used in calculating heating demand (Chapter 3); drought indices, which attempt to quantify the severity and length of a drought (Chapter 11); and comfort indices, which give a quantitative estimate of how cold or hot a human being feels (Chapter 11). These all play an important role in describing the climate for particular applications. Evaporation is a special case for a derived element. Direct measurement of evaporation is possible, but an accurate result is costly to produce and requires great care. Therefore direct measurements are restricted to a few locations. Evaporation, however, is very important in many aspects of water management and so techniques have been devised to derive estimates by calculation from other, more commonly and easily measured elements.

### 1.3.3 Proxy elements

Finally, there are elements which can be called 'proxy' elements. These are usually non-atmospheric indicators of atmospheric conditions, commonly used to infer past climatic conditions in the time before instrumental records. Tree rings, pollen frequencies, ice cores, rock types, and dates of grape harvests from diaries have all been used as proxy variables and are used in Chapter 12. Probably other such proxy elements will be discovered and utilised in the future.

## 1.4 The climate system

The key to the current development of climatology is found in the view of the Earth's climate as a complete system or, at a slightly more detailed level, as a set of interlocking subsystems (Figure 1.5). This systematic viewpoint is not entirely new, but the inclusion of links with the Earth's surface adds a new dimension and creates an integrated whole. It has been found impossible to understand the flows and cycles of energy and matter in the atmosphere without considering the material which underlies it.

We have already indicated that energy exchanges provide the starting point and general framework for the study of climatology, and we shall consider them in detail in Chapters 2 and 3. Any consideration of energy exchanges must incorporate surface effects. The atmosphere is heated from below as a response to surface absorption of solar energy, but different surfaces react in different ways to the receipt of this energy. Ice and snow reflect much of it, land surfaces heat rather rapidly, while the oceans store the energy without experiencing a significant temperature rise. This stored energy in the oceans may be moved about by ocean currents or may be taken to great depth only to be released at the surface after being stored for many years. Since the various types of surface have various response times, there is no single time scale for transfer of energy from the surface into the atmosphere. Further, the surfaces themselves may change with time on an equally broad range of time scales. Consequently the atmospheric heating rate is constantly varying both at individual places and from

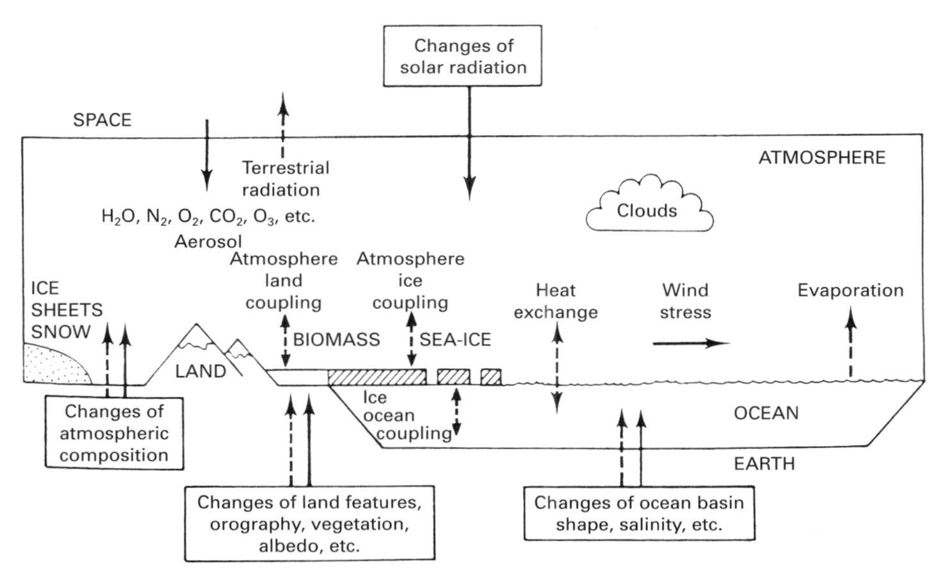


*Figure 1.5* Schematic illustration of the components of the climate system. Solid arrows are examples of external processes and dashed arrows are examples of internal processes. (From WMO, 1975.)

place to place. The causes and implications of these variations are currently the focus of much climatological research.

A similar systematic approach is also adopted for the study of the movement of matter within the atmosphere. Climatologically, water in its various forms is of most immediate concern. Many details of the water cycle are well known, and the systematic approach has been used for many years. We shall adopt it explicitly in Chapters 4 and 5. However, other materials are now receiving attention, most notably carbon dioxide. The main concern is with changes in the atmospheric concentration, and to estimate this it is necessary to consider the whole carbon cycle. While this is similar to the hydrological cycle, it is less well known, and involves much more than just the climate system.

### 1.4.1 Energy cascades and transformations

One way to view the climate system, with an emphasis on system energetics and their links to the cycling of matter, is through the concept of the energy cascade (Figure 1.6). This summarises the ways in which energy is transformed to produce atmospheric motions, and thus the climate. The vital role of the surface even when considering atmospheric motions in isolation is indicated by using 'strong surface heating' as the starting point. This is appropriate because the vast majority of the energy input from the Sun is absorbed at the surface, forming the fundamental energy source for all atmospheric motions. Spatial variations in this heating lead directly to both large-scale horizontal temperature gradients and to local convective instability. Any air mass, heated from below (input), tends to rise, thus increasing the available potential energy (uplift). Either this potential energy is released in convective activity or, through horizontal energy gradients, in large-scale horizontal motions and synoptic-scale weather patterns (circulation) (Chapters 6 and 7). As the resulting winds pass over the Earth's surface small-scale irregularities on the surface give rise to shear instabilities and other boundary effects (wind) (Chapter 10). The final fate of either energy path is dissipation into random molecular motions (output). All these energy transformations which constitute the energy cascade are degenerative, leading to lower level forms of energy such as noise and heat. Certainly some of the methods of final dissipation are interesting to consider. For instance, the blowing over of a dustbin creates movement and sound, while an electrical storm produces light and sound. On a somewhat larger scale, energy is imparted directly to the oceans through the induction of waves, while fluctuations in the length of the day are a result of the angular momentum interchange

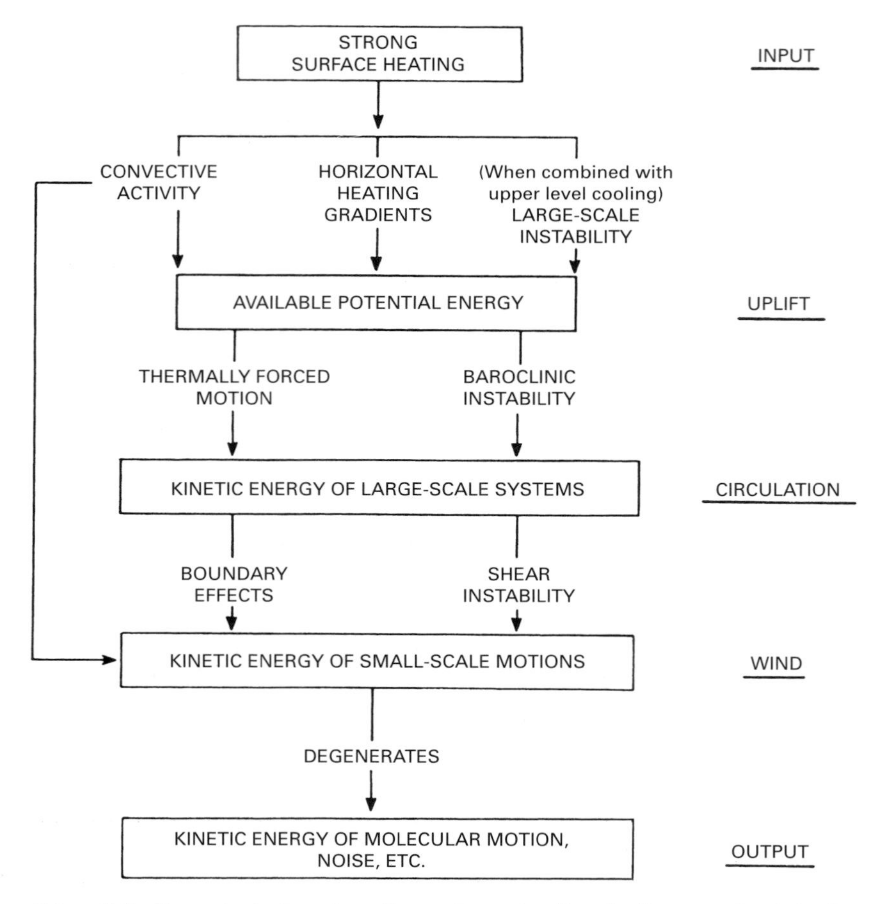


*Figure 1.6* Atmospheric transformations of energy. Almost all energy input to the atmosphere is the result of surface heating following absorption of solar radiation. Large-scale motions degenerate into smaller-scale movement and finally into kinetic energy of molecular motion.

between the atmosphere and the solid Earth. Wind and water power schemes 'tap' the atmospheric kinetic energy as do yachts and gliders and even commercial jet aircraft.

If we view the energy cascade in a slightly different way, we can consider the amounts of energy involved (Figure 1.7). There are tremendous variations in this, with a relatively small portion being stored in the system. Energy exchanges are going on all the time on a variety of time scales. Some processes are rapid enough to be seen. An example, which serves to emphasise the link between the energy and the water systems, is the formation of scattered clouds on a warm sunny afternoon after a cloudless morning. Energy from the Sun is used to heat a water surface throughout the day. Some of this energy is used for evaporation, which is a form of energy transfer. This creates humid air which, as the result of solar heating, ascends. The vigour of the ascent usually increases throughout the morning until, by afternoon, it is strong enough to force the water vapour to condense and form clouds. Energy is stored in these clouds as potential energy. This will soon be released if conditions are appropriate for the development of rain. Thus there is short-term energy storage. On a much longer time scale, when we heat our homes using coal or oil, we are experiencing the final stages of an energy exchange process that has taken millions of years to complete. Solar radiant energy was used directly to build plant and animal tissue which was subsequently stored when buried and converted to coal or oil. We now use this fossilised radiant energy to create heat.

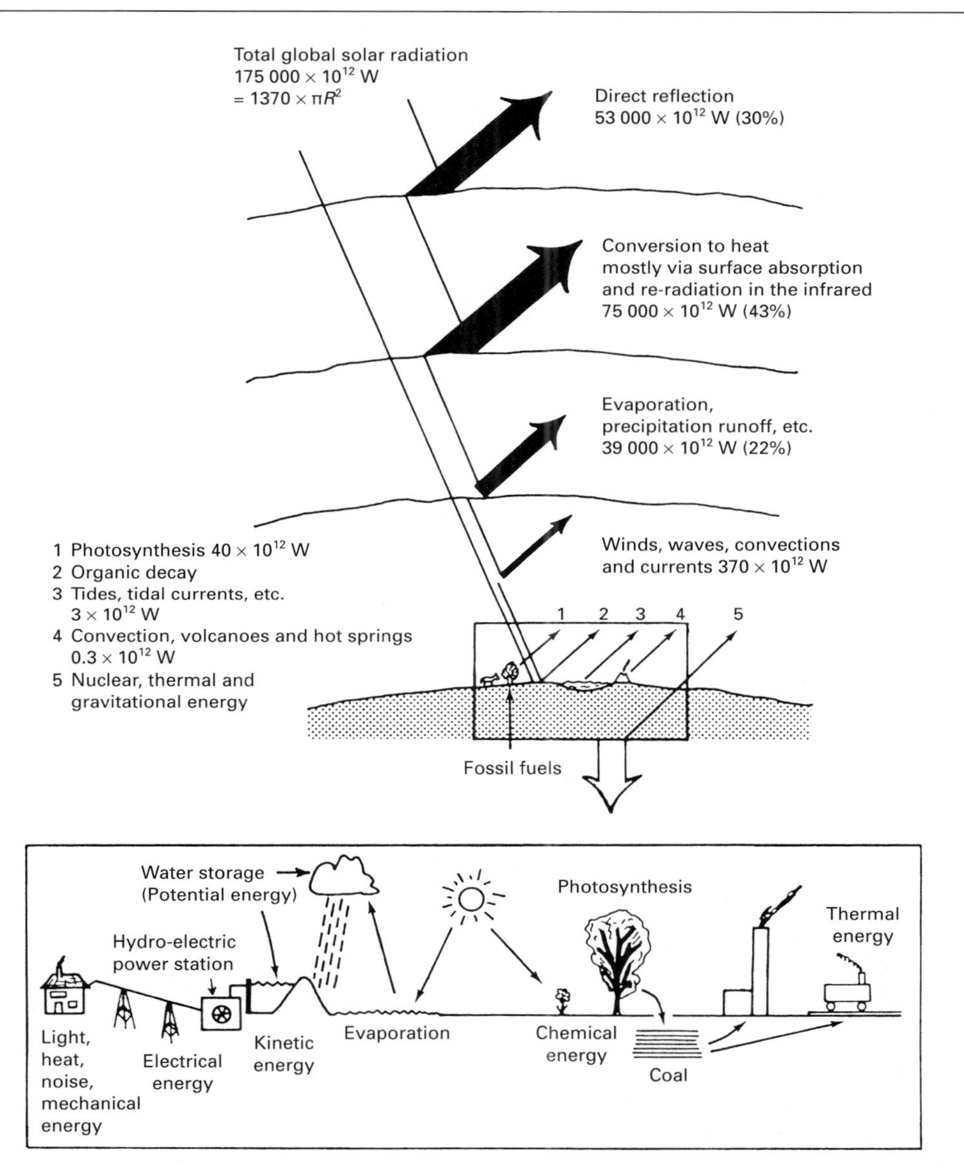


*Figure 1.7* The global energy cascade. Top-of-the-atmosphere irradiance is equal to the solar irradiance at the Earth's orbital distance multiplied by the instantaneous area of the planet illuminated, $\pi R^2$ (see Figure 13.1). The diagram shows schematically how solar energy, arriving at the top of the atmosphere, is absorbed by the surface and how this, when re-emitted, is distributed amongst the various features of the energy cascade. This cartoon complements the energy transfer systems shown in Figure 1.6. The lower box illustrates schematically some of the features of the energy cascade which most directly affect life on the surface of the Earth.

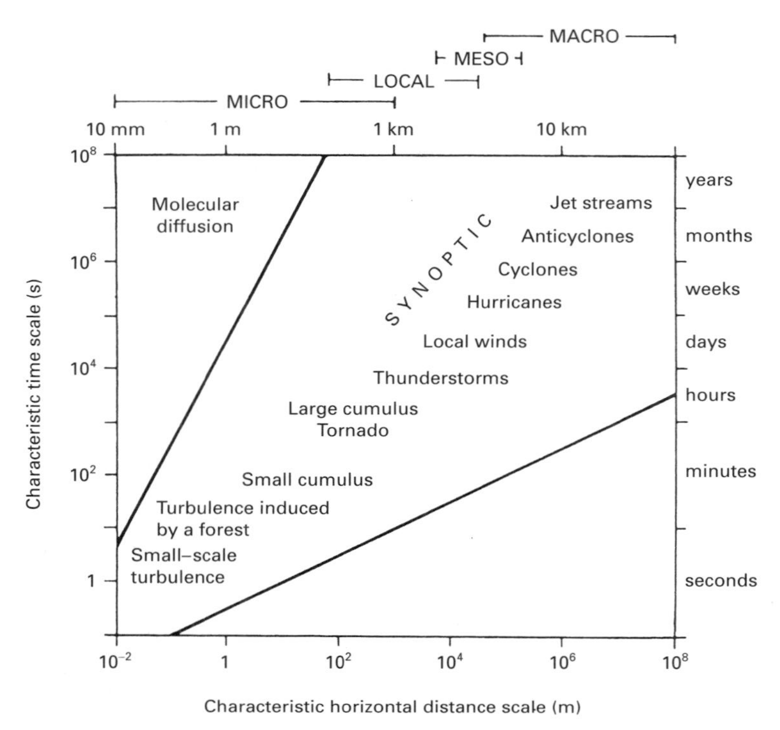


*Figure 1.8* Time and space scales of various atmospheric phenomena. There is an approximately linear relationship to be seen between the size of atmospheric features and their time scales. (From Hess, 1974.)

## 1.5 Climatological space and time scales

As indicated above, there are various scales of activity in the atmosphere. In fact, since the atmosphere is continually in motion, our scales of interest could vary from an individual tree or house to the whole globe, and from seconds to millennia. To attempt to describe and understand all the activity on all the scales would be a mammoth undertaking. Fortunately it is possible to simplify the situation. Traditionally the whole range has been divided into four general categories: micro, local, meso and macro (Figure 1.8). This applies to both space and time, since short time scale phenomena tend to occur on a small space scale, while long-term phenomena affect large regions. The division is based on a number of factors, notably the observational techniques needed to describe the phenomena on that particular scale and the analytical techniques required to obtain an understanding of those phenomena. In the first ten chapters of this book, where we consider the basic processes creating climate, we shall work from the large to the small scale. There are, however, many interactions between scales and we cannot treat any scale in complete isolation. Indeed, the last four chapters, as we look towards means of predicting climate, will take all scales into account.

All of the features indicated in Figure 1.8 have been studied in considerable detail. Probably best known are the features associated with meteorology, the core of which is associated with the scales marked as 'synoptic'. The synoptic scale deals with regions the size of nations in Europe or states in the USA and covers a time scale of a few days. The emphasis of meteorology has been on developing an understanding leading to site- and time-specific short-term forecasts, using techniques which are many, varied and frequently complex. The climatologist is more interested in an understanding which leads to longer term predictions, and many of the meteorological techniques need not greatly concern us here. Nevertheless, much of the information developed in meteorology must be incorporated into our understanding of climate. Indeed, as meteorology and climatology come closer and closer together, more and more interaction is required.

Although we must make great use of results from meteorological investigations, as climatologists we must take a wider view. Not only must we expand our time scale to encompass conditions from decades to millennia, but also we must incorporate the human dimension into our climatology. This involves both the impact of human activity on the atmosphere and the impact of the atmosphere on human activities. This is addressed directly in Chapter 11, and also occurs in many other places throughout the text.

### 1.5.1 Physical climatology

Certain basic processes go on at all scales, or irrespective of scale. These we can think of as the climate controls, and their study is usually called **physical climatology**. This is considered in Chapters 2–5. Since the atmosphere is driven by energy, we start, in Chapter 2, with a discussion of the global energy system. Chapter 3 then deals with some of the direct consequences of that system, notably temperature distributions in time and space. One of the most important consequences of energy exchanges is the development of a water cycle, which includes ice in the polar regions, liquid water in the oceans and in rain, and vapour in the atmosphere. Water in the atmosphere is the focus of Chapter 4. Chapter 5 completes our investigation of the water cycle by considering precipitation and some of its human consequences.

### 1.5.2 Dynamic climatology

As soon as we add consideration of horizontal air motions (i.e. wind) to our study of climate, not only do we need to change the name to **dynamic climatology**, but we also introduce the effect of scale. We have divided the dynamic climatology section into three scales in order to demonstrate more clearly how the climate controls generate what we now think of as the complete climate. Thus, after a general review of the processes creating winds in Chapter 6, in Chapter 7 we will be concerned with the largest scale: the whole globe or major portions thereof. The time scale will range from seasons to decades, a scale where we can use actual observations to develop our understanding. Following this we have, in Chapters 8 and 9, the regional scale – the synoptic scale treated in a climatological way. Chapter 10 is concerned with the smallest scale, from a field to a city and from minutes to a day or so. Each of these scales needs different techniques, and a differing emphasis, within the general framework of our understanding of the physical laws governing atmospheric activity.

The final four chapters have elements of both physical and dynamic climatology. After a review of the human dimension of climate and climatology in Chapter 11, we concentrate on the material necessary to address the questions of climate forecasting. Thus Chapter 12 treats past climates partly as a potential guide to the future and Chapter 13 looks directly at techniques of modelling the past, current and future climate. The final chapter then addresses the methods and possibilities for predicting future climates, the ultimate goal of contemporary climatology.

#### Box 1.A Climate applications and impacts: the use of climatological knowledge

Much of the main body of this book considers climatology as a scientific discipline where the main aim is to describe, understand and eventually predict the climate of an area. At the same time, however, climate has a profound role in human affairs. This can take two forms. First, climatic knowledge can be used to enhance human activities. Judicious use of the information can help to select the most productive crops to grow, the right size of a reservoir, the appropriate housing style, or the size of an electricity generating plant. Such investigations are the province of **applied climatology**. The second form of interaction is closely associated, but often seemingly takes the opposite viewpoint. This is the study of **climate impacts**, investigating ways in which climate influences, and sometimes dictates, the scope of human possibilities. Commonly, for a particular area climate constrains the range of crop types that can be grown, influences the diseases likely to develop, or determines whether or not a reservoir is feasible. Of particular concern are the changes in impact that may occur as a result of climate changes.

Each of the subsequent chapters in the book contains examples of climate applications. As with the present material, they are set off from the main text. The examples were chosen to demonstrate the range of human activities where climatic information might be useful and to suggest the types of analysis techniques likely to be used in generating that information, while also showing the links to the main narrative discussing the science of climatology. For any actual problem, of course, the information needs and the analysis techniques must be identified in close co-operation with those who will actually use the information.

One whole chapter, Chapter 11, is devoted explicitly to the links between humans and the atmosphere. In addition, much of the material in Chapters 12–14 implicitly involves climate impacts.

### Box 1.1 The climate observing system

We emphasised earlier that progress in understanding climate required both observational and theoretical advances. Much of the main narrative discusses the theoretical aspects. Each chapter, however, also includes a separate box containing comments about observations. Taken together, these boxes consider the **climate observing system**, the whole process from the actual observation to the provision of information useful for climatological analyses. Although the satellite and surface systems are very different in structure, the concept is identical for both. Hence, in this general overview, the surface observing system is emphasised.

Observations start, of course, with the choice, installation and use of a particular instrument at a single location. Commonly we are concerned with weather and climate variations over space and time, so a time series of measurements from a **network** of stations is needed. For valid intercomparisons, these observations must be made in a standardised way. Most national meteorological services operate networks which follow procedures outlined by the World Meteorological Organisation, and international comparisons are possible. Many of these networks are designed to meet the immediate operational needs for weather forecasting. Their climatological value comes as the record length increases and they supply more of the observational information necessary to develop or test theories or drive models. The climatological need for long-term and wide-area data requires that a **data archive** be created. An integral part of the archiving process is the **quality control** of the observed data. There are millions of separate observations taken every day, and errors, instrumental or human, are bound to occur. These must be corrected whenever possible, or we can never be sure whether our analyses are reasonable. The final product of the climate observing system is a series of quality controlled data sets which is available for climatological analysis, whether that analysis be for understanding climate processes, estimating the most extreme rainfall likely at a point in the next 100 years, or observing the trend in global temperatures over the last 100 years.

# CHAPTER 2 The Earth's radiation budget

All aspects of the climate system of the Earth – the winds, rain, clouds, humidity and temperature – are the result of energy transfers and transformations within the Earth/atmosphere system. These energy exchanges, which create and drive the climate, are the focus of this chapter. The whole process starts when energy from the Sun arrives at the top of the atmosphere in the form of radiant energy. Formally, this is **electromagnetic radiation**, but it is usually referred to simply as **radiation**. This energy interacts with the atmosphere as it is transferred downward, so that some energy is reflected back to space, some is absorbed and transformed into heat, and some is transmitted to the surface of the Earth. The radiation that penetrates to the surface and is absorbed can heat the surface, evaporate water, melt snow and heat the underlying soil. Thus radiant energy is transformed into a variety of other energy forms. This energy is eventually transferred to the atmosphere, where a transformation back to radiation occurs before the energy is finally returned to space. Variations in the amount of radiant energy received from the Sun and variations in the interaction with the Earth and atmosphere create the temporal and spatial variations in energy exchanges which lead to our climate.

### 2.0.1 Energy balance of the Earth

The globally and annually averaged values for the various energy exchanges are given in Figure 2.1. It shows that the surface loses as much energy as it receives. The same is true for the atmosphere and for the whole planet. Thus globally, and on a long-term basis, a radiation balance is maintained. Without this balance very rapid climatic changes would occur, although minor deviations occur on all time scales. On this global annual scale, all the horizontal motions in the atmosphere and oceans, which transfer energy, are omitted. The effects of vegetation, topography, day and night, and the seasons are ignored. The numbers thus represent a generalised picture of the energy exchange, a picture we can later refine. In Figure 2.1 radiation has been divided into two categories: the incoming short-wavelength solar radiation arriving from the Sun and the outgoing long-wavelength terrestrial radiation emitted by the Earth. This fundamental division between solar and terrestrial radiation arises directly from the nature and properties of electromagnetic radiation, which are described in the next section. This division not only has great implications for the Earth's climate system, but is also of practical significance for satellite observations of the Earth and its atmosphere, and for human use of the energy itself.

## 2.1 The nature of radiation

Radiation is a form of energy that is emitted by all objects having a temperature above absolute zero. It is the only form of energy that can travel through the vacuum of outer space. Thus energy coming to and leaving planet Earth must be in the form of radiation. This energy form is most conveniently considered as a stream of energy packets, photons, moving along a sinusoidal wave trajectory. For the moment we shall emphasise the wave-like aspects, but later we will need to consider the photon characteristics.

### 2.1.1 Electromagnetic spectrum

A fundamental characteristic of radiation is the wavelength of propagation. All bodies radiate at a large number of wavelengths. The complete set of possible

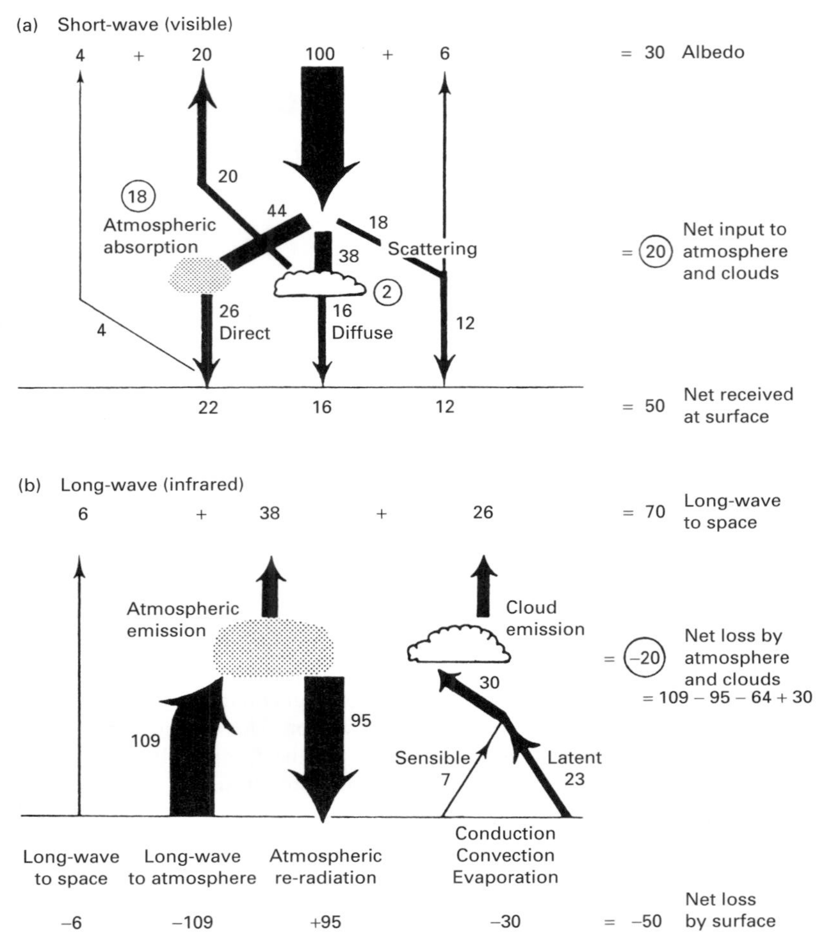


*Figure 2.1* The energy balance of planet Earth. The incident solar irradiance is shown as 100 units. Thus the reflected, transmitted and absorbed components of both incident solar and emitted terrestrial radiation are percentages of this value. For example, the diagram shows a globally averaged value of surface albedo of 4/26 or 0.15. The short-wave radiation (a) is balanced by the long-wave radiation (b) at (i) the top of the atmosphere (the planetary budget); (ii) the atmosphere; and (iii) the surface. Note the major contribution to the planetary albedo (20%) made by clouds, which also make a significant contribution to the emitted radiation. Despite the importance of these features the primary site of absorption of short-wave radiation is the surface.

wavelengths is called the **electromagnetic spectrum** (Figure 2.2). Climatologically important radiation is within the range 0.1 to 100 μm. The human eye responds to a very small portion of this region, which we call **visible light**. Our sense of colour depends on the wavelength of the light our eyes receive. Wavelengths around 0.40 μm give violet light. As wavelength increases we see the colours of the rainbow until at 0.70 μm we reach red light. Regions adjacent to this visible portion are given names associated with the nearest colour. The region with wavelengths slightly shorter than 0.40 μm is termed **ultraviolet**, while radiation with wavelengths longer than ~1.0 μm (and ≤1 mm) is termed **infrared** radiation.

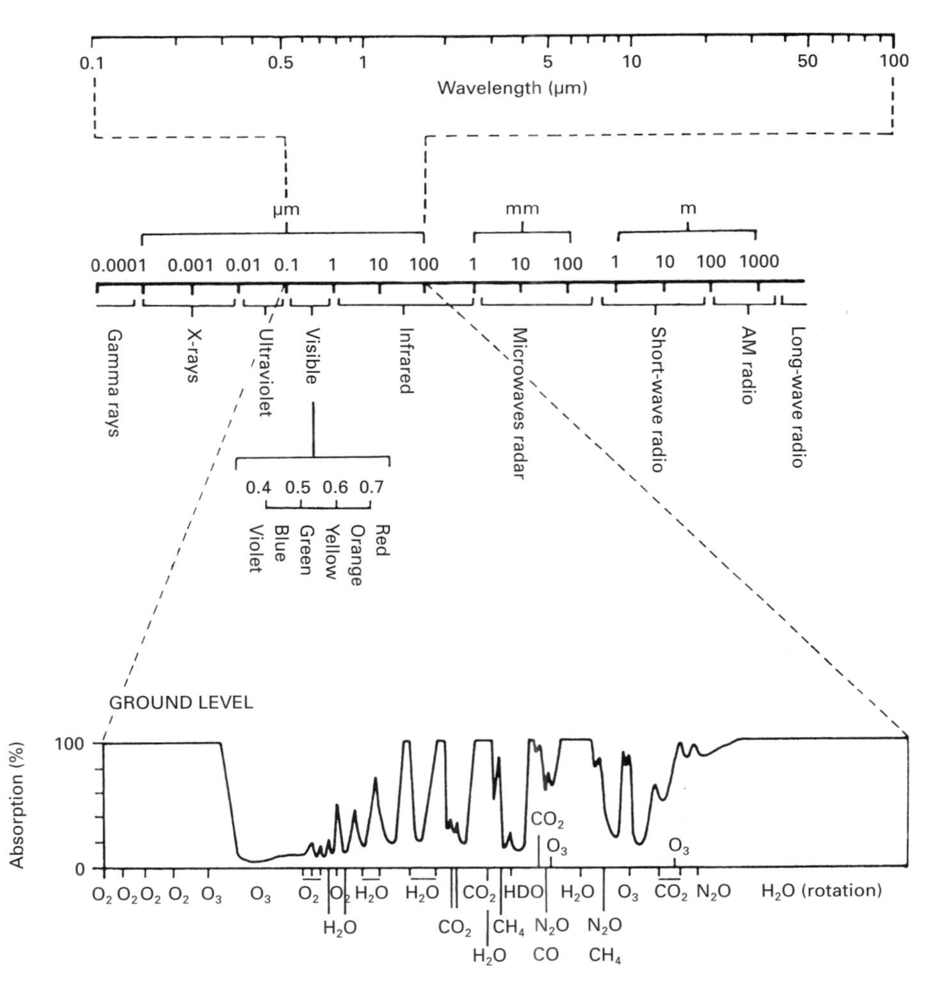


*Figure 2.2* The electromagnetic spectrum, emphasising the wavelength regions of importance for climatology, which range from ~0.1 to ~100 μm. The lower part of the diagram shows atmospheric absorption over this wavelength range. The atmosphere is seen to be transparent (little absorption) in the visible part of the spectrum but exhibits considerable absorption by $O_3$ in the ultraviolet and by $H_2O$ and $CO_2$ and other molecules in the infrared part of the spectrum.

## 2.1.2 Radiation laws

The wavelength of propagation depends on the temperature of the emitting body. The basic law of radiant emission is **Planck's Law**:

$$E^*_\lambda = c_1/\{\lambda^5[\exp(c_2/\lambda T) - 1]\} \qquad (2.1)$$

where $E^*_\lambda$ is the amount of energy (W m$^{-2}$ μm$^{-1}$) emitted at wavelength $\lambda$ (μm) by a body at temperature $T$ (K). The two constants $c_1$ and $c_2$ have values of $3.74 \times 10^8$ W μm$^4$ m$^{-2}$ and $1.44 \times 10^4$ μm K, respectively. This equation is displayed graphically in Figure 2.3 for perfectly radiating bodies that have temperatures that correspond approximately to those of the Sun (5800 K) and the Earth (255 K). For a given temperature, $T$, these Planck curves are uniquely defined and have a characteristic shape. The radiating temperature of the Earth is about 33 K less than the mean surface temperature because of radiation exchanges in the atmosphere, the difference being due mainly to the greenhouse effect, which will be explained more fully in Section 2.5.

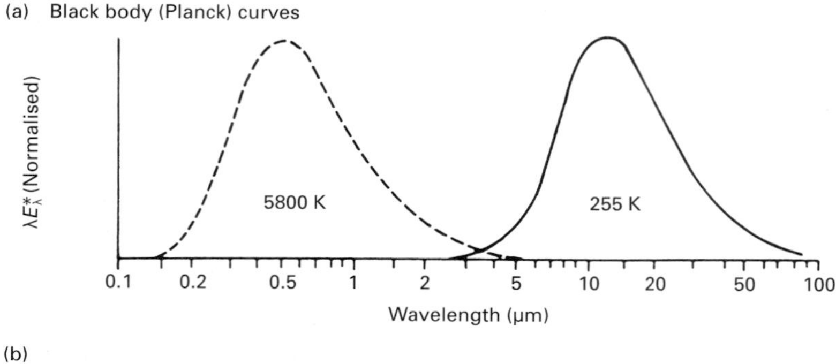


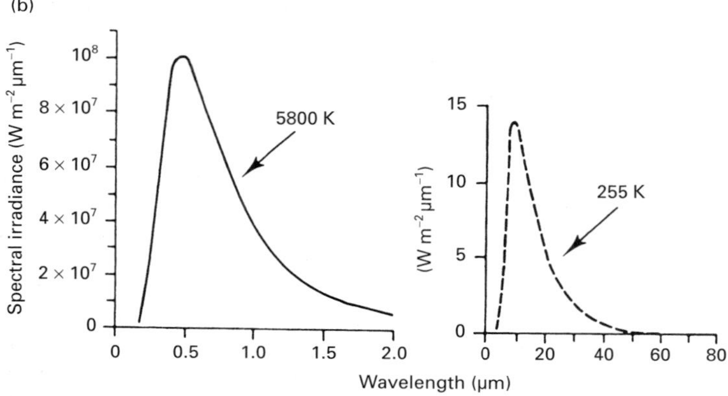


*Figure 2.3* (a) Planck (or black body) curves for the Sun (temperature approximately 5800 K) and the Earth (temperature approximately 255 K). The vertical scale is normalised so that the two curves have the same vertical range. The peak of the solar curve is at approximately 0.5 μm, while the peak of the terrestrial curve is at approximately 11.4 μm. The very small spectral overlap (at approximately 3.5 μm) means that there is little chance of confusing the sources of remotely sensed radiation. (b) Quantitative vertical scales (W $m^{-2}$ $\mu m^{-1}$) show the differences in total energy ($E^*$) emitted by the Sun and the Earth. The value of $E^*$ is given by the area beneath the Planck curves.

The wavelength of maximum emission for a body at a particular temperature is inversely proportional to the temperature and is given by **Wien's Law**:

$$\lambda_{max} = 2897/T \qquad (2.2)$$

This equation is obtained by differentiation of equation (2.1) and indicates that the maximum emission wavelengths for the Sun and the Earth are 0.50 and 11.4 μm, respectively.

The total energy emitted by a body, the area under each curve in Figure 2.3(b), can be found by integration of equation (2.1). This gives the **Stefan–Boltzmann Law**:

$$E^* = \sigma T^4 \qquad (2.3)$$

where $\sigma$ (= $5.67 \times 10^{-8}$ W $m^{-2}$ $K^{-4}$) is the Stefan–Boltzmann constant. Thus the amount of energy emitted increases with temperature, as indicated by Figure 2.3(b).

These three laws apply directly to bodies that are theoretically perfect radiators. Such bodies are called **black bodies**. The degree to which a real body approaches a black body is given by the emissivity of the body:

$$\varepsilon_\lambda = E_\lambda / E^*_\lambda \qquad (2.4)$$

where, for the wavelength $\lambda$, $\varepsilon_\lambda$ is the emissivity and $E_\lambda$ and $E^*_\lambda$ are the emissions of the real and black bodies, respectively. The value of the emissivity is dependent on wavelength. In general, solids and liquids radiate across a continuous spectral interval with a more or less constant emissivity, usually between 0.9 and 1.0. Gases, on the other hand, radiate only at specific wavelengths, and so have variable emissivities, a characteristic which complicates calculations of how much energy they emit at a given temperature but which can be exploited in satellite-based measurements of atmospheric properties.

Although, so far, we have been concerned with emission, there is a very simple relationship between emissivity ($\varepsilon_\lambda$) and absorptance ($\alpha_\lambda$):

$$\varepsilon_\lambda = \alpha_\lambda \tag{2.5}$$

This is known as **Kirchhoff's Law** and indicates that, at the same wavelength, good emitters are equally good absorbers. This equation refers to emissivity and absorptance, not to actual amounts of absorption or emission. An earthbound body with a high emissivity, and thus absorptance, at solar wavelengths will absorb a great amount of the solar radiation which falls upon it, but can emit very little at that wavelength because its temperature is such that Planck's law dictates that even a black body can emit little energy at that wavelength.

Energy that is not absorbed when it impinges on a body must be either reflected or transmitted. The proportion of each that will occur in a given situation depends on the wavelength of the incident radiation and the properties of the body.

The basic radiation laws emphasise that when considering radiation in the climate system it is advantageous to use two distinct radiation regimes: the short-wave (solar) radiation originating in the Sun and the long-wave (terrestrial) radiation emitted by the Earth and its atmosphere. Although the two overlap slightly, they are sufficiently distinct to be treated completely separately. This separation, indicated in Figure 2.3(a), is very convenient since the two regimes interact with the climate system in different ways, which we explore in the rest of this chapter.

## 2.2 Solar radiation reaching the Earth

Solar radiation is produced by nuclear reactions within the core of the Sun. This radiation, a good approximation to that of a black body, essentially follows the Planck curve for a temperature of 5800 K. Only in the high-energy region, at wavelengths shorter than about 0.3 μm, does emission fall significantly below the theoretical level. The wavelength of peak emission at 0.474 μm suggests a Sun with a blue-green colour. The observed yellow Sun is a result of the relative sensitivity of the human eye, interaction of radiation with the Earth's atmosphere and the relative intensities of emission because of the shape of the Planck curve. As a result of this shape, 99% of the energy is emitted in the wavelength region 0.15 to 4.0 μm, with 9% in the ultraviolet, 45% in the visible and 46% in the infrared (Figure 2.3b).

### 2.2.1 Solar energy reaching the top of the atmosphere

The top of the Earth's atmosphere receives about $4.5 \times 10^{-10}$ of the total energy output of the Sun. This is the energy available to drive the climate system. The amount is usually expressed in terms of the **solar constant**: the amount of energy passing in unit time through a unit surface perpendicular to the Sun's rays at the outer edge of the atmosphere at the mean distance between the Earth and the Sun. This constant, approximately 1367 W $m^{-2}$, is known to vary on a number of time scales.

The longest of the variations in the solar constant is caused by the evolution of the Sun itself. Over the lifetime of the solar system the temperature of, and hence the energy emitted from, the Sun is believed to have increased by between 20 and 40%. This increase has not only changed the solar constant, but also the spectral distribution of the energy. On a much shorter time scale, variations in solar energy output are associated with sunspots, dark regions on the Sun's surface, about 4000 km in radius, which result from convective activity in the upper layers of the solar mass. The number of spots varies in an approximately cyclic way, a complete cycle taking about 22 years. Activity within the sunspots leads to solar flares, giving enhanced ultraviolet and X-ray emissions. These variations influence the value of the solar constant and are one main reason for our uncertainty concerning its exact value.

In addition to the processes acting within the Sun itself, the energy received at the Earth also varies as a result of the astronomical relationships between the Sun and the Earth. These relationships are generally well known and many can be exactly specified. The shortest ones account for the systematic daily and seasonal changes in the solar energy received at the top of the atmosphere, and are considered here. Longer ones, which play a major role in climatic change, are considered in Section 12.3.

### 2.2.2 Earth's orbit around the Sun

The main characteristics of the orbital geometry of the Earth are shown in Figure 2.4. The Earth revolves

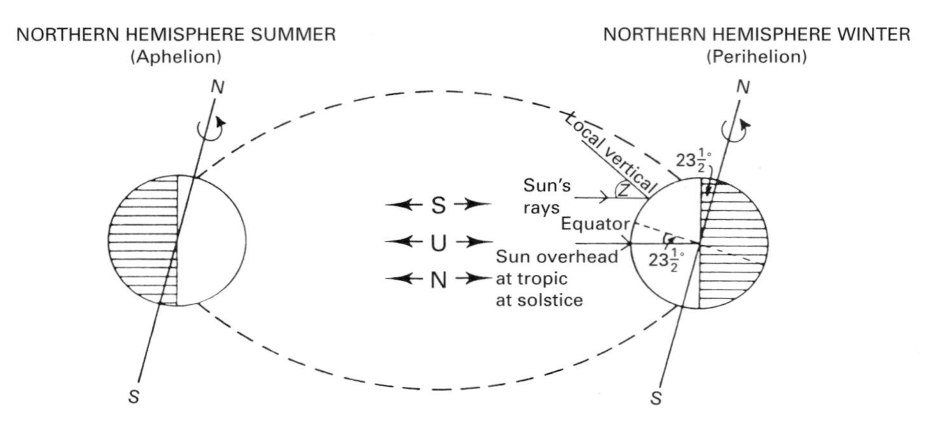


*Figure 2.4* The situation and orientation of the Earth at aphelion (June) and perihelion (December). The solar zenith angle, *Z*, is shown.

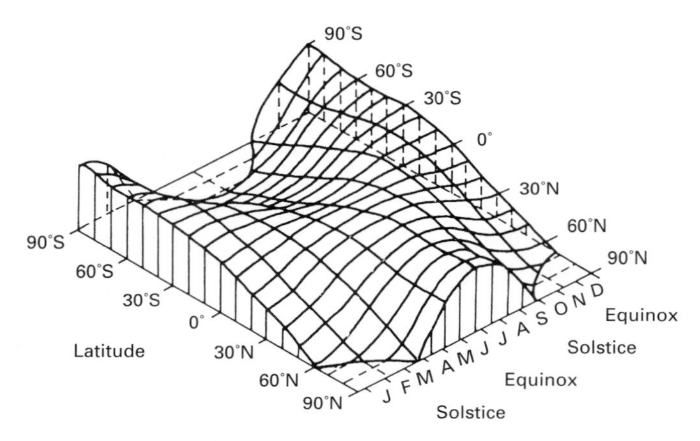


*Figure 2.5* The variation of insolation (at the top of the atmosphere) as a function of latitude and month for the whole globe. The very large amounts of radiation received at the poles in summer are the result of the 24 hours of daylight at that time. (From Strahler, 1965.)

around the Sun in an elliptical orbit with a period of one year. At present the Earth is closest to the Sun (perihelion) in the Northern Hemisphere winter, and farthest away (aphelion) in July. The **eccentricity** of the orbit, a measure of the degree to which the ellipse departs from a circle, is small. It, along with the actual dates of perihelion and aphelion, changes very slowly with time; so slowly that they have a minor impact on climate on the scale of a human lifetime.

Much more significant for seasonal variations is the inclination of the Earth's axis (Figure 2.4). The axis, the line through the centre of the Earth joining the two poles, is tilted by 23.5° from the perpendicular to the plane of the ecliptic, the plane of the Earth's orbit around the Sun. Since the axis retains the same orientation with respect to the galaxy, the effect is to create seasonal variations in the amount and intensity of radiation received at a given spot at the top of the atmosphere and, eventually, at the surface of the Earth.

### 2.2.3 Daily and seasonal variations in solar radiation

As a consequence of the **axial tilt**, both the duration of daylight and the height of the Sun in the sky vary with time (Figure 2.5). To a good approximation the length of daylight is proportional to the fraction of the latitude circle that is unshaded. At the extreme positions, the noontime Sun is overhead at either the Tropic of Cancer, giving the summer solstice, or the Tropic of Capricorn, creating the winter solstice. At the latter, in areas north of the Arctic Circle there is 24-hour night, while south of the Antarctic Circle there is 24-hour daylight. The equinoxes, at which there are 12 hours of daylight everywhere on the Earth, occur when the Sun is overhead at the equator on 22 March and 22 September.

The height of the Sun in the sky is usually given in terms of the solar zenith angle. This is the angular distance between the Sun's rays and the local vertical, and is given by:

$$\cos Z = \sin \theta \sin \delta + \cos \theta \cos \delta \cos h \qquad (2.6)$$

where $\theta$ is the latitude of the place, $\delta$ is the solar declination and $h$ is the hour angle. $\delta$ is a function of the time of year, while $h$ is zero at local noon and increases in magnitude by $\pi/12$, (15°), for every hour before or after noon. The 'half-day length', $H$, can be calculated from equation (2.6) by noting that at sunrise and sunset $\cos Z = 0$ (i.e. $Z = \pi/2$, (90°)). Thus:

$$\cos H = -\tan \theta \tan \delta \qquad (2.7)$$

The amount of incoming energy on a horizontal surface at the top of the atmosphere, $I_0$, is related to $Z$ by:

$$I_0 = S_F(d/d')^2 \cos Z \qquad (2.8)$$

where $S_F$ is the solar constant and $d/d'$ is the correction factor for the variable Earth–Sun distance, $d$ being the mean Earth–Sun distance and $d'$ its instantaneous value. Combining equations (2.6) and (2.8) and integrating from $-H$ to $+H$, the daily total radiation at the top of the atmosphere is:

$$I_{DT} = (86\,400/\pi) S_F (d/d')^2 (H \sin \theta \sin \delta + \cos \theta \cos \delta \cos H) \qquad (2.9)$$

where $H$ is expressed in radians.

The resultant variation in the incident solar energy, or insolation, at the top of the atmosphere as a function of latitude and month is shown in Figure 2.5. While the insolation above the equator is never as great as the maxima achieved above the poles, it is consistently high throughout the year. Polar insolation, on the other hand, is zero during the polar night, but peaks during the polar day when the Sun is permanently above the horizon. This spatial and temporal variation is vital for creating the climate system. In order to explore the full implications, however, we must first consider the radiation as it enters and interacts with the atmosphere.

## 2.3 Interaction of solar radiation with the atmosphere

As solar radiation passes down from the top of the atmosphere, scattering and absorption commences. These occur almost simultaneously, although they must be treated separately here. Indeed, they are separate processes with different consequences.

### 2.3.1 Scattering of solar radiation

Scattering of radiation occurs whenever a photon impinges on an obstacle without being absorbed. Within the atmosphere the radiation interacts with small particles, and the process is known as 'scattering'. When radiation impinges on a much larger body, such as the Earth's surface, there is a complete change in direction for the radiation and it is conventional to regard this special case of scattering as 'reflection'. The only effect of scattering is to change the direction of travel of the photon. The change can be in any and all directions, but it is usually sufficient to think of atmospheric scattering in just two directions: upwards and downwards relative to the Earth. Upward scattered radiation, unless it is bounced back downwards and thus takes part in multiple scattering, is lost to space and takes no further part in energy processes. Downward scattered radiation, on the other hand, is still in the system and susceptible to further interactions.

Scattering particles, such as atmospheric gas molecules, which are small compared to the wavelength of the incident radiation, produce **Rayleigh scattering** (Figure 2.6a). The amount of scattering is inversely proportional to the fourth power of the wavelength. Thus for visible radiation the scattering of blue light (wavelength approximately 0.4 μm) is approximately 10 times greater than for red light (wavelength approximately 0.7 μm). This type of scattering is characteristic of a cloudless atmosphere without contaminants. The major, and obvious, consequence of Rayleigh scattering is the blue colour of the daytime sky. The effect of the very efficient

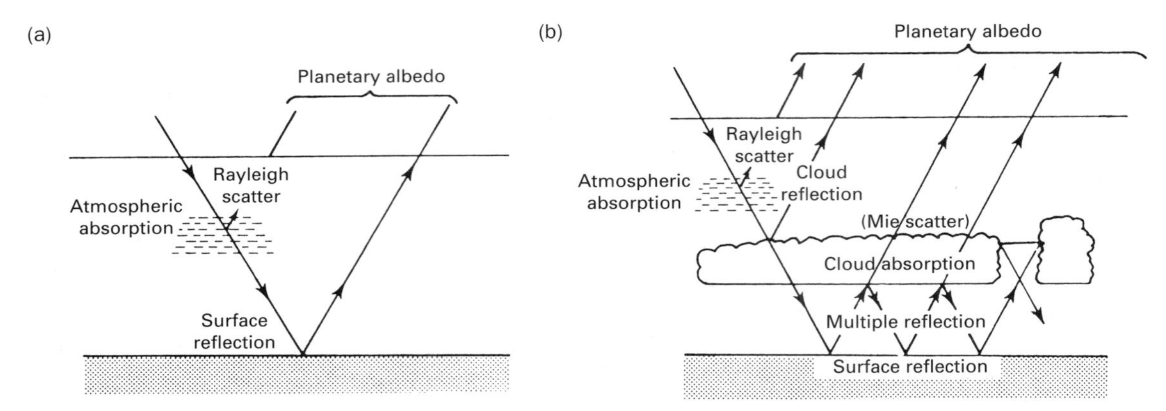


*Figure 2.6* (a) When a solar photon interacts with atmospheric gases the scattering process is well described by Rayleigh theory. (b) However, Mie theory is required to describe the more complex interaction which takes place between incident solar photons and water droplets and particulate material (especially in clouds).

scattering of blue light can be viewed on any evening at around twilight. The very long path length of the radiation through the atmosphere at that time causes most of the shorter visible wavelengths to be scattered many times and hence dispersed from the direct solar beam, leaving only the red component unaffected, creating a much redder Sun than during the rest of the day. The reddish sky often associated with this is a result of the scattering of red wavelengths slightly out of the direct beam.

Whenever clouds, pollution particles or water droplets are present in the atmosphere **Mie scattering** occurs (Figure 2.6b). This is effective whenever the scattering particles are of a similar size to the wavelength of the incident radiation. In this case scattering is almost entirely in the forward direction. The amount of scattering is a function of both the particle size and the wavelength of the radiation, but there is a tendency to scatter light of all wavelengths approximately equally. This is most readily seen when the atmosphere is **turbid** (i.e. laden with pollution) and creates a light blue/grey sky, since no wavelengths are preferentially scattered.

Although some photons are scattered only once as they pass through the atmosphere, multiple scattering is more common (Figure 2.6b). Scattering within clouds is always multiple scattering. A dramatic example of the effect near the surface occurs in high latitudes when a **whiteout** occurs. Here multiple scattering between a bright snow surface and the base of a low cloud layer makes it extremely difficult to see which is which, or where the horizon is located, and to distinguish any features on the surface.

As a result of scattering (whether single or multiple) and reflection, there are both upward and downward radiation streams. The ratio of the two is the **albedo**:

$$A = K{\downarrow}/K{\uparrow} \tag{2.10}$$

where $K$ is solar radiation and the arrows indicate the direction of travel. The albedo at any level of the Earth/atmosphere system can be determined from measurements of the two fluxes. However, two levels are of paramount importance: the Earth's surface and the top of the atmosphere. Knowledge of the value at the top of the atmosphere, the **planetary albedo**, for the whole globe or large regions thereof, is vital for an understanding of large-scale climate processes. Figure 2.1 indicates that the long-term average planetary albedo is about 0.3. The **surface albedo** is highly dependent on the nature of the surface, and variations between surface types have significant consequences for local climates, as considered in the next section. In addition, it is often useful to consider **cloud albedo** separately, since this is important for atmospheric heating and motion calculations and has an influence on the local climate.

### 2.3.2 Absorption of solar radiation

**Absorption** of short-wave radiation by atmospheric gases is relatively small and is usually considerably less important than scattering. The atmosphere is almost completely transparent near the peak of the solar Planck curve corresponding to the visible wavelengths (Figure 2.2). There is strong absorption of ultraviolet radiation by ozone in the lower stratosphere. This not only shields the surface from biologically harmful ultraviolet radiation but also leads to atmospheric warming some distance above the surface (see Figure 3.7). In addition, there is absorption in a few narrow regions of the spectrum, called 'bands', where absorption by oxygen, ozone and water vapour occurs. In total, however, only about 18% of the energy arriving at the top of the atmosphere is absorbed by gases before arriving at the Earth's surface (Figure 2.1). Absorption by clouds removes another 2%. This is much less than the amount of energy that is absorbed at the surface (~50%).

The net effect of the interactions between short-wave radiation and the atmosphere is summarised in Figure 2.1. Of the incoming energy at the top of the atmosphere, 26% is backscattered by the air and clouds and 20% is absorbed in the atmosphere. Thus 54% reaches the surface of the Earth. We now, therefore, consider the solar energy that actually reaches the Earth's surface.

## 2.4 Solar radiation at the Earth's surface

The energy which arrives at the surface of the Earth after traversing the atmosphere has both direct and diffuse components. The direct radiation is that which has travelled through the atmosphere without scattering. We see this as the direct solar beam. The diffuse component, the sky-light, is the portion which has been scattered. Energetically, they act in the same way once they arrive at the Earth's surface.

### 2.4.1 Global distribution of solar radiation at the surface

Global maps of radiation receipt can be constructed from the observational and empirical data. Some indications of the instruments and methods involved are given in Box 2.I. Although surface observational data are relatively sparse, particularly over the oceans, and a good deal of subjectivity is involved in map construction, radiation tends to be 'spatially conservative', not varying rapidly horizontally, so that broad regional patterns emerge (Figure 2.7). It is clear that the areas of maximum radiation receipt are the desert regions of the Earth, while minima are reached in polar regions.

### 2.4.2 Surface albedo

Radiation incident on the Earth's surface is either absorbed or reflected. The proportion of the incident energy that is reflected is the **surface albedo**, given by equation (2.10). This varies with surface type and has a profound role in determining the amount of surface heating and thus the local climate. The surface albedo is not highly wavelength dependent and can be treated as a single value for a given surface type (Table 2.1). Most natural land surfaces have albedos between 0.10 and 0.25, with surfaces created by human activity having slightly higher values. Snow and, to a lesser degree, frozen water surfaces have much higher values. Water is much lower, except at the end of each day, when low zenith angles lead to very high albedos for a short time. At the end of a day, for example, it is possible to see the Sun reflected by a calm water surface.

A high albedo indicates that much of the incident energy is reflected rather than absorbed. The high albedo of snow means that it reflects much of the incident energy, so that little is available for absorption and heating. Thus a clean snow surface can survive on a sunny day. Dirty snow includes material of a lower albedo within it and will absorb more energy and melt much faster. Similarly, albedo differences between vegetation and nearby artificial surfaces are one of the causes of the temperature differences between them and consequently are one of the factors to be considered in the establishment of local climates, as is discussed in Chapter 10.

The global average value of the surface albedo is around 0.15, mainly because water is the predominant surface type. Consequently the majority of the energy reaching the surface is absorbed. Indeed, Figure 2.1 indicates that approximately 50% of the solar energy reaching the planet is absorbed at the surface, compared with only 20% absorbed by the atmosphere. The result is that the major source of heating for the lower part of the atmosphere is the surface. This has profound implications for vertical air motions and cloud formation, for regional weather and climate differences, and for local differentiation of climate, all of which will be considered in several later sections.

### 2.4.3 Energy for photosynthesis

It is rarely necessary to subdivide short-wave radiation at the surface into specific wavelength regions when considering the energetics of climate. However,

**Box 2.I Instruments and methods for monitoring radiation**

*2.I.1 Instruments to measure solar radiation*
Surface-based instruments which measure radiation are commonly of three types:

1. **Pyranometers** (Figure 2.I.1a) measure both direct and diffuse solar radiation at all wavelengths coming from an entire hemisphere. They are usually mounted with the receiver surface horizontal, thus observing the total radiation from the Sun and the sky. A pair of pyranometers, one upward facing, one looking down, are usually used to determine surface albedo. With the use of appropriate filters, represented by the dome in Figure 2.I.1, it is possible to measure radiation in particular spectral bands. Commonly the filter is either of clear glass, which transmits all wavelengths and determines the energy in the total solar spectrum, or of a material that transmits in the 0.40–0.75 μm range, measuring the photosynthetically active radiation of interest in agricultural production.
2. **Pyrheliometers** measure direct solar radiation incident on a collector perpendicular to the Sun's

(a)

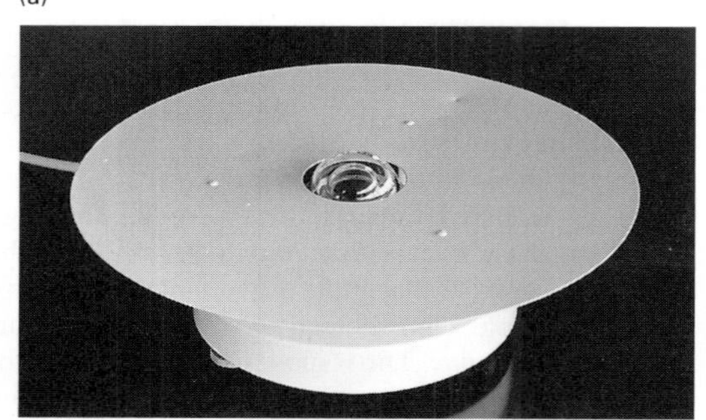

(b)

(c)

*Figure 2.I.1* Photograph of (a) pyranometer (Kipp and Zonen solarimeter) and (b) the method of mounting a shade ring above such an instrument so that only diffuse solar radiation is monitored. By subtracting the diffuse irradiance measured by one pyranometer from the total radiation measured by an unshaded pyranometer, a value of the direct solar radiation is achieved. (c) A Campbell-Stokes sunshine recorder. On bright days with little or no cloud the Sun's rays, concentrated through the glass sphere, burn the card encircling the recorder.

**Box 2.I (cont'd)**

rays. Again, these can observe the total spectrum or certain wavelength intervals.

3. **Diffusographs** observe radiation from the sky only. This instrument consists of a normal pyranometer surrounded by a 'shade ring', which obscures the direct beam of the Sun (Figure 2.I.1b).

Pyrheliometers are usually precision instruments, which must be carefully used and more or less constantly maintained. Consequently, they are commonly found only at specialised research sites. However, they are instruments which can measure radiation directly as an energy flux. Pyranometers, on the other hand, are designed to be more robust 'field' instruments. Numerous types have been developed, but most operate on a 'differential absorption' principle. Part of the sensor surface is coated with a paint with a high albedo, the rest with a low albedo paint. The temperature difference between the surfaces is usually measured electrically. The relationship between the observed temperature difference and the amount of radiation must then be established by 'calibrating' the instrument against an absolute instrument such as a pyrheliometer.

*2.I.2 Measuring sunshine duration*

From an energetic perspective, solar radiation is of major concern, while the separate visible component is of less direct interest. For many people, however, it is the duration of sunshine that is of more obvious interest. Two rather simple instruments have been developed to provide this information. The **Campbell-Stokes sunshine recorder** (Figure 2.I.1c) focuses the sunlight through a glass sphere to produce a burn on a graduated card whenever there is bright sunshine, while the **sunshine switch** opens an electric circuit whenever the sunlight is sufficiently intense to vaporise a bead of mercury inside a glass tube. The former is the preferred instrument in Britain; the latter is most common in the United States. Both instruments are rather inexact, but measure the duration of bright sunshine. When used directly, it is obviously an absolute value, but can be misleading since, as equation (2.7) indicates, the length of daylight varies with latitude and date. Thus the expression of the per cent of possible sunshine is often a more meaningful measure of the 'sunniness' of the day, particularly for holiday purposes.

*2.I.3 Empirical methods of obtaining solar radiation data*

There is a network of pyranometers over the land areas of the Earth making more or less continuous observations. Many have been operated for upwards of 50 years. They provide detailed information, for a specific location, on the long-term radiation climate and its variability. As suggested in Box 2.A, however, the network density is not always sufficient for particular applications, and solar radiation has been estimated for locations without measurements by using available sunshine duration and/or cloud amount data. The approach adopted is a typical example of an **empirical** method of obtaining data. It is clear that radiation amount must be related in some way to cloud amount and daylight duration. While detailed calculations of atmospheric transfer would be needed to provide the required information for a specific location for a specific day, it is possible to use statistical **regression** techniques if only long-term average values are required. For a station which measures both solar radiation and cloud amount (or sunshine hours), daily values of each are correlated to provide a regression relationship of the form $X = f(Y)$. Usually a linear relationship is assumed, so that the equation becomes:

$$X = a + bY \tag{2.I.1}$$

where $a$ and $b$ are empirical constants. For example, for some Nigerian stations a relationship was found such that $X = 0.19 + 0.60Y$, where $Y$ is the daily averaged percentage of possible sunshine and $X$ is the fraction of the extraterrestrial radiation received at the surface. The results illustrated in Figure 2.I.2 show that despite the fact that the regression coefficients vary spatially and temporally, a good empirical basis exists for extrapolation.

*2.I.4 Measuring and estimating long-wave and net radiation*

Measurements of long-wave and net radiation are commonly made using pyranometers with appropriate filters. It is rather difficult to devise a filter which transmits only long-wave radiation, so it is usual to measure solar and net radiation at the same site, and calculate the long-wave component from the radiation budget. Even for net radiation, it is difficult to find a filter material that will not absorb some energy at some wavelengths. Thus, while it is scientifically preferable to have no filter, in practice it is difficult to keep an exposed surface free from contamination. In general, they need almost constant care, so that they are found at research sites rather than as part of a network of field instruments. Thus they are much less numerous than solar radiation instruments. Surface values of long-wave and net

**Box 2.1 (cont'd)**

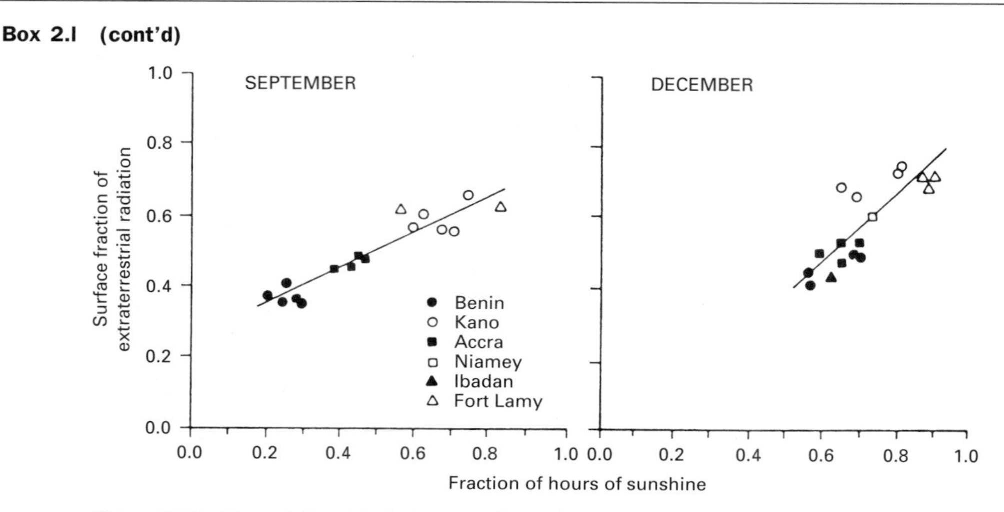


*Figure 2.1.2* The relationship between solar radiation at the Earth's surface and hours of sunshine for six stations in Nigeria for the months of September and December. The solid line indicates the relationship obtained by statistical regression techniques. The differing slopes between months indicate the temporal variation in the relationship. (From Davies, 1965.)

radiation are more likely to be obtained using empirical methods based on surface temperature and humidity than from direct measurements.

*2.1.5 Monitoring radiation from space*

Satellites observe the flux of radiation leaving the top of the atmosphere. They provide a wide variety of information and are complementary to the surface-based observations in that they take regular, repetitive observations of a large portion of the Earth (Figure 1.4). Satellites are generally inserted into one of two orbital configurations. In a **polar orbit** the satellite is usually between 500 and 1500 km above the surface of the Earth and the orbital path crosses the equator at approximately 90°. Characteristic orbital periods are of the order of 90 minutes, each orbit taking the satellite directly over or close to both poles. Usually such satellites are Sun synchronous: as the Earth rotates, each new orbit is over an area experiencing the same local time as the area viewed on the previous orbit. Half of each orbit, therefore, is over the dark or night side of the planet. A meteorological polar orbiter typically has a field of view sufficiently large that it takes about 15 orbits to cover the globe, and a specific location is seen approximately twice a day.

**Geostationary** satellites are very much higher than those of the polar orbiters. They must be approximately 35 400 km above the Earth's surface and in the plane of the equator so that they remain geosynchronous, remaining in the same position relative to the Earth at all times. As a consequence of their great height, they continuously view the full disc of the Earth and have much lower resolution than polar orbiting satellites. The curvature of the Earth makes it difficult to provide information for latitudes higher than about 45–50°. Commonly these satellites sample the radiation fluxes over the whole disc once every few minutes.

On board the satellites, radiance measurements are generally made in a series of spectral intervals, the number of which varies from satellite to satellite. Two relatively broadband wavelength regions, one within the solar spectrum (approximately 0.3–1.0 μm) and one within the atmospheric window (approximately 10–12 μm), are commonly used to identify cloud fields, derive winds and retrieve surface temperatures. In addition they provide the only source of information about planetary albedo and net radiation. Sensors responsive to a narrow spectral interval can provide more specific information. For example, a sensor centred at 6.3 μm, a water vapour absorption band, measures the amount of absorption in an atmospheric column, which is then related to the column water vapour content. Similarly, several narrow bands in the infrared and microwave regions of the spectrum can be used to obtain vertical temperature profiles which augment radiosonde data.

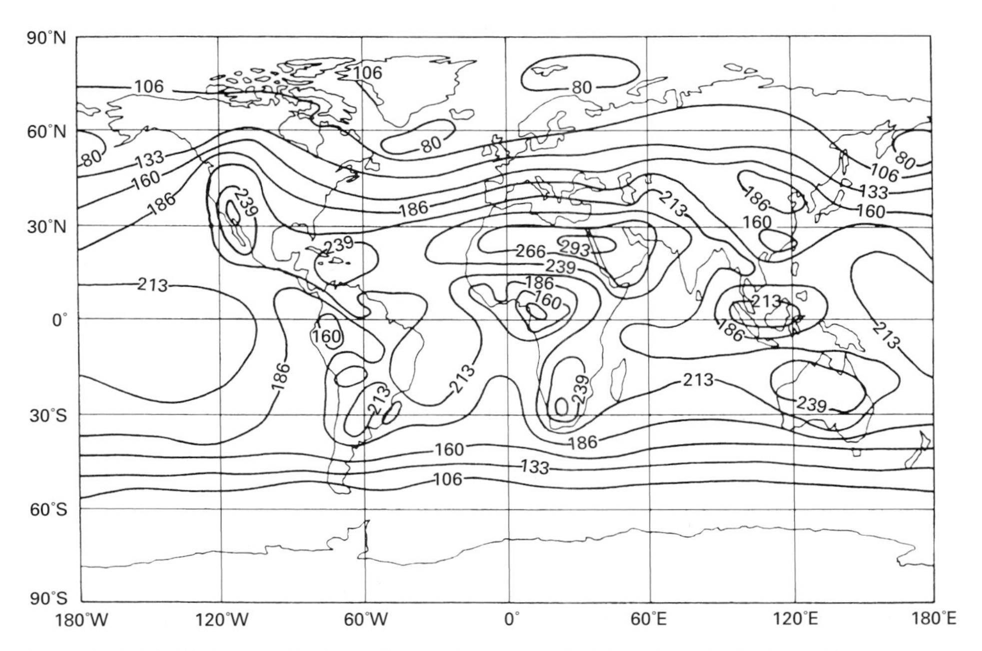


*Figure 2.7* World-wide distribution of annually averaged global (i.e. direct plus diffuse) surface-received solar radiation (W $m^{-2}$). (After Budyko, 1974; from Lockwood, 1979.)

*Table 2.1* Albedos and emissivities of common surface types (annual means)

| Type | Albedo (*A*) | Emissivity (ε) |
|---|---|---|
| Tropical forest | 0.13 | 0.99 |
| Woodland | 0.14 | 0.98 |
| Farmland/natural grassland | 0.20 | 0.95 |
| Semi-desert/stony desert | 0.24 | 0.92 |
| Dry sandy desert/salt pans | 0.37 | 0.89 |
| Water (0–60°)[a] | <0.08 | 0.96 |
| Water (60–90°)[a] | <0.10 | 0.96 |
| Sea-ice | 0.25–0.60 | 0.90 |
| Snow-covered vegetation | 0.20–0.80 | 0.88 |
| Snow-covered ice | 0.80 | 0.92 |

[a] The albedo of a water surface increases as the solar zenith angle increases. Ocean surface albedos are also increased by the occurrence of white caps on the waves.

for some applications such detail is needed. A pertinent example is the variability of plant response to the wavelength of incoming radiation (Figure 2.8). The leaf of the plant *Populus deltoides* (cottonwood tree) is reasonably characteristic of most leaves. There is strong absorption in the ultraviolet and at blue and red wavelengths, where the energy is used for photosynthesis. In the near-infrared, where the energy is not required for plant processes, most of the energy is transmitted or reflected, avoiding the overheating problems which would come with absorption. Variations in the spectral composition of the solar radiation, however, can influence plant growth. The short-term natural fluctuations from cloud-free to overcast conditions mainly decrease incoming radiation at wavelengths >0.6 μm, where transmission occurs, and have little influence on photosynthesis. However, increased atmospheric pollution or, on a longer time scale, changes in the nature of the Sun's radiant output, may decrease the amounts at shorter wavelengths and thus influence plant growth.

The amount of photosynthesis, and therefore the amount of growth, depends not only on the spectral composition of the radiation, but also on the amount. In general, other factors such as water availability and carbon dioxide concentration being non-limiting, the amount of photosynthesis increases approximately linearly with incident energy amount up to a certain point (Figure 2.9). Beyond this point the plant becomes light-saturated and the photosynthetic rate remains constant. This saturation value depends greatly on the

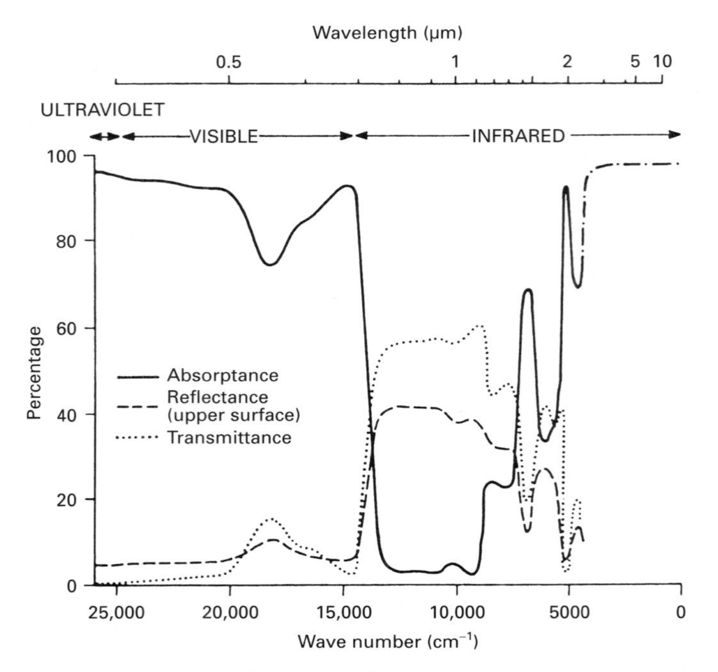


*Figure 2.8* Spectral distribution of reflectance, transmittance and absorptance of the leaves of *Populus deltoides* (cottonwood tree). The sudden increase in reflectance at approximately 0.7 μm is easy to see. Note the terms reflectance, transmittance and absorptance are applied when radiation impinges on a partially transparent substance. In the case of a completely opaque body (e.g. the Earth's surface) the terms reflectivity and absorptivity can be used. (Gates, 1965)

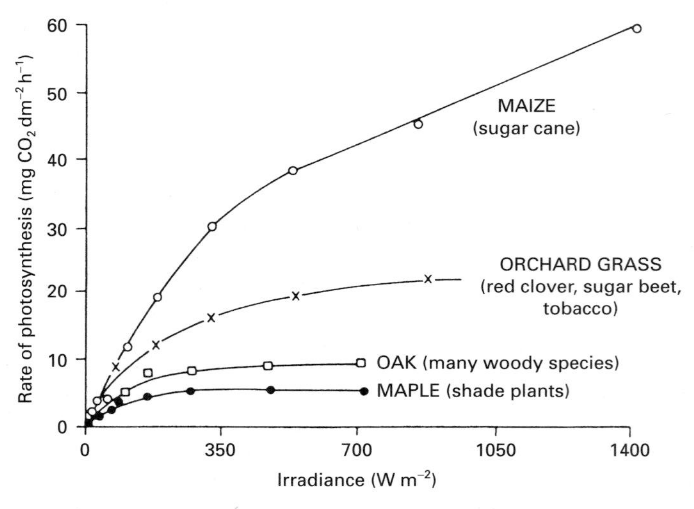


*Figure 2.9* The effect of irradiance upon the photosynthetic rate for four green plants which are typical of the plant groups indicated in parentheses. (From Hesketh and Moss, 1963.)

species of plant, some shade-loving plants having low values while others rarely receive sufficient energy to reach saturation. At extremely high levels of irradiance, however, the plant's temperature-regulating mechanism may break down and wilting and death occur.

## 2.5 Terrestrial radiation

As indicated in Figure 2.1, some solar radiation is absorbed directly in the atmosphere but most passes through to be absorbed at the Earth's surface. The absorption of solar radiation leads to heating which, in turn, leads to the emission of long-wave radiation. The amount of energy emitted is given by the Stefan–Boltzmann Law (equation (2.3)) modified for the effects of emissivity, $\varepsilon$. Thus:

$$E^* = \varepsilon \sigma T^4 \qquad (2.11)$$

Since the variation of emissivity with wavelength is small for solid objects, but large for gases, it is convenient to treat long-wave emission separately for the Earth's surface and for the atmosphere.

### 2.5.1 Surface long-wave emissivity

Although various types of surface on the Earth have different emissivities (Table 2.1), almost all equal or exceed 0.90. Artificial surfaces tend to be slightly below this value, vegetated surfaces just above it, and water considerably above it. In detailed calculations of the energy balance of a particular surface type, or when comparing contrasting surfaces, these differences must be incorporated. Nevertheless, for many calculations it is possible to assume a uniform emissivity for land surfaces, and a uniform one for water. Indeed for many purposes it is possible to assume that the Earth's surface acts as a black body for long-wave radiation.

Such simple assumptions are not possible for the atmosphere. Values of absorptivity and emissivity vary greatly with wavelength. The value also depends on the amount, temperature and pressure of the emitting gas.

### 2.5.2 Gaseous absorption and emission

Each gas absorbs radiant energy in a series of very narrow wavelength intervals, called **spectral absorption lines**. Commonly these lines are grouped together forming **absorption bands**. The location of the bands, and the strength with which they absorb, depends on the molecular structure of the gas. Further, as the amount of gas, its temperature, and the total atmospheric pressure increases, these bands are broadened and the amount of absorption increases.

The major absorbing gases which occur naturally in the Earth's atmosphere are water vapour, carbon dioxide and, to a much smaller extent, ozone and methane. Human activity is adding chlorofluorocarbons (CFCs) and various oxides of nitrogen to this list (Table 1.1). Most of the major absorption bands occur in the infrared portion of the spectrum (i.e. $\lambda \geq 3$ μm) and within this portion there is some absorption at virtually all wavelengths where significant amounts of energy are emitted (Figure 2.2). The major exception is the region between 8 and 14 μm, which is the so-called **atmospheric window**.

As a result of these absorption characteristics most of the long-wave radiation emitted by the surface of the Earth is absorbed by the atmosphere. Only a small portion is transmitted through the atmospheric window and escapes to space. The portion that is absorbed combines with the absorbed short-wave energy to heat the atmosphere. The heating stimulates the atmosphere to emit radiation. This emission is in all directions, but again we can simplify matters and note that a portion is emitted upwards and eventually is returned to space, while a portion travels downwards and is received at the surface as **incoming long-wave radiation**. The complementary term, **outgoing long-wave radiation**, can refer either to the energy emitted from the Earth's surface or to the energy leaving the atmosphere for space. The context makes it clear which of these is being discussed.

### 2.5.3 Effect of clouds

In a cloudy atmosphere the same basic radiation exchanges take place. However, complications arise because the presence of cloud modifies the wavelength dependence of emissivity. The most important effect is that clouds tend to 'close' the atmospheric window because of their very much greater absorptivity in the infrared region. This effect can readily be seen by comparing temperatures on a cloudy night to those on a cloudless one. In the former the closing of the atmospheric window prevents the significant loss of long-wave radiation that occurs in the latter. The result is that in normal circumstances the overnight temperature decrease on a cloudy night is much less than on a cloudless one.

### 2.5.4 Greenhouse effect

Since the atmosphere is almost transparent to solar wavelengths but absorbs terrestrial radiation strongly, an analogy was drawn long ago between the operation of the atmosphere and that of a greenhouse. The term **greenhouse effect** has passed into the literature to denote the process in the atmosphere whereby solar energy passes almost unimpeded to the surface, creates a warm surface which emits long-wave radiation, which in turn is absorbed in the atmosphere only to be re-radiated back to the surface. The net effect is to maintain the surface at a higher temperature than would occur if the atmosphere were as transparent to long-wave radiation as it is to short-wave energy. While the glass of a greenhouse displays similar radiative transfer characteristics, we now know that a greenhouse maintains its higher internal temperature largely because the shelter it provides reduces the wind speed and thus the turbulent transfer of energy away from the surface. This mechanism, vital for understanding local climates, is discussed in Section 10.1.

The greenhouse effect, albeit mislabelled, is a naturally occurring phenomenon, and it keeps the planet approximately 30 °C warmer than it would be without an atmosphere. However, one of the prime concerns of modern climatology is with an increasing greenhouse effect, and the consequent potential for elevated planetary temperatures. Atmospheric concentrations of the gases responsible for the effect, collectively known as the **radiatively active trace gases**, are increasing as a result of human activity. Some, such as carbon dioxide and nitrous oxide, are the result of industrial activity; others, including methane, are partly a consequence of increased agriculture. All, whatever their source, become mixed with the other atmospheric gases and have a world-wide effect. The result is that the magnitude of the greenhouse effect is increasing. This increases the energy content of the planet. The detailed consequences of this are far from clear, and indeed much of the rest of this book is devoted to exploring the climatic consequences of changes in energy regimes. Nevertheless, a reasonable assumption is that planetary temperatures will rise. Much of Chapter 14 is devoted to expansion of that statement.

All of the terrestrial radiation exchanges discussed above are summarised, as before, by Figure 2.1. This indicates clearly the small proportion of energy that passes through the atmospheric window directly to space and the large amount of long-wave radiation exchange between the Earth and the atmosphere. These incoming and outgoing fluxes are of the same order as the incoming short-wave flux. However, they are the result of continuous interactions without the marked diurnal variations characteristic of the solar flux.

## 2.6 The global radiation budget

On a time scale of several years there is approximate equality between the amount of solar radiation received from the Sun at the top of the atmosphere and the amount of long-wave radiation emitted to space. Hence there is a global radiation balance, as implied by the values given in Figure 2.1. Minor variations, such as those resulting from fluctuations in solar output, will occur on various time scales, including the annual one. Since radiation is the only commonly occurring form of energy that can be transferred through space, this planetary radiation budget is also the planetary energy budget and the overall balance must be maintained if the climate, again on a long time scale, is to remain stable. Certainly the presence of even a small imbalance, if maintained for a considerable time, would lead to climatic variations. Indeed, such variations may be one of the causes of climatic change.

Within the global and long-term energy balance are local and short-term imbalances. In fact, it is these imbalances that drive the climate as we know it. Although they can exist anywhere in the atmosphere, we are particularly concerned with two levels: the top of the atmosphere and the surface of the Earth. The former is the focus of this section; the latter will be discussed in the next.

At the top of the atmosphere the energy flows can be expressed in terms of a simple energy budget:

$$Q_t^* = K_t^* - L_t\uparrow \qquad (2.12)$$

where $K_t^*$ is the absorbed solar radiation below the point (i.e. the difference between the incoming and reflected energy), and $L_t\uparrow$ is the outgoing long-wave radiation from below the point. $Q_t^*$, the difference between the two, is the **net radiation**. Only when the two are equal do we have an energy balance.

### 2.6.1 Latitudinal and seasonal variations in radiative fluxes

The radiation imbalance at each latitude is readily illustrated by comparing latitudinally averaged values

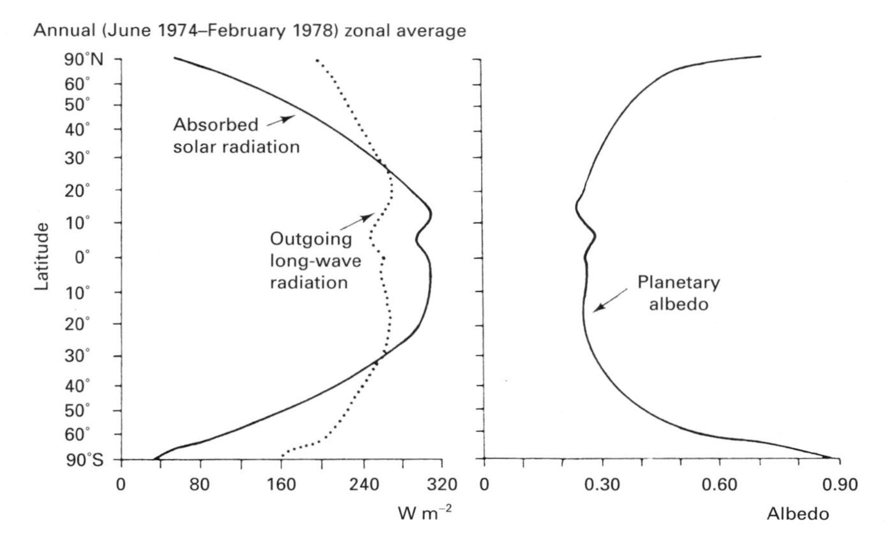


*Figure 2.10* Latitudinally averaged elements of the planetary radiation budget as observed by the scanning radiometers on board NOAA polar orbiting satellites in the period June 1974 to February 1978. Annually averaged values of absorbed solar radiation, outgoing long-wave radiation and albedo are shown. (Data from NOAA-NESS, Washington, DC.)

of the solar radiation absorbed and the infrared radiation emitted by the system (Figure 2.10). The amount of solar energy absorbed is affected by both the total amount incident and the albedo. Thus in high latitudes the prevalence of ice and snow further reduces by reflection the already low level of solar radiation. Similarly the increased albedo just north of the equator, caused by the large amount of cloud common in the area, leads to a decrease in the solar radiation absorbed there.

In contrast to the considerable gradient between equator and poles for the absorbed solar radiation, there is only a slight gradient for the emitted long-wave radiation. This suggests that the equator-to-pole temperature gradient is considerably less than would be expected from considerations of solar radiation alone. In fact, the mean annual climatic conditions at every latitude are considerably more hospitable for humans than they would be if each zone were in radiative equilibrium. The atmosphere acts, very efficiently, to redistribute energy. The tropical latitudes are cooled as they export energy to the mid- and high-latitude regions, which thus gain energy and are warmed. The redistribution of energy is a direct result of the equator-to-pole temperature gradient, which is itself a consequence of the latitudinal radiation imbalance. The transport between latitudes is accomplished by horizontal energy transfers using both the atmospheric and the oceanic circulation. These processes operate in such a way that the whole system, driven by the radiation imbalances, tries to achieve equilibrium.

Seasonal variations in the radiative fluxes must be added to obtain a complete picture of the average global radiation budget. This seasonal budget is a strong function of albedo variations (Figure 2.11a). The Northern Hemisphere albedo reaches a maximum in winter, underlining the importance of the seasonal snow and ice cover variations. The smaller, but often significant, mid-latitude and tropical variations result mainly from changes in vegetation cover. Seasonal changes in the Southern Hemisphere are much less marked. The emitted long-wave radiation (Figure 2.11b), for similar reasons, has a similar pattern.

### 2.6.2 Spatial variations in radiative fluxes

While these generalised results for seasonal and latitudinal variations are of great importance in understanding the dynamics of climate, it is frequently necessary to consider the actual spatial distribution of the components of the radiation balance. Small

(a)

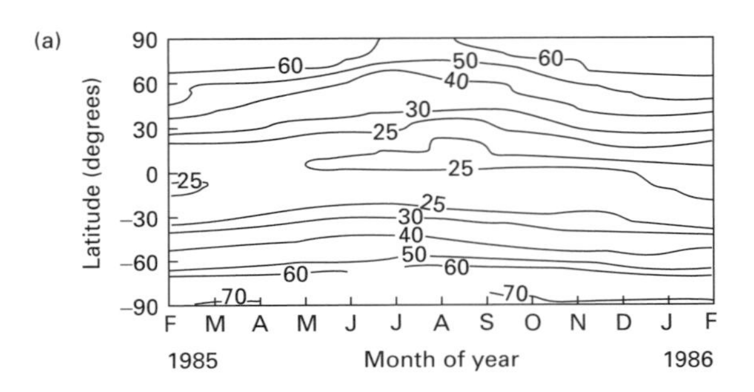


(b)

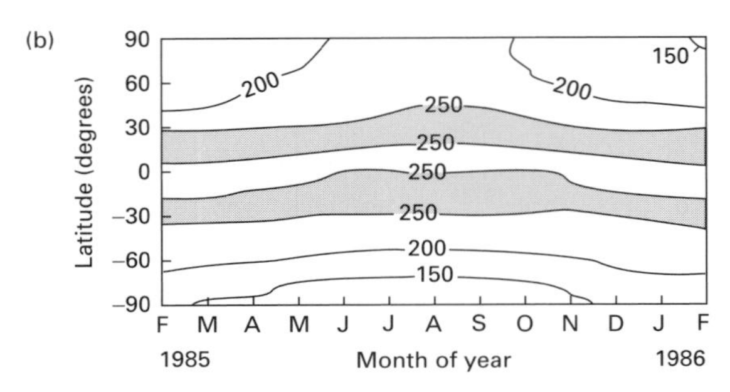


*Figure 2.11* Annual cycle of (a) planetary albedo (%) and (b) outgoing long-wave radiation (W m$^{-2}$) as a function of latitude. (From Harrison *et al.*, 1993.)

changes in one component for one area may, for example, be compensated by changes in the opposite direction at another place along the same latitude circle, thus leaving the zonal average unchanged. The change, however, may have tremendous consequences for the atmospheric dynamics both locally and globally. There are numerous short-term variations in energy exchanges at the top of the atmosphere, comparable to the stream of daily weather changes near the Earth's surface. We illustrate this using monthly average conditions for a typical month (Plate 1) as a refinement of the general values of Figure 2.11.

Regions of high albedo (Plate 1a) occur in the polar regions throughout the year. The high reflectivity of the North African desert is also permanent, but other areas of high albedo in the tropics and subtropics, such as in the eastern Pacific and South American regions, vary in position with season. The spatial variation is echoed in the outgoing long-wave radiation distribution (Plate 1b), with values decreasing poleward throughout the year, but with marked seasonal variations in the area between 30°N and 30°S. The most striking characteristic of the net radiation maps (Plate 1c) is the seasonal reversal in signal almost everywhere; except for the persistent negative values over both poles and the generally positive values near the equator.

## 2.7 Surface radiation budgets

The surface of the Earth is akin to the top of the atmosphere in that it also experiences a global, annual energy balance and local and short-term energy imbalances. Here, however, it is inappropriate to think solely of radiative fluxes. As Figure 2.1 indicates, the fluxes of latent and sensible heat, which are not radiative, must be incorporated. Thus, even on the global scale, there is not a true surface radiation balance. Instead it is useful to consider the total amount of radiation at all wavelengths that is absorbed at the surface. Expanding the concepts of equation (2.12), the surface **net radiation** is given by:

$$\begin{aligned} Q^* &= K\downarrow - K\uparrow + L\downarrow - L\uparrow \\ &= (1 - A)K\downarrow + L\downarrow - \varepsilon\sigma T_s^4 \end{aligned} \qquad (2.13)$$

where $Q^*$ is the net radiation; $K$ and $L$ are the short-wave and long-wave fluxes, respectively, and the arrows indicate the direction of the radiation streams. The first form of the equation emphasises that the net radiation is the sum of all the fluxes, while the second form emphasises the role of the surface characteristics (surface albedo, $A$, and emissivity, $\varepsilon$) in determining the amount of radiation absorbed.

### 2.7.1 Daily cycle of surface radiation budget

Although the net radiation and its component fluxes vary on a global and seasonal scale in much the same way as we have already discussed for the planetary case, we can here consider variations on a daily basis. In general, the short-wave element, $K$, is the component of the net radiation that is most variable. It varies with latitude, season and time of day, as indicated in equation (2.9). A typical diurnal variation for a mid-latitude location is shown in Figure 2.12. Figure 2.12(a) represents cloudless conditions, Figure 2.12(b) a day upon which cloud was intermittent

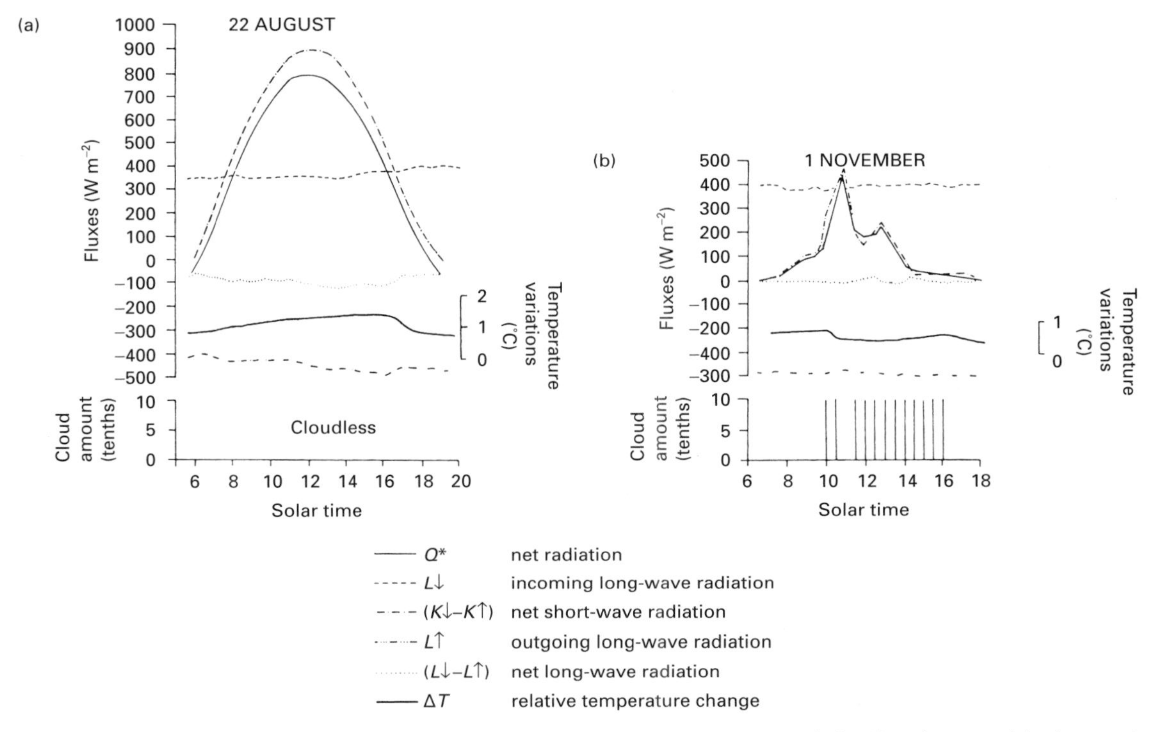


*Figure 2.12* Diurnal variation in the surface radiation budget components over Lake Ontario on a (a) clear and (b) partly cloudy day.

in the morning and formed a complete overcast in the afternoon. The net solar component simply echoes the incoming component, the magnitude of the variation being damped by the effect of the surface albedo. The incoming long-wave radiation is much less variable. The amount, depending on the temperature and humidity of the overlying air, will change through such effects as horizontal movements associated with winds. Thus while the net radiation follows the net short-wave component quite closely, especially in the cloudless case (Figure 2.12a), no diurnal pattern is completely predictable from radiative considerations alone. If there is relatively little air movement, an increase in $L\downarrow$ is to be expected in the afternoon, as the atmosphere is heated by direct absorption of solar energy and by transfer of heat from the underlying surface. The surface itself will be heated by the absorption of radiation and $L\uparrow$ will vary in response to that heating. The time of maximum $L\uparrow$ (N.B. negative and hence a 'dip' in Figure 2.12a) will be later than the time of maximum $K\downarrow - K\uparrow$, because heating will be occurring throughout the period with positive net radiation. During the night $Q^*$ is likely to be negative. Only the long-wave components are active and $Q^*$ will then depend on the difference between the radiative temperatures of the atmosphere and the surface. Again, for the atmosphere, this will partly depend on horizontal motions and on cloud conditions. There is likely to be greater long-wave loss on cloudless nights than when clouds close the atmospheric window.

Although the surface radiation budget, with the net radiation representing the energy absorbed at the surface, plays a major role in the determination of surface temperatures, other energy fluxes are also active. Thus to understand how temperatures are created and how they vary spatially and temporally, we need to consider both the relationship between energy and temperature and the whole surface energy budget. The following chapter considers these in detail.

## Box 2.A Application of solar radiation information

Throughout this chapter solar radiation has been treated as a source of energy. Indeed, it is often put to practical use as an alternative energy source for domestic use. Solar energy collectors on a roof are used to convert the radiation to heat for space or water heating (Figure 2.A.1). A fundamental piece of information required by a system designer is how much energy will be produced, and how reliably, by a collector of a given size. This obviously incorporates the question of how much solar radiation is received by a collector at a given location, and thus appears to be a climatological problem. Indeed, we can use this problem as the first practical example of applied climatology, developing information for a facility such as that of Figure 2.A.1, and using North Carolina information.

### *2.A.1 Information requirements for sizing solar energy collectors*

Discussion between climatologists and solar engineers indicated that climatological information was needed for sizing a collector at a particular location. The available design criteria dictated that long-term average monthly total solar radiation on a horizontal surface be used. Although the daily solar radiation data were available, they could not be used because the engineers had no method of incorporating them into the sizing calculations. The development of a more refined sizing method using daily data would be possible, of course, if warranted by anticipated performance and economic gains. Certainly, the daily data were needed to evaluate the performance of a particular system once it was installed. However, we are concentrating here on the creation of information directly and immediately useful for system sizing. Hence monthly averages are exclusively of concern.

The need for information at the collector site raised a typical climatological problem. Data are rarely available for the site of interest, so that some data manipulation and interpolation will be required. Thus in many instances, although there may be a specific target location, it is advantageous to provide maps which can be used at any site. In the current example, the solar engineers preferred such maps.

### *2.A.2 Developing maps from point data*

The network density for direct measurements of solar radiation in North Carolina is low, with only two stations in an area of 126 500 km$^2$. However they, along with four other stations, measure sunshine duration. Regression relations between solar radiation and sunshine (Box 2.I) were established for these two stations, and then used with the other four stations. Hence, although values must be treated with caution, there are six stations available for the spatial analysis of the data.

Although six stations is a small number for an area as large as North Carolina, the stations are well scattered through the state and reflect most of the local climate variations, particularly cloud amount, which are likely to influence radiation receipt. Further, radiation is spatially **conservative**, indicating that usually it does not vary very rapidly with distance. Hence it is possible to use these data with some confidence to construct statewide maps of radiation receipt. The development of the maps incorporated a mixture of objective statistical techniques and climatological insights. Objectively, several methods of interpolating values between data points, and extrapolating beyond the data points, were tested before a simple linear scheme was adopted. This was then subjectively modified to account for the known general climatology. In particular, along the coast the summer sea breeze gives a cloud regime which does not penetrate far inland. Hence the values for the coastal stations were restricted to the immediate coastline, and the influence of the inland sites was extended much farther eastward than would be usual. Similarly, the influence of the one mountain station in the west was extended to the whole mountain region, which runs NE–SW through the western quarter of the state. The resulting maps (Figure 2.A.2) reflect these modifications, but represent the best estimate of the monthly total solar radiation available. Hence they provide the information required by the solar heating system designers.

**Box 2.A (cont'd)**

(a)

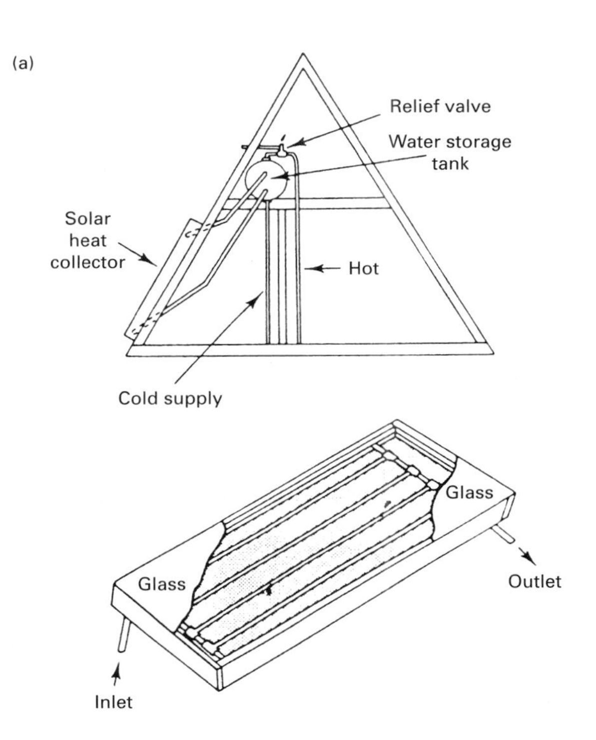


(b)

*Figure 2.A.1* (a) Layout and installation mode of a solar collector. (b) This method is used to heat this dwelling in Jacksonville, North Carolina. (Courtesy of John J. Busenberg, Astron Technologies Inc.)

**Box 2.A (cont'd)**

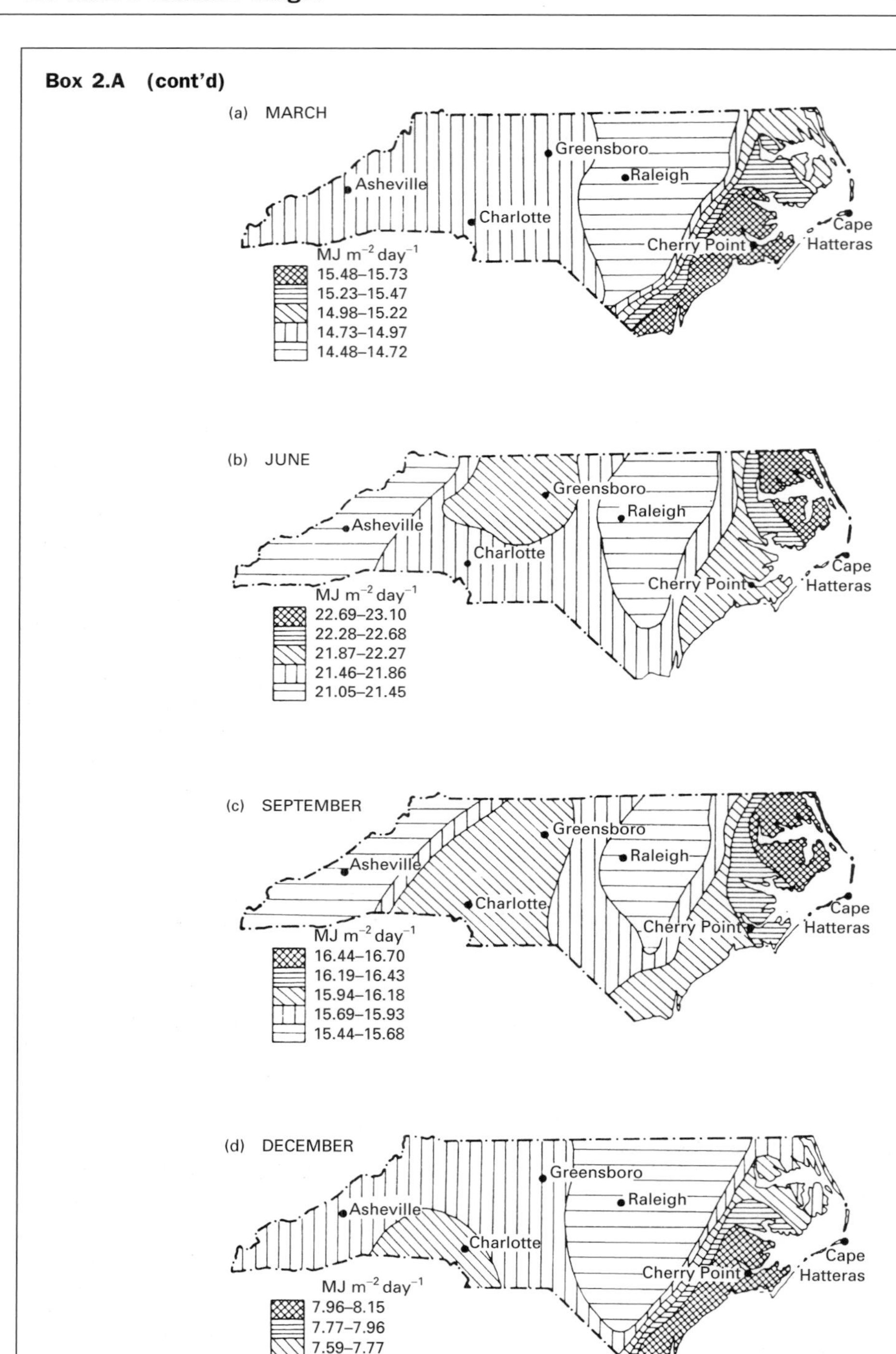


*Figure 2.A.2* Spatial distribution of monthly averaged daily total solar radiation on a horizontal surface in North Carolina for (a) March, (b) June, (c) September and (d) December. The six named stations are the complete observational network used. (After Robinson and Easterling, 1982.)

# CHAPTER 3 Energy and temperature

We have already indicated in the previous chapter that when a body absorbs energy, its temperature increases. In fact there is a simple relationship between the change in energy of a body and its temperature change:

$$\Delta E = \rho C \Delta T \qquad (3.1)$$

where $\Delta E$ is the change in energy and $\Delta T$ the change in temperature of a body of density $\rho$ and specific heat $C$, with $\rho C$ being termed the heat capacity of the body. The relationship here relates to changes for a unit volume in unit time, so that both $\Delta T$ and $\Delta E$ are proportional to rates of change. Thus the equation enables us to calculate heating rates directly since $\Delta T/\Delta t = (\Delta E/\Delta t)/\rho C$.

If we consider a situation where only radiative energy is involved, we can perform a simple analysis. As a body absorbs solar radiation its temperature will rise in accordance with equation (3.1). This will lead to an increase in the amount of long-wave energy emitted, in accordance with the Stefan–Boltzmann law (equation (2.3)). Neglecting non-radiative energy fluxes, the temperature will increase until the absorption rate is equal to the emission rate. The net change in energy within the body will thus be zero, there will be no further temperature change and the body will be in **radiative equilibrium**. The actual temperature at this point will depend, for a given incoming radiation stream, on the albedo of the body, controlling the amount absorbed, and on its emissivity, controlling the amount emitted, for a given temperature. If either the emissivity or the albedo is increased, the equilibrium temperature is decreased. Note that the heat capacity of a body is not important in determining the equilibrium temperature. It is very important, however, in determining the time needed to reach that temperature, the length of time increasing as the heat capacity increases.

This rather straightforward relationship between radiative exchanges and heating rates is sometimes directly applicable to the free atmosphere, but for most times and places radiation is rarely the sole determinant of temperature. The other forms of energy, and their own exchanges, must be considered for a full analysis of the causes of the temporal and spatial distribution of temperatures over the Earth. It is most convenient to discuss conditions at the surface of the Earth before considering the atmosphere. Hence the first three sections of this chapter will develop the concept of energy flow as the process controlling temperature change, leading to a description and explanation of surface temperatures on the global scale. Thereafter the conditions in the atmosphere will be discussed.

## 3.1 Surface energy budgets

The temperature at the surface of the Earth is a response to all of the energy fluxes affecting the surface. Thus the energy responsible for temperature changes is given by the **energy budget** equation

$$\Delta E = Q^* - (H + LE + G) \qquad (3.2)$$

where $Q^*$ is the net radiation; $H$ and $LE$ are, respectively, the sensible and latent heat fluxes into the air; and $G$ is the heat flux into the underlying surface. Sensible energy flows from high- to low-temperature areas predominantly by the vertical movement of air warmed by surface contact. Latent energy is associated with the movement of water vapour molecules, and the exchanges are the result of evaporation and condensation. $G$, commonly called the ground heat flux, moves heat to and from lower levels by conduction from hot to cold layers in a manner similar to heat flow along a heated rod. When the underlying surface is water, $G$ can also be by convection.

As written in equation (3.2), the energy budget equation emphasises the contrast and links between the radiative and non-radiative fluxes. It also indicates that there is an imbalance between them which drives temperature changes. As with the simple radiative situation discussed at the beginning of this chapter, the surface can be conceived as striving to achieve energetic, and thus temperature, equilibrium. Given the constantly varying atmospheric conditions, including those associated with the seasonal and diurnal cycles of solar and net radiation considered in the last chapter, this is achieved only in a transient manner. However, on the global annual scale indicated by Figure 2.1, $\Delta E = 0$, which represents the planetary **energy balance**.

As an example of typical energy exchanges we can again consider the diurnal energy cycle. Surface temperatures start to rise as soon as the net radiation becomes positive. In most cases the surface becomes warmer than the overlying air and a sensible heat flux upwards is initiated. The net radiation is also likely to provide energy needed for evaporation, and thus latent heat transfer begins. At the same time heat is transferred from the warm surface to lower layers of the underlying medium. This situation is likely to continue throughout the period of positive net radiation, although changes in the air above the surface may disrupt the simple pattern. A particular airflow may bring in air that is warmer than the surface and create a sensible heat flux towards the ground. If the underlying medium is water, internal currents may create a similar effect. If the ground dries, there can be no upward latent heat flux. Such influences obviously are outside the immediate and direct control of the near-vertical energy exchanges we are considering here. However, if we continue our idealised diurnal cycle, once the net radiation becomes negative and radiative cooling dominates, the non-radiative fluxes tend to be directed towards the surface, decreasing the rapidity of cooling. Thus, in general, the non-radiative energy transfers tend to minimise the diurnal temperature changes that would result from radiative exchanges alone.

### 3.1.1 Heat transfer from the surface

The transfer of energy away from a surface into the underlying medium and the overlying air is, at least close to the surface, primarily by conduction. This can be considered by analogy with heat flow along a rod. When one end is heated the heat will start to flow from the hotter to the colder region. The rod is thus progressively heated, with a maximum temperature change at the heated end, the change being gradually damped at larger distances from that end. The rate of heat penetration is dependent on the **thermal diffusivity**, $K^*$, of the material (or equivalently dependent on the **thermal conductivity**, $K = \rho c_p K^*$, where $c_p$ is the specific heat at constant pressure). At any instant after the start of heating, the depth of penetration, which can be defined as the point where the temperature rise is a small fraction, say 5%, of that at the heated surface, is proportional to $\sqrt{K^*}$. If we have a heating cycle, as with the diurnal cycle, rather than steady heating, temperature waves will spread vertically downwards with their amplitude diminishing as they progress (Figure 3.1). Eventually a point will be reached where the diurnal cycle will be sufficiently damped to be negligible. Values of $K^*$, together with depths of penetration for various surface types and for the atmosphere, are given in Table 3.1. Although values differ for various land surfaces, the major difference is between solid land, stirred water, and stirred air. Penetration rates are more rapid and energy reaches a greater depth in water than in land, while penetration in air is greater than both. Heat transfer in solids can only be through molecular interactions, the true **conduction** process. However, both air and water can transfer heat through the mass movements associated with stirring: **turbulent transfer** and **convection**. These will be considered in more detail in Chapters 10 and 4, respectively.

Since different substances conduct heat away from the surface at different rates, the surface temperature resulting from a particular amount of energy input will also differ. The volume over which the heat is effective is proportional to $\sqrt{K^*}$ and the temperature rise is similarly found to be proportional to $\rho C\sqrt{K^*}$. This is often called the **conductive capacity**, $C^*$ (Table 3.1). At the interface between two substances the heat will be shared in proportion to their respective conductive capacities. The temperature range at the interface must be the same for both media and will be approximately given by the inverse of the sum of the conductive capacities.

For a moist or wet surface the latent heat flux provides an additional means of energy transfer. Available energy at the surface, notably that associated with positive net radiation, is used to stimulate the phase change of water from the solid or liquid state into a gas. This provides to individual water molecules

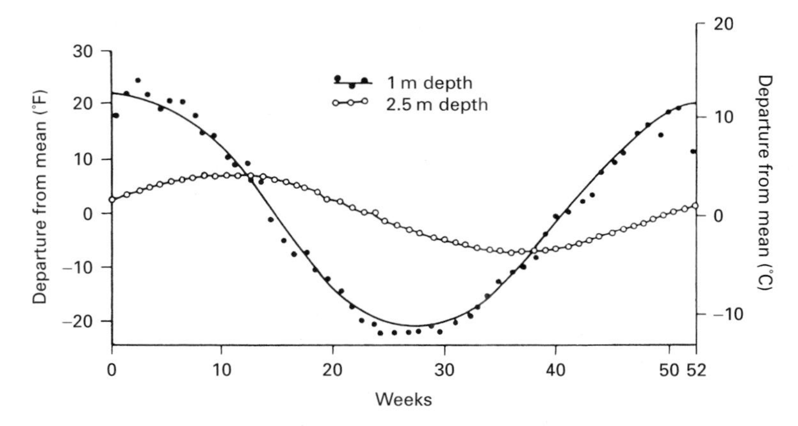


*Figure 3.1* The annual temperature wave at soil depths of 1 and 2.5 m measured at Griffith, Australia with fitted sine curves. Temperature departures from the mean value are plotted as a function of weeks of the year. (From Rose, 1966.)

*Table 3.1* Thermal properties of air and various surfaces

| Substance | Heat capacity, $\rho C$ ($J\ m^{-3}\ K^{-1}$) | Thermal diffusivity, $K^*$ ($m^2\ s^{-1}$) | Thermal conductivity, $K$ ($W\ m^{-1}\ K^{-1}$) | Conductive capacity, $C^*$ ($J\ m^{-2}\ K^{-1}\ s^{-1/2}$) | Penetration depth | |
|---|---|---|---|---|---|---|
| | | | | | Diurnal (m) | Annual (m) |
| Ice | $1.89 \times 10^6$ | $1.2 \times 10^{-6}$ | 2.27 | $2.1 \times 10^3$ | 0.6 | 10 |
| Dry sand | $1.26 \times 10^6$ | $1.3 \times 10^{-7}$ | 0.16 | $4.5 \times 10^2$ | 0.2 | 4 |
| Wet soil | $1.68 \times 10^6$ | $1.0 \times 10^{-6}$ | 1.68 | $1.7 \times 10^3$ | 0.5 | 9 |
| Still water | $4.2 \times 10^6$ | $1.5 \times 10^{-7}$ | 0.63 | $1.6 \times 10^3$ | 0.2 | 4 |
| Stirred water | $4.2 \times 10^6$ | $5.0 \times 10^{-3}$[a] | $2.1 \times 10^4$ | $3 \times 10^5$[a] | 40[a] | |
| Still air | $1.26 \times 10^3$ | $2.0 \times 10^{-5}$ | $2.5 \times 10^{-2}$ | 5.6 | | |
| Stirred air | $1.26 \times 10^3$ | 10.0[a] | $1.3 \times 10^4$ | $4 \times 10^3$[a] | 1500[a] | Troposphere |

[a] These values are not determined, as are the others, by molecular properties and so cannot be measured precisely as in laboratory experiments.
*Source*: from Petterssen (1969).

sufficient kinetic energy to escape the attraction of the surrounding molecules. Once free of the surface, the water vapour molecules participate in the turbulent transfer and convection of the atmosphere itself. The energy in these molecules remains 'latent' until condensation occurs, commonly through cloud formation well away from the Earth's surface. This will be treated as part of the atmospheric water cycle in Chapter 4.

The energy available for the sensible, latent and ground heat fluxes is not partitioned equally between them. For many land surfaces the soil heat flux is small and can be neglected, so that the major partitioning of energy is between sensible and latent heat. The ratio of the two ($H/LE$), the **Bowen ratio**, is an indication of this partitioning. In general, surfaces act to keep the ratio at a minimum. Indeed a moist surface will increase little in temperature while evaporation is occurring, but there will be a rapid temperature rise once it has dried out and the sensible heat flux takes over as the major energy transfer mechanism. Hence the Bowen ratio links water availability and temperature. On the broadest scale, when considered on an annual basis, it plays a major role in the energetics of regional climates, and will be considered further in Chapters 8 and 9.

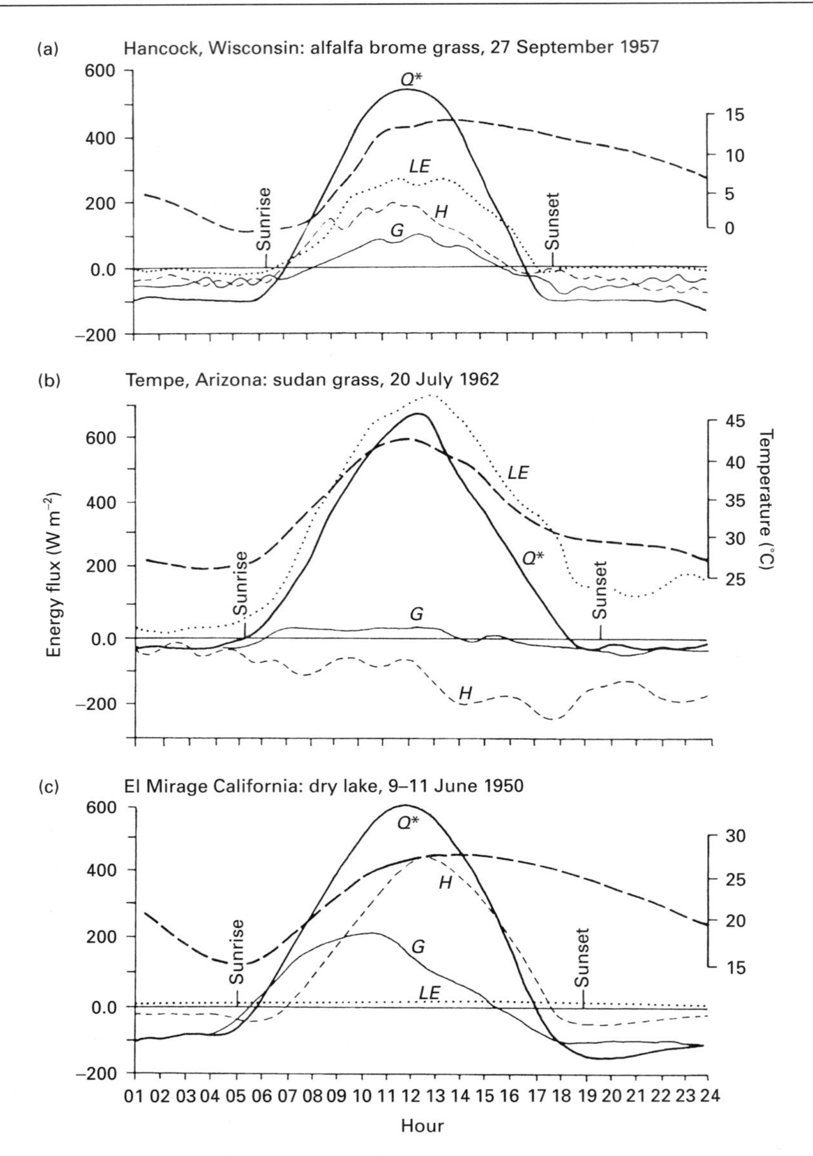


*Figure 3.2* Diurnal variation of the surface energy budget components over grass at (a) Hancock, Wisconsin and (b) Tempe, Arizona; and (c) over soil at El Mirage, California. $Q^*$ is net radiation, $H$ and $LE$ are sensible and latent heat fluxes, and $G$ is the heat flow into the surface. The dashed curve shows the variation in surface temperature. (From Sellers, 1965.)

### 3.1.2 Annual and diurnal energy cycles

The diurnal cycle of the various components of the energy budget depends greatly, through the influence of the Bowen ratio, on surface type. Figure 3.2 illustrates the diurnal variations in the components for three contrasting surfaces. All observations were made under clear skies (as compared with Figure 2.12, which shows the effect upon the radiation in terms of cloudy skies) in light to moderate wind conditions. The contrast in *LE* between the grass surfaces and the bare desert soil is immediately apparent and not

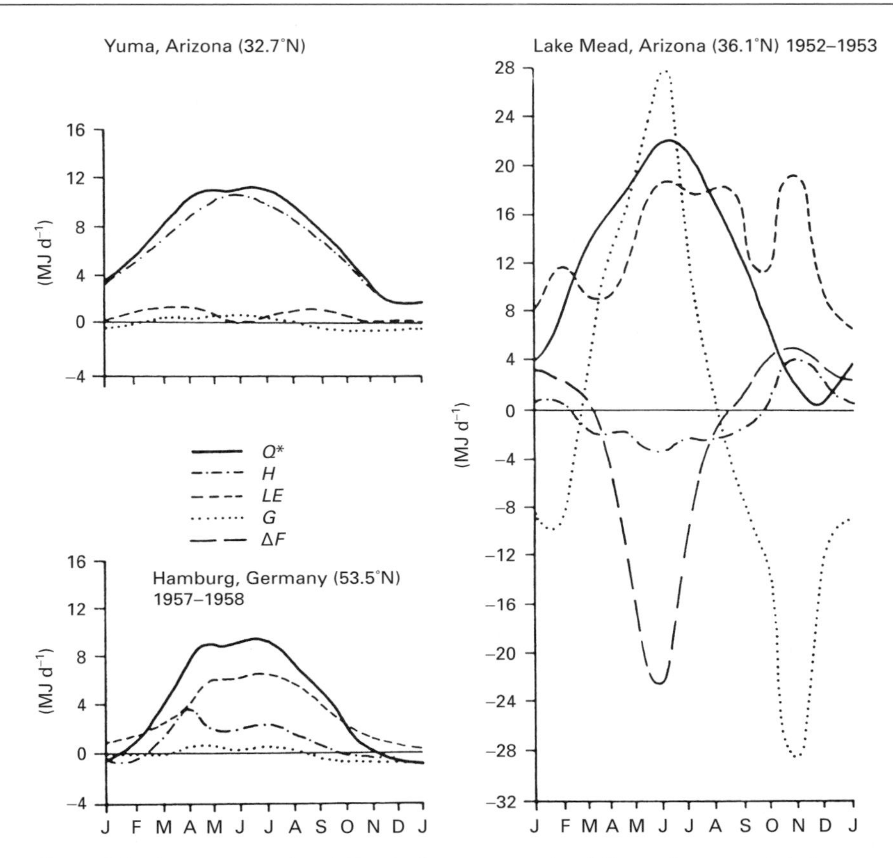


*Figure 3.3* Average annual variation in the components of the surface energy balance at Yuma, Arizona; Lake Mead, Arizona and Hamburg, Germany. $Q^*$ is net radiation, $H$ and $LE$ are sensible and latent heat fluxes, $G$ is the heat flow into the surface and $\Delta F$ is the net subsurface flux of heat out of the water column. (From Sellers, 1965.)

surprising. The dry desert atmosphere above irrigated grass in Arizona promotes rapid evaporation (maintains a high humidity gradient) and hence a high *LE*. This flux is much smaller into the more humid air (a lower humidity gradient) at Hancock, Wisconsin. In both California and Wisconsin the sensible heat flux is upward during the day, and downward at night. This is the most common situation. However, in Arizona the flux is always directed towards the ground, indicating that the energy needed for evaporation is being drawn from the warm air as sensible heat as well as from the radiation fluxes. The diurnal temperature changes right at the surface, calculated from the energy fluxes, clearly indicate the role of the surface in creating local temperatures.

Contrasting surface types also influence the components of the energy balance on an annual basis (Figure 3.3). The moist conditions at Hamburg lead to a flux of *LE* that almost always exceeds *H*. The dry Arizona surface has a very low *LE* and almost all of the net radiation absorbed by the surface is removed by the sensible heat flux. For both stations the ground heat flux is low throughout the year. Over the water of Lake Mead high evaporation rates occur all year while the sensible heat flow is always relatively small. In individual months there is a large energy exchange between the surface and the underlying water. In addition, major horizontal energy movements are indicated for the lake. Throughout this chapter we virtually ignore such movements because in most cases they are negligible compared to the vertical flows. In some cases, as here for the lake water, and as considered in Chapter 10 for air, horizontal motions have a great impact on local energy budgets and thus on local climates. The larger scale horizontal winds, of course, play a vital role in vertical energy exchanges, as considered in the next section.

The ground heat flux, $G$, is commonly a rather small component of the energy budget. Indeed, in many places the daily or annual total flux equals zero. Consequently it is often possible to ignore it when investigating the energetics of the atmosphere. However, complete neglect can be very misleading. In practical terms, $G$ is responsible for temperature changes in the soil (Figure 3.1), which lead to the springtime rise in temperatures and the germination of winter-dormant seeds. More broadly, $G$ provides a major means of energy input to the oceans. This energy, whether reflected in the pervasive effects of continentality to be considered in the next section, or in the fluctuating values of sea surface temperatures and the changes in ocean current direction and strength considered in many later sections, have a profound effect on climate and climate variability.

## 3.2 Major surface types and their consequences

The preceding discussion indicates that temperature has a direct relationship with the energy budget. It also indicates that the type of surface plays a major role in determining that relationship. Hence, as we work towards an understanding of the spatial distribution of temperature, we must consider the role of the surface in more detail. This has two major aspects: differences arising directly from the interaction of different surfaces with the energy streams; and differences which arise when different surface types interact. In order to consider these, however, we have also to consider the spatial scales which are of concern to us. No two places on the surface of the Earth have exactly the same type of surface. Changes from one side of a field to another may be vital for an agriculturalist, but of little significance to someone interested in global climate change. Although as climatologists we may be interested in both, it is advantageous to separate them. Here we consider the global scale, leaving the local scale until Chapter 10.

### 3.2.1 Land, water, snow and ice

The vital role of albedo in energy exchanges was emphasised previously. Considering the values of surface albedo (Table 2.1) on the broadest scale, the surface of the planet can be divided into three types: water with low values; land with intermediate ones; and snow and ice with a high albedo. Snow and ice, of course, are largely confined to high latitudes, where their role in absorption has already been considered (Figure 2.10).

Outside the polar regions land and water are intermixed. Albedo differences dictate that, for a given incoming amount of solar energy, more will be absorbed by the ocean than by land. The oceans are thus a great storehouse of energy and, as the surface currents and vertical overturnings slowly move that energy around, they have a profound influence on our climate. The long-term implications of this will be treated in detail in Section 7.3 and throughout Chapter 13. Here, however, we are concerned with the more immediate influence on the global distribution of temperature, and the major concern is with the contrast in temperature range between the two surface types. Because of the differences in heat capacities, transparency and ability to mix, land and water have very different conductive capacities (Table 3.1). At the surface, where both are in contact with air, the temperature range is proportional to the inverse of the sum of conductive capacities. For the interface between air and land this is about 7; for water and air it is close to 0.14. Consequently the temperature range is about 50 times larger over land than water. It therefore follows that land surfaces heat and cool more rapidly, and have a greater temperature range, than do water surfaces. This result applies on both diurnal and seasonal time scales. The most direct impact for the global distribution of temperature is on the continental space scale, where it is expressed as the concept of **continentality**, indicating that continental interiors heat (cool) more rapidly, and to a higher (lower) temperature, than oceanic or coastal locations (Figure 3.4). If we use the values in Table 3.1 with fairly typical values for the radiation fluxes and calculate the annual range of temperature, we find that over the oceans it is typically only a few degrees, while for land the annual range can be many tens of degrees. These are somewhat larger values than those observed. The differences arise partly because such a calculation neglects the effects of heat fluxes, and partly because interactions between surfaces, fostered by horizontal motions of energy, are ignored.

### 3.2.2 Interactions between surfaces

Almost all of our discussions of energy and temperature so far have implicitly considered only vertical

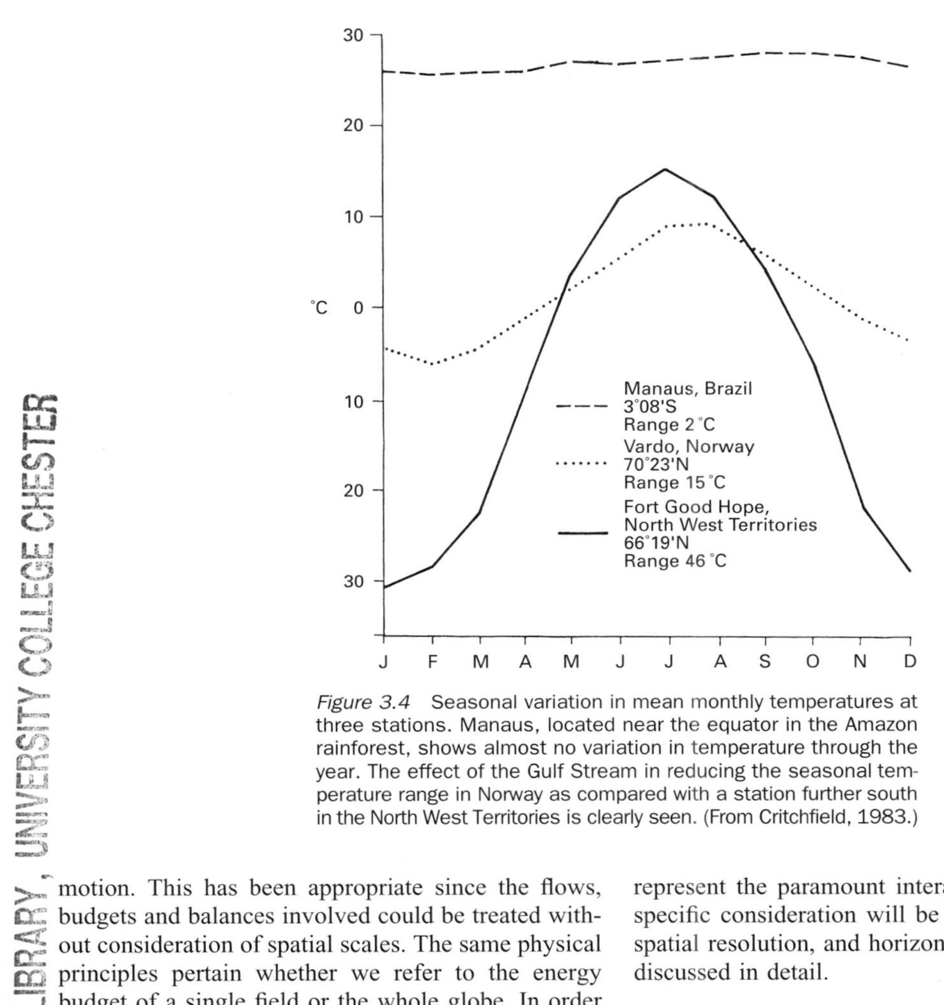


*Figure 3.4* Seasonal variation in mean monthly temperatures at three stations. Manaus, located near the equator in the Amazon rainforest, shows almost no variation in temperature through the year. The effect of the Gulf Stream in reducing the seasonal temperature range in Norway as compared with a station further south in the North West Territories is clearly seen. (From Critchfield, 1983.)

motion. This has been appropriate since the flows, budgets and balances involved could be treated without consideration of spatial scales. The same physical principles pertain whether we refer to the energy budget of a single field or the whole globe. In order to complete the discussion of global temperatures, however, we have to consider horizontal motions. The continentality effect indicates, for example, that oceans are cooler than land in the summer. Our energy analysis also indicates that the air above the ocean must be cooler than that over land. Hence, if any oceanic air blows onto land it will, at the very least, modify the sensible heat flux. Indeed, any time there is horizontal motion, i.e. wind, there is going to be interaction between surfaces.

All components of the energy budget will be involved in these horizontal interactions. However, their relative roles will be dependent on the nature of the surfaces involved and the spatial scale being considered. In this and the following six chapters we will be concerned primarily with the global and regional scales, where considerations of continentality represent the paramount interactions. In Chapter 10 specific consideration will be given to a much finer spatial resolution, and horizontal interactions will be discussed in detail.

## 3.3 Temperatures at the Earth's surface

On the global scale, the factors responsible for the distribution of temperature over the Earth's surface are latitude, continentality and altitude. The first two of these are a direct response to the energy budgets we have been considering, and so we discuss them first, and then add the effect of altitude.

### 3.3.1 Global patterns of mean sea-level temperature

The global distribution of sea-level temperature (Figure 3.5) clearly shows the effect of latitude. Using Northern Hemisphere nomenclature for convenience, over the oceans in both winter (December–February,

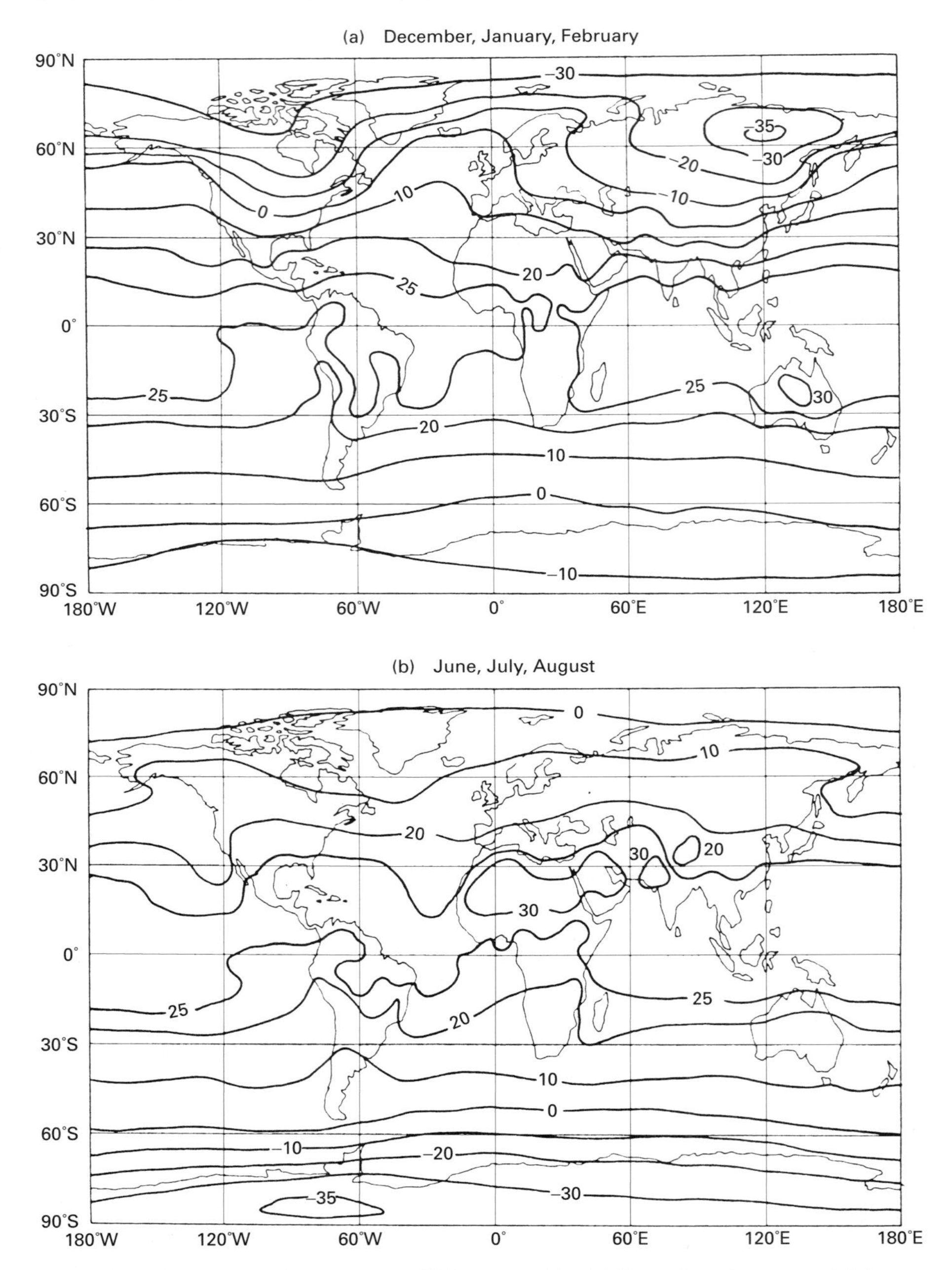


*Figure 3.5* Mean sea-level temperature (°C) averaged for (a) December, January and February (1963–1973) and (b) June, July and August (1963–1973). (From Oort, 1983.)

Figure 3.5a) and summer (June–August, Figure 3.5b) seasons the highest temperatures lie in a belt close to the equator, while over land they are displaced slightly northward in summer, and southward in winter. Maximum temperatures exceed 30 °C over portions of these land areas. Temperatures decrease poleward, the latitudinal dependence being particularly clear in the Southern Hemisphere in winter. The superimposition of the continentality effect is displayed by the minimum temperature regions of the Northern

Hemisphere winter, with the lowest values (below –30 °C) in north central Asia. There are sharp temperature contrasts between land and sea, particularly on the western sides of continents. In winter, at any given latitude the Northern Hemisphere land is colder and the Southern Hemisphere land warmer than the adjacent ocean. A similar phenomenon, with the hemispheres reversed, occurs in June–August. In this season Antarctica is the coldest region of the globe while the warmest areas are those continental areas just north of the equator.

A close comparison of the conditions in the two seasons clearly indicates that seasonal changes in ocean surface temperature are relatively minor, but that mid-latitude continental interiors suffer a much greater range. Figure 3.4 indicates the annual course of mean monthly temperatures for three stations: a continental interior, a coastal location and a tropical situation. These results are largely as expected from our previous considerations of energy exchanges.

### 3.3.2 Effects of altitude

While Figure 3.5 clearly shows the dependence of sea-level temperatures on latitude and continentality, much of the land surface of the Earth is not at sea level. In general, there is a decrease in temperature with altitude, and a map of actual surface temperatures tends to reflect the continental topography. In some mountainous regions the great local relief makes temperature generalisation difficult – a problem we shall consider in Section 7.4. In other areas, notably the high plateaux of many continental interiors, the temperatures of whole regions are modified. Four stations in Kenya, representing a cross-section from sea-level Mombasa on the Indian Ocean coast westward and upward through Garissa and Nairobi to the interior highlands at Eldoret (Figure 3.6) shows a highland interior some 10 °C cooler than at sea level. Reference to Figure 3.5 suggests that this is roughly equivalent to a poleward movement of about 10° latitude.

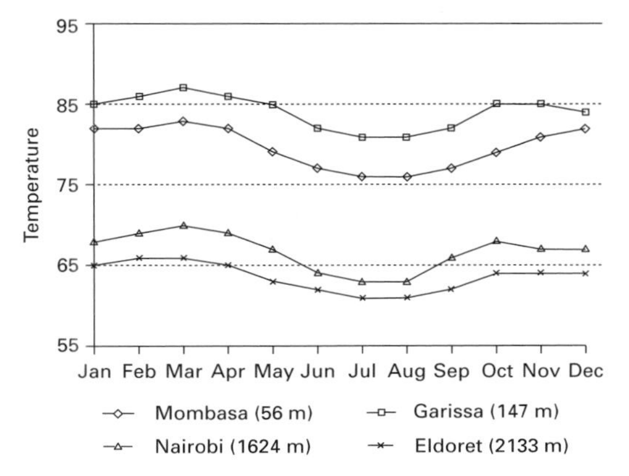


*Figure 3.6* Monthly mean temperature (°C) for four stations in Kenya, showing the effect of increasing altitude on temperature. Station heights (in metres) are given in the key.

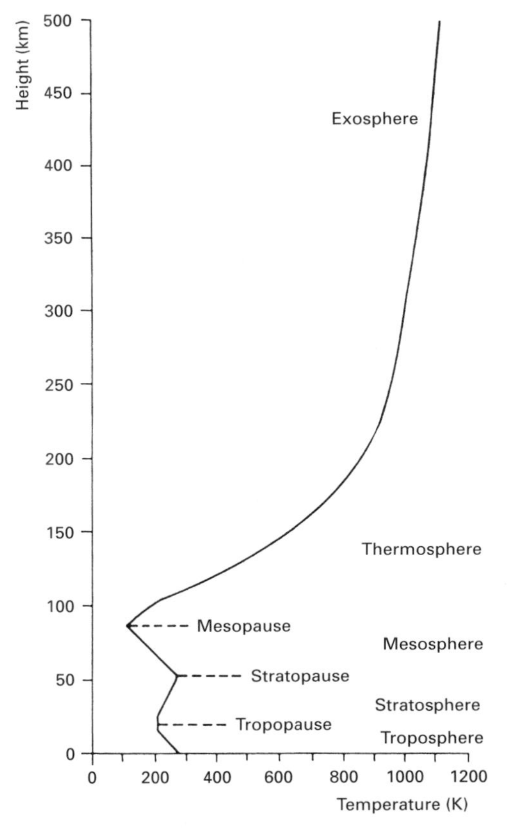


*Figure 3.7* Temperature variation with height in the atmosphere (sometimes termed the 'lapse rate'). Note the three regions of increased temperature where absorption occurs: (i) the surface (visible and near-infrared radiation); (ii) lower stratosphere (ultraviolet absorption by $O_3$); (iii) thermosphere (high energy absorption causing photoionisation).

## 3.4 The temperature structure of the atmosphere

One important consequence of energy exchanges is the development of a distinct temperature structure in the atmosphere. An averaged temperature profile of the atmosphere (Figure 3.7) indicates that there

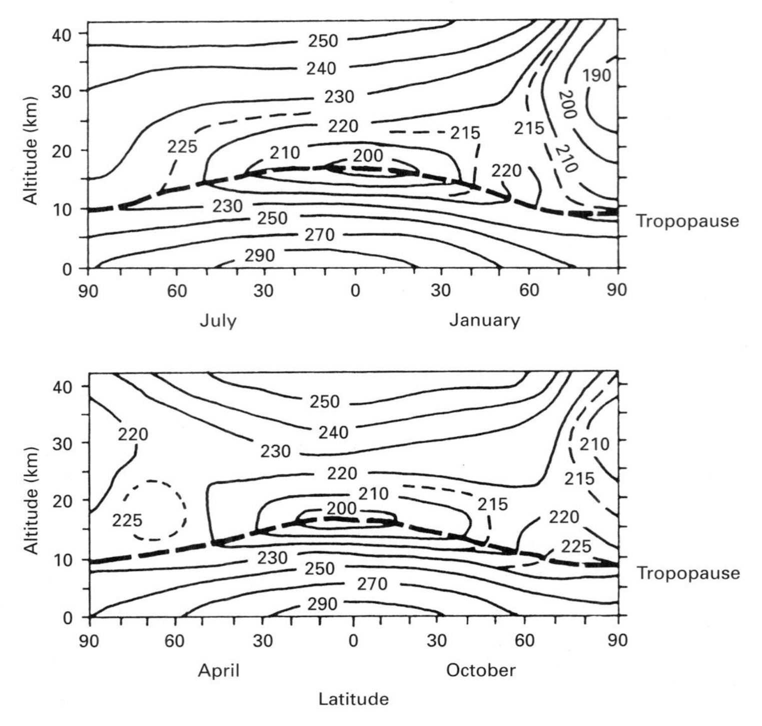


*Figure 3.8* Seasonal and latitudinal temperature (K) structure of the Northern Hemisphere atmosphere for the four mid-season months. Note the variation in height of the tropopause and the lower temperatures in the polar stratosphere in winter. (From Manabe and Strickler, 1964.)

are three major 'warm' regions where radiation is absorbed directly: the surface, the stratosphere and the thermosphere. These regions of absorption create a series of distinct layers with distinct temperature gradients within the atmosphere. These layers are termed 'spheres'. The spheres are separated by levels of demarcation called 'pauses' at which the temperature gradient reverses. The cause of the heating at the surface, the surface absorption, has been noted already. Heating in the lower stratosphere is caused by absorption of high-energy ultraviolet radiation by ozone. In the thermosphere the heating is also the result of interaction between radiation and atmospheric constituents, through a process known as photoionisation.

### 3.4.1 Importance of the troposphere

Most of the activity that we usually associate with climate occurs in the layer between the lower stratosphere and the surface. This region accounts for approximately 80% of the atmospheric mass and is the region where the atmosphere is characterised by a decrease in temperature with elevation. The dominant mechanism for heating is through surface absorption of solar radiation and re-radiation from the surface. The height of the top of the layer, the tropopause, varies with latitude and season, being generally higher at the equator than at the poles and higher in the summer hemisphere than in the winter one (Figure 3.8). The detailed structure of the troposphere at a given time and place is dictated by the levels at which radiation absorption occurs and is influenced by vertical motions in the atmosphere.

**Box 3.A Applications of energy and temperature information**

Surface temperature information has been abundant for many years. Further, temperature also has an impact on many human activities, notably in the health, agriculture and energy fields. Consequently there has been a great deal of use of temperature information for practical applications and numerous analyses of the relationship between temperature and human activities. We have selected as examples a statistical analysis for practical application in agriculture, an impact and potential application in the energy production sector and a use of energy information in agriculture.

*3.A.1 Temperature and the length of the growing season*

One prime agricultural concern is the length of the growing season. A farmer's definition of growing season will vary with, among other things, the climate of the region. In tropical regions temperatures rarely vary much from month to month (Figure 3.6) and it is the availability of water that is important. While water is often a critical concern in other areas, in many mid-latitude regions the growing season is defined as the length of time during which the temperatures remain above a particular threshold value. Here we concentrate on the latter aspect, although the basic ideas are equally applicable to growing seasons dictated by precipitation.

Crop growth is initiated when temperatures in spring rise above a certain threshold value and ceases in autumn when they fall below this value. From the temperature record it is possible to determine for a given place and year the length of time between these events and thus the length of the growing season. Since this length will vary from year to year it is not possible to use data for one year to provide a forecast for another. Long periods of data are required to produce such forecasts. A network of measuring sites is used to prepare a map of the average length of the growing, or in this case, freeze-free, season (Figure 3.A.1), enabling the agriculturalist to obtain a general idea of the growing season length at a particular location. Further information can be gained by analysing the long-period record to determine the probability with which the growing season will have a particular length (Table 3.A.1). Using the same method, the probability that the season will start on or after a given date in spring and will end on or before a given date in autumn is calculated. Thus a knowledge of planting and harvesting dates, as well as the time available for crop growth, can be acquired. This information therefore provides a probability forecast for a particular season. The individual farmer can use it to balance his strategy for maximum yields against the likelihood of crop failure for climatic reasons.

The temperature values chosen for Table 3.A.1 are rather low for most crops. Nevertheless the figure and table are illustrative of the type of information that can be produced for agricultural purposes and are suggestive of applications in other industries, such as the construction industry, where temperatures can dictate work schedules.

*3.A.2 Energy demand as a function of temperature*

The energy industry commonly uses temperature observations in a way which is similar to that described above. Demand for energy, particularly electricity, natural gas and home heating oil, is temperature dependent. This is primarily the result of the increase in building heating needs as the ambient temperature decreases or, in regions where air conditioning is common, an increase in demand as

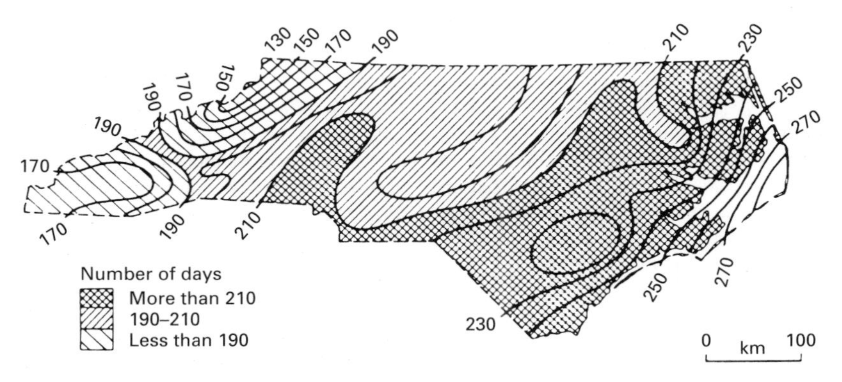


*Figure 3.A.1* Average length of the freeze-free season in North Carolina (number of days). (From Clay *et al.*, 1975.)

**Box 3.A (cont'd)**

*Table 3.A.1* Freeze date probabilities

| Temperature (°C) | Probability | | | | | | | | |
|---|---|---|---|---|---|---|---|---|---|
| | 0.10 | 0.20 | 0.30 | 0.40 | 0.50 | 0.60 | 0.70 | 0.80 | 0.90 |
| *Probability of longer than indicated period (days) with temperatures continuously above indicated value* | | | | | | | | | |
| 0 | 276 | 266 | 258 | 251 | 245 | 239 | 232 | 224 | 213 |
| –2.2 | 318 | 303 | 292 | 283 | 274 | 266 | 256 | 245 | 230 |
| –4.4 | >365 | 329 | 304 | 304 | 295 | 286 | 278 | 268 | 254 |
| –6.7 | >365 | >365 | >365 | >365 | 340 | 330 | 322 | 313 | 303 |
| –8.9 | >365 | >365 | >365 | >365 | >365 | >365 | >365 | >365 | 345 |
| *Probability of the occurrence of the listed temperatures at a later date in spring (month/day) than the date indicated* | | | | | | | | | |
| 0 | 4/1 | 3/25 | 3/20 | 3/16 | 3/12 | 3/8 | 3/4 | 2/27 | 2/20 |
| –2.2 | 3/31 | 3/19 | 3/10 | 3/3 | 2/24 | 2/9 | 2/9 | 1/31 | 1/19 |
| –4.4 | 3/17 | 3/5 | 2/25 | 2/18 | 2/11 | 1/27 | 1/27 | 1/18 | 1/3 |
| –6.7 | 2/17 | 2/8 | 2/1 | 1/25 | 1/19 | 1/12 | 1/3 | | |
| –8.9 | 2/1 | 1/23 | 1/14 | | | | | | |
| *Probability of the occurrence of the listed temperatures at an earlier date in autumn (month/day) than the date indicated* | | | | | | | | | |
| 0 | 10/25 | 11/1 | 11/5 | 11/9 | 11/13 | 11/16 | 11/20 | 11/25 | 12/1 |
| –2.2 | 11/7 | 11/13 | 11/18 | 11/22 | 11/26 | 11/29 | 12/3 | 12/8 | 12/14 |
| –4.4 | 11/12 | 11/21 | 11/28 | 12/4 | 12/9 | 12/15 | 12/21 | 12/28 | 1/9 |
| –6.7 | 11/30 | 12/10 | 12/17 | 12/24 | 12/30 | 1/6 | 1/16 | | |
| –8.9 | 12/24 | 1/8 | 1/25 | | | | | | |

temperatures rise. This demand, of course, is superimposed on the more or less fixed need for energy by machine operations. However, the demand caused by temperature variations is responsible for the transient peaks in demand that dictate a power company's generation or storage requirements.

One application of temperature information is for energy demand forecasting. For any specific utility company there will be a unique relationship between demand and temperature. This will depend on the population of the service area, on the mix of domestic, commercial and industrial users, and on the level of technology in the region. It will also depend on where the temperature stations are located in relation to the places where the use occurs. The relationship will almost certainly change from year to year as any and all of these factors evolve. Even when the effect of this long-term development is removed, there will be uncertainty in the relationship (Figure 3.A.2). Nevertheless, the information can be used with temperature forecasts to estimate future demand. This approach has been used for many years with the short-term weather forecast to estimate the demand for the following day. The potential use of the relationship with the newly developed long-lead climate forecasts is discussed in Box 8.A.

A broad assessment of the impact of temperature changes on energy demand cannot use the results of Figure 3.A.2 directly, since they are for a specific utility at a specific time. Nevertheless, their form has been found to be typical of most utilities, and a generalisation suitable for such impact assessment has been developed. This is based on the concept of the **heating degree day** (HDD), defined as the number of degrees by which the average daily temperature falls below a threshold, or base, temperature. On a daily basis HDD has units of degrees, and values can be summed for any selected number of days to give, for example, monthly or annual total heating degree days. The HDD represents what might be termed the climatic energy demand, expressed in climatic, not energy, units, and allows comparisons between regions and times irrespective of a particular utility or population. The warm season equivalent to the HDD is the **cooling degree day** (CDD), which is concerned with the number of degrees by which the temperature falls above a selected threshold.

**Box 3.A (cont'd)**

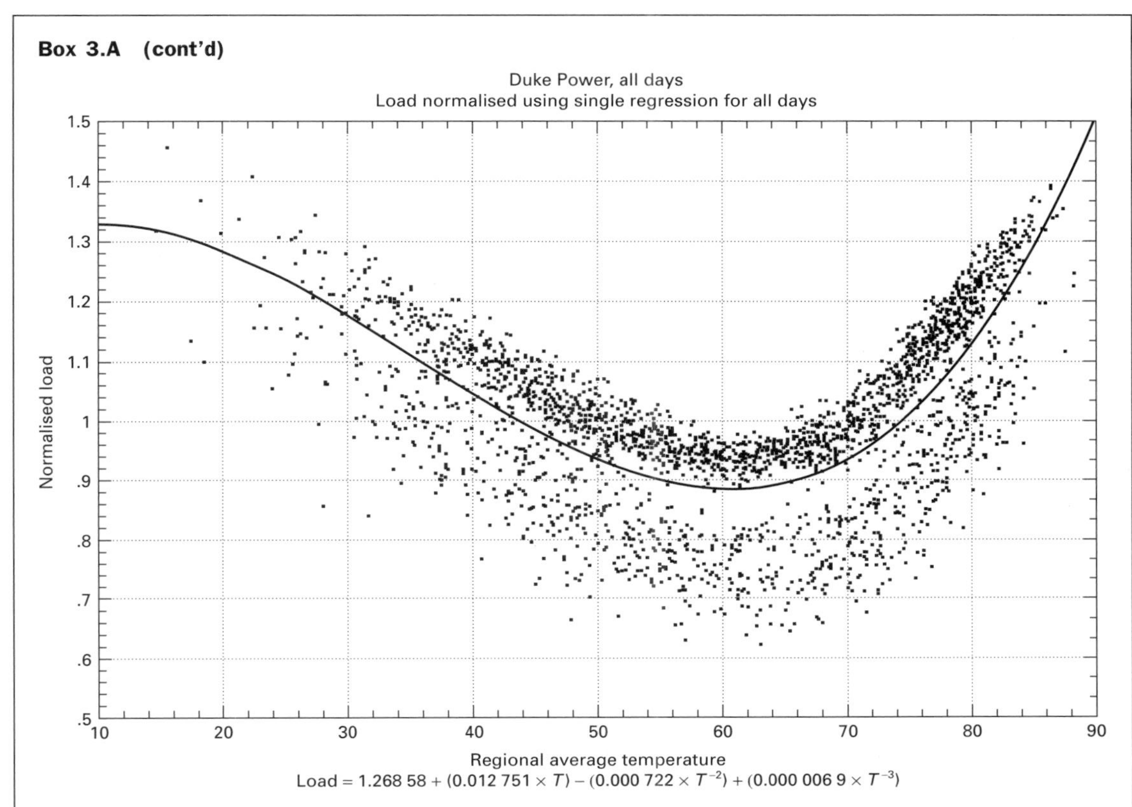


*Figure 3.A.2* Relationship between temperature and energy demand for the period 1985–1991 for a single power utility. The effect of population and technology changes during the period of record have been removed. The solid line indicates the least-squares regression, representing the best general indication of the relationship. In general, the observation points near and above that line represent weekday values, those below are for weekends or holiday periods. The scatter indicates the uncertainties in any energy demand forecast made using temperature alone. Note that it is possible to show the temperature–demand relationship as two linear segments with a minimum near 17 °C, analagous to the generalisation associated with heating and cooling degree days. (After Robinson, 1997.)

HDD and CDD values are readily available for many locations. In North America the threshold temperature traditionally used is 18.3 °C (65 °F). If a house is heated to this temperature it will, because of the heating provided by lights, appliances and people, eventually attain a temperature around 22 °C, which is felt to be 'comfortable' for living. Europeans tend to live with a comfort level 3–4 °C lower. Consequently they may use a lower threshold value, and their published values are not always directly comparable with those from North America.

The degree day concept has proved sufficiently powerful in the analysis of climatic impacts on energy use that it has been expanded into the field of agriculture, where analyses of **growing degree days** for specific crops, using appropriate threshold temperatures, have been developed.

*3.A.3 Temperature modification for frost protection*

A knowledge of the surface energy flows and balances has been used to develop frost protection techniques for agriculture. The objective is to prevent the temperature of valuable crops falling below freezing. Several methods are available when a radiation frost, often called a ground frost, is forecast. This is a frost associated with a temperature fall as a consequence of loss of long-wave energy, often on a calm clear night. Maximum cooling occurs at the surface, so that temperatures increase with altitude, a situation known as an **inversion** (see Section 4.4). One method is to use smoke-generating heaters within the area to be protected. These warm the air directly and effectively, since the heat is unlikely to be transported to higher levels in these inversion conditions. However, the amount of heat needed to prevent the

**Box 3.A (cont'd)**

frost is too great to allow success by heating alone. It is supplemented by the smoke emitted by the heaters. This smoke rises somewhat, but tends then to spread laterally because of the inversion, and creates an elevated surface for radiation exchanges above the crop being protected. The top of the smoke layer is radiatively cooled, while the bottom acts in the same way as a cloud deck for the actual surface and minimises the radiation loss from it. In these radiative frost conditions frost protection can also be obtained by utilising fans or propellers (Figure 3.A.3). These are mounted, vertically or horizontally, above the crop, stirring the air and bringing to the surface the warmer air from aloft, which replaces the cold surface air. The continual stirring prevents the continued temperature drop at the surface.

Both these methods are restricted to radiative cooling conditions and are not effective when the cold air extends for a great depth within the atmosphere, giving a freeze rather than a ground frost. Alternative methods can be used for low-growing crops, such as strawberries, which can tolerate temperatures one or two degrees below freezing but not lower temperatures. The crop is sprayed with water prior to the onset of the freeze, so that the fruit is completely surrounded with water. As temperatures fall the water freezes and releases latent heat. Provided the fruit is completely enclosed, some of this heat enters it, raising its temperature slightly. Thereafter the encircling ice acts as an insulator impeding heat loss from the fruit and minimising the internal temperature drop.

With a crop of high enough value, or small enough areal extent, or with the crop of a backyard gardener, it is of course possible to cover the crop physically. This will provide good protection from radiative frosts and some protection from those associated with deep cold layers. The covering material and method may range from simple Styrofoam cups placed over the susceptible parts of individual plants to permanent glasshouses.

*Figure 3.A.3* A frost protection 'wind machine' constructed from an aeroplane propeller and driven by an automobile engine. In winter this machine is used to mix warmer air downwards to prevent frost damage to the apple trees. (Courtesy of Greg Johnson, USDA – National Water & Climate Center, Portland, Oregon.)

Whichever method is suitable for a particular area or type and size of operation, the decision whether or not to invest in frost protection equipment is largely a climatological one. Indeed, it is very similar to that discussed in Section 5.A.1 concerning budgets for snow removal. Further, the use of the equipment can readily be related to the forecasting considerations given in Section 6.A.1.

**Box 3.I The measurement of temperature**

*3.I.1 The meaning of 'air temperature'*

The temperature of concern throughout Chapter 3 could be called 'air temperature' or 'surface temperature'. In climatology the terms are often used rather loosely, interchangeably, and even in combination as 'surface air temperature'. Perhaps the last is the clearest, if least appealing, term since we are usually talking about a measurement by a thermometer exposed to the air in a screen not too far above the surface. If we think of this temperature being representative of the surface layer where we humans live and breathe, the term 'surface temperature' appears reasonable. However, anyone walking on concrete on a sunny summer day knows that the true surface is at a very different temperature from the air only a short distance above it. Sometimes, to avoid further confusion, this true surface is called the skin, and has a 'skin temperature'. For a concrete surface, the concept of a skin appears sound, but if we think instead of a forest, the concept becomes much more

**Box 3.I (cont'd)**

vague, and begins to be mixed in with our original definition of 'surface air temperature'.

With this terminology, the skin is where the energy exchanges take place. Hence we determine the skin temperature when we calculate temperatures from energy fluxes. Most important from a practical standpoint, satellites observe upwelling radiation in selected channels in the atmospheric window. The results, after corrections for atmospheric transmission losses and surface emissivity, are skin temperatures integrated over the field of view of the sensor. Hence satellites and surface-based thermometers do not measure the same thing. If we need a general indication of the surface temperature of a forest or a city, for example because we need to determine energy exchanges for use in climate models, the satellite information is probably most suitable. For an ocean surface there is often no choice, because only satellite information may be available. Over a single surface type, such as a concrete car park, both may be available and then care is needed. In most cases the various names can indeed be used interchangeably. Only when satellite and ground-based observations are being used together is it important to state exactly what is meant by the terms being used.

*3.I.2 The measurement of surface temperature*

The basic instrument for measuring temperature in the atmosphere near the Earth's surface is the **thermometer**. For many years the most common type used by most national meteorological services has been the mercury-in-glass thermometer (Figure 3.I.1). Mercury in the bulb expands when heated, the expanding liquid being forced along the tube. The temperature is thus indicated by the length of the mercury column in the tube. The instrument must first be calibrated to provide the relationship between length and temperature. Thereafter the length can be expressed directly in temperature units.

There are numerous variations on this basic design. Two common ones are used in the maximum and minimum thermometers (Figures 3.I.1 and 3.I.2).

(a)

(b)

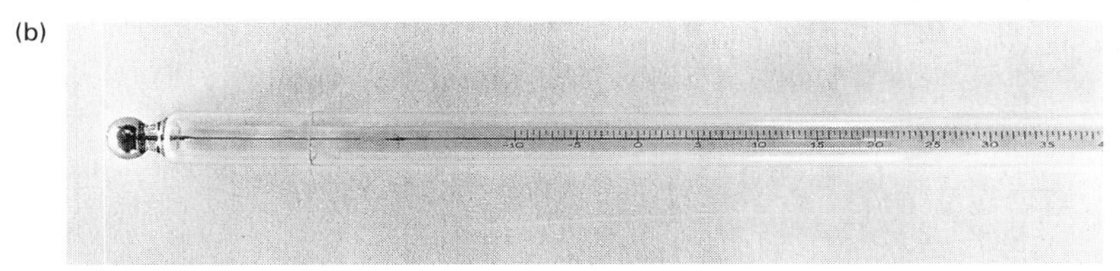

*Figure 3.I.1* (a) Orientation of wet and dry bulb thermometers (vertical) and maximum and minimum thermometers (horizontal) in a Stevenson screen (see Figure 3.I.3). (b) The maximum thermometer operates in the same way as a clinical thermometer: a constriction close to the bulb forces the complete column to remain extended showing the maximum temperature until 'shaken down' when reset.

**Box 3.I (cont'd)**

(a)

(b)

(c)

*Figure 3.I.2* (a) Minimum thermometer, showing the barbell shaped index which is drawn back by the (colourless) alcohol as temperatures decrease but remains in place when the alcohol column again increases in length (here the thermometer is installed with its bulb just touching the blades of short grass. The recorded temperature is termed the 'grass minimum'). (b) An earth thermometer which is installed in a steel tube driven vertically into the ground to standard depths of 30 and 100 cm. (c) Set of soil thermometers installed so that their bulbs are at depths of approximately 5, 10 and 20 cm (usually 2, 4 and 8 inches).

**Box 3.I (cont'd)**

In the former a constriction is placed in the tube close to the bulb. The expanding mercury is able to force its way past this, but when cooling and contraction take place the constriction does not allow the mercury to return to the bulb. The mercury is thus left in the tube to record the maximum temperature. A rapid shaking of the tube can restore a continuous thread of mercury across the constriction, thus resetting the instrument for use again. In the minimum thermometer the mercury is usually replaced by alcohol. A small rod called an index and usually barbell-shaped, is placed within the alcohol in the tube. As temperatures fall and the liquid contracts, this index is drawn towards the bulb by the surface tension of the alcohol surface. The index is left behind when the liquid again expands, the end of the index farthest from the bulb thus indicating the lowest temperature during the period. Tilting the instrument so that the index slides to the liquid meniscus resets it.

Recent advances in instrument technology have allowed some meteorological services to replace the mercury-in-glass thermometers with **electrical resistance** instruments, which relate resistance to temperature. These have advantages in durability and ease of maintenance, but from a climatological perspective, a change from one instrument to another is fraught with pitfalls. Response times, sensor exposure, and ageing characteristics are all different. So if a long observational record is involved, it is often difficult to compare temperatures before and after the instrument change.

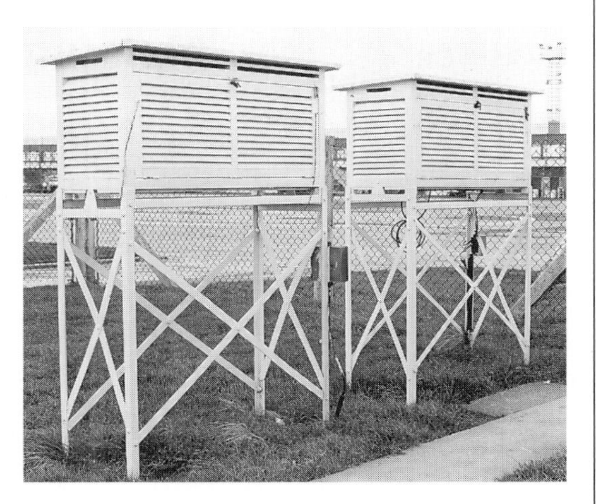

*Figure 3.I.3* A Stevenson screen houses meteorological instruments, particularly thermometers and thermohydrographs, so that they are well ventilated but protected from direct radiation. The screen is a double-louvered white-painted wooden box erected in an open position so that the thermometer bulbs are 1.25 m above the ground. (Two screens are shown in the photograph.)

Thermometers must be in some form of shelter, which serves to shield the instrument from direct radiation from the sun and from upwelling energy from the ground, while providing ventilation so that there is a free flow of air past the thermometers. The Stevenson screen is a typical example (Figure 3.I.3). Most nations, encouraged by the World Meteorological Organization, have established standard conditions for their screen placements so that comparisons between sites are possible. The standard usually requires the screen to be of the order of 1 m above a particular type of surface. The surface is commonly grass, but for states with a variety of climatic conditions the selection of an appropriate surface type is not simple. There is the danger of requiring a grass surface in a desert region, where the land under the screen may be the only patch of grass in the whole area. Sensor heights must also be standardised. In general, the higher the placement, the greater the integration of the effects of the underlying surface types, since skin temperatures can change very rapidly over short distances. A low level placement would force the result to depend on the exact location of the sensor, and comparison between sites would be difficult. However, too high a placement allows energy exchanges between the surface and the sensor, so that it no longer represents the conditions in the life-layer, which is the level of major interest. Consequently a compromise is reached and most sensors are placed at a standard level close to 1 m above the surface.

Non-standard instrument types and exposures may be needed for special applications. Nevertheless, unless detailed metadata are available, such observations are difficult to compare with data from a standardised network. Without a screen, for example, it is easy to allow direct solar radiation to fall on a thermometer bulb, with a consequent high temperature. Any use of data taken in non-standard conditions, therefore, must be done with extreme caution.

# CHAPTER 4 Moisture in the atmosphere

Clouds and precipitation, along with temperature and wind, are the most striking elements of weather and climate, elements that can change very rapidly with time and space. However, water in all its forms and in all its various activities in the atmosphere plays an important role in sustaining not only the climate but also life itself. In order to understand and predict the actions of water in the climate system it is useful to think of the water as being part of a distinct system, sometimes called the **hydrological cycle** (Figure 4.1). A complete understanding of this system would, as the figure implies, require excursions into geomorphology, pedology, botany, glaciology, oceanography and, if human structures are included as part of the Earth's surface, civil engineering. Further, the implications of water and water supply would require a knowledge of public health, water supply engineering, agricultural practices and, eventually, appreciation of public policy development and implementation. While such considerations may indicate the broad range of disciplines that are of interest to the climatologist, we must obviously concentrate on those aspects of water that are directly connected with the climate, drawing on the other disciplines only when needed and indicating some areas where climatology can contribute to these other disciplines. In this chapter we concentrate on water leaving the surface and forming clouds, while the next chapter will be

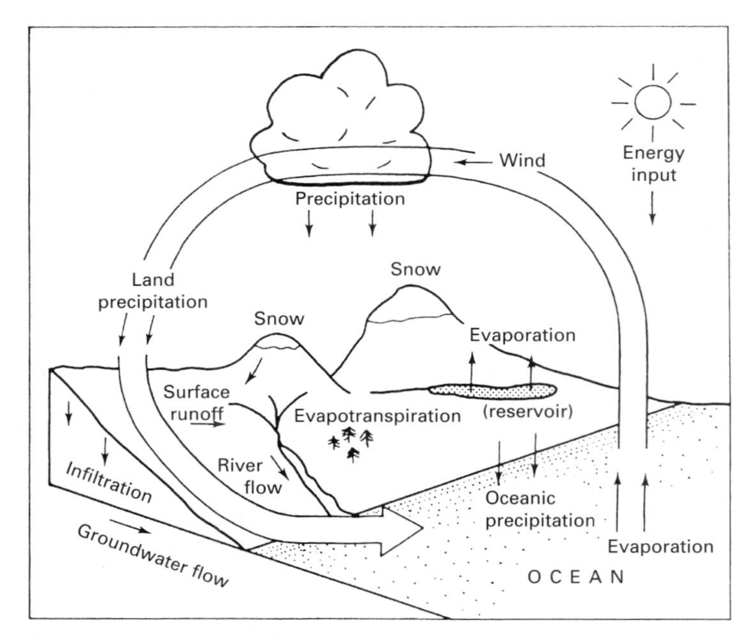


*Figure 4.1* The global hydrological cycle.

concerned with precipitation returning the water to the surface.

Since we are thinking of water moving in a cycle, there is no one best place to start our discussion of the climatic role of water. However, the previous chapters have emphasised energy exchanges, so it is convenient to start our discussion at the air–Earth interface, where these exchanges are vital.

## 4.1 Evaporation

The concept of evaporation was introduced in Chapter 2, but there it was treated as a consequence of energy exchanges. Here we are concerned with evaporation as it influences the amount of water at the surface and in the atmosphere. Nevertheless, there is a simple relationship between the energy used for evaporation and the amount of water evaporated, as was demonstrated by our use of the symbol $LE$, where $L$ is the latent heat of vaporisation and $E$ is the depth of water evaporated, for the latent heat flux.

### 4.1.1 Evapotranspiration

Water is removed from the surface into the atmosphere by two distinct processes. **Evaporation** occurs when there is a 'free' water surface, be it water in an ocean or between individual soil particles. **Transpiration** occurs when water is removed from the interior of plant leaves through their stomata. Although these two processes are different and distinct, it is possible in most climatological contexts to treat them together, since the forcing mechanisms are similar and the end result is the same. The combined process is called **evapotranspiration**. We shall be dealing with this combined process unless it is specifically noted to the contrary.

The rate of evapotranspiration at any instant, and thus by integration the amount of evapotranspiration over a particular time interval, is controlled by four factors: (1) energy availability; (2) the humidity gradient away from the surface; (3) the wind speed immediately above the surface; and (4) water availability. The varying influences of these four components were considered briefly, from an energetic perspective, in the discussion of Figure 3.2 in the previous chapter.

Viewing the evaporative process on the molecular level, evaporation occurs whenever the flow of individual water molecules away from the water body exceeds the return flow to the water from the atmosphere. Dew is formed if the return flow exceeds the amount leaving the surface. The flow from the surface is largely dependent on the energy available, as discussed in Chapter 2, and on there being a supply of water at the surface. The rate of return flow is highly dependent on the humidity of the air immediately above the surface. As the humidity increases, this return flow increases and the net rate of evaporation decreases. Thus for rapid evaporation we need a steep humidity gradient away from the surface, in exactly the same way that a steep temperature gradient is needed for rapid heat flow. In near-calm conditions the surface evaporation will lead to a rapid build-up of the moisture content of the air layer immediately adjacent to the surface and a decrease in the evaporation rate. In the presence of wind, with its propensity for turbulent mixing, the low-level moist layer will be removed and replaced with less humid air. Usually this replacement is associated with large-scale horizontal advection, which introduces a new, drier air mass to the surface. The net result is that moderate wind speeds tend to maintain steep humidity gradients and high evaporation rates.

High rates also depend on a continuous supply of water at the surface. For an open water surface this is no problem. However, for a land surface, water movement from depth, whether through the soil or through plants, requires a considerable time. There is, for example, an upper limit to the rate at which plants can transpire. Often on a sunny midsummer day plants can wilt in the afternoon simply because water vapour is being removed from the leaves faster than it can be brought in from the roots. Similarly the surface layers of a soil may become dry although plenty of water remains at depth.

Not all of the energy absorbed at the surface is used to evaporate water, since the other energy fluxes, sensible and ground, will be maintained. For many land surfaces, however, the minimisation of the Bowen ratio (Section 3.1) means that sensible heat flows and temperature changes are small while evaporation is occurring, but there will be a rapid temperature rise once the surface dries and the sensible heat flux takes over as the major energy transfer mechanism.

The action of the four factors controlling evapotranspiration, any of which can be limiting, has led to the development of two concepts of evapotranspiration for practical applications. The first is **potential evapotranspiration** (PET), which is the rate that will occur from a well-watered, actively growing, short green crop completely covering the ground surface.

This is essentially identical with the values that would be obtained over a large open water surface. It represents the rate controlled entirely by atmospheric conditions and is the maximum possible in the prevailing meteorological conditions. The second concept, **actual evapotranspiration** (AET), is the amount that is actually lost from the surface given the prevailing atmospheric and ground conditions. Both are important, since PET provides some measure of possible agricultural productivity if, for example, irrigation is initiated, while AET provides information vital for the determination of soil moisture conditions and the local water balance.

The relative sparseness of observations, together with the difficulty of reconciling various estimation methods (see Box 4.I), makes it difficult to construct

**Box 4.I Monitoring water in the air**

*4.I.1 Evapotranspiration measurement and estimation*

The measurement of evapotranspiration, potential or actual, is difficult. Many techniques have been devised. The most common method of measuring PET directly is using **evaporation pans** (Figure 4.I.1). These are simply containers of a standard size containing water freely exposed to the atmosphere. The water depth is measured at the beginning of the time period of interest and again at the end. The difference, after correction for any precipitation received, is the evaporation. Energy transfer through the sides of the pan, together with turbulence created by the pan itself, make it difficult to relate the results to evaporation from natural open water surfaces. Usually a correction factor, a 'pan coefficient', is employed before the results are used. Rather more sophisticated instruments, **lysimeters** (Figure 4.I.2), are available, but these require very careful installation and maintenance if they are to give useful results. They are thus restricted mainly to a few agricultural research establishments. A section of the land surface is removed, a pan placed in the cavity and the land replaced in the pan with as little alteration to its initial structure as possible. Usually the pan is placed on a weighing mechanism which is used to record the change in weight and thus the amount of evaporation.

As it is difficult to measure evapotranspiration directly, it is usual to estimate it from more commonly measured parameters. Numerous methods are available. Almost all start by estimating PET from atmospheric measurements. Some simple ones, requiring only air temperature as input, can be used for the determination of monthly average values. Others are much more complex and incorporate solar radiation, wind speed, air temperature and humidity measurements, and can be used to estimate daily values. Once PET is determined, some form of book-keeping method is used to track the amount of moisture in the soil and to relate this to the amount that will be available for actual evapotranspiration.

*4.I.2 Instruments for measuring atmospheric moisture content*

Several types of instruments have been devised to measure humidity. Only those commonly used currently are discussed here. The first is the **dew point hygrometer** (dew cell), which measures the dew point temperature, $T_d$, directly. A mirrored surface is cooled electrically until dew forms. As soon as this occurs a photoelectric detector senses a change in surface reflectance and switches the cooling circuit to a heating one. The heater remains operative until the dew is evaporated, at which point the cooling cycle is again initiated. The cycling is repeated until a stable

*Figure 4.I.1* An evaporation pan. The amount of evaporation is established by measuring the depth changes and compensating for the input of precipitation.

**Box 4.I (cont'd)**

*Figure 4.I.2* A weighing lysimeter being lowered into position. When *in situ*, the lysimeter should be undetectable. The amount of evapotranspiration taking place over a fixed period is calculated by weighing the soil plus biomass at the start and finish of the experiment and carefully monitoring rainfall. The sides and base of the soil container are carefully sealed before emplacement. (Courtesy of John Stewart.)

temperature is reached, representing the dew point of the air above the mirror surface. Figure 4.I.3 illustrates another type of hygrometer which senses relative humidity directly.

A less sophisticated instrument is the **psychrometer** (see Figure 3.I.1). This consists of two thermometers. One, called the **dry bulb**, is unmodified. The bulb of the second, the **wet bulb**, is kept moist by being encased in a wick connected to a water reservoir. Water is evaporated from around the wet bulb. The energy required for this evaporation is extracted from the bulb itself and thus its temperature is lowered. The amount of cooling depends on the evaporation rate, and thus on the humidity of the air. Provided an even flow of water comes from the reservoir, allowing neither flooding nor drying of the bulb, the temperature will become stable. The amount of moisture can then be calculated from the dry bulb or 'air' temperature together with the difference between the dry and wet bulb temperatures ($T_a$ and $T_w$), the latter being known as the wet bulb depression. Note that the wet bulb temperature is not the same as the dew point but that $T_d \leq T_w \leq T_a$, the equalities holding only when the air is saturated. To obtain reliable estimates of humidity with a psychrometer it is necessary to ensure some airflow over the two thermometers, which are usually held so that they lie parallel to each other a few centimetres apart. This airflow is created either by placing the instrument in a tube and drawing air across it, forming an aspirated psychrometer, or by whirling it through the air manually, as in a sling psychrometer.

Humidity in the free atmosphere is measured by both radiosondes and satellites. Radiosondes utilise **resistance hygrometers**. These use the property of some materials (e.g. carbon black) of having an electrical resistance that varies with relative humidity. Although relative humidity is thus measured directly, radiosondes also measure temperature, so that other expressions for humidity can be calculated. The instrument is fairly easy and cheap to manufacture but in general the results are rarely accurate to more than ± 10%. Increasing use is therefore being made of humidity determined by satellites. Use is made of the radiances observed in a number of infrared and microwave channels to establish the humidity profile. Simultaneous retrievals of the atmospheric temperatures, using measured radiances in other channels, allow determination of the moisture content at various levels in the atmosphere.

*4.I.3 Measurements of clouds*

Cloud amounts and distributions are measured from the surface and from satellites. The traditional surface-based method, routinely used at a large number of stations for many years, has been simply to divide the

**Box 4.I (cont'd)**

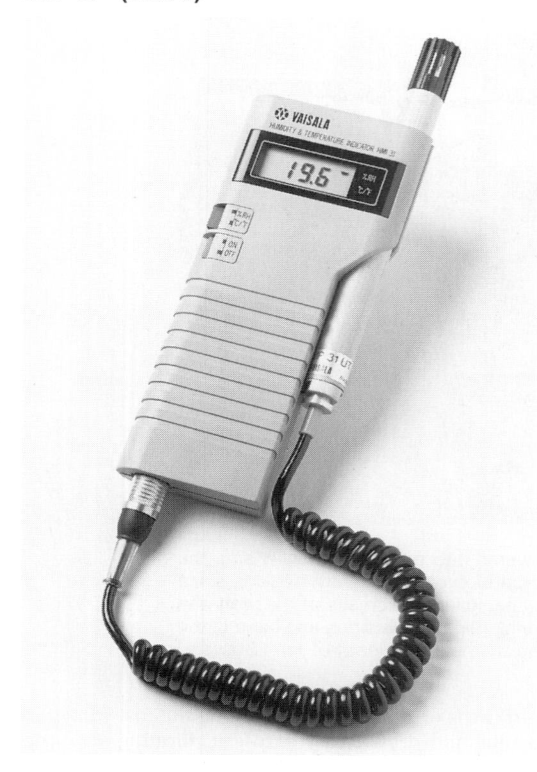


*Figure 4.I.3* A humidity sensing instrument. The sensor is similar to that used for humidity measurement in radiosondes. It is based on a polymer thin-film capacitor contained in the probe which can be held at a distance from the display. A 1-μm-thick dielectric polymer layer absorbs water molecules through a thin metal electrode. This causes a capacitance change proportional to the relative humidity. This particular instrument permits display of both temperature and relative humidity. (Courtesy of Vaisala.)

sky into eight or ten parts by eye and estimate the fraction covered (oktas or tenths) by clouds. In some cases the observation is refined to include cloud type and the amount at each of the three levels indicated for horizontal clouds in Figure 4.4. This approach is rather subjective, and when estimates are made for several levels it is impossible to determine whether higher cloud exists but is hidden by lower layers.

Cloud images derived from satellites are familiar to television viewers and are an important tool in weather forecasting. However, extracting cloud information for climatic purposes is not always easy, since the information is based on analysis of simultaneous measurements of albedo and outgoing long-

*Table 4.I.1* Albedos and absorptances of four cloud types

| Cloud type | Low | Middle | High | Cumuliform |
|---|---|---|---|---|
| Albedo | 0.69 | 0.48 | 0.21 | 0.70 |
| Absorptance | 0.06 | 0.04 | 0.01 | 0.10 |

wave radiation. The short-wave albedo depends on the 'optical thickness' of the cloud. This is largely a function of the water content, which in turn depends on the thickness of the cloud. It also depends on the type of cloud, since liquid and frozen water behave differently. The long-wave radiation from a cloud top consists of radiation emitted by the cloud and radiation transmitted through it. Both depend on cloud absorptance, and thus on cloud water content and thickness, although a cloud thicker than 1 km will absorb virtually all the radiation it receives from below. Since the amounts also depend on the radiating temperature, and there will be a decrease in temperature through the cloud, the amount of radiation finally emitted upwards is largely a function of the cloud top temperature. Thus values of albedo and absorptance for individual clouds vary tremendously, but generalised values are given in Table 4.I.1. Despite these inherent problems, many clouds can be readily identified. The towering cumulonimbus of the belt near the equator and the predominantly cloudy regions of the mid-latitude depression belts are readily detectable. Other cloud areas are less easy to detect. Low-level stratus, common along the western coasts of North and South America and Africa, appears similar to the underlying surface, since both have high albedo and relatively high outgoing radiation. Studies of the characteristic spatial patterns associated with these clouds are helping to resolve ambiguities.

The surface and satellite cloud observations differ both in the type of cloud they see in any multilayer situation and in the total cloud amount observed. Hence it is very difficult to use them together (Box 7.1). Although a single climatology might be intrinsically desirable, it is preferable to treat surface- and satellite-based cloud observations as different facets of the same phenomenon. Surface values are useful in specifying the local climate and in considering what a person or plant 'sees', particularly in terms of the way in which clouds interact with solar radiation. Also surface observations can be used to estimate cloud base heights, a vital consideration for aviation. On the other hand, the satellite observations are much more useful when considering global energy flows, climate models, and the possible causes and effects of climatic change.

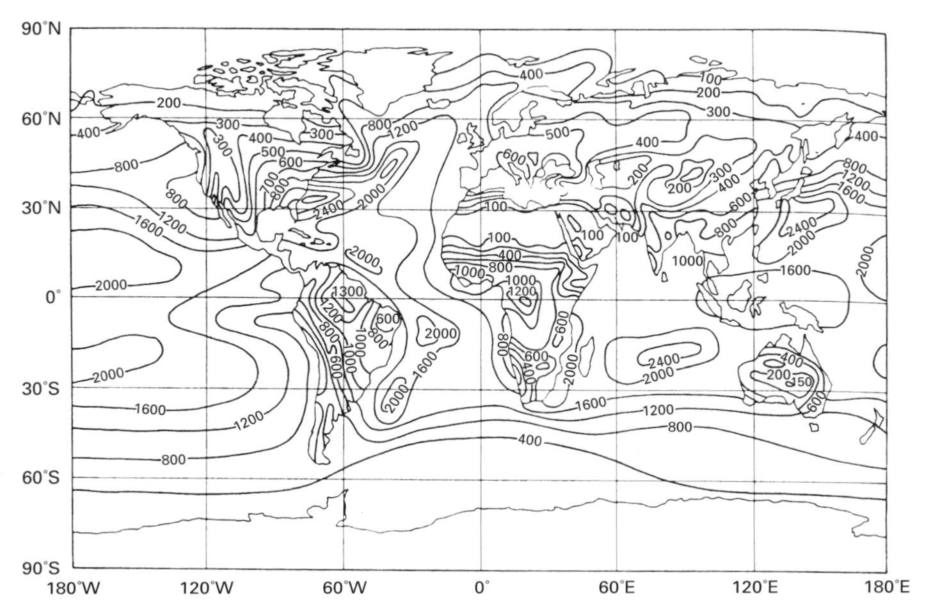


*Figure 4.2* Global distribution of evaporation (mm year$^{-1}$) based upon a number of recent estimates. Two features are immediately obvious: (i) the discontinuity between adjacent land and ocean regions, which might be anticipated as the oceans are an almost infinite moisture source; and (ii) the dependence of oceanic evaporation upon ocean temperatures (especially clear in the North Atlantic). (From Brutsaert, 1982.)

global maps of evaporation. Nevertheless, various attempts have been made and, although the results differ in detail, the general picture is well established (Figure 4.2). Maximum rates of actual evaporation occur over the subtropical oceans with a general decrease in amounts poleward. Land values are lower than oceanic ones, the isopleths making a sharp break at the coasts. A rough extrapolation of the oceanic values over the continents gives an idea of the PET of land. This serves to emphasise the considerably lower rates of AET over the land, most marked over the desert regions of the Earth.

## 4.2 Moisture in the atmosphere

Evaporated water enters the atmosphere as individual energetic water vapour molecules. There are numerous ways of expressing the resulting moisture content of the atmosphere, each appropriate for particular applications. Prior to discussing these, however, it is necessary to introduce consideration of saturation, a fundamental concept vital for the understanding of atmospheric processes.

### 4.2.1 Saturation of the air

There is an upper limit to the amount of water vapour that the air can hold. This is the point at which air becomes **saturated**. A rigorous definition of saturation, which will be important later, is that it is the maximum water vapour content of air in equilibrium with a plane surface of pure water or of pure ice at the same temperature as the air. Values of saturation differ between the two types of surface. The saturation vapour pressure is smaller over ice than over water at the same temperature because the latent heat required for the solid to vapour transition ($2.834 \times 10^6$ J kg$^{-1}$) is greater than that for the change from liquid to vapour ($2.501 \times 10^6$ J kg$^{-1}$). At the point where saturation occurs, the interaction of the large number of high energy water vapour molecules with the lower energy air molecules becomes sufficient to cause the vapour molecules to give up some energy and become liquid droplets (or solid crystals) as they revert to the lower energy state. Latent heat is released in the process.

The saturation value varies with temperature. Expressing humidity in terms of the **vapour pressure** – the force per unit area created by the motions of

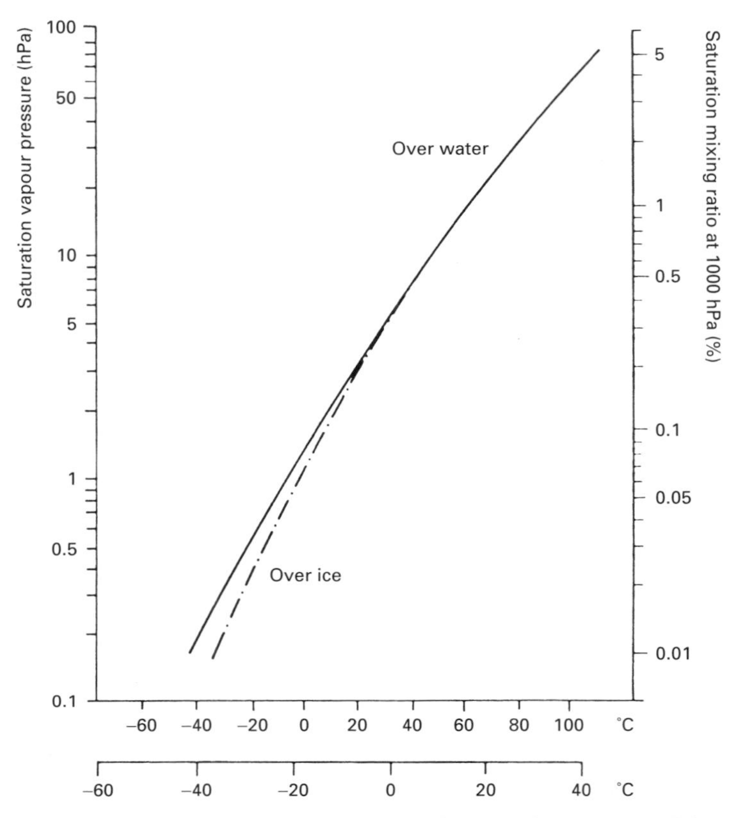


*Figure 4.3* Saturation vapour pressure as a function of temperature. Below 0 °C there are two curves: one for supercooled water, the other for ice. (From Lockwood, 1979.)

the vapour molecules treated in isolation from all the other gases of the atmosphere – we can quantify the relationship. The relationship between the saturation vapour pressure, $e_s$, and temperature, $T$, can be derived from the second law of thermodynamics. The result is the **Clausius–Clapeyron** equation:

$$de_s/dT = L/T \times 1/(V_2 - V_1) \qquad (4.1)$$

where $L$ is the latent heat of vaporisation and $V_2$ and $V_1$ are the specific volumes (volume occupied by unit mass) of water vapour and liquid water, respectively. The value of $V_1$ is usually negligible in comparison with the value of $V_2$. The saturation vapour pressure increases approximately logarithmically with temperature (Figure 4.3). Since $e_s$ differs between liquid and solid water surfaces, at temperatures below freezing the saturation vapour pressure depends on whether condensation occurs as liquid water droplets or as ice crystals.

### 4.2.2 Evaluation of atmospheric moisture content

So far the moisture content has been expressed in terms of the vapour pressure. Alternative means of expression, which are used in various places later, can be collected and defined here.

The **absolute humidity** is defined as the mass of water vapour per unit volume of air and is expressed in grams per cubic metre. The **specific humidity** is the ratio of the mass of water vapour to the mass of moist air in the same volume.

The **mixing ratio**, strictly the 'water vapour mixing ratio', is the ratio of the mass of water vapour to the mass of dry air in a specified volume. It is expressed in grams per kilogram. The saturation mixing ratio, $w_s$, is frequently used, being directly analogous to and having the same shaped curve as the saturation vapour pressure, $e_s$. Indeed, it can be shown that they can be related in the normal atmospheric situation, where $e_s$ is very much less than the total atmospheric pressure, $p$. The relationship takes the form:

$$w_s \approx 0.622 e_s / p \tag{4.2}$$

The **relative humidity** with respect to water is the ratio of the actual mixing ratio, $w$, to the saturation mixing ratio, $w_s$. It is usually expressed as a percentage such that:

$$RH = (w/w_s) \times 100\% \tag{4.3}$$

The relative humidity of a parcel of air therefore depends upon its temperature since $w_s$ is functionally dependent upon temperature. Consequently the value will alter whenever the temperature changes without a change in moisture content. For instance, the relative humidity may decrease by as much as 50% between morning and noon as temperatures rise.

The **dew point** temperature, $T_d$, is the temperature at which an air parcel would become saturated if it were cooled without a change in pressure or moisture content. Since there is a unique relationship between saturation and temperature, the dew point temperature also has a unique value for any mass of air. It must be emphasised that although $T_d$ is a temperature, it is only of interest as a measure of humidity.

### 4.2.3 Global distribution of precipitable water

Although atmospheric moisture is derived from surface evaporation, the spatial distribution cannot be determined by consideration of the source strength alone since horizontal and vertical mixing almost invariably occur. Nevertheless, knowledge of the spatial distribution is vital for understanding and modelling the dynamics of the climate system and for understanding and quantifying the processes of precipitation formation. The atmospheric vapour content is expressed as the **precipitable water vapour** (PWV): the total amount of water vapour in the atmospheric column above a point on the Earth's surface, or in a specific atmospheric layer. The amount between any two pressure levels $p_1$ and $p_2$, is given (in kilograms per unit area) by:

$$\mathrm{PWV} = 1/g \int_{p_1}^{p_2} q \mathrm{d}p \tag{4.4}$$

where $g$ is the acceleration due to gravity and $q$ is the specific humidity.

The PWV represents, as the name indicates, the amount of water that is available to fall as precipitation. It also indicates the amount of moisture available in the cloud-free atmosphere to interact with the various radiation streams and thus influences atmospheric heating rates.

PWV can be determined directly from moisture profiles obtained by radiosonde ascents or satellites. It can also be estimated from surface observations alone, since most vapour is concentrated in the lowest atmospheric layers, close to its source. The distribution of PWV is such that at any time there is considerable moisture throughout the atmosphere (Plate 2). This is true even for the desert regions. Although this map could be combined with one showing the wind field to indicate the moisture flux over an area, it is clear that the distribution of clouds and precipitation is dependent on much more than simply the availability of moisture. Mechanisms are needed to convert the atmospheric water vapour into liquid or solid form to produce clouds and to cause this moisture to fall as precipitation. These mechanisms are treated in subsequent sections.

## 4.3 Cloud types and distribution

When air becomes saturated, water vapour condenses and clouds are formed. The type of cloud produced depends greatly on the process by which the air is brought to saturation (Table 4.1). Prior to our

*Table 4.1* Cloud types and formation processes

| Process | Common cloud types (abbreviations) |
|---|---|
| A. Without cooling: | |
| air mass mixing | Stratocumulus (Sc) |
| B. Cooling without vertical motions: | |
| radiative cooling | Radiative fog |
| advective cooling | Advective fog |
| C. Cooling with vertical motions: | |
| orographic uplift | Stratus (St), altostratus (As) |
| frontal uplift | All types |
| airflow confluence | Cirrus (Ci), stratus (St), altostratus (As) |
| convection | Cumulus (Cu); cumulonimbus (Cb) |

discussion of the cloud-forming processes, therefore, it is useful to consider the main types of clouds.

### 4.3.1 Classification of cloud types

The first systematic attempt to classify clouds was made by Luke Howard (1772–1864). Although changes have been made subsequently, the names he invented – at a time when it was usual to name scientific phenomena in Latin – have been retained. The infinite variety of individual clouds are classified into four main 'families' (low, middle, high and vertically extended), which are themselves intermixed to form subgroups (Figure 4.4). **Cumulus** (Cu) represents the family with predominantly vertical development. They can range from the small white fluffy clouds of a summer afternoon (Figure 4.5a) to the towering black threatening cumulonimbus (Cb) – the thundercloud (Figure 4.5b). The term 'nimbus' means rain- or snow-producing and usually refers to a very well developed member of a particular cloud family, such as nimbostratus (Ns). Layer clouds with predominantly horizontal development are divided into three families, their distinctive appearance being created because of height and temperature differences. **Cirrus** (Ci) clouds are high and often wispy, being composed of ice crystals (Figure 4.5c). The lowest layered clouds are **stratus** (St), composed of liquid water droplets and giving dull overcast, often drizzly, conditions (Figure 4.5d). Stratocumulus (Sc) clouds (Figure 4.5e) have a distinct cellular structure. The **alto** always appears in conjunction with another family. Altostratus (As) is similar to stratus except that it tends to be less dense and less likely to give precipitation. Such clouds are

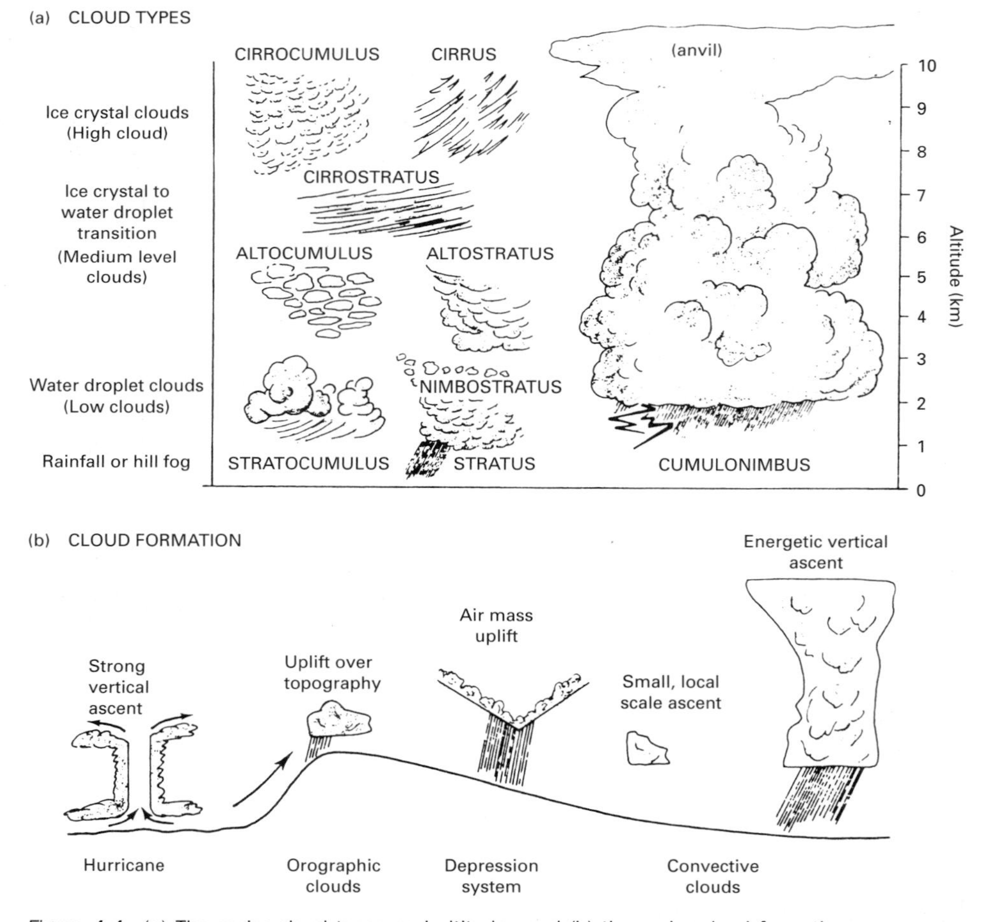


*Figure 4.4* (a) The major cloud types and altitudes and (b) the major cloud formation processes.

*Figure 4.5* continued on next page

*Figure 4.5* Photographs illustrating (a) cumulus (Cu); (b) cumulonimbus (Cb); (c) cirrus cloud (Ci); (d) stratus cloud (St); (e) stratocumulus (Sc) viewed from above; (f) altocumulus (Ac) overlying lower level cloud.

commonly composed of **supercooled** water droplets with temperatures below 0 °C. Altocumulus (Ac) is a cumulus type cloud, having significant vertical development but with a base sufficiently high to be composed of supercooled water (Figure 4.5f). Finally, it is convenient when considering cloud-forming processes to treat **fog** as a cloud at ground level.

### 4.3.2 Cloud distribution

A knowledge of the global cloud distribution is vital whether we consider clouds primarily in their role as precursors of precipitation, the main focus in this chapter, or as major participants in energy exchanges, as considered previously. We consider methods of measurement in Box 4.I, but for the global view which is of concern here, most of our knowledge comes from satellite observations. Although the observational record is still relatively short compared with that of surface observations, and not yet long enough to consider long-term trends or fluctuations, we can now provide global average distributions with some confidence (Plate 3). Although this new information is most directly beneficial to the modelling efforts described in Chapter 13, aspects of this cloud climatology will be used throughout our discussion of regional climates and their origins.

## 4.4 Cloud formation

### 4.4.1 Cloud formation processes

In our discussion of the processes leading to cloud formation we shall follow the order outlined in Table 4.1, starting with the only process that can create clouds without a cooling of the air. In some circumstances when two air masses with different temperatures and moisture contents converge the two mix together. As a result of the non-linear relationship between saturation vapour pressure and temperature (Figure 4.3), the mixture may be saturated with respect to the new temperature although both initial air masses were unsaturated. The types of cloud created depend on the level of the mixing, but commonly this process leads to stratocumulus cloud. The process is rarely seen in its pure form, since airstream convergence commonly leads to the widespread vertical motions or frontal uplifts discussed below, and these overshadow the effect of mixing. Nevertheless, the latent heat released during mixing is frequently an important source of energy for sustaining the vertical motions and the cloud formation.

The radiative cooling mechanism for condensation depends directly on radiation exchanges at the surface. When net radiation is negative, particularly on a calm, clear night, the air in contact with the ground is cooled. If there is sufficient moisture in the air, and the cooling is sufficient, this air will be cooled below its dew point and **dew** will be formed. As the energy loss continues through the night, it is possible for an increasingly deep layer to cool below the dew point. A ground fog, usually called a **radiation fog**, will result. This will begin to form, very close to the ground, an hour or two after midnight and will gradually thicken and deepen as the night progresses. Soon after sunrise net radiation will become positive, heating of the air will commence and the liquid water droplets of the fog will evaporate back into the air.

Another cooling mechanism producing fog is associated with horizontal movement of air (**advection**). If a warm airstream starts to blow over a cooler surface the air itself rapidly adjusts to the temperature of the new surface. Again, given sufficient cooling, or sufficiently moist air, a fog will result. This **advective fog** will persist as long as the moist airstream blows over the cooler surface.

Although these cooling mechanisms may be locally important, by far the most common mechanisms for cloud formation occur as a result of vertical motions in the atmosphere. In order to appreciate these it is first necessary to examine the consequences of allowing air to rise vertically.

### 4.4.2 Vertical motion in the atmosphere

To develop the concepts associated with vertical motions we consider a 'parcel' of air. Although such a parcel can be of almost any size, it is useful initially to visualise it as being about the same size as a small cumulus cloud. We make the assumption, which is very good in practice, that the parcel's vertical motion is sufficiently fast to prevent energy exchange with its surroundings. Such a process is called **adiabatic**. In this case, as the parcel rises through the atmosphere, its pressure decreases along with that of the surrounding environment; so the parcel expands and cools. The reverse also holds. In an adiabatic descent, a parcel of air would warm as it contracts. A familiar example of this descending condition would be the increase in temperature of the valve on a bicycle pump when used to inflate a tyre. The relationships can be quantified through the first law of thermodynamics, which states that:

$$\text{energy added} = \text{increase in internal energy} + \text{work done} \qquad (4.5)$$

In the adiabatic case the heat added is, by definition, zero. The increase in internal energy is proportional to the temperature change in the parcel, while the work done is represented by the effort needed for the parcel to expand against the outside pressure. Thus equation (4.5) can be restated for our parcel as:

$$0 = c_p dT - (1/\rho)dp \qquad (4.6)$$

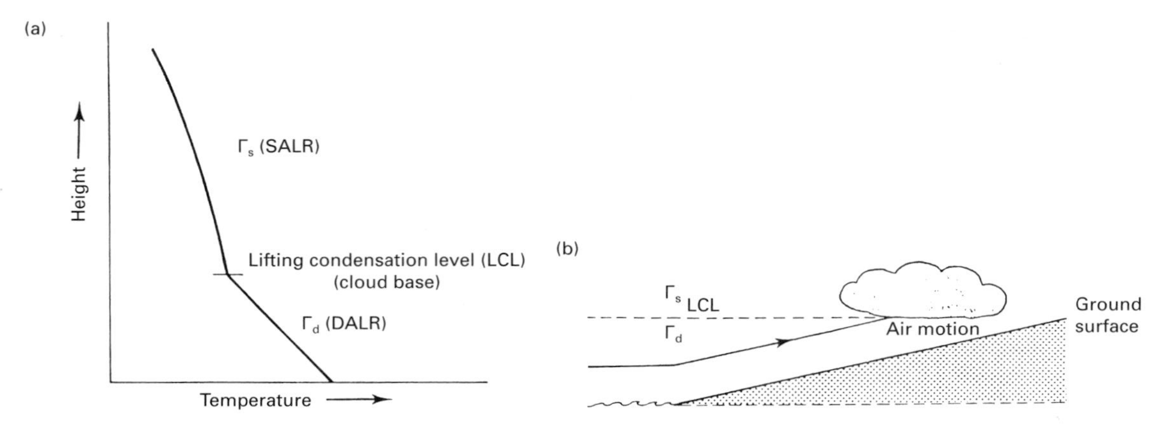


*Figure 4.6* (a) Simplified thermodynamic diagram illustrating the variation of temperature with height as the parcel is lifted over a raised ground surface as shown in (b).

where $c_p$ is the specific heat at constant pressure. Using the **hydrostatic equation** which relates changes in pressure to changes in height in the atmosphere such that:

$$dp/dz = -g\rho \quad (4.7)$$

we have:

$$dT/dz = -g/c_p \quad (4.8)$$

i.e. the rate of change of temperature with height has a constant value. This is the **dry adiabatic lapse rate** (DALR), $\Gamma_d$. Numerically it is equal to 9.8 K km$^{-1}$, and is a constant. Note that this value is a **decrease** of temperature with height.

If the air is saturated when lifting occurs, this rate must be modified to account for the latent heat released during condensation. In this situation the first law of thermodynamics leads to:

$$-L dw_s = c_p dT - (1/\rho)dp \quad (4.9)$$

where $w_s$ is the saturation mixing ratio. Inclusion of this term leads to:

$$\begin{aligned} dT/dz &= -g/c_p - [(L/c_p) \times (dw_s/dz)] \\ &= -\Gamma_d - [(L/c_p) \times (dw_s/dz)] = -\Gamma_s \quad (4.10) \end{aligned}$$

This **saturated adiabatic lapse rate** (SALR), $\Gamma_s$, is slightly variable with height because $dw_s/dz$ (which has a negative value) varies with height. Nevertheless, when considering the lower and middle troposphere a value close to 5.0 K km$^{-1}$ for $\Gamma_s$ is sufficient for most purposes.

### 4.4.3 Cooling rate changes during ascent

Usually when a parcel starts to rise in the atmosphere it contains some moisture but is not saturated. Hence it cools initially at $\Gamma_d$ (Figure 4.6). Eventually the dew point temperature is reached and condensation commences. This is the **lifting condensation level** (LCL) and is the level of a cloud base. Continued uplift is accompanied by a temperature decrease at the saturated adiabatic lapse rate. Simple calculations of the height of the lifting condensation level are thus possible, provided it is borne in mind that there is a dew point lapse rate. This is approximately 1 K km$^{-1}$ and results from the influence of decreasing pressure on the condensation process.

There are a number of ways of creating the ascent of a parcel of air. Whenever the topography of the land surface dictates that air must rise, **orographic** uplift occurs. The angle of slope of the surface largely determines the size of the parcel that is uplifted and the type of cloud that results. Air flowing over a gently rising coastal plain, for example, will slowly be uplifted *en masse*, leading to stratus development. If the air reaches a larger obstacle, say a mountain, it may be forced to rise steeply and, if there is additional surface heating, cumulus clouds may result.

**Frontal uplift** occurs when two air masses of different temperatures come into contact. Although some mixing may occur as described above, it is usual for the warmer, less dense and often moist, air to override the colder, thus being forced to ascend and create clouds. The dynamics and characteristics of fronts, including the clouds associated with them, will be considered in detail in Chapter 9.

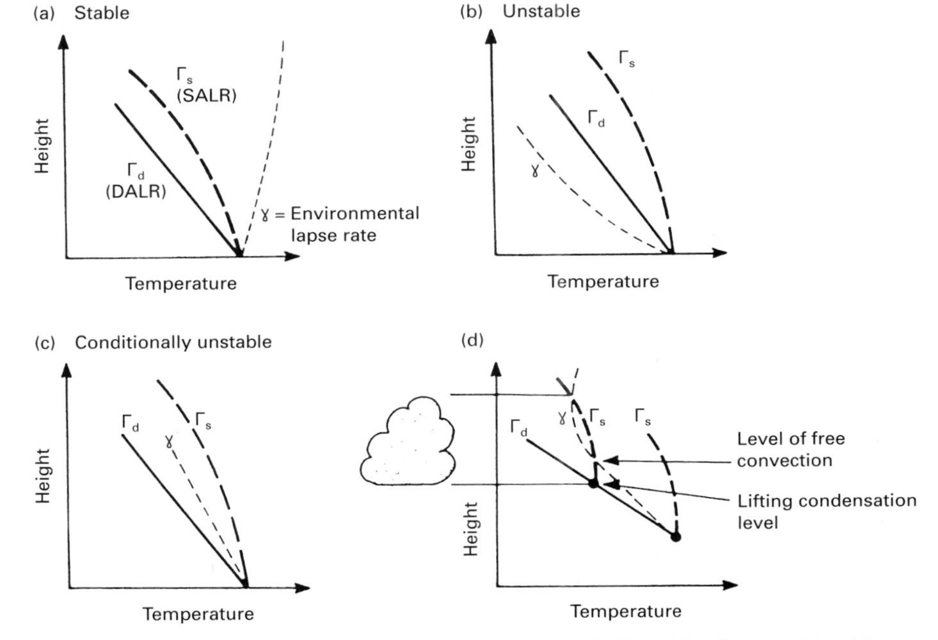


*Figure 4.7* Stability diagrams illustrating the conditions of (a) stable, (b) unstable, (c) conditionally unstable atmospheres and (d) the formation of cloud at the lifting condensation level.

Uplift as the result of **confluence** between two airstreams generally results in widespread ascent. Such uplift is commonly associated with depressions in mid-latitudes, where air spiralling inwards towards the low pressure centre constitutes the confluence. This will be treated along with fronts in greater detail in Chapter 9.

So far we have considered parcels that are forced to rise. However, vertical motions leading to cooling can be created in a much more spontaneous manner through the process of **convection**. This process, which leads to cumulus clouds, is the result of **hydrostatic instability** in the atmosphere.

## 4.5 Hydrostatic stability

When an air parcel, at any level in the atmosphere, is warmer than the surrounding environment the parcel will be less dense than its surroundings, more buoyant, and thus will rise through the atmosphere, while continuing to cool at the appropriate adiabatic lapse rate. This simple statement summarises the concept of **hydrostatic stability**: the property of the atmosphere which controls the small-scale vertical motions within it.

Prior to considering hydrostatic stability in more detail we must introduce another lapse rate, the **environmental lapse rate**, $\gamma$. This is the observed temperature distribution with height at a given time and place. It is variable both in time and space. It must be clearly distinguished from the adiabatic rates since it bears no relation to parcels rising or falling. Nevertheless, the contrast between the environmental rate and the adiabatic rates controls the stability of the atmosphere.

### 4.5.1 Stability conditions

Idealised stability conditions are illustrated in Figure 4.7. In Figure 4.7(a) it is assumed that a parcel, initially at the same temperature as the environment, rises at the saturated adiabatic lapse rate. Immediately upon uplift it becomes cooler than the environment and sinks back. Note that if it were a dry parcel, cooling at the DALR ($\Gamma_d$), the effect would be even more marked. These conditions in which the environmental lapse rate is such that it forces air to return to its original level are called **stable** conditions.

Figure 4.7(b) indicates the opposite, **unstable** conditions. Here a parcel, whether saturated or dry, immediately becomes warmer than the environment and continues to rise. A third case (Figure 4.7c) illustrates **conditional instability**. Here the stability condition depends on whether the parcel is dry or saturated.

Most parcels starting at or near the surface of the Earth contain some moisture but are not saturated. Frequently the atmosphere is conditionally unstable. These common conditions are illustrated in Figure 4.7(d). The process starts with the parcel being forced upwards from its starting point. It initially cools at the dry rate until saturation is reached at the lifting condensation level. Further upward movement is accompanied by cooling at the saturated rate. Usually this slower cooling rate in the conditionally unstable atmosphere eventually causes the parcel to change from being cooler to being warmer than the environment. At this point, the **level of free convection**, instability is established and vertical motions are continued without the necessity of the forced uplift needed to get the parcel to the level of free convection. Ascent will continue until the parcel once again becomes colder than the environment. At this point stability will be established and the cloud top reached.

When a deep layer of air is uplifted, the stability condition can change as it ascends. If the base of the layer is very moist while the top is comparatively dry (Figure 4.8a), the bottom will reach saturation before the top. Thereafter it will cool more slowly than the top, steepening the lapse rate within the layer and eventually creating unstable conditions. This is **convective instability** (Figure 4.8b).

The environmental lapse rate is rarely constant with height. On the largest scale, as illustrated in Figure 3.7, the various atmospheric layers have differing stabilities. In particular, the troposphere is conditionally unstable while the stratosphere is stable. Consequently few convective motions can penetrate into the stratosphere, and the tropopause can be regarded as the practical upper limit of cloud formation. Not all clouds, of course, reach this height. However, a well-developed cumulonimbus top is frequently very close to the tropopause.

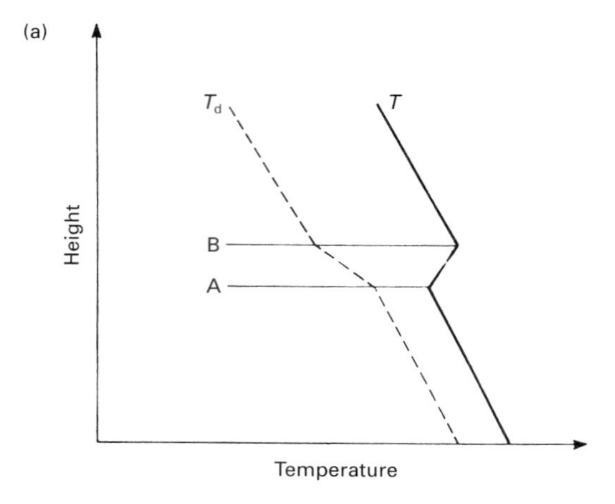


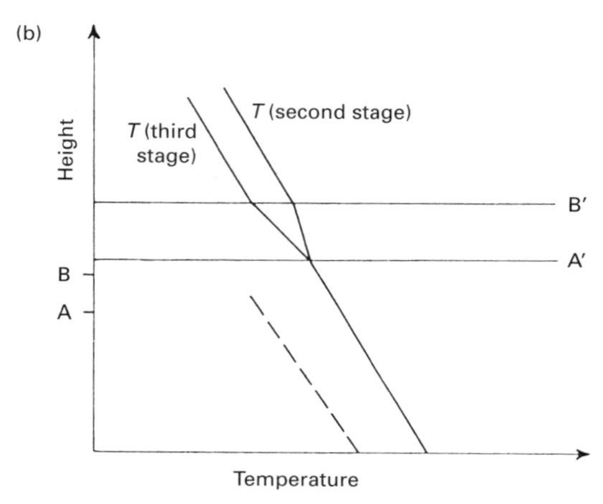


*Figure 4.8* Simplified thermodynamic diagram illustrating convective instability in which the lifting of a large mass of moist air results in less rapid cooling of the base as it reaches saturation first, i.e. at height A before height B. This results in a slower cooling of the base (second stage) so that the lapse rate within the layer is steepened, finally creating the unstable conditions between heights A′ and B′. ($T_d$ and $T$ are the dew point and air temperatures, respectively.)

### 4.5.2 Diurnal variations in the environmental lapse rate

Within the troposphere the environmental lapse rate can change with location, time and height. At a given time and place several layers with differing stability may occur above each other (Figure 4.9). As an example of temporal changes, we can consider what might occur near the surface during a cloudless day (Figure 4.10). The results are an extension of the effects of the diurnal variation in the surface energy balance. Commencing soon after sunrise the heating of the ground causes a steep, unstable lapse rate in the lowest air layers. This persists and deepens throughout the morning and early afternoon. Once surface cooling is established after sunset, the lapse

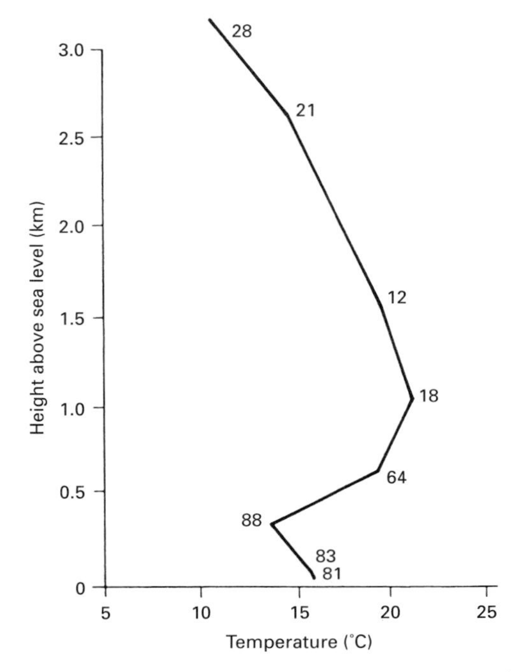


*Figure 4.9* Results from a typical radiosonde ascent taken on the Californian coast in summer (July 1957). The numbers beside the sounding curve give the relative humidity in per cent. Note the change in sign in the lapse rate at about 0.35 km above sea level. (From Neiburger *et al.*, 1982.)

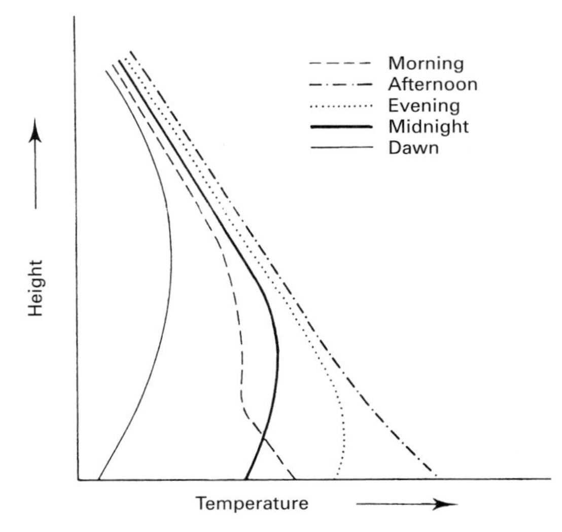


*Figure 4.10* Typical diurnal variation in atmospheric stability. Note the breakdown of inversion conditions established during the preceding cloudless night. The morning heating is seen to proceed more rapidly for the surface than for the atmosphere and the night-time cooling is also more rapid for the surface than for the air.

rate in the lowest layers is reversed. During the night this layer of reversed lapse rate deepens, only to be replaced by more normal conditions as the cycle starts again after dawn.

An environmental temperature profile where the temperature increases with height is an **inversion**. Such layers are very stable and allow little vertical mixing, a very important consideration in air pollution control (see Section 11.3). Inversions can be formed in several ways in addition to the radiative cooling illustrated in Figure 4.10. Many, such as that in Figure 4.9, are the result of large-scale atmospheric motions which have advected warm, stable air over a cooler, less stable near-surface air mass.

### 4.5.3 Triggers of convective processes

In an unstable atmosphere convection will occur spontaneously, redistributing heat, and tending to create a **neutral** atmosphere, where $\gamma$ approaches $\Gamma_d$. If saturation occurs during convection cumulus clouds are formed, while convective activity which does not lead to saturation is known as **clear air turbulence** (CAT). In a stable atmosphere any tendency towards convection is rapidly damped and the stability retained.

Even with an unstable atmosphere there must be some 'trigger' mechanism to initiate the convection process. Rarely is there a shortage of such triggers. The turbulence inherent in any wind flow is almost always sufficient to move a parcel and thus create a temperature difference between it and the environment. Nevertheless, there are certain situations where there are well-defined triggers which enhance the likelihood of convection at certain times or places. The orographic, frontal and confluence mechanisms treated above as forcing air to rise can all act to initiate convection. Similarly, differences in surface characteristics can start the process. A simple example would be the case of an island in a lake. During the day the island heats much more rapidly than the water, forming a heated air parcel over the island. Afternoon convective showers during the rainy season over the Lake Victoria coast of Uganda, for example, can frequently be traced to a source over the islands in the lake.

### 4.5.4 Concept of potential temperature

The relatively simple diagrams of stability we have used so far provide only a qualitative picture of the atmospheric processes. Quantitative diagrams, called thermodynamic diagrams, however, can be developed. These depend for their basic construction on the concept of potential temperature. There are many atmospheric processes in which during ascent and descent the temperature changes at the dry adiabatic lapse rate. When this occurs a quantity called the **potential temperature**, $\Phi$, is conserved. The potential temperature of a parcel of dry air is the temperature it would have if brought dry adiabatically to a surface pressure of $10^5$ pascals. $\Phi$ is a function of $T$ and $p$ only. The second law of thermodynamics and the ideal gas law can be combined and rearranged to give:

$$(c_p/R) \times (dT/T) = dp/p \quad (4.11)$$

Integrating upwards from surface conditions $p_o$ and $\Phi$ to parcel position conditions $p$ and $T$ leads to **Poisson's equation**:

$$(T/\Phi) \times (c_p/R) = p/p_o \quad (4.12)$$

This equation may be solved numerically but traditionally was solved graphically using a **thermodynamic diagram**. These, for a particular situation measured by a radiosonde ascent, allow exact specification of condensation levels and are used for short-term forecasting in meteorology. As such they are not of immediate concern here, but they also allow determination of the energetics associated with the motions, linking with the aspects of the energy cascade considered in Chapter 1.

In our consideration of hydrostatic stability so far, the effects of horizontal motions have been virtually ignored. These effects, which may transform a rather benign cumulus cloud into a severe thunderstorm or even a tornado event, are considered in Chapter 9. In addition, a strong **wind shear**, i.e. a change of wind speed or direction with height, can also destabilise the atmosphere. This 'dynamic' instability, which can occur even in an atmosphere that is statically stable, is mainly responsible for clear air turbulence (CAT) and is also frequently associated with the jet stream, and is also described in Chapter 9.

**Box 4.A Application of moisture information**

*4.A.1 Moisture and transportation*

There are numerous links between moisture and transportation, but here we use two examples; one an application of climate information which incorporates knowledge of the energy budget as well as atmospheric moisture, the other an example of a seemingly simple direct impact of climate on transportation.

The climatological challenge for the application example is to define the frequency with which frost or ice is likely to form on a particular bridge. This information could be used to decide whether or not to install frost protection equipment on the bridge. For simplicity, we will assume that frost will occur when the bridge surface cools both below the dew point of the immediately overlying air and below freezing. Cooling is driven by the energy budget, and we can use the information from Chapter 3 to identify qualitatively the conditions when the required cooling is likely to occur. In Section 13.1 we shall deal with a simple energy budget model which, with minor extensions, could be used to provide a quantitative assessment. The major complicating factor for a bridge is the interaction between the bottom and the air and surface beneath it. Whether or not we use a formal model, the basic energy balance and moisture concepts involved allow us to identify the climatological factors that are likely to be important. These will certainly include cloud amount, wind speed and humidity. We shall assume that there are long-period observations of these at a station not too far away from our bridge. We can then define the combinations of values of these elements which have in the past led to bridge freezing. There are probably many different combinations. A statistical analysis, providing estimates of the most likely combinations, is possible. In some types of problems this may be the only possible approach. However, here it might be advantageous to use the energy budget model, since it has the potential to provide more precise detail than the statistical approach. The final step, whether modelling or statistics is used, is to use the relationships established with the observational data, to deduce the required frequency of bridge freezing.

The second example explores the impact of fog on aircraft delays. Initially, this seems to be a simple exercise of relating late arrivals to the occurrence of fog. However, there are numerous complicating factors. Just a few need be mentioned. How do we incorporate a flight with a completely clear path from origin to destination airport, but which was delayed because the aircraft itself was late arriving at the origin as a result of a fog delay? A non-weather-related delay may prevent an aircraft leaving before fog forms, and so create a much longer delay. How do we count

**Box 4.A (cont'd)**

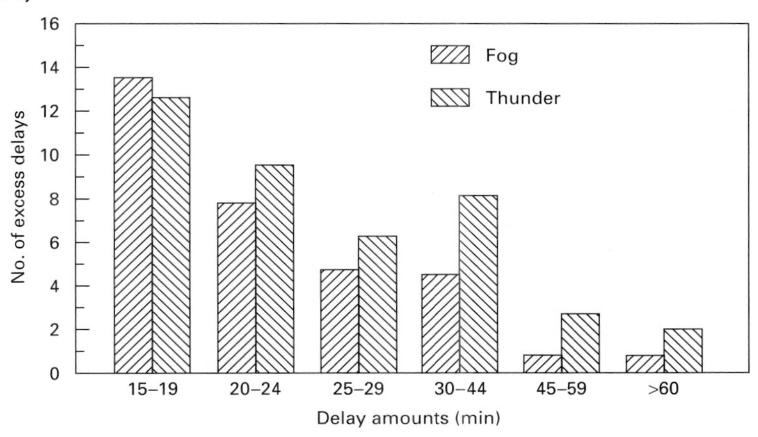


*Figure 4.A.1* The increase in the number of flights delayed daily in fog and thunder conditions over the number that would be expected on clear days, for delays of various lengths (minutes). These values refer to delays at Atlanta Hartsfield International Airport for one airline with approximately 260 daily operations. (From Robinson, 1989.)

that? How long is a delay when a flight is cancelled, or diverted to a completely different airport? How, in general, do we separate weather-related delays from all others in a traffic flow system where aircraft are involved in an almost continuous sequence of flights? It is virtually impossible to give an absolute value to the delay caused by a particular fog event. Instead, it is possible to give a relative answer, by comparing the total delays on a day with fog to those likely on a perfectly normal day (Figure 4.A.1). As is the case with many climate impacts, there are clearly some direct relationships but the many other factors involved make it impossible to isolate the direct ones. As a result, estimates of the climate impact relative to that unrelated to climate must be created and used.

*4.A.2 Extraction of atmospheric water*

Although for most of the world fog and low cloud are regarded as a hazard, in certain regions they are a water source. Wherever there is a situation where fog or very low cloud frequently passes through an area, it is possible to place an obstacle in the flow so that cloud droplets impinge on it and flow towards the ground to be collected. The phenomenon is familiar to anyone who has walked through a fog. In some highland regions, such as the Andes Mountains in Central and South America, orographic uplift of very moist air provides an almost constant water source by such a mechanism. The region is in cloud virtually all of the time and there can be no precipitation in the sense of water falling from a cloud base. Instead, this is the water source for vegetation growth. In natural conditions the moisture is intercepted by trees, where it drips off leaves or runs down the trunk into the soil, providing some moisture for continued tree growth. Humans have created obstacles to intercept the cloud water for their own use. There have been numerous experiments to find the best collector configuration. The most successful instruments use an aspirator motor to draw air over thin metal collector wires, but this requires a power source. Without this it appears that thin wires arranged in a tubular configuration, and capable of collecting whatever the wind direction, are most effective (Table 4.A.1).

In other areas with scarce precipitation, but where dew forms with some regularity, it is possible to conserve and use that dew. For example, **dew traps** are constructed in the vine-growing regions of the island of Lanzarote in the Canaries. Volcanic ash, an abundant resource there, is used as the building material. This, as a layer on the soil, both insulates the soil from the direct effects of solar radiation during the day and cools rapidly during the night, thus enhancing dew production. Even in the complete absence of rainfall the vines can thrive on the water from the trapped dew alone.

**Box 4.A (cont'd)**

*Table 4.A.1* Configurations and moisture collection efficiency of various cloud water collection systems in the South American Andes

| Shape[a] | Dimensions (m) | | | Mesh | Area, $A$ ($m^2$) | Altitude (m) | Date | Collected volume ($l\ m^{-2}\ day^{-1}$) | |
|---|---|---|---|---|---|---|---|---|---|
| | $a$ | $b$ | $h$ | | | | | Maximum | Mean |
| I | 0.3 | 1.2 | ? | VNS[b] | 0.38 | 1050 | Sep–Oct 1962 | | 3.0 |
| I | 0.3 | 0.6 | ? | VNS | 0.21 | 1050 | Sep 1963–Dec 1964 | 34.2 | 6.2 |
| II | 0.7 | 2.0 | 0.9 | VNS | 1.4 | 850 | Dec 1961–Dec 1963 | | <4.5 |
| III | ? | ? | ? | MM[c] | ? | 850 | 1965 | | <4.5 |
| I | 0.3 | 0.6 | ? | VNS | 0.21 | 850 | Aug 1963–Aug 1964 | | |
| | | | | | | | Sep 1964–Oct 1965 | | |
| | | | | | | | Aug 1967–Feb 1968 | 7.6 | 0.7 |
| IV | 0.4 | 1.5 | 0.9 | VNS | 0.7 | Several | Sep 1968–Aug 1972 | 12.3 | 3.0 |
| I | 1.0 | 1.0 | ? | MM | 1.0 | 3 heights | Aug 1970–Feb 1971 | | 2.7 |
| | | | | | | | | | 5.5 |
| | | | | | | | | | 13.4 |
| ? | ? | ? | ? | ? | 0.02 | 950 | Sep 1970–May 1972 | | 9.0 |

[a] Shape:

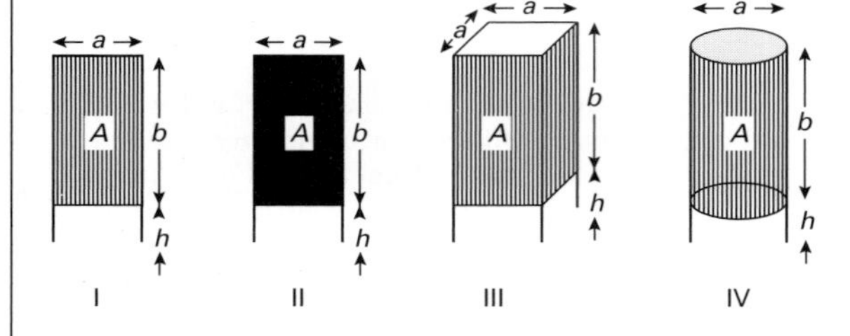


[b] Vertical nylon strings.
[c] Mosquito mesh.
*Source*: after Schemenauer *et al.* (1988).

# CHAPTER 5 Precipitation

In this chapter we consider ways in which precipitation forms, and discuss the types, amounts and distribution of precipitation at the surface of the Earth. Thus the chapter represents the final stages in the hydrological cycle (Figure 4.1) which was introduced in the previous chapter. Conditions at the surface are investigated through the concept of the water budget. This is complementary to the energy budget considered earlier.

## 5.1 Precipitation formation

In our consideration of cloud formation in the previous chapter we implied that as soon as the air reached saturation, condensation and cloud formation occurred. However, rarely does condensation occur immediately. Indeed if a parcel were rising in pure, particle-free air, relative humidity could reach several hundred per cent before spontaneous condensation occurred. However, even clean air is rarely entirely free of particles, and certain particles act as **cloud condensation nuclei** (CCN), promoting condensation at relative humidity at or close to 100%.

Common CCN include dust, clay and organic particles derived from land surfaces, salt crystals derived from sea spray, and particles created in the atmosphere by chemical actions usually called gas-to-particle conversions. An idea of the size and concentration of these particles can be gained from Figure 5.1. Here clouds are divided into two groups, marine and continental. The division depends primarily on the origin of the air and the particles it contains, not on the location of the clouds themselves. Thus marine clouds, which tend to have a smaller concentration but a larger range of droplet sizes when compared with continental clouds, can occur over continental interiors, although they are likely to be more common over the oceans and the continental margins.

### 5.1.1 Growth of cloud droplets

The size of the particle has a great influence on its propensity for growth. Recalling that saturation is defined with respect to a planar pure water surface, water that condenses on a CCN is neither plane nor pure. Two effects come into play. The first is the **solute effect**. If a CCN is dissolved by the water, the air surrounding it may be saturated with respect to the resulting droplet at relative humidity <100% and further condensation is possible. Also, because we have a spherical droplet, the surface is not planar and surface tension creates the **curvature effect**. This second effect requires that the air be supersaturated before further growth can occur. In practice, these two effects occur simultaneously, so that droplet growth follows curves of the type shown by curves X and Y in Figure 5.2. Curve Y starts with condensation on a CCN smaller than does X, and requires a higher supersaturation in order to continue growth. The value of supersaturation reached depends on the concentration and size of nuclei present and on the speed of the cooling process. Rapid ascent usually gives high values, although rarely do they exceed 101%. If there are a large number of small CCN competing for water vapour, as with the continental cloud suggested by Figure 5.1, the supersaturation may not achieve a high enough value to allow the droplets to pass over the maximum of the curve of Figure 5.2. The droplets remain small and precipitation is unlikely. If this happens at low levels, **haze** is formed. With a variety of CCN sizes, as found in marine clouds, some droplets will be able to pass over the 'hump' and continue to grow. They are then said to be **activated** and may continue to grow into raindrops.

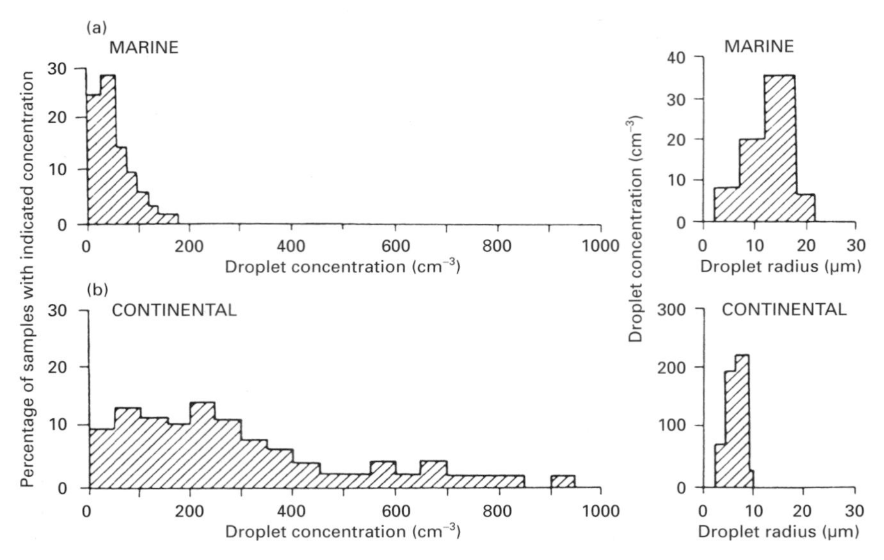


*Figure 5.1* Droplet distribution for (a) marine and (b) continental cumulus clouds showing (left-hand side) percentages of samples with given concentrations (droplets per cubic centimetre) and (right-hand side) concentrations of different-sized droplets. Note the scale change in droplet concentration between the two cloud types. The very much higher droplet concentration in continental cumulus clouds suggests a much higher concentration of CCNs in the air mass from which they are formed. (From Wallace and Hobbs, 1977.)

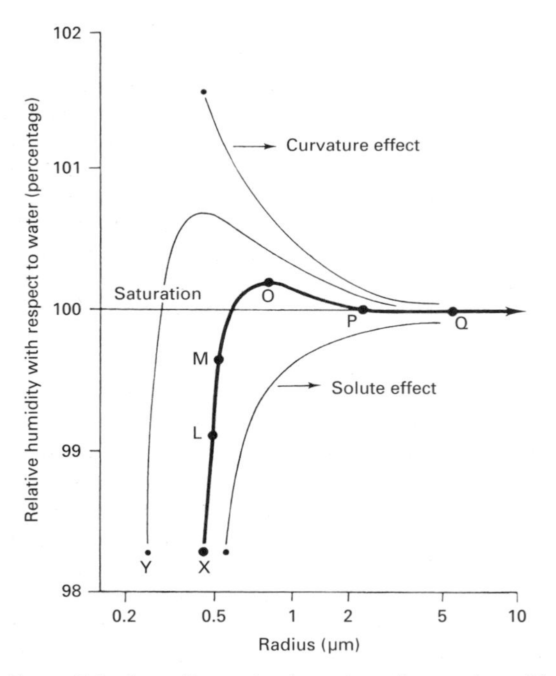


*Figure 5.2* An ordinary cloud condensation nucleus (X) will grow as a compromise between the solute and the curvature effect, the path of this process following the curve LMOPQ. The smaller condensation nucleus (Y) requires a large supersaturation in order to continue growth. (From Petterssen, 1969.)

### 5.1.2 Growth of raindrops

Growth of activated droplets will continue by direct condensation. However, this process is very slow, much slower than the observed times needed for precipitation formation. Hence other mechanisms must be involved. To explore these mechanisms it is necessary first to consider the motion of drops within clouds.

Particles suspended in the atmosphere fall under their own weight according to **Stokes' law**:

$$V = 2g(\rho_p - \rho_a)r^2/9\eta \qquad (5.1)$$

where $V$ is the terminal velocity, $g$ is the acceleration due to gravity, $r$ is the particle radius, $\rho_p$ and $\rho_a$ are the densities of the particle and the atmosphere, and $\eta$ is the viscosity of air. The **terminal velocity** is the velocity a particle attains in free fall through still air, being the result of the balance between the gravitational attraction downwards and the frictional drag of the air itself. Stokes' law indicates that this velocity increases rapidly with particle size. For example, a 10 μm droplet will have a terminal velocity of about $10^{-3}$ m s$^{-1}$, while a 1 mm drop will have a terminal velocity of approximately 10 m s$^{-1}$. When clouds are formed, the air is rarely motionless, and in particular, there is likely to be an updraught in the cloud. This will mitigate against the downward movement so that the speed of the final movement towards the Earth's surface, the **fall velocity**, is given by:

$$\text{Fall velocity} = \text{terminal velocity} - \text{updraught velocity} \quad (5.2)$$

Only if the terminal velocity is greater than the updraught velocity will the drop fall. In practice, the small cloud condensation droplets will have a terminal velocity that is smaller than the updraught velocity and will remain suspended in the cloud. Indeed, they may never grow to sufficient size to have a positive fall velocity and thus may never leave the cloud. In order for precipitation to occur, therefore, some mechanism is needed to increase the size of a cloud droplet to that of a raindrop. The mechanism which becomes active depends greatly on the temperature of the cloud.

### 5.1.3 Collision and coalescence

In **warm clouds**, where the ambient cloud temperature is above 0 °C, all the condensation products are liquid water. The larger droplets will have a higher terminal velocity than smaller ones and will fall through them, collecting them and increasing in size through the **collision and coalescence** processes. For collision to occur the smaller droplet must be close to the axis of fall of the larger drop, otherwise it will follow the air currents around the falling drop and there will be no impact. Generally the larger the falling drop, the more efficient is the collection, and drops with a radius greater than 40 μm collect most of the droplets they encounter. Even with a collision, growth will only occur if the two drops coalesce. This will occur most readily if the drops are of considerably different sizes. These factors thus re-emphasise the importance of a wide range of sizes for the original CCN for precipitation production.

### 5.1.4 Bergeron–Findeisen mechanism

In **cold clouds**, where the ambient temperature is below freezing, the condensation products can be both liquid water and ice crystals. At temperatures below about –40 °C all the products are ice crystals and the cloud is said to be **glaciated**. Between 0 and –40 °C ice and water coexist, giving a **mixed cloud**. In glaciated clouds there is some crystal growth by processes analogous to the direct condensation and the collision/coalescence processes of warm clouds, but rarely do these high, cirrus type clouds yield precipitation that reaches the surface. In mixed clouds the initial growth phase depends on the coexistence of ice and water. The process is known as the **Bergeron–Findeisen** process. At temperatures below 0 °C the saturation vapour pressure with respect to water is greater than that with respect to ice (Figure 4.3). Thus, in a mixed cloud, the air which is close to saturation with respect to water is supersaturated with respect to ice. Consequently ice crystals can grow much more rapidly than, and at the expense of, water droplets.

The individual ice crystals will grow and collect together to form snowflakes. The shape of the snowflakes that are created depends on the temperature at which the condensation occurs. Since a crystal may move about in a cloud and experience a variety of temperatures, the shape of the snowflake which grows as a result of the accumulation of crystals can be very complex. In addition, the growing ice crystals may come into contact with supercooled liquid water droplets which will freeze onto or around the crystal immediately on contact, a growth process known as **riming**. Riming is a primary mechanism of hail formation. Crystals may also grow by **aggregation**, a process similar to collision/coalescence. Aggregation is most marked at ambient temperatures above –5 °C, when crystal surfaces become moist and sticky.

An individual cloud need not be exclusively warm or cold. The lower portions of a cloud may be warm while higher parts may be mixed or even glaciated. Thus precipitation, created as snow, may melt before falling from the cloud. Furthermore, clouds can seed themselves naturally, when ice particles falling from a cirrus cloud, or from the glaciated portion of a towering cumulus, act as nuclei. In the same way, ice particles or snowflakes from the mixed portion of a cloud can fall into the warm portion and provide the large nuclei needed for further growth.

### 5.1.5 Cloud seeding

The action of the Bergeron–Findeisen process is used in artificial **cloud seeding**. Tiny particles of silver iodide or dry ice (solid carbon dioxide) are dropped into the tops of mixed clouds. Silver iodide particles serve as CCN with the correct crystal structure to act as a large freezing nucleus and initiate the growth of drops. Dry ice serves to cool the ambient air locally, enhancing the difference in saturation vapour pressure between ice and water and thus initiating the whole process. Cloud seeding can only be undertaken when a cold cloud is already present, and so serves as a means of precipitation augmentation, not 'rainmaking' in an area with no naturally occurring chance of rain. Due to this restriction it is extremely difficult to demonstrate rigorously that a specific cloud seeding project has led to increased rainfall. Nevertheless, the balance of evidence suggests that the technique is useful in areas where a small increase in precipitation can lead to significant economic gain. Hence, for example, it is widely practised in the somewhat marginal agricultural areas of the American Midwest and in South Africa. It is also used to augment winter snowfall in the American Rockies, helping to enhance the amount of water in that frozen reservoir which will become available in the subsequent growing season.

Attempts are being made to use cloud seeding techniques to modify the microstructure of clouds as an aid to hail suppression. If a large number of small hail stones can be produced to replace a small number of large ones, there is likely to be a decrease in the damage the hail causes. Although theoretically cloud seeding could produce this effect, there is not at present any reliable and reproducible observational evidence that this has been successful. A similar conclusion can be drawn from the attempts to modify hurricanes, either by decreasing their intensity or altering their direction, by cloud seeding.

Cloud and fog dissipation schemes use a very similar technique but here the aim is to overseed. Solid carbon dioxide, with a temperature around −78 °C, dropped into a cloud from an overflying aircraft locally supercools the cloud and produces ice crystals by spontaneous nucleation. A dry ice pellet 10 mm in diameter falling through a cloud at −10 °C produces about $10^{11}$ crystals before evaporating. This effectively overseeds the cloud, giving a very large number of small crystals which rapidly evaporate into the unsaturated air and thus the cloud dissipates. Silver iodide introduced near the top of a cloud behaves like ice crystals and again overseeding can cause dissipation. This technique is often used for the dispersal of cold fog at airports.

## 5.2 Precipitation at the ground

The type and size of precipitation leaving a cloud base depends on the conditions within the cloud, but the precipitation that actually reaches the ground is modified by conditions in the air layer between the cloud and the ground. In general, the temperature structure determines whether the precipitation will arrive as frozen precipitation or as liquid water, while the humidity of the layer determines the amount of evaporation that will occur and hence the ultimate size of the precipitation particles (Figure 5.3). In both

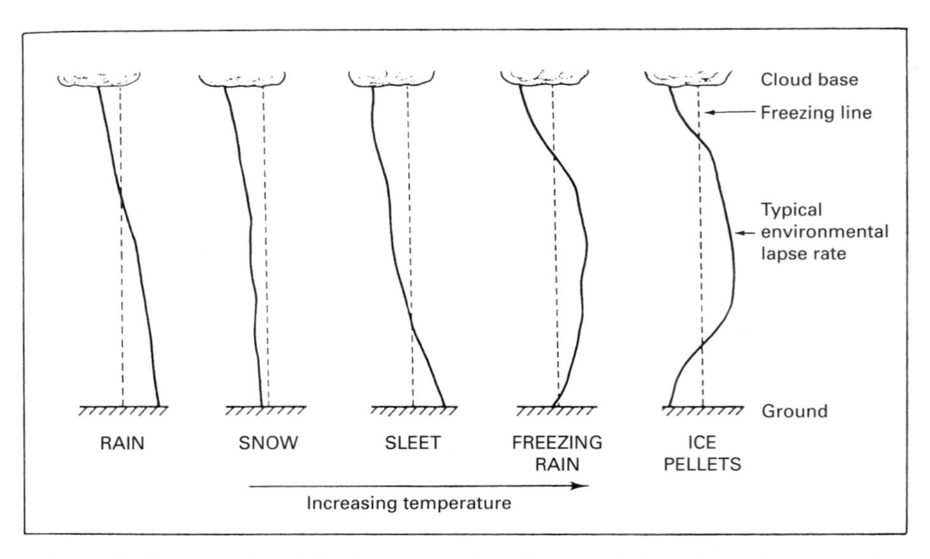


*Figure 5.3* The type of precipitation that reaches the ground depends on the temperature structure of the air layer between the cloud base and the ground. In the situations illustrated, precipitation falls from a mixed cloud. The resultant precipitation type depends on the relation between the environmental lapse rate and the freezing temperature. In all cases typical profiles are shown and temperature increases from left to right.

cases the fall velocity will dictate the time over which the processes can act and hence how complete they will be. The major types of **hydrometeors** that can occur are summarised in Table 5.1.

### 5.2.1 Precipitation type and duration

The intensity and duration of precipitation is determined largely by the type of cloud system involved. This in turn is intimately connected with the cloud formation processes considered above. In general, cumulus type clouds involve vigorous vertical motions, giving large drops and intense precipitation for a short period. Usually their influence is restricted to a fairly small geographical area. Stratus and altostratus, in contrast, involve more persistent, less vigorous vertical motions over a much wider area. Hence prolonged, steadier, and usually less intense, precipitation results. This can be illustrated by reference to Figure 5.4. Miami, Florida, USA, is in an area dominated by cumulus type clouds. Short-duration rainfalls are likely to be much more intense than in Seattle, Washington, USA, where precipitation from depressions is predominant. The difference in intensity decreases as the duration increases, but is still clearly discernible for 24-hour rainfall totals.

Figure 5.4 also indicates that intensity decreases as duration increases. World rainfall statistics suggest that intensity is approximately proportional to the inverse square root of the duration, but that there are many regional variations.

### 5.2.2 Precipitation measures

Although the preceding sections indicate that there are a wide range of types, durations and intensities of precipitation, we know relatively little about the frequency or distribution of most of them. The only common world-wide observational networks, using instruments described in Box 5.I, measure either total precipitation or total snowfall. Total precipitation is usually given as the liquid water equivalent value derived by melting frozen forms and including them with the liquid ones. There are no standardised observational methods for the specific forms of precipitation given in Table 5.1. Many national weather organisations record the frequency of days when they occur, but these are usually based on visual observation. Hence observations are restricted to a few locations and tend to be rather subjective. Certainly there are no routine measurements of the amounts of the various precipitation forms.

*Table 5.1* Major types of hydrometeors (or precipitation)

| Hydrometeor | Description | Normal clouds from which precipitation can fall and reach the ground |
|---|---|---|
| Rain | Drops with diameter >0.5 mm, but smaller drops are still called rain if they are widely scattered | Ns, As, Sc, Ac castellanus, Cu congestus |
| Drizzle | Fine drops with diameter <0.5 mm and very close to one another | St, Sc |
| Freezing rain (or drizzle) | Rain (or drizzle), the drops of which freeze on impact with the ground | The same clouds as for rain or drizzle |
| Snowflakes | Loose aggregates of ice crystals, most of which are branched | Ns, As, Sc, Cb |
| Sleet | In Britain, partly melted snowflakes, or rain and snow falling together | The same clouds as snowflakes |
| Snow pellets (also known as soft hail and graupel) | White opaque grains of ice, spherical, or sometimes conical with diameter about 2–5 mm | Cb in cold weather |
| Snow grains (also known as granular snow and graupel) | Very small, white, opaque grains of ice. Flat or elongated with diameter generally <1 mm | Sc or St in cold weather |
| Ice pellets | Transparent or translucent pellets of ice, spherical or irregular, with diameter <5 mm. There are two types: | |
| | (a) frozen rain or drizzle drops, or largely melted and then refrozen snowflakes; | Ns, As, Cb |
| | (b) snow pellets encased in a thin layer of ice (also known as small hail) | Cb |
| Hail | Small balls or pieces of ice with diameters 5–50 mm or sometimes more | Cb |
| Ice prisms | Unbranched ice crystals in the form of needles, columns or plates | St, Ns, Sc (sometimes falls from clear air, when it is just an advanced stage of ice fog) |

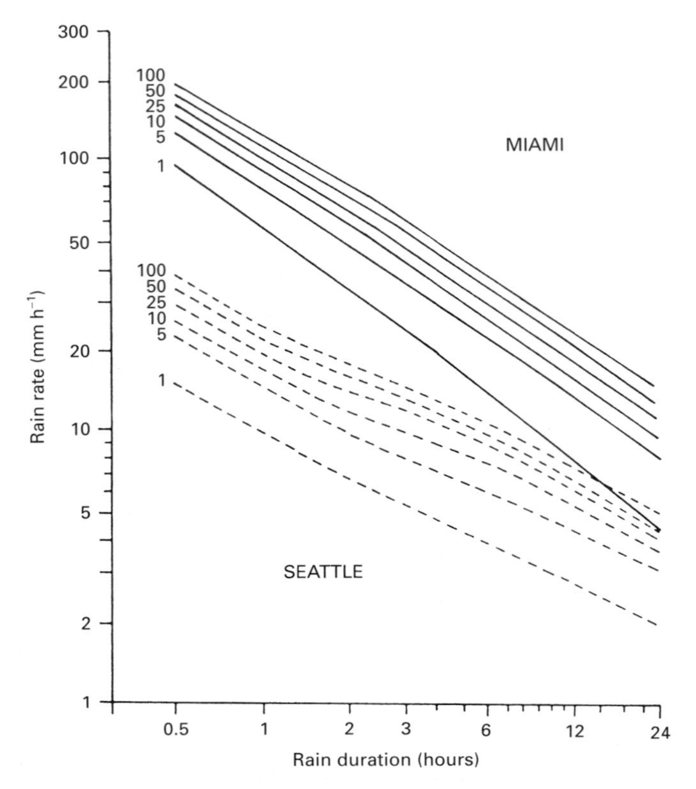


*Figure 5.4* Intensity–duration rain curves for various return periods (1–100 years) for Miami, Florida (solid lines) and Seattle, Washington (dashed lines). The curves indicate that only once in 10 years is Seattle expected to have rainfall averaged over 12 h that will reach or exceed 5.5 mm $h^{-1}$. (From Barrett and Martin, 1981.)

**Box 5.I The measurement of precipitation**

*5.I.1 Liquid precipitation measurement*

Rainfall can be directly measured using a **rain gauge** (Figure 5.I.1). This is essentially a bucket with a horizontal orifice of known size exposed just above the surface of the Earth. The water is caught in the bucket and funnelled into a measuring cylinder, and the depth of water collected at the end of a period of interest is then noted. The cylinder is then emptied and re-installed. In order to maintain uniformity, most nations have established standards for the size, exposure and measurement times. There are several problems inherent in this simple measurement technique. In windy conditions turbulence in the airflow is created by the gauge itself. In particular this increases the speed of the flow across the top of the orifice, decreasing the catch and leading to unrepresentative results. This underestimate in measured precipitation is much greater for snow than for rain because the fall velocities and momenta of snowflakes are much less than those of raindrops. The decrease varies with wind speed and, in sloping terrain, with wind direction. Some gauges are equipped with wind

*Figure 5.I.1* (a) Example of one type of rain gauge and (b) the chart recorder of a tipping bucket rain gauge. (c) A baffled surface erected to try to eliminate splashing into the gauge of rain that has fallen onto the ground. Rain gauges are sited near the centre of the meteorological enclosure, not less than 3 m away from the Stevenson screen (Figure 3.I.3). The gauge should be installed in horizontal ground in a vertical hole so that the rim is a specified height above ground (usually 30 cm or 12 inches). Recording rain gauges are used to provide information about the time and duration (and hence intensity) of precipitation. The mechanism records each time the complete filling of a small water trough occurs. The trough tips when full, causing the water to be lost and permitting another cycle of filling to occur.

**Box 5.I (cont'd)**

(a)

(b)

(c)

**Box 5.I (cont'd)**

shields, but this only partially corrects the problem. Since wind speed decreases rapidly as the ground is approached, a gauge close to the ground is desirable. However, the closer to the ground it is, the more likely it is to receive precipitation that is splashed into it in addition to the falling rain (cf. Figure 5.I.1c). Hence it is difficult to measure with great accuracy the amount of rainfall that actually arrives at the Earth's surface. Nevertheless it remains the only instrument giving absolute measurements of precipitation, and there are numerous records of precipitation available from many locations around the world.

Rain gauges can be constructed so that a continuous record of receipt is made (Figure 5.I.1b). A mechanism is inserted within the bucket which senses the change in water level as precipitation occurs and translates this into a trace on a chart located on a revolving drum. Such instruments allow not only the determination of total precipitation but also its intensity and duration.

Although these measurements give a good idea of the spatial distribution of precipitation, especially when long-term averages are needed, they are spot measurements. Rainfall amounts, especially from a single storm, can have a very great spatial variation. Unless there is a very dense network of rain gauges these variations will not be identified. This can be a very serious consideration in areas where intense storms that may cause flooding are common.

The lack of spatial coverage by rain gauges can be overcome to some extent by using information obtained by **radar** (Figure 5.I.2). A radar emits a pulse of electromagnetic radiation that is reflected off any

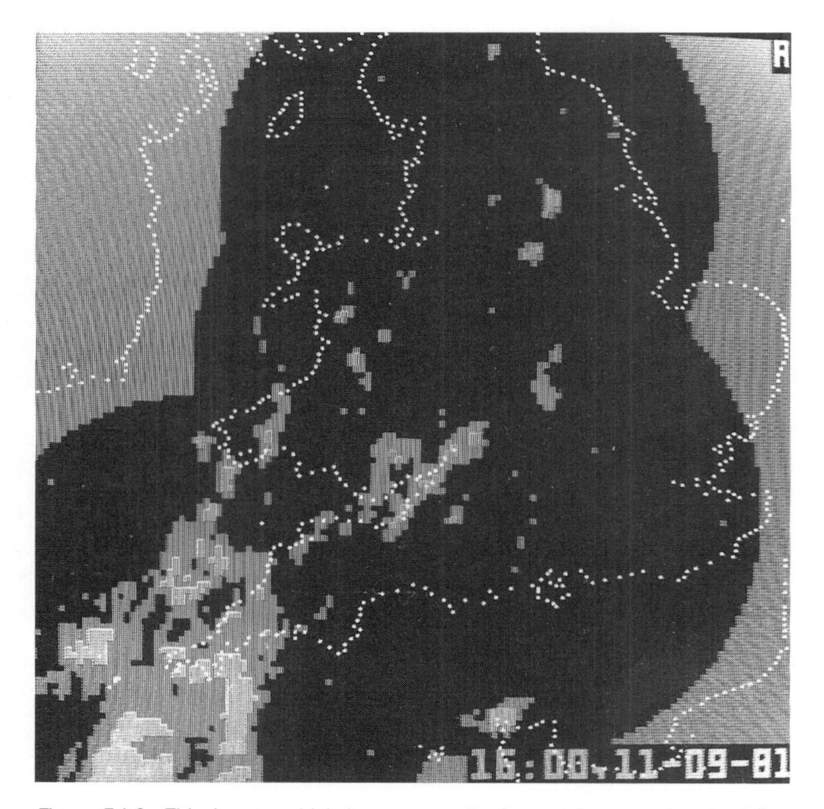


*Figure 5.I.2* This image, which is a composite formed from a network of four weather radars, shows where rain was falling in England and Wales at 1600 hours on 11 September 1981. An unstable southwesterly airstream covered the British Isles, producing heavy showers over Wales and the Midlands. An area of thundery rain can be seen approaching from the southwest, ahead of an upper trough. In this black and white photograph of a colour display, the dark grey represents light rain and pale grey heavy rain. Each picture element is a 5 × 5 km square. (Courtesy of RSRE Malvern: © British Crown copyright/MOD.)

**Box 5.1 (cont'd)**

obstacle in the emission path and returned to the source. By measuring the time between emission and receipt, the distance of the obstacle from the source can be deduced. By choosing an appropriate wavelength for the emission, raindrops can become the obstacles and the number of drops in the path deduced. By calibration with conventional rain gauges, the amount of rainfall can be deduced. By using a full radar sweep, it is possible to determine rainfall over a wide area. In sophisticated systems the **Doppler effect**, which leads to a shift in the wavelength of the returned signal, can be used to determine the direction and speed of motion of the cloud droplets or the raindrops, and has the potential to provide a great deal of information about rainfall rates.

It is also becoming increasingly common to deduce rainfall amounts from satellite observations. Measures of cloud top height and optical thickness for specific cloud types can be related to the amount of precipitation at the surface below the clouds. At present this is proving most useful in oceanic areas where surface-based observations, whether with conventional gauges or radar, are not practical.

*5.1.2 Snow measurement*

Another feature of precipitation that is routinely measured is snow depth. The procedure is again simple. A site where the depth of snow appears to be representative of the area, usually in the centre of a region of open ground, is chosen and the depth measured with a normal measuring stick. In some countries measurements are taken on a **snow board** in an attempt to provide as much observational standardisation as possible. Results are commonly reported both as the current depth of snow on the ground and as the amount of snowfall since the last observation.

For some applications this is the important information, but for others it is more useful to determine the water content of the snow, which is related to snow density. This is particularly important when the snow-pack is being used as a water reservoir, so that the spring meltwaters can be used for downstream water supply, a common practice in the western part of the United States. Water content can be determined by extracting a snow core of known diameter, melting it and weighing the meltwater. A more common, and much simpler, method is to assume that the snow is of average density and compaction and may be converted to a liquid water equivalent using the rule of thumb that 10 units of snow correspond to 1 unit of water. However, actual values range from about 6 : 1 to 30 : 1.

A set of measurements related to frozen conditions are those recording the areal extent of ice and snow cover. This has practical implications in polar regions for navigation and for overland transport. However, since small changes in the extent of these high albedo surfaces can have a marked influence in energy exchanges, continuous monitoring is important. Routine satellite surveillance gives daily information, and now we have a clear idea of the interannual variability of polar snow and ice extent. This enhances our understanding of climate processes and in some cases helps to provide climate forecasts.

Even for total precipitation or total snowfall, the observational network allows only certain generalisations. Precipitation, depending on the formation mechanism, is highly variable from place to place. A network of point observation sites, therefore, can easily miss significant variability and give misleading results. In general it is possible to assume that the longer the period involved in creating a rainfall total, the less dense the observational network should be. Thus for most of the land areas of the world the existing network is perfectly adequate to give annual total, or even monthly normal, values. These will be considered in the next section, and in the next several chapters. On a finer time scale, as used for Figure 5.4, special observations are needed. Similarly, the finer spatial scale considered in Chapter 10 requires not only special observation networks, but also careful statistical analyses of the results, which includes application of the principles of cloud and precipitation formation as discussed here.

### 5.2.3 Rain amounts and rain days

The spatial distribution of rainfall can also be viewed in terms of the number of **rain days** per year. A rain day is usually defined as a 24-hour period, usually starting at 0900Z, during which 0.2 mm or more of precipitation falls. The climatic average of rain days varies from over 180 per year in humid coastal regions, such as Washington State in the USA, to less than one per year in very arid regions, such as west central Australia. In general there is a close relationship between the number of rain days and the total precipitation. For example, the bases of the Hawaiian Islands are in a very different climatic regime from the mountain peaks. While the base has low precipitation and few rain days, the peak of Mount Wai-ale-ale on the island of Kauai could claim to be the wettest place in the world with an average annual total precipitation of 11 455 mm and an average of 335 rain days per year. However, seasonality can

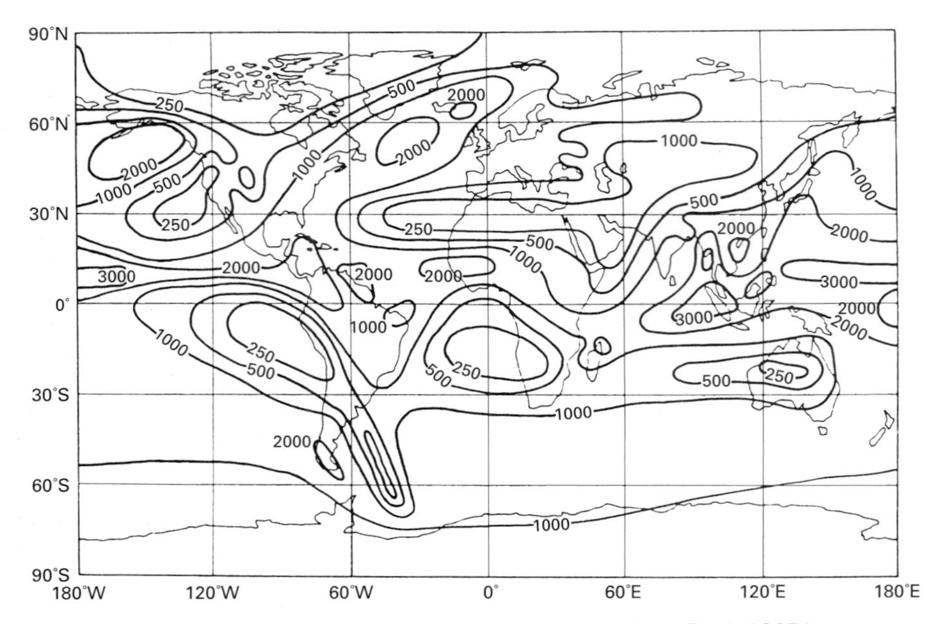


*Figure 5.5* Global mean annual precipitation (in millimetres). (From Riehl, 1965.)

influence the relationship between rainfall totals and rain day numbers. Places influenced by the Asiatic monsoon may have annual totals approaching that of Wai-ale-ale, but, because of distinct wet and dry seasons, only half as many rain days.

The relationship between rainfall and rain days therefore depends strongly on the climatic regime and on the nature of the precipitation producing systems. For many purposes the total rainfall in a given period is the most useful measure of precipitation, but in some cases the number of rain days is more appropriate. It is perhaps unfortunate that most travel brochures and travel agents, when they consider climate, provide monthly average rainfall statistics; whereas the average number of rain days in a particular month would be of much more use to the prospective holidaymaker.

## 5.3 Global precipitation distribution

We have seen that both the intensity and duration, and thus the amount, of precipitation, in an individual event depend on the processes acting to create the precipitating clouds, and that the areal extent of the precipitation depends on the same factors. Since particular processes tend to dominate particular areas of the globe, we can make several pertinent generalisations about the global precipitation, leaving detailed discussion for subsequent chapters.

The area of maximum annual precipitation, over 2000 mm per year, extends in a band through the equatorial regions (Figure 5.5). The subtropical deserts and the polar regions have values below 250 mm. The mid-latitude regions have intermediate values, being in general about 1000 mm per year.

### 5.3.1 Tropical precipitation

Precipitation in much of the tropics is associated with convective activity. Strong vertical motions occur in a fluctuating band near the equator. These release the abundant water vapour to create a regime of intense, short-lived storms from cumulus clouds. Rainfall rates in excess of 100 mm $h^{-1}$ are not uncommon. Although the location of the storms is partly controlled by local topographic features, storms tend to recur sporadically, so that precipitation does not occur at a particular place every day even though there may be a storm in the area each day.

More widespread uplift is associated with monsoonal circulations. Such circulations are particularly well developed over tropical Asia. Although this is a strongly seasonal precipitation regime, the effects of convective uplift, dynamic uplift and topographical forcing combine to produce high annual rainfall totals. Locally rainfall rates may be very high but generally the monsoonal condition is characterised by longer lasting, less intense precipitation.

### 5.3.2 Mid-latitude precipitation

In mid-latitudes much of the precipitation production is associated with depressions and fronts. The result is widespread uplift giving extended periods of gentle rain over a broad area. Rainfall rates can vary greatly, although 1–2 mm $h^{-1}$ can be regarded as a typical value. The intensity is partly controlled by the amount of water vapour available, which in turn depends on the source of the air that is being uplifted. Air derived directly from the subtropical oceans, where evaporation rates are high, is likely to lead to higher precipitation rates. If the source is the tropical deserts, the air is likely to be much drier and it is not uncommon in these conditions for dust and sand particles to form the condensation nuclei and hence be deposited in large quantities with the rain.

Convective activity in the mid-latitudes is primarily a summer phenomenon. It can be as intense, but is usually less regular, than in the tropics. A rainfall of 31 mm in 1 minute (1860 mm $h^{-1}$), was recorded in Maryland, USA, in 1956, and a fall of 126 mm in 8 minutes (945 mm $h^{-1}$) in Bavaria, Germany, in 1920. The strong upwelling of air which must be associated with such phenomena can lead to some surprising results. On 9 February 1859 an area of about 1000 $m^2$ centred on Aberdare, Wales, received precipitation filled with small fish (sticklebacks and minnows). Similarly, the Yachting Olympics in October 1968 at Acapulco, Mexico, experienced a heavy storm which deposited live maggots from 5 to 25 mm in length.

### 5.3.3 Regions of low precipitation

The regions of low precipitation in the subtropics result mainly from a lack of mechanisms for creating uplift and bringing the air to saturation. Certainly over the oceans, and to a large extent over the land deserts as well, there is no lack of moisture in the atmosphere. On those rare occasions when the mechanisms are present, thunderstorms, frequently intense ones, occur. In contrast, over the polar regions the low precipitation totals are as much associated with a lack of atmospheric moisture (Plate 2) as with a lack of uplift mechanisms. Here the infrequent precipitation events are more likely to be rather long periods of light snow, although often associated with high winds.

### 5.3.4 The thunderstorm

The one precipitation-producing feature that is common to virtually all regions of the globe is the **thunderstorm** (Figure 5.6). This is a mesoscale feature whose frequent occurrences make it vital in any consideration of global precipitation distributions. All that is needed is an unstable, preferably a highly unstable, humid atmosphere. This triggers a rising bubble of air which forms a cumulonimbus cloud (see Section 4.5). A thunderstorm consists of a series of these clouds or cells, each going through a sequential development, often called a life cycle, in about half an hour (Figure 5.7a). Rising air dominates during the initial, or cumulus, stage. When precipitation starts to fall the cell has reached maturity. This precipitation creates a downdraught by entrainment of the adjacent air (Figure 5.7b). The downdraught initially occurs close to the leading edge of a moving cell and brings snowflakes or hail pellets below the freezing level. This downdraught becomes increasingly dominant and locally decreases the instability, so that eventually the updraught is suppressed completely. Meanwhile the cloud top is approaching the tropopause and beginning to spread laterally. Maximum upward velocities are now found in the middle of the cloud. Once the updraught is removed the cell enters the dissipating stage. Without the updraught the cloud droplets cannot grow. Hence precipitation soon ceases and the remaining cloud droplets evaporate back into the air. However, another cell may have formed ahead of the dissipating cell. This is frequently the result of the cold downdraught air undercutting warm surface air ahead of the cell and releasing the latent instability there. Consequently the storm may continue as another cloud. Usually subsequent clouds have somewhat less vigour than the initial one, simply because each cloud serves to redistribute the vertical temperatures and thus decrease the instability.

In the type of thunderstorm just described it has been assumed that there is little vertical variation in wind speed. This is typical of **air mass thunderstorms**. These are the common type in the tropics and are the isolated thunderstorms typical of mid-latitude summer afternoons. However, when there is significant vertical variation in wind speed, thunderstorms can occur in organised groups and become severe, forming **mesoscale convective complexes**. Our understanding of these is rapidly increasing, stimulated by the advent and use of **Doppler radar** (see Section 5.I.1), but is still far from complete. Here

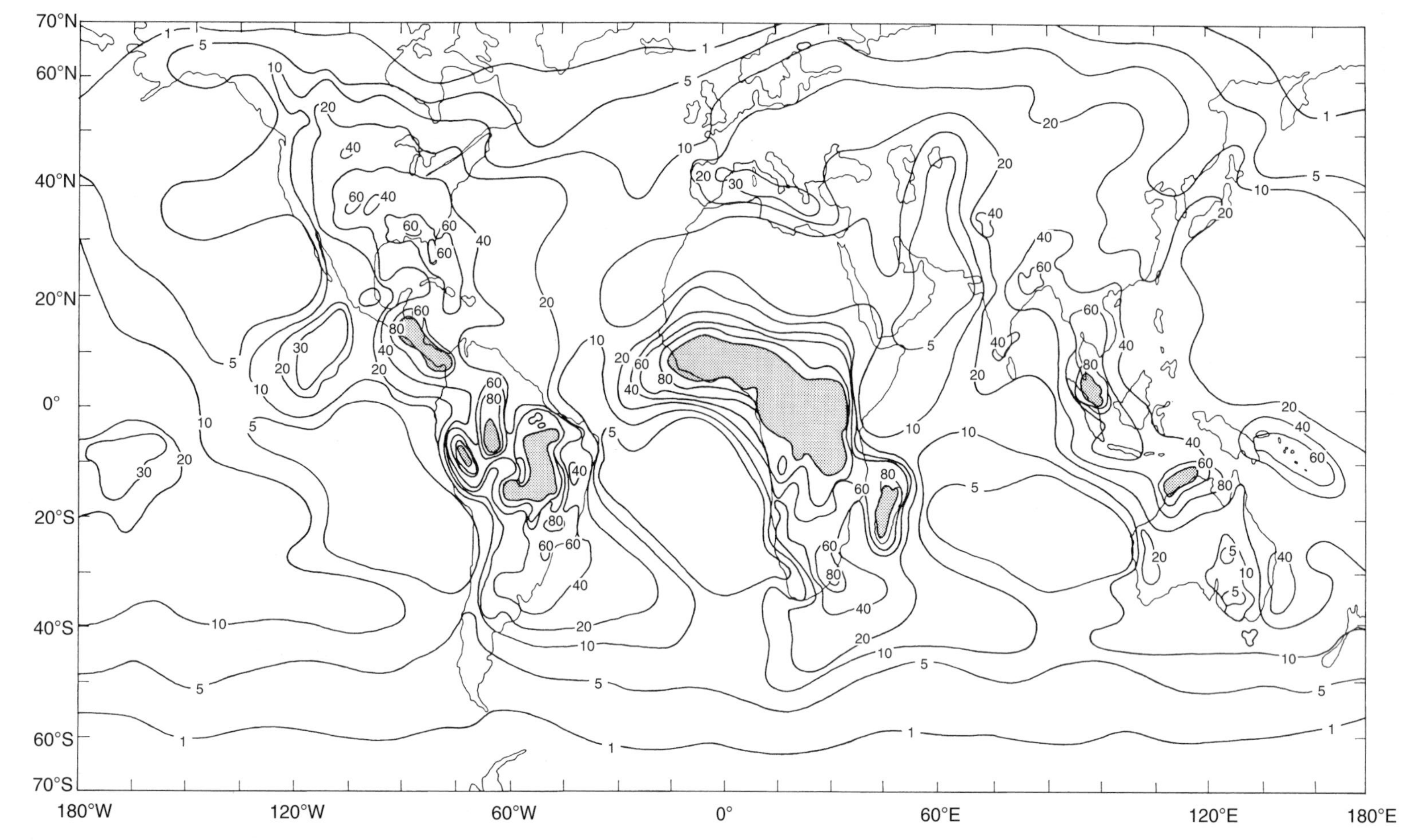


*Figure 5.6* Global distribution of the annual number of thunderstorms. The tropical continents dominate in terms of numbers, although these may not be the most intense storms. Over mid-latitude land areas the eastern portions of the continents have the highest values. Many subtropical desert regions may have several storms per year, but occurrence in the polar regions is rare. Areas with >100 storms year$^{-1}$ are shaded. (From WMO, 1953.)

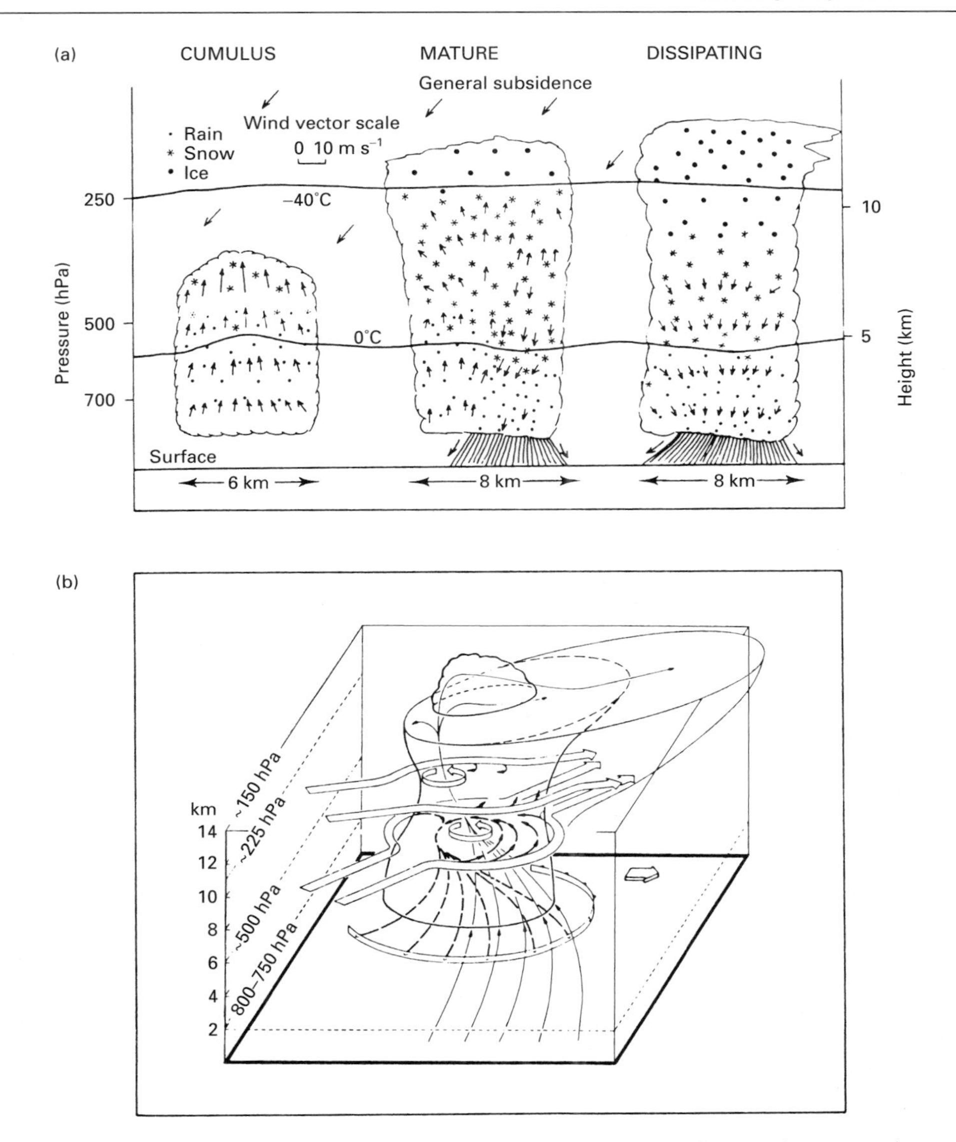


*Figure 5.7* (a) Typical life cycle of a thunderstorm cell showing the cumulus stage, the mature stage and the dissipating stage. (b) Three-dimensional depiction of a cumulonimbus cloud. The flat arrow on the upper right of the surface shows the direction of travel. The thin, solid inflowing and ascending streamlines represent the trajectory of moist air originating at low levels (surface to approximately 750 hPa). The heavy dashed streamlines show the entry and descent of potentially cold and dry middle level (700–400 hPa) air feeding the down-rushing and diverging downdraught. The surface boundary between the inflow and downdraught is shown as a barbed band. The internal circular arrows show the net updraught rotation.

we provide a descriptive approach, using the **squall line** as a representative example.

A squall line is a series of thunder cells aligned at right angles to the direction of storm motion. Although individual cells go through a life cycle akin to that for the air mass thunderstorm, the results are different because the wind speed of the environment increases with altitude, and the cells move at a speed approximately equal to the wind speed in the middle troposphere. Thus at low levels ambient air is being

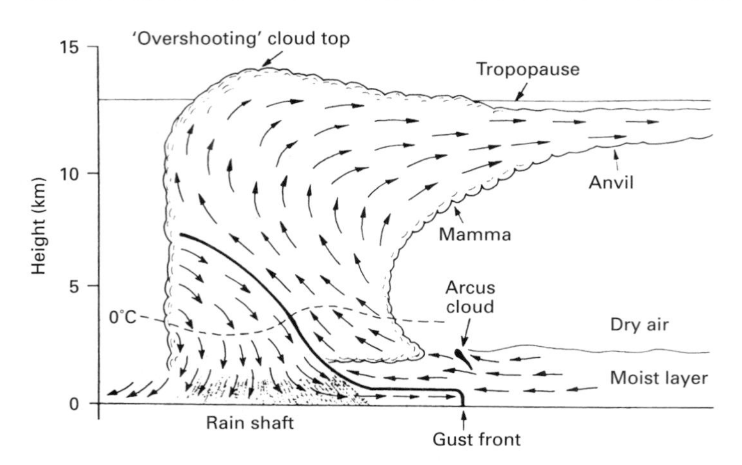


*Figure 5.8* Cloud outline and air motions relative to the movement of a squall-line storm (from left to right). (From Wallace and Hobbs, 1977.)

overtaken by the cell and incorporated into it (Figure 5.8). Outside the storm, the ambient air is usually capped by a weak inversion, which prevents any spontaneous convection occurring outside the cell. The incorporated air is undercut by the cold downdraught air and lifted to the level of free convection. It then moves upwards towards the tropopause. At these higher levels the wind is moving faster than the cell and the cloud top is drawn out into an 'anvil'. The updraught creates the precipitation, which now falls to the ground through the trailing edge of the moving storm, creating the downdraught area. Part of the downdraught travels along the ground under the updraught, creating a **gust front**. This gust front undercuts the warm air, forcing it to rise and join the updraught. Thus in the squall-line storm the updraughts and the downdraughts are complementary, not in opposition as in the air-mass storm. This means that the storm can persist much longer and can reach much higher intensities, and many of the most intense short-term precipitation events ever recorded have been associated with this type of storm. In special circumstances, considered in Section 9.2, these storms can generate tornadoes.

Lightning is the result of charge separation within a thundercloud. Although the mechanism of charge separation is not well understood, the effect is to produce a gradient of electrical potential within the cloud and between the cloud and the ground. Once this gradient exceeds a threshold value a lightning stroke results. A stroke is initiated by a 'stepped leader' moving in a jerky, but rapid, motion. This is usually directed from the cloud to the ground but may be in the opposite sense from an upstanding object such as a tall building. This stepped leader is met by a 'travelling spark', with the 'return stroke' flowing in the channel created by the two. This gives the bright flash that is visible as the lightning stroke, which may be followed by one or two 'streamers' and by a second return stroke. In fact most lightning flashes are made up of three or four strokes about 50 ms apart. The return stroke raises the temperature of its channel by about 30 000 K. This occurs so fast that the air has no time to expand, and the pressure within the channel increases to 10, or sometimes 100, times its normal value. This sets up a shock wave emanating from the channel, which is converted into the sound wave we hear as thunder. While lightning can be seen for long distances, thunder can usually only be heard up to about 25 km from the lightning site, partly because of the damping effect of the air itself and partly because the temperature structure of the air usually ensures that the sound wave is refracted upwards away from the Earth's surface.

## 5.4 The surface water budget

It remains for us to complete the cycle that we started with evaporation from the surface. In the evaporation section we noted that rates of evaporation were partly controlled by water availability, which, over most land surfaces at least, is largely controlled by precipitation. The difference between precipitation and evaporation in any area can loosely be called the **water budget**

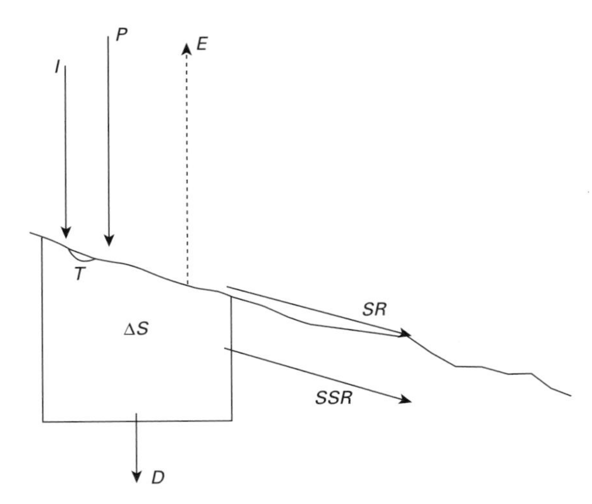


*Figure 5.9* Schematic representation of the surface water budget, emphasising those components which are likely to be pertinent to an analysis of the change in soil water content for an agricultural enterprise. *P*, precipitation; *I*, irrigation (if any); *E*, evapotranspiration; *SR*, surface runoff; *SSR*, subsurface runoff; *D*, deep drainage; $\Delta S$, change in soil moisture; *T*, surface storage.

of the area. In many ways this is directly analogous to the energy budget concept treated in Chapters 2 and 3. It too applies on various scales and for various time periods, while there must be a long-term global **water balance**. Indeed, sometimes the concept is simply referred to as the water balance rather than the water budget.

The water budget concept is of most immediate interest over land areas, since any excess of precipitation over evaporation represents water that runs off the surface or is stored in the soil or lakes. It therefore represents the water that can be used for human activities. For ocean areas, however, the main concern for the water budget is as an important component of air–sea interactions, which have a profound effect on climate.

The way in which we consider the water budget depends greatly on the space and time scales and application of interest. The most comprehensive analysis is commonly needed for the smallest scales (Figure 5.9). The method of formulation depends on the application. Thus, for example, we can emphasise agricultural interests and consider the budget in terms of changes in soil water content:

$$\Delta S = (P + I) - (E + R + D + T) \qquad (5.3)$$

where $\Delta S$ is the change in water content of the soil; $P$ and $I$ are inputs of precipitation and irrigation, respectively; while $E$, $R$, $D$ and $T$ are, respectively, outputs in the form of evapotranspiration, runoff (surface and subsurface), deep drainage vertically out of the root zone of the soil, and storage in puddles, ponds and lakes on the surface. Equation (5.3) forms the basis for some of the discussion of the impact of climatic changes in Sections 11.1 and 11.3.

On a regional scale the land surface water budget has a significant impact on the distribution of natural vegetation. This relationship has been expressed in numerous ways. A typical example relates potential evapotranspiration and a moisture index to specific vegetation types (Figure 5.10). The moisture index used here is equal to 100[(P/PET) – 1] where P and PET are annual average values of precipitation and potential evapotranspiration, respectively.

### 5.4.1 Latitudinal hydrological imbalance

Ultimately, on a global scale there must be a balance between the water that is evaporated and that which is redeposited as precipitation. The total mass of water vapour in the atmosphere is approximately equal to one week's rainfall over the globe. Since this mass does not change appreciably with time it seems that the average time for a complete water cycle is about one week. Although we have emphasised the vertical component of the cycle, there are also, of course, significant horizontal motions for water vapour during the time that it resides in the atmosphere. In fact, there must be horizontal motions in order to maintain the global water balance (Figure 5.11). There is a large movement of water vapour from the primary source area in the subtropics both into the equatorial regions and towards the middle and high latitudes.

These latitudinal motions involve not only the transport of water vapour, but also the transport of energy. Energy contained in the water vapour that is evaporated in the subtropics is released during cloud formation in the equatorial regions. Indeed, this energy transfer is vital for driving the circulation in the tropics. Similarly the poleward transfer from the subtropics plays a significant role in creating the circulation in mid-latitudes.

In order to understand and describe these circulations and, hence, the various climatic regimes of the globe, we have to combine the hydrological cycle we have discussed in these past two chapters with the energy exchanges considered in Chapters 2 and 3. This will be the focus of the next two chapters, and will be an integral part of all subsequent ones.

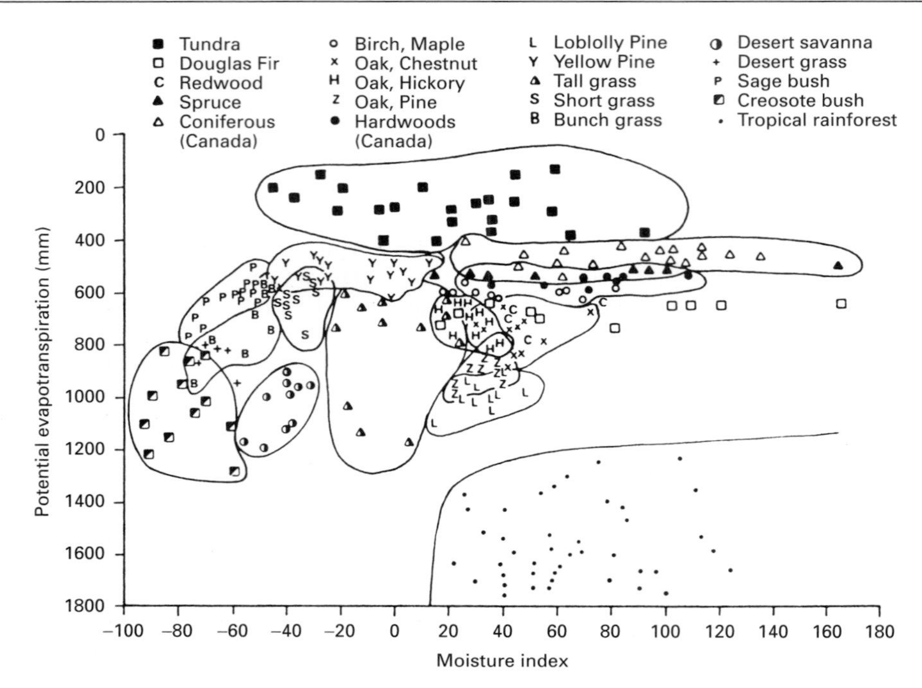


*Figure 5.10* Relation between natural vegetation and the climatic factors of potential evapotranspiration and a moisture index at selected stations in the United States, Canada and the tropics. The moisture index is 100[(P/PET) – 1], where P and PET are the annually averaged precipitation and potential evapotranspiration, respectively. (After Mather and Yoshioka, 1968.)

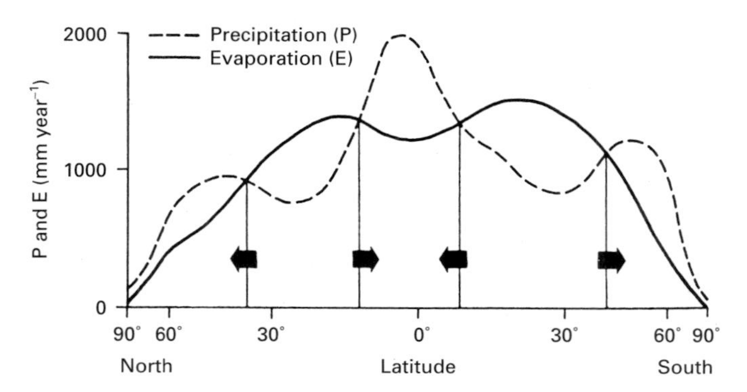


*Figure 5.11* Annual average evaporation and precipitation per unit area in millimetres per year as a function of latitude. The arrows show the sense of the required water vapour flux in the atmosphere. (From Sellers, 1965.)

**Box 5.A Applications of precipitation information**

Three examples have been chosen to illustrate various aspects of the use of precipitation information. They illustrate a class of problem where the required information is the **probability of occurrence of an event**. These probabilities, determined from the climatic record, give no information about when the event will occur, only about its chance in a certain period. A short-term weather forecast would have to be used for guidance concerning the time of the actual event.

*5.A.1 City snow clearance strategies*

The first example emphasises the distinction between using climate probability information to develop a long-term strategy and using weather forecast information to put it into action. The problem is snow removal from city streets. Long-term planning for snow removal is needed, since equipment must be purchased and funds put aside to ensure adequate resources to operate the equipment and pay the crews. The city authorities must first determine how much snow they can tolerate lying on the streets before initiating cleaning operations. Thereafter the climatologist can use the historical record to determine how frequently snow above this threshold value occurs, in terms of the number of times per winter. Standard statistical methods can then be used to assess the probability of a certain number of snowfalls in the coming season. Armed with this climatological information the appropriate authorities can plan their strategies accordingly.

Frequently the determination of the number of snowfalls from the historical record is not easy. Long-term observations are likely to be available only for the local airport, usually some distance from the city centre. Estimates of how much snow fell in various parts of the city therefore have to be made. Since precipitation is highly variable even on a small spatial scale, these estimates have to be based on inferences concerning local topographic lifting, the possible increased convection caused by the heat of the city, and on possible lack of time for melting during fall at higher elevations, all topics discussed in various parts of this chapter. Further, the time of occurrence of a snowfall may be important, since an early morning rush-hour fall may require a different response from one which occurs in the late evening. Again, the data may not provide an answer directly and estimates based on knowledge of local climatic processes may be needed. Nevertheless, the climatologist should be able to provide to the city authorities an estimate of the probability of the number of snowfalls likely in any winter. Thereafter it is the job of the local weather forecaster to provide the short-term warnings that are needed to put the city's snow clearing strategy into operation.

*5.A.2 Power-line icing*

The second example involves a situation where no appropriate measurements are available, and concerns the probability of ice formation on overhead power lines. Such icing can create loads of several tonnes per kilometre on the lines, leading to their subsequent collapse. This results not only in loss of service for the customers, but also repair costs and lost revenue for the utility. A light coating of ice can be overcome by boosting the power to create self-heating within the lines, but regular, severe icing can only be overcome by laying the cable underground: a costly undertaking. Hence the climatological problem is similar to that for the snow removal: how frequently and over how wide an area will icing occur, and how severe will it be? Experiments and theory indicate that accumulation depends on precipitation type and rate, wire and air temperature, wind speed and direction, and humidity. These have rather sparse measurement networks and it has proved impossible to link observational data with the power companies' own records of line failure to produce a statistically based predictive equation. Modelling techniques are helping to identify more precisely the range of meteorological conditions creating icing on a wire, and thus helping to identify the conditions at routine observing stations which might lead to icing. However, the conditions are highly uncertain and the problem, containing the same need to incorporate process considerations as the snow-removal problem, has no precise solution. Thus this example is deliberately left as one where a rather unsatisfactory climatological answer can be given and where further investigation to gain insight into the problem is needed.

*5.A.3 Urban drainage design*

The final example of a climatic probability approach to problems associated with water concerns urban drainage design. Storm drains are generally designed to be large enough to accommodate the largest storm runoff expected during the lifetime of the drain. The obvious approach of searching records for the largest storm on record is not adequate, since the record is only a sample of all possible conditions and there is no guarantee that it will contain the largest possible event. Fortunately there are established statistical methods, called extreme value analyses, which allow one to estimate the 'largest possible' event from a set of records. Thereafter a set of intensity/duration curves can be developed (e.g. Figure 5.4) for a variety

**Box 5.A (cont'd)**

of return periods. (An event with a return period of $n$ years has a probability of occurrence in any one year of $(100/n)$%.) The drainage designer can then choose the appropriate return period, making a reasonable compromise between construction cost and size that ensures that the drain is adequate for all but the most unusual, extreme storm event. The designer will also be concerned with the areal variation of the intensity/duration curves. As with the snow removal example, it is unlikely that there will be records from a dense network of rain gauges available to provide the required information. Hence a knowledge of the precipitation-producing mechanisms must be combined with the records that are available to produce the types of estimates that the designer needs.

# CHAPTER 6 Winds and pressure

So far our discussion of climate has said very little about one of the most important climate characteristics: the wind. This deliberate omission has allowed us to consider the set of processes leading to the energy and water budgets, and eventually controlling climate itself, without any consideration of the spatial scale of the discussion. The budgets apply whether we are talking about the whole planet or a single field. However, as soon as we start to consider the horizontal climatic component represented by the wind, spatial scale becomes very important. Thus, in this chapter we shall investigate the basic processes creating wind, and then the next few chapters will look on a global, regional and local scale at the influence of wind.

## 6.1 Atmospheric pressure

Horizontal air motion is a response to horizontal variations in atmospheric pressure. We shall, in the next few sections, consider pressure, pressure patterns and the forces creating motion. Then we shall return to the wind itself. Such an approach not only fosters an understanding of climate processes, but also allows some consideration of atmospheric forecasting. Most weather forecasts are based on predictions of pressure changes and calculations of wind, and some of the same ideas will underlie our discussion of future climate in Chapters 13 and 14. More practically, although it is easy to measure wind near the surface, it is extremely difficult at higher levels, so alternative observational methods must be used, as considered in Box 6.I.

### Box 6.I Measurement of pressure and winds

*6.I.1 Atmospheric pressure*

Surface pressure is measured by a **barometer**: an evacuated tube sealed at one end which is inverted into an open dish of mercury (Figure 6.I.1). The level of the mercury in the tube adjusts itself until the pressure it exerts on the mercury in the dish exactly balances that of the atmosphere. The height of the mercury column, after adjustment for expansion and contraction caused by temperature variations, is thus a measure of atmospheric pressure. This, of course, has led to pressure being expressed in terms of 'millimetres of mercury', or even just in terms of height, rather than the fundamental pressure unit (Pa) used here.

In the free atmosphere, pressure measurements are made during radiosonde ascents. Although the method adopted depends on the particular type of radiosonde used, it is common to employ an **aneroid barometer**. This is a partially evacuated semi-rigid metal bellows, which flexes as pressure changes. The change in box configuration is transmitted by a mechanical linkage to a pointer or recorder, which has been calibrated to relate the mechanical changes to pressure changes. Similar instruments, calibrated somewhat differently, are the basis of many aircraft altimeters. The instruments are also familiar to most people, since they are the type of barometer displayed in many homes. Such home observations of pressure changes, being for a single point, will not reveal all the complexities of the atmospheric motions. However, because pressure is so intimately linked to winds and weather, such observations remain the single most useful measurement, which, if judiciously used, can reveal a great deal about the current weather and its likely future course.

**Box 6.I (cont'd)**

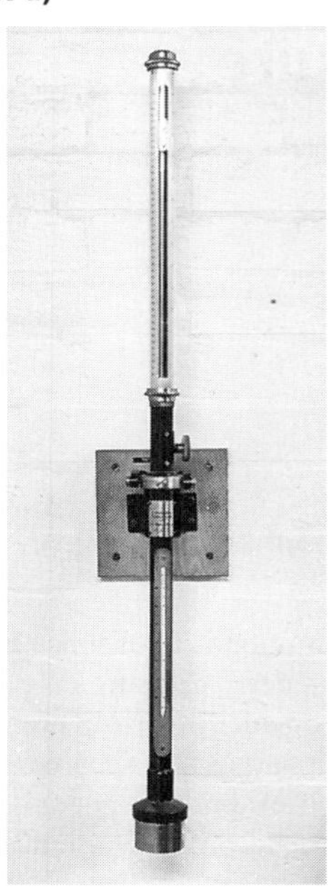

*Figure 6.I.1* Photograph of a barometer showing the mercury column and the vernier scale used to read the height of the column. The instrument reading is usually scaled to sea-level pressure.

*Figure 6.I.2* Anemometers must be installed at a number of heights in the lower atmosphere if the effect of surface friction on wind speed is to be measured. In complex terrain, such as an urban centre, a large array of anemometers may be required as wind speeds can vary considerably over short distances.

*6.I.2 Wind speed measurements*

The standard instrument used for wind measurement by most national weather services is the **anemometer** (Figure 6.I.2). This consists of a set of (usually three) rigid arms with cups attached, radiating from a central spindle. The speed of rotation of these arms is a function of the wind speed. The instrument is calibrated so that the number of rotations in a given time can be translated directly into wind speed. The mechanism used to measure this number has an influence on the quantity that is reported as 'wind speed'. Some instruments accumulate the number of rotations for a given period, such as an hour or a day, leading to a measurement of **wind run**, the amount of air which has passed over the instrument during the period. This is roughly equivalent to the average wind speed during that period. Other instruments provide virtually instantaneous readout, showing second-by-second fluctuations. Some stations average these for a short period, typically 2 minutes, and report this as the wind speed at a particular time. Others note the highest speed indicated during a specific time, such as an hour, and then report the hourly **peak gust speed** as the wind speed. Hence care must be taken in using wind speed information.

Usually associated with the anemometer will be a **wind vane**, measuring and recording the wind direction at the same location as the speed. Even here, there is often some ambiguity, since direction is sometimes recorded directly in degrees, proceeding clockwise with north being 0°, and sometimes in terms of the eight (and sometimes even 16) cardinal directions. However, all are consistent in indicating the direction from which the wind is blowing.

The values obtained from the anemometer/vane combination are extremely sensitive to the instrument siting, perhaps more so than for any other meteorological element. Wind speed and direction vary rapidly with height near the surface of the Earth (see Chapter 10), so that observations at a standard height are vital if inter-station comparisons are to be made. A location 10 m above the ground on a level site free from major obstructions for several kilometres in any direction is highly desirable, and very difficult to achieve.

As with other elements, there are a variety of other wind-measuring instruments for special purposes. In particular, very fast response instruments which measure the components of the wind (north–south, east–west, vertical) are required for studies of turbulence and other small-scale climate variations of the type considered in Chapter 10. Moreover, methods of estimating winds from satellites, such as those relating the nature of ocean whitecaps to speed, are increasingly common in remote areas.

### 6.1.1 Sea-level pressure

Pressure is created because the atmospheric gas molecules are in constant motion. They therefore exert a force whenever they impact upon a surface. The total force produced per unit area is the pressure. Variations in pressure, both in time and space, result from changes in the energy and number of molecular impacts, which are primarily determined by the density and temperature of the gas. The most fundamental density variation is its decrease with height in the atmosphere. Thus at sea level the pressure is commonly around 1013.2 hPa, which is taken as standard sea-level pressure. At 3000 m it is about 70% of this value and at 10 000 m the pressure is around 300 hPa. Near the surface a change in height of about 100 m leads to a decrease in pressure of about 10 hPa. Such rapid vertical changes can mask horizontal changes unless all pressure observations are taken at a uniform height. Since over land this is a practical impossibility, observations are modified using known relationships between height and pressure, so that all refer to sea level. This modification process is known as **reduction to sea level**.

Once reduced to sea level, observations from a network of observing stations reveal distinct horizontal **pressure patterns**: discrete regions or bands of high and low pressure. These are readily seen whenever a pressure, or **isobaric**, map is inspected (e.g. Figure 6.1). An isobar is simply a line joining places with equal atmospheric pressure. In the atmosphere the surface pressure is likely to be low in regions of high temperature, in accordance with the universal gas laws. It is also likely to be low in regions of ascending air where molecules are being removed from the surface. Although the geographical distribution of pressure need not concern us until later in the chapter, Figure 6.1 clearly indicates certain regions of persistent high and low pressure. It is these features that are responsible for the **primary circulation features** of the atmospheric circulation. Certain zones have frequent frontal activity and it is in these regions that the **secondary circulation features** are particularly important.

### 6.1.2 Pressure variations higher in the atmosphere

Horizontal pressure variations also occur in the free atmosphere away from the surface. It is possible to represent these on a map in two ways. The first is directly analogous to the isobaric chart already introduced, but with the surface replaced by a specific level in the atmosphere. Pressure changes at a constant altitude are thus displayed. However, it is usually much more convenient to use an alternative approach and display changes of height over a constant pressure surface. The resulting constant pressure surface map looks rather like an isobaric chart, but the lines are height contours rather than isobars (Figure 6.2). When we discuss winds associated with pressure variations, it will be seen that, in fact, the two types of maps can be interpreted in a similar way.

## 6.2 Air movement around a rotating planet

### 6.2.1 Pressure gradient force

The driving force for all air motions is variations in atmospheric pressure. For horizontal variations in pressure, a force is created acting from high to low pressure (Figure 6.3). This **pressure gradient force**, $P_f$, is given, per unit mass, by:

$$P_f = -(1/\rho) \times (\Delta p/\Delta x) \qquad (6.1)$$

where $\Delta p$ is the pressure change over distance $\Delta x$. Thus $\Delta p/\Delta x$ is the pressure gradient. The negative sign indicates that the force operates from high to low pressure, i.e. that $p$ decreases as $x$ increases. The atmospheric situation is illustrated by analogy in Figure 6.3. If the valve separating the two liquid columns is opened, the fluid attempts to readjust to equalise pressure.

Although there is a similar pressure gradient force in the vertical, acting upwards, this is almost exactly balanced by the downward acting force of gravity. Only in certain circumstances, such as during cloud formation, is vertical motion important. Generally the horizontal force exceeds the vertical one by about three orders of magnitude for most scales of atmospheric motion and the horizontal wind speed is very much greater than the vertical speed. For our present analysis it is a very convenient simplifying assumption to regard wind as having only a horizontal component.

On a non-rotating planet, air would flow under the influence of the pressure gradient force towards low pressure, a simple result of Newton's second law of motion. Its speed would depend on the magnitude of the force, which in turn is proportional to the spacing of the isobars or height contours on a chart such as Figure 6.2.

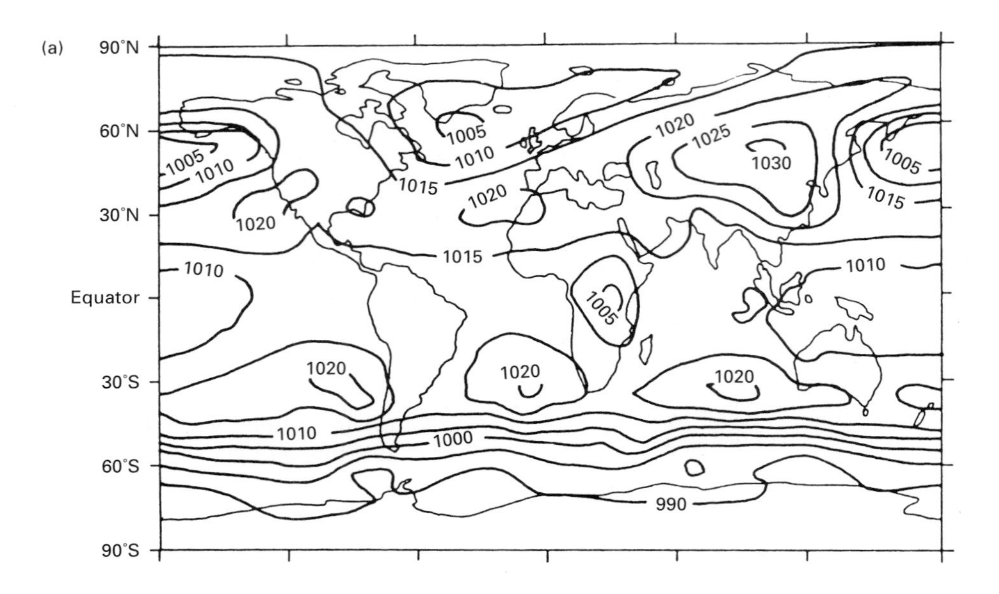


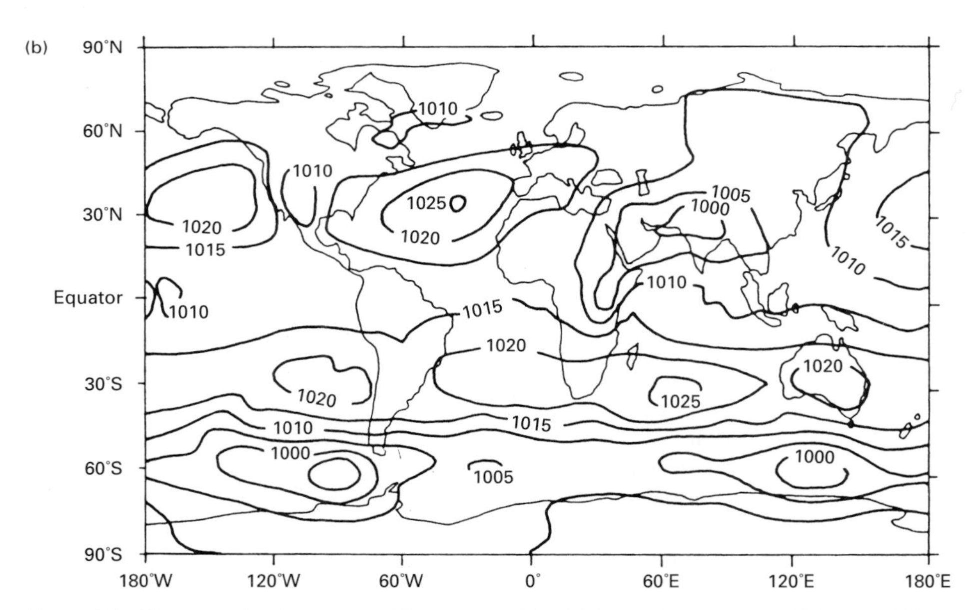


*Figure 6.1* Mean sea-level pressure (hPa) averaged for (a) December, January, February (1963–1973); and (b) June, July, August (1963–1973). Note the movement of the position of the Intertropical Convergence Zone (the belt of lower pressure near the equatorial belt) and the subtropical high pressure areas. (From Oort, 1983.)

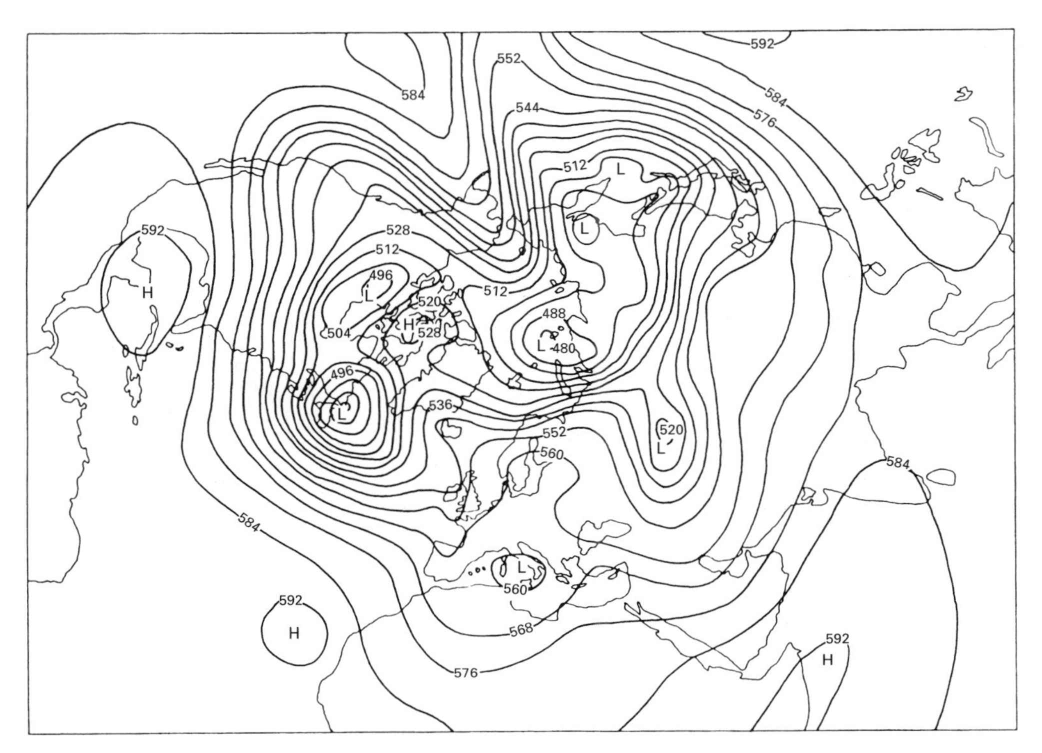


*Figure 6.2* 500 hPa chart for 20 January 1982 (0000Z). Heights are shown in decametres (dm). This should be compared with the associated surface chart shown in Figure 9.8. In particular, note the close association between the position of Rossby waves over the UK and the North Atlantic and the family of depression systems in Figure 9.8. The lows over the Mediterranean and south of Greenland are seen on both the surface and the upper air charts. (From *European Meteorological Bulletin*.)

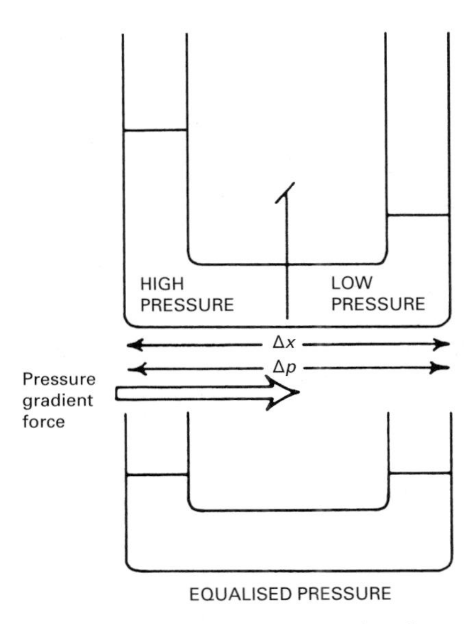


*Figure 6.3* Schematic diagram illustrating the pressure gradient force. As the valve between the two sides of the container is opened the pressure gradient force operates to move the liquid until the pressures are equalised.

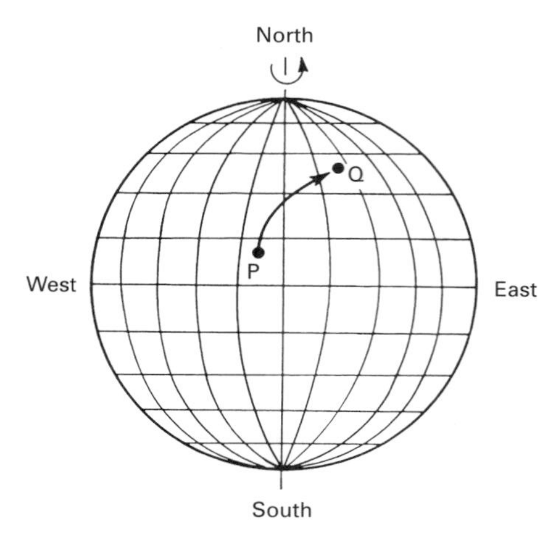


*Figure 6.4* Schematic diagram illustrating the direction (P to Q) which a projectile will take due to the rotation of the Earth, given an initial velocity towards the North Pole. When apparently stationary at point P the projectile already has a large angular velocity about the axis of rotation of the Earth, as does the piece of land on which it lies. Points near the poles have a smaller angular velocity and so, once moving, the projectile appears to turn towards the right as it travels poleward.

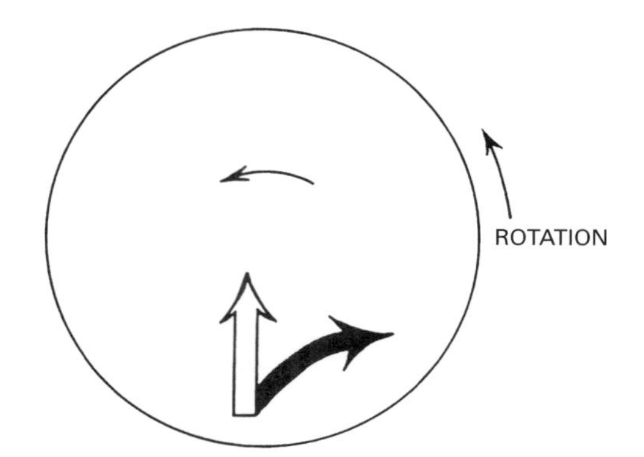


*Figure 6.5* Deflection (black arrow) to the right of a projectile (whose initial direction is shown by the open arrow) moving across the face of a disc which is spinning anticlockwise.

### 6.2.2 Coriolis force

On a rotating planet the speed of the wind is still governed by this pressure gradient force, but the rotation causes a change in the direction of the flow. In terms of Newton's second law, another force, created by the rotation, acts on the moving air parcel. This is the **Coriolis force**. Any projectile given a velocity across the surface of the Earth will experience an apparent force, which tends to turn it to the right in an anticlockwise rotating frame (e.g. the Northern Hemisphere as viewed from space) (see Figure 6.4) and to the left in a frame that is rotating in a clockwise direction (e.g. the Southern Hemisphere as viewed from space). This additional force acts upon all projectiles moving within the rotating frame of the Earth. The deflection can be most easily understood by thinking of the surface of the Earth as being approximated by a disc at the location of interest. This disc has a spin about the centre (the local vertical) of $\Omega \sin \theta$, where $\Omega$ is the rotation rate of the Earth and $\theta$ the latitude of the location. As seen in Figure 6.5, any projectile appears to move to the right because of the rotation of the disc. Thus to a stationary observer standing at the centre of the disc the projectile coming from the edge, which is moving more rapidly around the centre than the inner disc, appears to be deflected to the right. The extra apparent acceleration given to all air motions around the Earth is given by:

$$\text{Coriolis acceleration} = (2\Omega \sin \theta)u \qquad (6.2)$$

where $u$ is the speed of the projectile. The quantity known as the Coriolis parameter, $f$, where:

$$f = 2\Omega \sin \theta \qquad (6.3)$$

is constant for a given latitude. Since the angular speed of the Earth is $2\pi$ radians, in 24 hours we have:

$$\Omega = 2\pi/(24 \times 60 \times 60) = 7.27 \times 10^{-5}\,\text{s}^{-1} \qquad (6.4)$$

Hence $f$ is always small, varying from $1.5 \times 10^{-5}\,\text{s}^{-1}$ at the poles to zero at the equator. However, this force imparts the east–west component to meridional atmospheric motions.

### 6.2.3 Geostrophic flow

When isobars are straight and parallel and we are considering motion in the free atmosphere away from the effects of surface friction, only the pressure gradient force and the Coriolis force act on a parcel of air. The pressure gradient force initiates the motion and immediately the Coriolis force commences its deflecting action. The two forces rapidly come into equilibrium and there is a balanced flow with two equal forces. Thus, using equations (6.1) and (6.2) and rearranging:

$$V_g = -(1/\rho f) \times (\Delta p/\Delta x) \qquad (6.5)$$

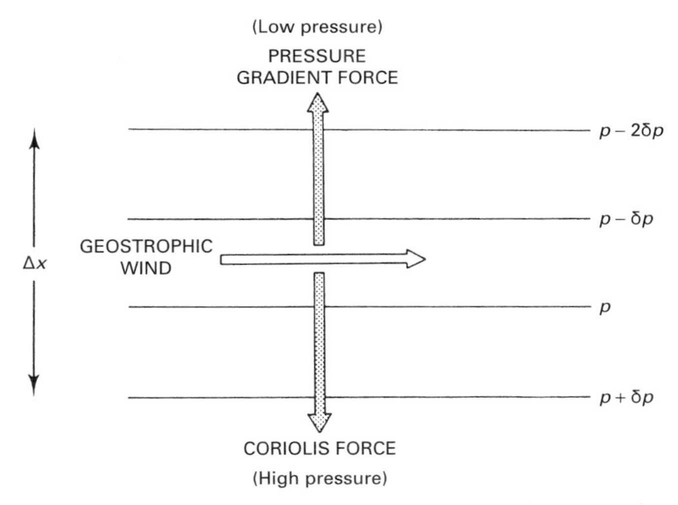


*Figure 6.6* Diagram showing the geostrophic wind, which is the balanced result of the two opposing forces (pressure gradient and Coriolis).

where $V_g$ is the **geostrophic wind**. This wind therefore blows parallel to the isobars with low pressure on the left-(right-)hand side as you stand with your back to the wind in the Northern (Southern) Hemisphere (Figure 6.6). Its speed is proportional to the isobaric spacing. The geostrophic wind, determined entirely from observations of the pressure distribution, is identical to the actual wind when isobars are straight and parallel and when the motion is far enough removed from the surface. In reality, of course, the actual situation rarely completely fulfils these constraints. For much of the time, however, the isobars are not highly curved and the geostrophic wind provides a very good approximation to the real wind. Certainly in these conditions it enables us to get a much better idea of the wind speed than can be obtained from the rather sparse and expensive direct observations. The geostrophic approximation can only be used with confidence poleward of about 30°, since in equatorial regions the Coriolis force tends towards zero and there is no strong deflection of the winds.

The major drawback to the use of equation (6.5) is that it includes the air density, which varies with temperature and height. This difficulty can be overcome by introducing the hydrostatic equation (equation (4.7)). Substituting for the density, the geostrophic wind equation becomes:

$$V_g = (g/f) \times (\Delta h/\Delta x) \qquad (6.6)$$

where $(\Delta h/\Delta x)$ is the rate of change of height of a constant pressure surface. This can be obtained directly from a constant pressure chart, such as Figure 6.2. This equation is much simpler to use than equation (6.5), because only the Coriolis parameter, the acceleration of gravity and the slope $(\Delta h/\Delta x)$ are needed. The equation, however, can be interpreted in a similar way to equation (6.5), with the geostrophic wind blowing parallel to the height contours with low height on the left-hand side (for the Northern Hemisphere) and at a speed proportional to the contour spacing. This is just one example of the simplification possible when constant pressure, rather than constant height charts are used to depict conditions in the upper air.

Isobars are rarely straight. In most cases there is some degree of curvature, either **cyclonic**, with air having a counter-clockwise motion around a low pressure area, or **anticyclonic**, with clockwise motion around a high pressure system. (NB These directions are given for the Northern Hemisphere. They must be reversed when the Southern Hemisphere is being considered.) However, except for small-scale motions or motions associated with severe storms, the geostrophic approximation holds for such curved motions.

### 6.2.4 Gradient wind

When there is marked curvature in the isobars a third force, the **centrifugal force**, must be introduced. This

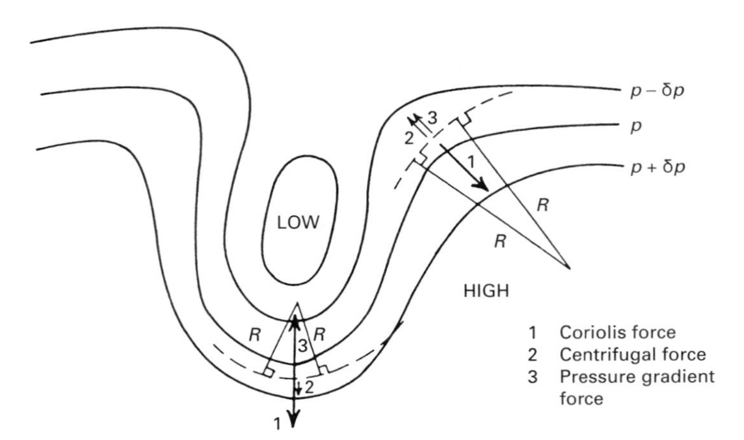


*Figure 6.7* The three-way balance between the horizontal pressure gradient force (3), the Coriolis force (1) and the centrifugal force (2) in atmospheric flow along curved trajectories (dashed) with radius of curvature *R*. The resulting winds are gradient winds. Only three isobars are shown for simplicity ($p - \delta p$, $p$, $p + \delta p$), although the concept of the gradient wind only becomes important in situations of large pressure gradients. (From Wallace and Hobbs, 1977.)

acts outwards from the centre in any curved motion. It can readily be demonstrated by whirling a stone on a string. The tension felt in the arm during this activity is an expression of this force. In the case of rotation around a high pressure area this force is in the same direction as the pressure gradient force and hence leads to an increase in wind speed over that calculated for the geostrophic wind. Around a low pressure centre the centrifugal force opposes the pressure gradient force and decreases the wind speed (Figure 6.7). The wind that results from a balance of the three forces is known as the **gradient wind**. The direction of this wind, like the geostrophic wind, is parallel to the isobars. The speed of the gradient wind differs significantly from that of the geostrophic wind only when there is severe curvature at very large pressure gradients. For instance, in the case of a tropical hurricane the calculated geostrophic flow might be 500 m $s^{-1}$, but the gradient wind speed is only 75 m $s^{-1}$.

### 6.2.5 Near-surface wind

As we approach the surface of the Earth the influence of surface friction is increasingly felt. This **frictional force** acts directly against the airflow, leading to a reduction in wind speed. Since the Coriolis force is a function of wind speed it is also reduced and the flow, even with straight parallel isobars, is no longer balanced (Figure 6.8). A cross-isobaric flow, directed towards low pressure, is induced. The angle at which the air crosses the isobars depends on the magnitude of the frictional force. A fairly smooth water surface rarely produces inflow to low pressure at an angle greater than about 8° to the isobars, while a land surface of rolling terrain may lead to an angle somewhat in excess of 25°. Very rough terrain is much more likely to create its own circulation patterns in the lowest air layers than simply to modify the geostrophic wind. The frictional force is at its maximum right at the surface and gradually decreases with height until it becomes insignificant and the geostrophic wind approximation holds. This decrease with height also leads to a clockwise change in wind direction with height, which is sometimes called the **Ekman spiral**. The layer where friction is effective is usually known as the **friction layer**, with the free atmosphere above it.

So far, when considering the free atmosphere we have assumed that the motions have been horizontal, or at least quasi-horizontal. Certainly for any given level the geostrophic approximation gives a good indication of wind flow patterns. However, interactions between levels, particularly when spatial temperature variations are present, can lead to vertical motions that modify the simple analysis presented here.

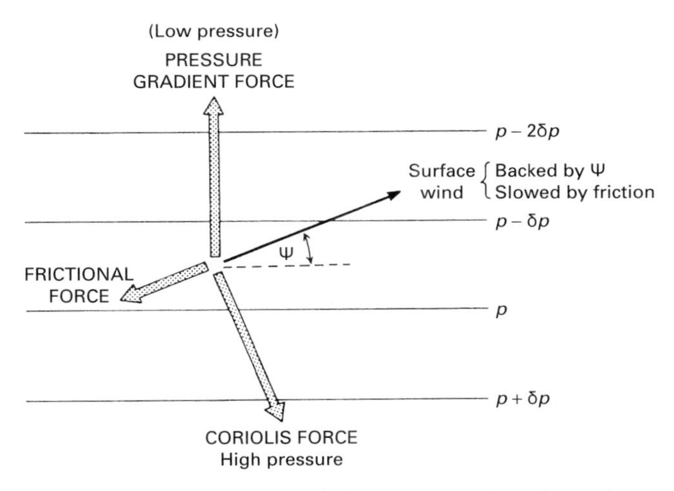


*Figure 6.8* The effect of surface friction is to 'back' (move in an anticlockwise direction) the surface wind compared with the direction of geostrophic flow and to reduce its speed.

## 6.3 Barotropic and baroclinic conditions

We can begin our discussion of factors which influence the vertical variations of horizontal wind, and sometimes themselves lead to vertical motions, by considering the relationship between temperature and pressure in the vertical. The height difference between any two given pressure levels is known as the **thickness**. Since density decreases as temperature increases, it follows that a warmer layer must cover a greater geometrical height to embrace the same mass of gas, so that thickness varies directly with temperature.

### 6.3.1 Barotropic atmosphere

Using this relationship, the simplest situation we can envisage is one where there are horizontally uniform temperatures at all levels. Hence there is no spatial change of thickness. We can introduce a pressure gradient and therefore allow horizontal motion. This situation, where there is a pressure gradient but no temperature gradient, is a **barotropic** situation. There is no change in wind speed or direction with height and there is no chance for disturbances to grow in the airflow.

### 6.3.2 Equivalent barotropic state and the thermal wind

If we introduce a temperature gradient such that the isotherms are parallel to the isobars, we generate an **equivalent barotropic** atmosphere (Figure 6.9). If we assume, as in Figure 6.9(a), that the low pressure area is cold and the high pressure area is warm, the increasing thickness as we move into the warmer air will lead to a steepening of the pressure gradient with height. This will cause an increase in wind speed with height, but the wind direction will not change. Figure 6.9(b) indicates the opposite conditions, with the warm air over the near-surface low pressure region. Now the wind speed decreases with height, eventually becomes calm, and then increases with height, but moving in the completely opposite direction. Since we can treat this change as simply the wind blowing in the same direction, but with a negative velocity, we can state that in an equivalent barotropic atmosphere there is no change in wind direction. As with the true barotropic situation, there is no chance for disturbances to grow. The difference in wind speed between the top and bottom, the vertical shear in the calculated geostrophic winds, is proportional to the horizontal gradient of the mean temperature of the intervening layer. Such a thermally induced gradient in wind speed is termed a **thermal wind**.

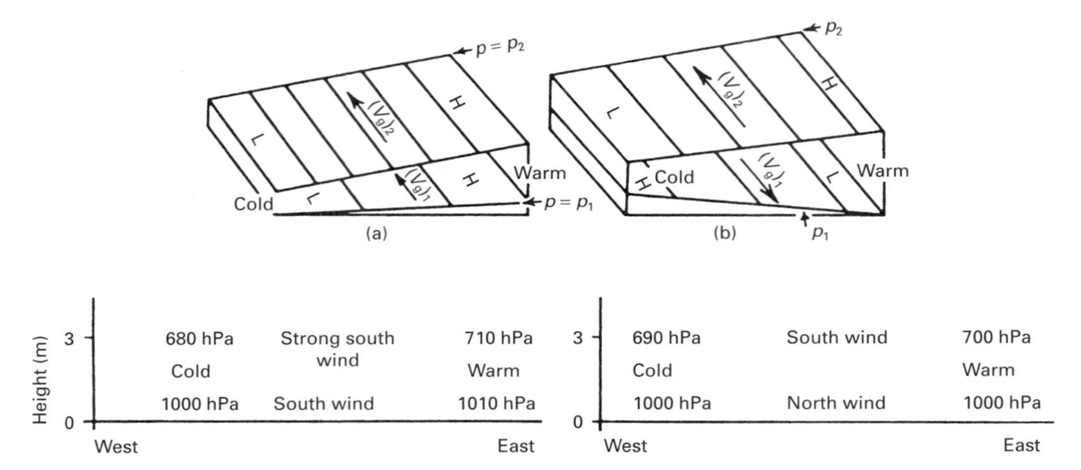


*Figure 6.9* The change of the geostrophic wind with height in an equivalent barotropic flow in the Northern Hemisphere. (a) $V_g$ increasing with height and (b) $V_g$ reversing in direction with height. (From Wallace and Hobbs, 1977.)

### 6.3.3 Baroclinic atmosphere

When the isotherms are not parallel to the isobars a **baroclinic** situation occurs (Figure 6.10). In this case the temperature, and thus the thickness, varies along an isobar. Consequently the pressure pattern changes with height, as does the wind speed and direction. The vector difference between the speed and direction at the lower and upper levels is also a thermal wind. It has a direction parallel to the isotherms and a speed proportional to the isotherm spacing, again blowing with low temperature to its left (in the Northern Hemisphere). The governing equation is similar to that of the geostrophic wind, so that its value can easily be calculated. The thermal wind is a very useful concept in meteorology, since it allows the calculation of the actual wind at any level once the pressure distribution at one level and the horizontal and vertical temperature distributions are known. In practice the sea-level pressure distribution is obtained from the numerous surface-based observations, while the temperature distributions can be obtained from upper air soundings (radiosonde ascents) or from satellite measurements.

In baroclinic conditions the wind is blowing across the isotherms, which leads to **advection** of energy into or out of an area. Thus a flow from a colder to a warmer region leads to cold advection. When the wind is **backing** (changing in a counter-clockwise direction) with height, cold advection is occurring, while a **veering** wind indicates warm advection. Figure 6.10 indicates a veering wind and warm advection. Backing or veering can often be observed when several different cloud layers occur simultaneously, although the phenomenon can easily occur without the presence of clouds.

This energy advection associated with baroclinicity plays a vital role in the creation of disturbances in the atmospheric flow patterns. We can start with an equivalent barotropic situation which has west to east, or **zonal**, airflow and isotherms parallel to the isobars. If, for some reason, such as the presence of a topographic barrier, the zonal flow is disturbed, a baroclinic situation will be generated (Figure 6.11). The flow at point A is carrying cold air southward, while at point B warm air is being moved northward. The latitudinal temperature contrasts will continue to increase as the advection continues. Eventually the contrasts will be so large that spontaneous vertical motions will be generated. In energy terms, this implies that the increase in baroclinicity increases the potential energy in the system, which is then released as the kinetic energy of motion. Southward-moving cold air is associated with sinking motions, northward-moving warm air with rising ones. These motions are induced by the dynamics

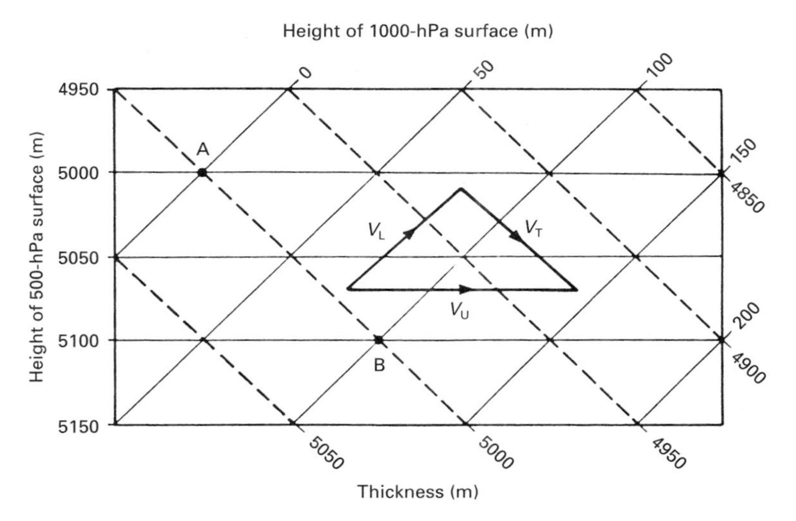


*Figure 6.10* Schematic chart showing the contours of 1000- and 500-hPa surfaces and the thickness of the layer. Lines of constant values for the height of the 500-hPa surface are horizontal and those of the height of the 1000-hPa surface are skewed (both unbroken in the diagram). Thus, for example, at point A, the height of the 500-hPa surface is 5000 m and the height of the 1000-hPa surface is 0 m, giving a thickness of 5000 m. Similarly, at point B the thickness equals 5100 − 100 = 5000 m. The locus of such points is thus a line of constant thickness and a set of such lines is shown dashed in the diagram.

The lower level wind, $V_L$, blows parallel to the lines of constant values of the height of the 1000-hPa surface; and similarly the upper level wind, $V_U$, is parallel to the lines of constant values of the height of the 500-hPa surface. It can be seen from the diagram that the vector difference between upper and lower level winds, $V_T$, blows parallel to the lines of constant thickness and is known as the thermal wind. (From Petterssen, 1969.)

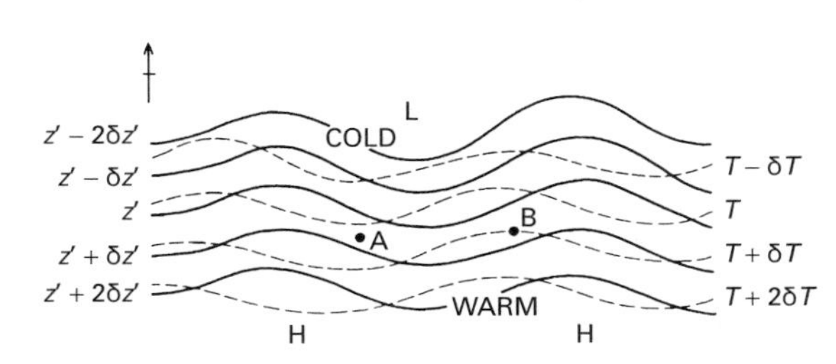


*Figure 6.11* Plan view (on a constant pressure surface) of the variation of temperature, $T$ (dashed), and geopotential height, $z'$ (solid), in a developing baroclinic wave in the Northern Hemisphere. (From Wallace and Hobbs, 1977.)

of the flow and will occur whether or not the atmosphere is hydrostatically stable in the sense introduced in Chapter 4.

The vertical motions associated with waves in the horizontal flow are enhanced by the curvature of the waves themselves. Recalling from the discussion of the geostrophic and gradient winds that for a given pressure gradient the wind speed will increase as the anticyclonic curvature increases, there will be accelerations in the airflow as it moves through a wave. In particular, as the air passes through point A (Figure 6.11) its curvature is becoming more cyclonic and the air is decelerating, while at point B the opposite is occurring. Assuming that there is no significant lateral motion, this leads to further vertical motions.

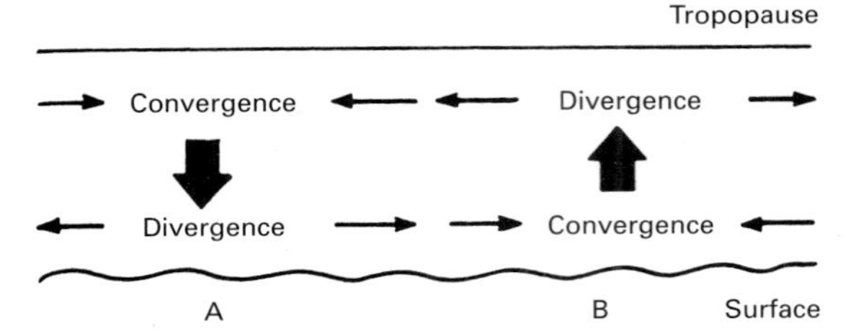


*Figure 6.12* Schematic illustration of the balance of surface convergence/divergence by divergence/convergence aloft.

### 6.3.4 Divergence, convergence and vorticity

It is a simple extension of the necessity to conserve mass that if a constant volume of fluid extends its horizontal dimensions, or **diverges**, its vertical extent must decrease. Similarly a vertical column experiencing **convergence**, for example through a decrease in downstream velocity, is constricted in the horizontal and must therefore be extended vertically. Figure 6.12 is a schematic representation of the vertical velocities associated with the situations of convergence and divergence at the surface and aloft.

In the case of our baroclinic wave occurring in the free atmosphere, convergence aloft at point A (Figures 6.11 and 6.12) reinforces the sinking motion, while high level divergence near point B enhances the upward motions. Thus the near-surface conditions are of sinking and diverging air at A and converging and rising air at point B.

The rising air near the surface at point B tends to be associated with a low pressure region at the surface. Air converges across the isobars within the friction layer. This increases the **vorticity**, defined as twice the angular velocity of the rotation, of the air in this region. Figure 6.13 illustrates the principle. It is also clearly displayed by a skater who draws in arms and legs closer to the body, causing personal 'convergence' and an increase in vorticity as expressed by the increase in spin rate. Hence at point B in Figure 6.12 a cyclonically curved spiral is generated as the air moves in towards low pressure.

The effects of the baroclinic instability, divergence and vorticity, all combine to create disturbances within the baroclinic zone. Detailed consideration of the processes involved is vital for day-to-day weather forecasting, and is important in computer modelling of the general circulation, but is not warranted in the context of climatology. It must be emphasised, however, that baroclinic conditions leading to disturbances occur only in restricted regions of the globe and even in these regions only for part of the time. Equivalent barotropic conditions are by far the dominant conditions over the globe.

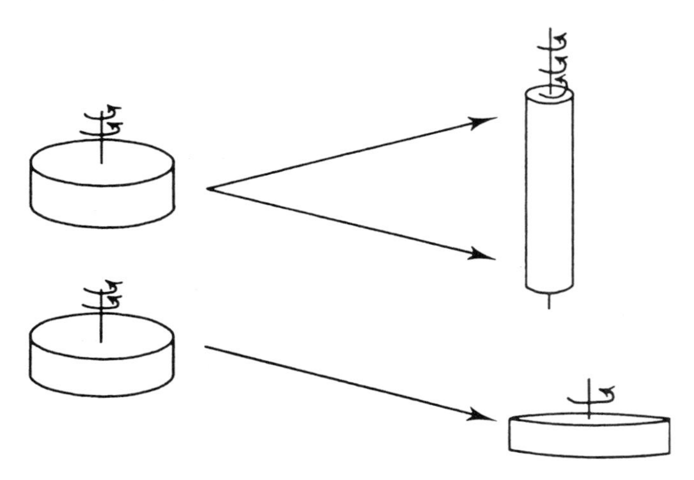

*Figure 6.13* Schematic illustration of the variation of vorticity caused by the conservation of angular momentum. As the upper cylinder decreases its radius, vorticity increases, while for the lower cylinder the increased radius leads to a vorticity decrease.

**Box 6.A Applications of wind information**

Wind is a source of energy for human use. In this section we explore this in two ways: first as the motive power for transportation, and second by considering means of conversion to more useful energy forms. Both concerns show rather ancient applications of wind information, which are now being used in thoroughly modern ways.

*6.A.1 Wind and sailing: three-stage forecasting for applications*

The wind has provided the motive power for sailing vessels for centuries. The plans needed to undertake long transoceanic voyages demanded a knowledge of the wind climatology of the oceans. The establishment of the major trade routes indicated a knowledge of that climatology, picking out areas with persistent favourable winds and avoiding, as far as possible, those regions with calm or light variable ones. Indeed, much of our early observational information about climate came from sailing experience, and is now reflected, as we shall see in the next chapter, in names we give to several areas of the general circulation of the atmosphere.

The rise of oceanic sailboat racing as a modern sport has provided new uses for climatological information. Here wind forecasts are vital, and it is common to use a three-stage forecasting technique. This can be regarded as a refinement of the two-stage method introduced for snow removal in the last chapter. Prior to an event, climatological wind information, in the sense of long-term averages or probabilities of speeds and directions over the region of the race, is developed. This allows creation of the basic suite of possible routes and sailing options. A short time prior to the actual event, the long-range weather forecast is consulted to narrow the route selection from this climatological suite, and to take associated actions, such as stowing appropriate sails. Finally, during the race the short-term forecasts, including 'nowcasts' of actual weather conditions, are used to refine the actual course taken.

Rarely is the process as simple and straightforward as the above statements sound. For example, in the case of a series of yacht races off Cuba, each starting at the same time of day, a major uncertainty was the time of onset of the sea breeze and how it would interact with the existing wind field. With such complications it is not always certain that the shorter-term forecasts will be more 'accurate' (if defined as being closer to the actual observed value) than the longer-term ones. In the Cuban case it could be suggested (Table 6.A.1) that the 18-hour wind speed forecast was no better than climatology, although the directional information was more accurate. The forecasts issued immediately before the race, when there was more information about the sea breeze onset, were considerably closer to the actual conditions.

The use of the three-stage forecasting process is common in many fields where operations are dependent on the weather. Within the transportation field, for example, the routeing of ocean-going cargo ships is virtually the same as that for sailing vessels, although wave heights are usually of more direct concern than winds themselves. Transoceanic airline routes also follow the scheme, with the climatological upper wind information being used for planning, usually to ascertain the seasonal changes in the general route patterns. For aircraft, of course, the short-term forecasts needed are shorter term than for sea-borne carriers. Indeed, when it comes to making operational decisions for several of the activities considered in

*Table 6.A.1* Forecast and observed sea-breeze winds for the site of the 1991 Pan American Games yacht races, off Havana, Cuba (after Powell, 1993)

| Date (in August 1991) | Climatology | | 18-hour forecast | | Race forecast | | Observed | |
|---|---|---|---|---|---|---|---|---|
| | Direction (degrees) | Speed (m s$^{-1}$) | Direction (degrees) | Speed (m s$^{-1}$) | Direction (degrees) | Speed (m s$^{-1}$) | Direction (degrees) | Speed (m s$^{-1}$) |
| 7 | 045 | 7 | 080 | 5 | 045 | 6 | 060 | 8.5 |
| 8 | 045 | 7 | 070 | 5 | 045 | 6–8 | 043 | 6.5 |
| 9 | 045 | 7 | 065 | 3 | 015 | 4–6 | 039 | 4.5 |
| 10 | 045 | 7 | 095 | 3 | 020 | 3–5 | 040 | 4.0 |
| 11 | 045 | 7 | 070 | 5 | 035 | 5 | 040 | 5.1 |
| 12 | 045 | 7 | 050 | 5 | 040 | 3–5 | 032 | 4.5 |
| 13 | 045 | 7 | 045 | 3 | 030 | 5 | 043 | 5.0 |

**Box 6.A (cont'd)**

earlier sections, whether it is providing enough sand and grit to minimise the impact of bridge freezing, the purchase of sufficient fuel to generate the required electricity, or to ensure that agricultural crops are planted, irrigated or harvested at the right time, the three-stage approach is vital.

*6.A.2 Energy from the wind*

It is increasingly common to convert the kinetic energy of moving air, the wind, into electrical energy and to feed it, along with the energy created from other sources, into the national distribution system. Again, the concept is not new, with windmills having been a significant part of the landscape in some areas for centuries. However, the sight of wind farms, with high-tech propellers of futuristic appearance, is new.

Climatologically the challenge is to provide the wind information needed to find the best location for these wind farms. Several factors must be taken into account. The power contained in the wind is proportional to the third power of the velocity. Hence simple wind-speed averages are very misleading, and the frequency of winds in various speed categories is needed. Furthermore, there are practical limits for the propellers. Winds below a specific threshold will not contain enough energy to overcome the inertia of the system and the propeller will not turn. At the other extreme, winds above a specific threshold may lead to rotational speeds dangerous to the system, and the propeller must be shut down. Finally, if the machine faces a fixed direction, which is commonly the case for engineering and efficiency reasons, the wind direction becomes important.

Given those problems, how do we go about 'wind prospecting' to find the best sites? Mountain or ridge tops are obvious candidates (Figure 6.A.1), but the wind may be too strong. Valley bottoms often channel winds, and may be ideal sites for continuous strong winds. However, some valleys may be highly prone to stagnant, near-calm conditions, and thus highly unsuitable. Existing wind data must therefore be scrutinised very carefully. As with most applied climatological problems, the observations are unlikely to be at sites where the information is needed, and knowledge of how the atmosphere works on the local scale (see Chapter 10) must be used to make preliminary estimates of candidate sites. Local experience, albeit usually non-quantitative and anecdotal in nature, can be extremely useful in identifying 'windy' sites.

In many cases, once candidate sites are identified, a measurement programme must be developed to test the hypothesis whether is suitable. The programme will probably be a short-term one lasting a few months at most. However, it should be long enough to cover the range of synoptic conditions likely to be encountered, so that relations between the short-term observations at the site of interest and the long-term observations at nearby weather stations can be developed. These relations will give the required performance information. Again, the results must be treated with care, and in particular the representativeness of the period of dual observation must be carefully checked. In addition, wind speed is likely to vary greatly with small shifts in location, so short-term spatial variations as well as temporal ones must be considered.

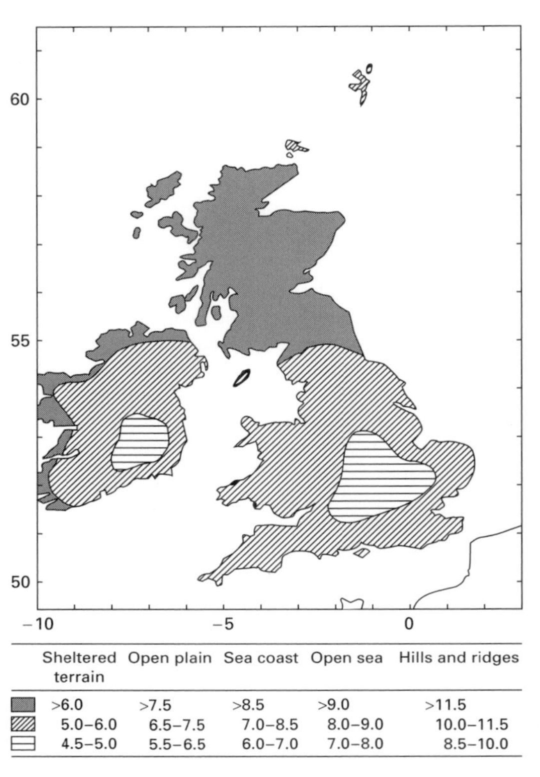


*Figure 6.A.1* The distribution of mean annual wind speed 50 m above ground for the British Isles. The basic distribution was determined from observations and then modified to give a range of values by consideration of topography and vegetation. This creates an indication of the potential for wind energy generation across the nation. (From Palutikof *et al.*, 1997; after Troen and Petersen, 1989.)

CHAPTER 7

# The general circulation and global climate

In the present chapter we are primarily concerned with the large-scale air flow pattern of the whole globe: the **general circulation of the atmosphere**. In addition to providing our winds, the general circulation serves to redistribute energy and moisture, modifying their latitudinal imbalances and thus establishing the climate. In this chapter we shall emphasise the processes acting to create and control this circulation, and then consider the global-scale climate regions which result. Detailed discussion of these regions is deferred to the subsequent chapters. Throughout, we shall also refer to the companion to the atmospheric circulation, the **general circulation of the ocean**. This is also involved in energy redistribution and climate production. Interactions between the two circulations, by no means fully understood, are proving to be vital for a full appreciation of climate and climate variability.

It is frequently very convenient to divide the atmospheric general circulation into two components: the **primary circulation features**, i.e. the persistent large-scale features which cover large areas of the globe, and which, while varying in detail, exist at all times; and the **secondary circulation features**, i.e. the short-lived, rapidly moving cyclones (depressions) and the much slower moving anticyclones, which are superimposed on the former and are responsible for day-to-day weather changes for much of the Earth.

## 7.1 The function of the general circulation

The latitudinal imbalance of absorbed and emitted radiation discussed in Chapter 2 and latitudinal variations in the components of the atmospheric water system detailed in Chapter 5 both indicate that horizontal motions are necessary to maintain the present climate. Thus a prime role of the general circulation, both atmospheric and oceanic, is to provide the necessary redistribution. Almost all information we have about past climates indicates that the general circulation has been operating in the same way for millennia, possibly aeons, with climate changes being modifications of the basic pattern rather than radical departures from it. Hence we can take the present-day circulation as a model which, once understood, not only explains the present climate system but also provides insights into the past and possible future climates.

### 7.1.1 Role of the general circulation

The profound role of the general circulation in our climate can be illustrated by comparing the latitudinally averaged observed temperatures with those calculated by considering vertical energy exchanges alone (Figure 7.1). Without the horizontal motions, summers would be rather warmer than observed for much of the globe. Winter temperatures in the tropics would also be higher, but would drop lower than the actual ones very rapidly as the poles are approached. The implication of the seasonal difference is that the required poleward energy transport is much greater in winter than in summer. This is observed in practice.

The fluxes that modify these radiative temperatures are shown in Figure 7.2. Here we have combined the effects of the latitudinal imbalance of radiation (Figure 2.10) with the imbalances of the hydrological cycle (Figure 5.11), added the effects of oceanic transport and expressed them all in energy units. Although the individual fluxes have seasonal variations, as is implied by Figure 7.1, we shall temporarily concentrate on the annual conditions.

The major role of the oceanic flux is movement of some of the sensible heat away from the equator

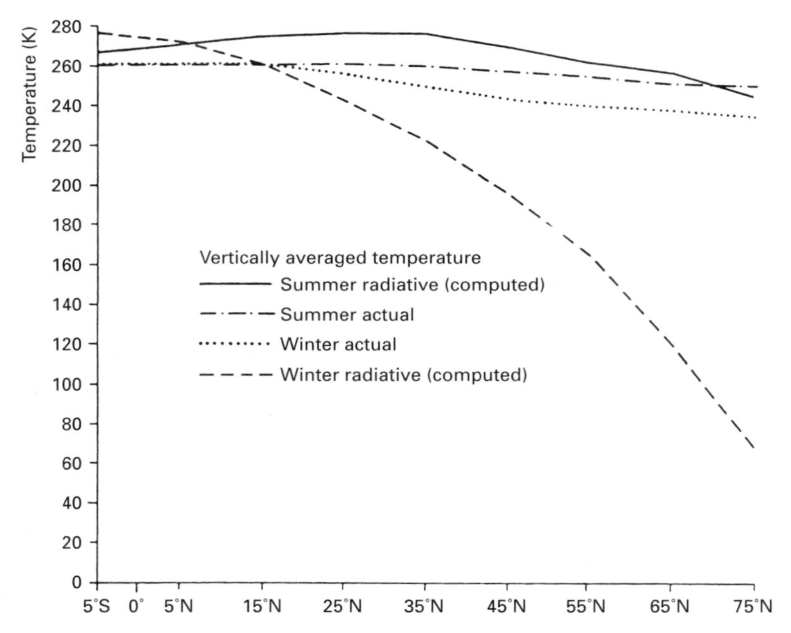


*Figure 7.1* Comparison of theoretically derived radiative equilibrium and observed vertically averaged temperature profiles for the summer and winter. Without energy transfer from low to high latitudes the equilibrium temperatures in mid and high latitudes are extremely low.

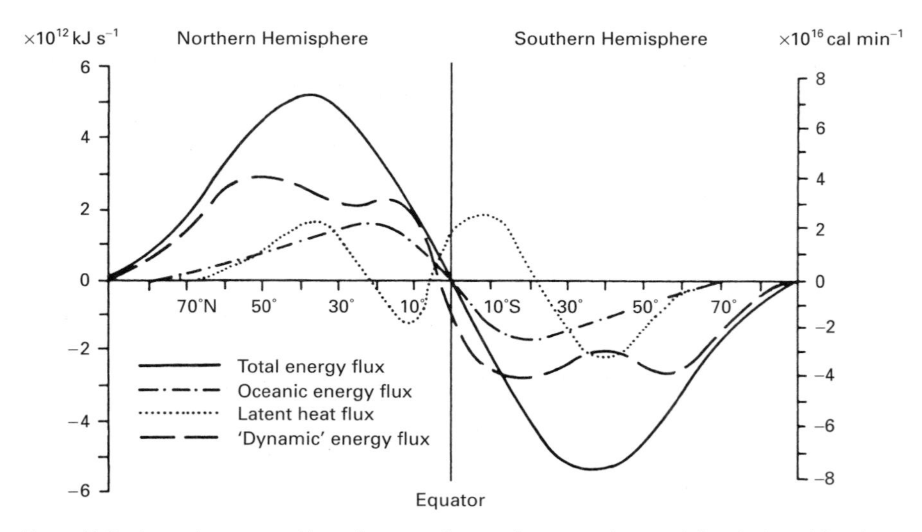


*Figure 7.2* Annual mean northward energy fluxes shown as the total flux (—) and its three components; (– · –) oceanic; (······) latent; and (— —) dynamic (sensible). (From Sellers, 1965.)

through the action of the ocean currents. The remaining sensible heat is transferred by the atmospheric circulation. Again this is mainly a simple equator to pole transfer, although reaching a maximum rate farther poleward than the oceanic flux. The latent heat flux is rather more complex. The major source regions for the water vapour are the subtropical oceans, where net radiation is high, so that the flow is both equatorward and poleward from source regions around 10°. The relative magnitude of the three fluxes is very roughly as follows:

| | |
|---|---|
| oceanic flux | 25% |
| atmospheric sensible heat (dynamic) flux | 60% |
| atmospheric latent heat flux | 15% |

All three combine to produce a flux close to zero at the equator and a maximum near $5 \times 10^{12}$ kJ s$^{-1}$ at about 40°N and 35°S.

### 7.1.2 Controls and constraints on the general circulation

Although this discussion emphasises that the general circulation is primarily a response to the energy imbalances, the introduction of consideration of water vapour movement indicates that there are 'constraints' under which the general circulation must act if the present climate is to be maintained. The atmosphere must act to maintain the global water balance, preserving in an approximate way the present distribution and amounts of precipitation and evaporation. Similarly, it must maintain a balance of atmospheric mass. Finally, it must also maintain the planet's angular momentum balance. Since there is frictional coupling between the atmosphere and the rotating Earth there is a possibility that the rotation rate can be altered by the general circulation. In essence, a constant angular momentum requires an approximate equality between eastward and westward components of the wind. These simple constraints must be borne in mind as we develop our description and understanding of the general circulation.

## 7.2 The nature of the general circulation of the atmosphere

### 7.2.1 The three-cell model

Characterisation of the general circulation of the atmosphere has been a central problem in both meteorology and climatology since the subjects emerged as sciences. Observation and theory have here gone hand in hand. Once surface-based observations became sufficiently widespread to give approximately global coverage, the **three-cell model** of the general circulation was developed as a theory fitting the known facts (Figure 7.3). Although this model is now seen to be a vast oversimplification, it still provides a useful conceptual tool.

The model was developed from the observation that there were zonal belts of low pressure around the equator and, in a more diffuse form, around 60° latitude. High pressure dominated around 30° and at the poles. Since low pressure is associated with

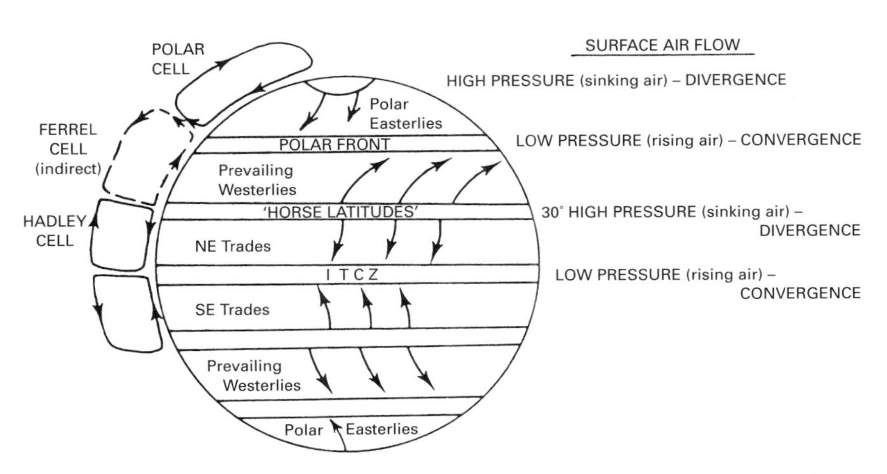


*Figure 7.3* The three-cell circulation of the atmosphere showing the resultant meridional pattern: high and low pressure areas and the direction of the surface wind flow.

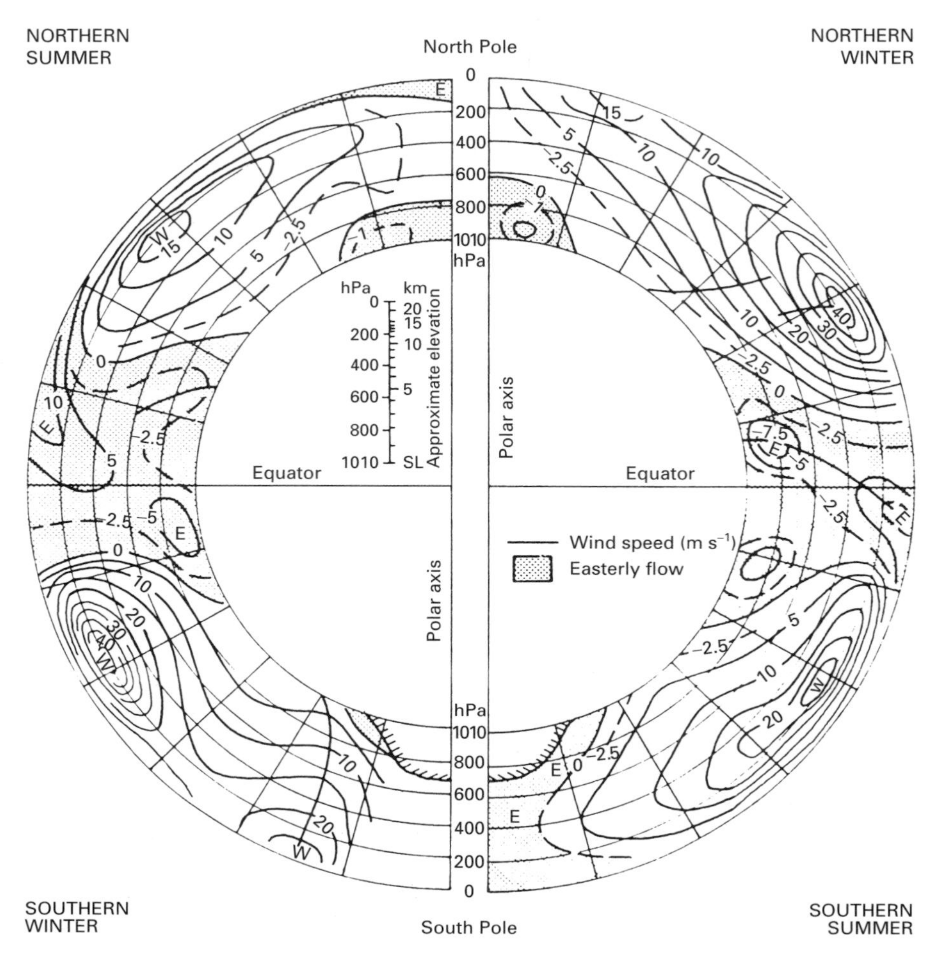


*Figure 7.4* Mean zonal wind (east–west) averaged over latitude circles for northern summer (left) and northern winter (right). Winds are in metres per second and easterly winds are shaded. Note the very strong upper westerly flow, which contradicts the three-cell model shown in Figure 7.3. A conversion scale between pressure (hPa) and height (km) co-ordinates is inserted. (From Goody and Walker, 1972.)

convergence and ascending air, and high pressure with descent and surface divergence, it was a relatively simple matter to create the three cells. The tropical and polar cells were supposedly being driven by the effects of surface heating and were called 'thermally direct' (i.e. ascent over a warm area and descent over a cold area), while the mid-latitude one was a response to them and thus was 'thermally indirect'. Ascent creates clouds and precipitation, while descent gives dry, cloudless conditions, in rough accordance with observations. The air motions at the surface, after incorporation of the Coriolis effect, also correspond reasonably well with observations. It can also be seen that the model fulfils the basic functions of the general circulation to redistribute energy and moisture without changing the angular momentum or mass balance of the planet.

### 7.2.2 Empirical refinement of the three-cell model

As further observations were made, particularly in the upper atmosphere, it became obvious that the three-cell model was incomplete in all areas, and incorrect in at least one. The mid-latitude upper flow is westerly (Figure 7.4), not easterly as the model

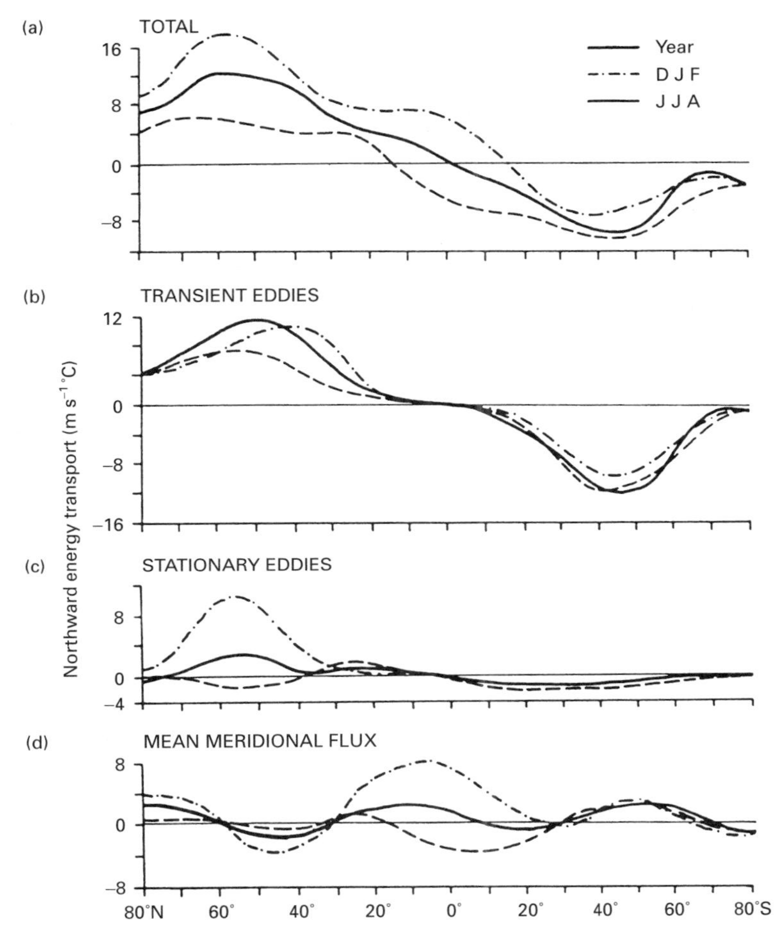


*Figure 7.5* Meridional profiles of northward transport of energy in m s$^{-1}$ °C for the cases of (a) total energy transfer, (b) transfer by transient eddies, (c) transfer by stationary eddies, in which the longitudinal component of the motion is responsible for the northward flux, and (d) the mean meridional flux. Note that the meridional flux is largest near the equator, while transient eddies perform most of the energy transport in the mid-latitudes. (From Oort and Piexoto, 1983.)

predicts. Indeed, the upper level winds in both hemispheres and both seasons are characterised by extremely high velocity **westerlies**, centred on a core, or **jet**, which changes position during the year. These westerlies take on a wave-like motion, the waves being known as the **Rossby**, **planetary** or **long** waves. When viewed from above either pole, they well justify their other appellation as the **circumpolar vortex** (see, for example, Figure 6.2). They are a function of the zonal temperature gradient and the angular velocity of the motion and are influenced by the characteristics of the underlying surface. Thus the thermally indirect mid-latitude cell is much more complex than was supposed, and it is inappropriate to call it a cell. The two thermally direct cells remain, although their great complexity is acknowledged and notions of their driving mechanisms have changed.

Much of the difference between cells can be attributed to the means used to accomplish the necessary poleward energy transports. In the mid-latitude cell the transport is performed by stationary eddies, the horizontal wave-like motions of the Rossby waves, and by their embedded disturbances, the transient eddies (Figure 7.5). In the tropics the eddies are small, and meridional transport dominates. For the relatively small polar cell there is a mix of all of them.

The three-cell framework, or perhaps better, the three-division framework (Hadley cell, Rossby waves, polar cell), is evident in our diagrams of the general circulation and is clearly reflected in the various regional climates. Our regional discussions will imply that mid-latitudes seem to be more complex than the tropical or polar regions. This is almost certainly the case, although it should be borne in mind that we know more about the mid-latitudes than the other regions and so are more likely to recognise complexity. It should also be borne in mind during these discussions that, despite the regionalisation, the whole of the planetary circulation operates as a single system.

### 7.2.3 The three regimes

The thermally direct **Hadley** cell generally exists within an equivalent barotropic atmosphere. At these low latitudes, without a distinct thermal winter and summer, continentality effects are minimal and so the near-surface temperature distribution is roughly uniform zonally, with a temperature decrease poleward. The basic pressure distribution is also zonal. The surface, heated by solar radiation absorption, provides a heat source for the lower atmosphere. Latent heat released by convective motions within the ascending air of the **intertropical convergence zone** (ITCZ), where the cells in each hemisphere meet, provides a second heat source. The heat sinks occur at the top of the atmosphere and at the poleward limb of the cell. The horizontal components of the cellular flow are only slightly affected by the Coriolis deflection at these low latitudes, and the resulting **trade winds** have a marked cross-isobaric component, blowing equatorward at low levels and poleward in the upper troposphere. Thus the wind patterns allow a direct meridional transfer of energy away from the equator. Within this general scheme, however, variations in surface characteristics lead to the superimposition of other circulation patterns.

In theory, the polar cell is similar to the Hadley one. A surface covered with snow and ice produces a uniform temperature distribution, which leads to equivalent barotropic conditions and cellular motions. Here the winds are completely deflected to zonal flow by the Coriolis force. The resultant surface easterlies are relatively weak throughout the year in the high latitude Arctic but are a fairly strong feature over Antarctica in the southern summer (Figure 7.4). There is general atmospheric subsidence. During winter in both hemispheres there is evidence for a local velocity maximum at 75–80°, which has a jet-like structure that is particularly well developed in the Southern Hemisphere. This feature, which can be regarded as an extension of the mid-latitude westerlies, dominates the polar circulation. Hence the 'pure' polar cell is rather weakly developed and there is relatively little meridional transport (Figure 7.5). In any event its geographical extent is very small and in both hemispheres it can be regarded as the area to which energy must be transported and hence is not itself required to play a great role in energy transport.

The mid-latitude region is also frequently a region with an equivalent barotropic atmosphere. Thus there is a tendency for cell-like meridional overturning and meridional energy transport. However, the unique characteristic of mid-latitudes, creating complexity in both the energy transfer mechanisms and the weather and climate, is the frequent establishment of baroclinic conditions. The causes of this complexity will be explored in the next section.

The present configuration of the general circulation of the atmosphere appears to be a relatively stable one. This has been established both by numerical model experiments and through laboratory simulations of atmospheric motions using **dishpan** experiments. The dishpan is a fluid-filled annulus which can be heated at the bottom and outer edges and rotated at various rates (Figure 7.6). Thus it is a type of scale model of the Earth's atmosphere. With a small amount of heating at the base and outer edge and a relatively low rotation rate, a single cellular circulation is established very similar in nature to the Hadley circulation. As the rotation rate and heating gradient are increased to simulate more closely the existing atmospheric conditions, this single system breaks up. The Hadley circulation becomes confined to the 'tropics', near to the outer edge, and a Rossby-like circulation develops in middle and high latitudes. In some cases a very weak polar circulation is also established. This configuration is maintained through a relatively wide range of rotation rates and heating gradients, suggesting that the present atmospheric configuration is 'stable', at least for the rotation rates and energy gradients likely to be encountered in the foreseeable future. The results, of course, do not indicate an unchanging climate, but rather one that can change within the limits imposed by the configuration of the general circulation of the atmosphere.

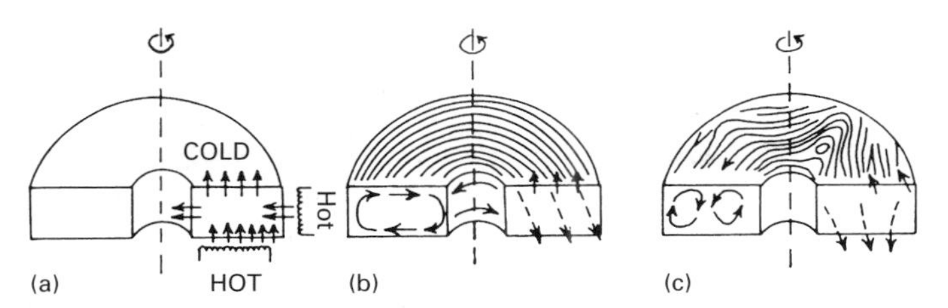


*Figure 7.6* A differentially heated rotating annulus experiment showing (a) the heating regime; (b) the symmetric flow, i.e. a Hadley circulation, in the case of a low rotation velocity; and (c) the Rossby wave regime set up in the case of a higher circulation velocity. (From Wallace and Hobbs, 1977.)

## 7.3 Large-scale effects of the surface boundary

The Earth's surface forms a highly discontinuous boundary to all atmospheric processes and therefore its effect on the general circulation must be taken into account. We have already shown that practically all of the energy input to the atmosphere comes from the surface, while surface friction changes the near-surface wind. The surface serves as a sink for kinetic energy, since the energy extracted from the wind as its speed decreases is transferred to the surface. This exchange, while much smaller than the sensible and latent heat exchanges, is significant in maintaining the global angular momentum balance and in producing motion in the oceans. Hence both the thermal and the frictional character of the surface have a profound influence on the atmosphere. In this context we can divide the Earth into three basic surface types – water, snow and ice, and land – and look at each in turn.

### 7.3.1 Oceans

A vital characteristic of the ocean for atmospheric circulation and climate is the large heat storage capacity it represents. By comparison, atmospheric storage of heat is an order of magnitude smaller (Figure 7.7). In terms of mass, the atmosphere is equivalent to a layer of water approximately 10 m deep. If we consider only the **mixed layer** of the ocean, that top layer where vertical stirring is possible, which is about 70 m deep, then, since the specific heat capacity of water is 4.2 times greater than that of air, this upper layer of the water can store approximately 30 times more heat than can the atmosphere. There is, in addition, a much slower movement of energy to and from oceanic storage at greater depth, which is associated with the deep overturning of the oceans which may occur on the order of several thousand years. An equally vital ocean characteristic is their ability to transport sensible heat. Although smaller on average than the atmospheric flux (Figure 7.2), a monthly and latitudinal comparison between them (Figure 7.7) indicates that the oceanic transports are comparable to those of the atmosphere in many places at many times.

Momentum is transferred to the ocean whenever the wind blows over it. The kinetic energy in the wind is effectively transformed into the kinetic energy associated with ocean currents. On a small scale, waves are produced. On a larger scale, when strong winds blow, the drag on the ocean surface is sufficient to permit the movement of the warm surface waters, which are then replaced by upwelling colder water from greater depth. Thus a specific sequence of events can be initiated by a warm water surface. The warm water tends to create unstable atmospheric conditions, which are favourable to the development of storms and strong winds. These in turn lead to upwelling of cold water, which serves to stabilise the atmosphere and dampen the winds: an example of a 'negative feedback' within the climate system. The major example, El Niño, is considered in the next chapter.

### 7.3.2 Cryosphere

The Earth's cryosphere consists of snow on continents and ice over both land and sea. The existence and persistence of the cryosphere depends upon subfreezing temperatures and thus it occurs mainly at high latitudes and at high altitudes. In addition there must be adequate precipitation to maintain the snow

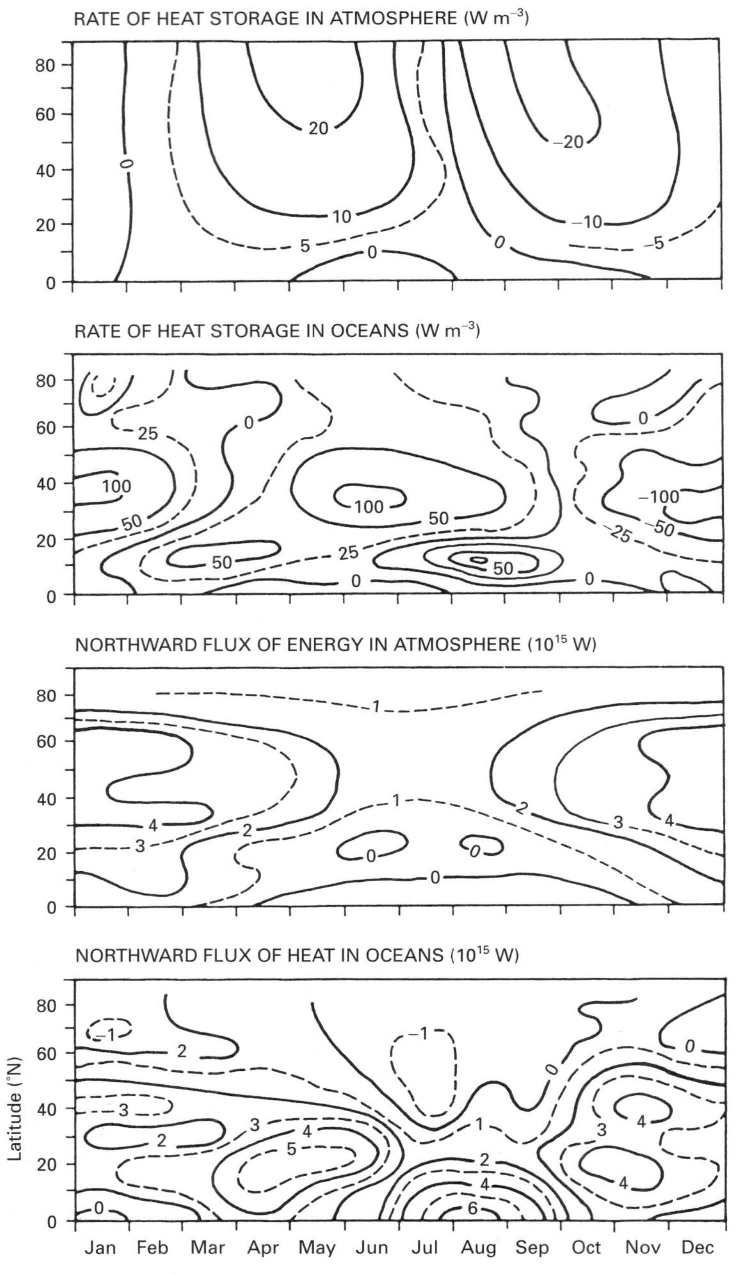


*Figure 7.7* Rate of heat storage in the atmosphere and oceans and the northward transport of energy in the atmosphere and oceans as a function of latitude and time of year. The atmospheric heat storage peaks in high latitudes in early summer and is matched by a heat loss in autumn. The oceanic heat storage, by contrast, is greatest at approximately 30° in midsummer. The fluxes of energy are similarly mismatched with the largest atmospheric flux being in mid-latitudes in winter while the maximum oceanic flux is in lower latitudes in summer. (From Oort and Vonder Haar, 1976.)

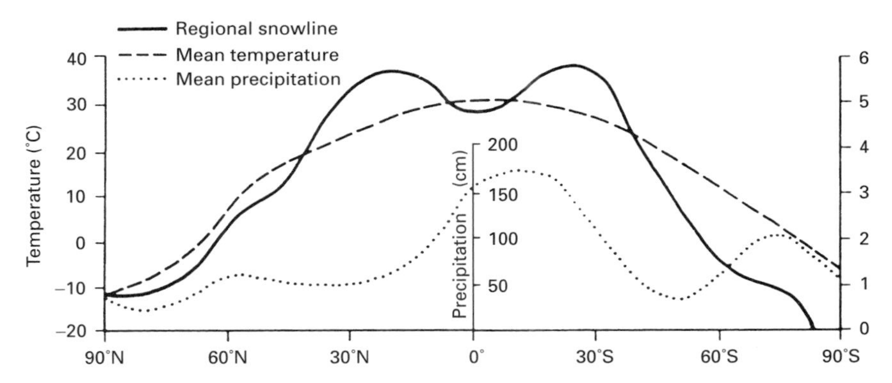


*Figure 7.8* Generalised curves showing the fluctuation of temperature, precipitation and regional snowline with latitude. Note that in the equatorial zone the snowline falls, despite increased temperature, apparently because of increased precipitation. (From Flint, 1971.)

and ice supply (Figure 7.8). At present the perennial cryosphere covers 8% of the Earth's surface (Figure 7.9). The very large seasonal fluctuations in the climate of the high latitudes are underlined by the fact that the seasonal cryosphere covers an additional 15% of the surface in January and an additional 9% in July. Antarctica has a complete cover, but the Arctic Ocean, although frozen all year, has an ice mass which is not complete. It consists rather of numerous large ice floes, with an average thickness around 4 m, which are in continuous motion within the Arctic Basin. The year-to-year fluctuations in seasonal sea-ice extent are considerable in both hemispheres.

The cryospheric interaction with the overlying air is primarily through the stabilising effect that the cold surface creates. This is really significant only in polar regions, where the cryosphere has its major extent. It tends to reinforce the high pressure region at the poles, and foster low level inversions and relatively calm conditions.

The extent of the cryosphere exerts an important control on the planetary albedo and, through the **ice–albedo feedback mechanism**, on planetary temperatures. It is hypothesised that the onset of glacial periods results from a trigger, such as a fluctuation in solar luminosity, causing an increase in overall glacial extent. If it is assumed that the areas which remain unglaciated do not suffer a significant change in cloud amount, the resultant increase in mean global albedo may lead to a new, colder climate which is sufficiently stable to persist for a long period of time. At present this feedback may be working in the opposite direction because of the potential warming associated with anthropogenic activity.

### 7.3.3 Continents

The continental surface is characterised by a vast array of topographic features and surface types, all of which, to varying degrees, influence the general circulation of the atmosphere. The most obvious and pervasive influence arises because of the contrast in thermal properties between land and sea leading to the continentality effect. The thermal contrasts created give rise to pressure contrasts, which influence the secondary circulation features of the general circulation and eventually create distinct regional variations in continental climates.

The regions are further differentiated as a result of the differences in the Bowen ratio associated with water availability. The Indian Ocean is typical of the tropical oceans in that about 90% of the available energy is used for evaporation. In arid land regions, such as parts of Asia and Australia, most of the energy goes directly into warming the air. In the moist mid-latitudes and the tropical jungles both sensible and latent heat are removed from the surface, but most of the available radiative energy is used for evapotranspiration. Finally, in the polar regions the average energy flux is from the air to the surface in the form of sensible heat flow. In each case the fluxes influence the amount of cloud and, to some extent, the amount of precipitation in the area. They are all, of course, influenced by the horizontal air motions mentioned earlier. These motions are themselves influenced by topography, which varies from continent to continent, leading to the different distributions of climate regions over each continent.

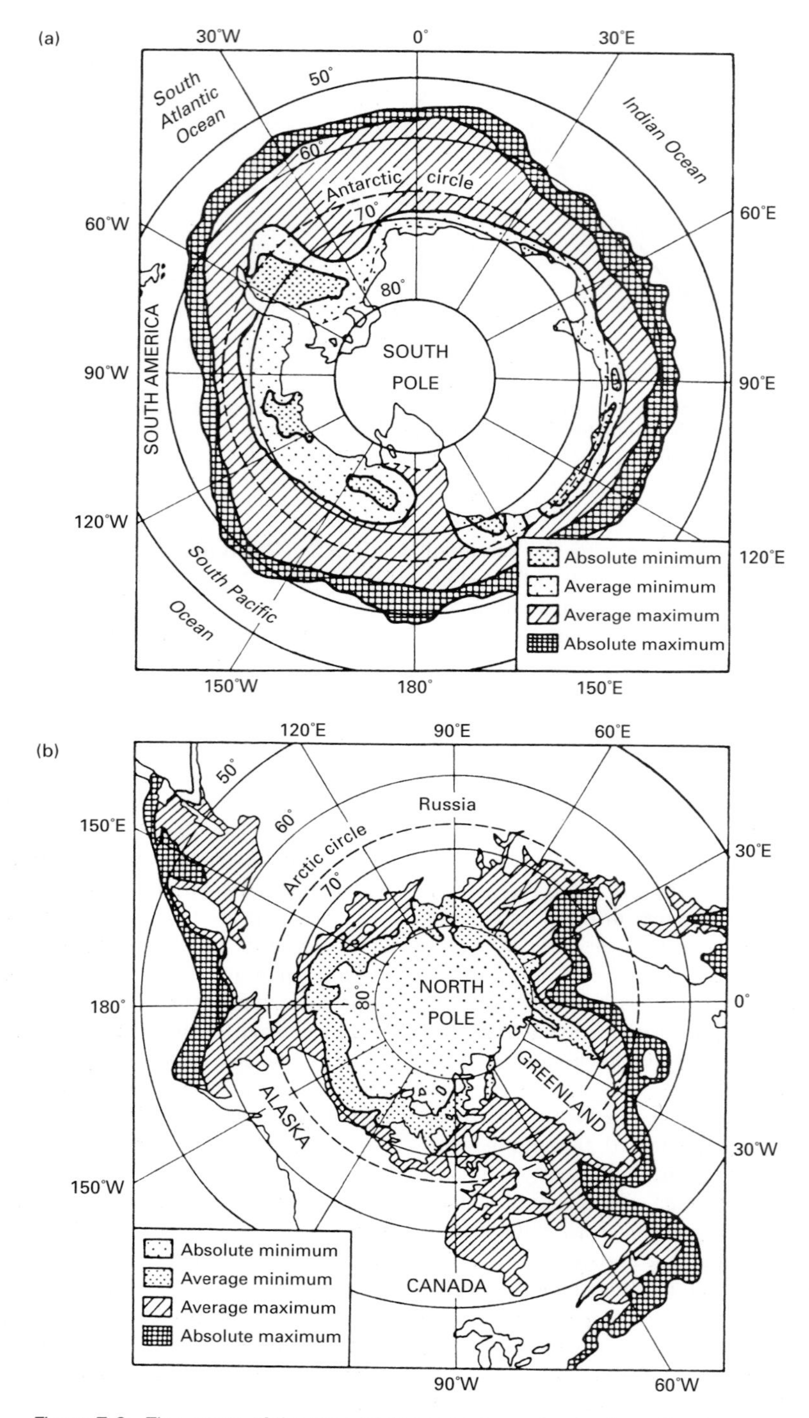


*Figure 7.9* The extent of the present-day cryosphere: (a) Southern Hemisphere; (b) Northern Hemisphere; (c) The extent of the Northern Hemisphere cryosphere at the height of the Pleistocene. (From Flint, 1971.)

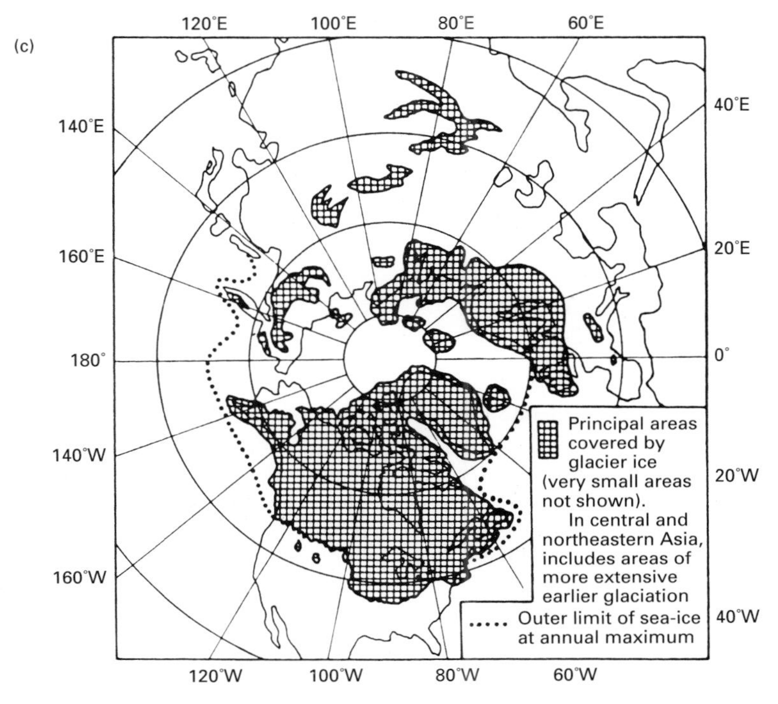


*Figure 7.9* continued

## 7.4 Climate classification and regions

One long-established aim of climatology has been to develop a regional climatology of the Earth. The objective is to develop succinct descriptions of the climatic conditions likely to be encountered at any point on Earth, together with an explanation of their causes and indications of their stability and variability with time. This serves two major purposes. First, the information is useful for anyone with more than a passing interest in a particular place. It can provide information, for example, pertinent to the types of crops that could possibly be grown, or the heating and cooling requirements of housing in the area. Thus it provides an estimate of the climatic 'resources' of an area. Second, the identification of regions is also important for climatology itself. If patterns in the spatial distribution of regions are found, they may provide insight into the processes that are acting to create those regions. Thus the development of regionalisation schemes has historically gone hand in hand with the understanding of climatic processes. The early recognition that several widely divergent areas on the Earth's surface have similar climates stimulated a search to identify the processes acting to create them. This search played a vital role in advancing our understanding of the nature of the general circulation of the atmosphere and thus stimulated study of most of the processes we have so far discussed in this book. Conversely, as understanding of the basic processes increased, we were able to refine our understanding and definition of the regions. This interaction is continuing. For example, improved observations of sea surface temperatures and energy budgets, mainly from satellites, have led both to better specification of processes associated with El Niño in the Pacific Ocean and to the recognition of links between that region and many others.

There are an infinite variety of climates over the Earth, every place being slightly different in some aspect from all others. Consequently the first step in developing a regional climatology must be to develop a classification scheme which allows us to identify major, or in some aspect 'significant', differences in climate. As with classification in any branch of science, such a scheme must aim to simplify and clarify the variations in order to enhance comprehension and understanding. The classification scheme produced automatically leads to the creation of a series of

**climate types**. Provided the scheme leads to a manageable number of these, they can be mapped to produce the **climate regions**. The major problem in developing a climate classification scheme is in defining climate. Many elements are involved. If only one is used, it hardly qualifies as a 'climate classification', although it might provide much useful information. On the other hand, if we try to use all elements, the resulting complexity defeats the purpose of classifying for simplicity and clarity. Hence, as a compromise, usually two or three elements are used.

The elements are usually chosen because they are perceived to be important in the context of the use to which the classification scheme is to be put. It is also within this context that the method of expression of the elements must be chosen. It is possible to express precipitation, for example, in terms of the number of rain days or the total rainfall, while various averaging periods, such as monthly or annual, could be used. Some considerations of problems inherent in this are explored in Section 7.A.1.

**Box 7.A Succinct and accurate climate descriptions**

Previous chapters have demonstrated that climate is a complex set of interactions between a variety of elements, with a variety of forcing functions. The section on classification indicated the difficulty of creating a simple scheme giving a succinct but realistic description of climate. The classifications noted there represent what have become 'standard' in climatology and provide a common body of information for all climatologists. They do not, and cannot, provide complete climatic descriptions serving all purposes for all places. There is, in fact, an almost infinite variety of ways of describing, comparing and classifying climate. We cannot consider them all. Rather, in this section we introduce some ideas and comments which suggest lines of approach to the development of appropriate descriptions and classifications for a variety of purposes.

*7.A.1 Popular climatic descriptions*

The commonest, and probably most popular, description of climate is to express it as the normals of average temperature and total precipitation each month, without necessarily invoking the formality of the Köppen system. The information is readily available for many stations, and in the next chapter such descriptions for selected stations are used to introduce discussion of the various climate types. However, here we can ask whether these elements really provide the best description of climate for a particular purpose in a particular area.

Taking as an example the common request to a climatologist or a travel agent for expected conditions at a proposed holiday destination, it is not clear that they do. As suggested in Section 5.2, information about rain days may be more useful than monthly totals, since, for example, a long-lived slow-moving depression may take five days to give the same amount of precipitation as a 5-hour convective storm. Similarly, our simple description for temperature only gives a monthly average. That might be a balmy 25 °C, which in reality is commonly composed of 15 days around 15 °C, and 15 days around 35 °C. Thus, the creation of misleading information is possible even when using only the common elements. If the effect of wind is added when low temperatures are experienced, or humidity considered during hot spells, a more meaningful description may be developed, but complexity increases.

These examples could be multiplied many times. Certainly the lesson they teach must be borne in mind throughout the rest of the book. In particular, they emphasise that the climatic descriptions in Chapters 8 and 9 can themselves be misleading, and must be treated with caution. Indeed, it may be useful as you read Chapters 8 and 9 to consider for each of the regions the simplifications that have been introduced, and to deduce from the information in the text the additional information needed before embarking on, say, a holiday in the area. That provides a good test of understanding. In many cases it is now possible through the resources of the Internet to obtain the data and information which allows you to check your deductions.

*7.A.2 Objective classification techniques*

The popular descriptions considered above treat individual climatic elements largely in isolation. In reality, the elements occur in combination so that, for example, the coldest days in winter may be the sunniest days of the season, and may be virtually calm. These relationships have been recognised for many years and there have been attempts to use them to create classifications at the other end of the complexity spectrum from the previous popular descriptions. They concentrate on rigorously defined combinations of elements and specific criteria for separating the various climatic types. They are powerful tools for identifying regional and temporal variations in some key atmospheric components and have the potential to provide much useful information. However, their development tends to require a deep understanding of climatic processes and, very

**Box 7.A (cont'd)**

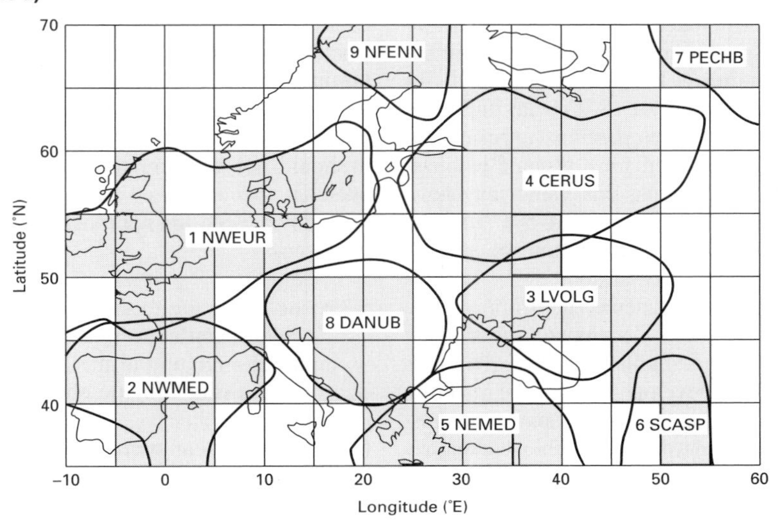


*Figure 7.A.1* A drought regionalisation of Europe. The solid lines encompass the areas where each of the principal components were strongest. Shaded areas were used to determine regional average drought conditions for each of the regions. The names represent symbolic abbreviations for the core areas. (After Briffa *et al.*, 1994.)

frequently, an appreciation of advanced statistical techniques. As such they tend to be highly complex, and a full analysis is beyond the scope of this book.

Nevertheless, a brief example can provide a few pointers to the techniques of much modern climatic analysis. A knowledge of drought severity, extent and frequency is vital for many aspects of agricultural and water resources planning. Drought, however, is a complex phenomenon (considered in Box 11.I) involving temperature, precipitation and soil water and their deviations from normal. It is usually expressed as a statistically derived 'drought index'. While maps of the index for a particular time are relatively easy to construct, drought extent and severity is constantly changing. To identify and portray those areas which seem to act in concert is extremely difficult, and involves analysis of a large amount of data and many variables. One statistical method commonly used in climatology to overcome this is **principal component analysis**. This technique assesses the variability in each of the original observed elements and seeks to find a mathematical combination of the elements (approximately equivalent to the 'principal components') which retains most of the variability information but which provides a smaller, and more manageable, number of variables. Figure 7.A.1 represents the result of such an analysis, where drought index information for the three summer months for 100 years at 120 points is reduced to nine principal components, which are then mapped, indicating a rather simple drought regionalisation of Europe. This regionalisation captures about 60% of the variability which is contained in the original data. Obviously this means that for individual points there has been some generalisation and loss of information. The map emphasises the corresponding gains, clearly demonstrating that certain areas act together as a drought unit for much of the time, thus providing the required information for agricultural and water resource planning.

The particular example used here demonstrates a second point associated with regionalisation. The drought analysis was carried out by dividing Europe into cells based on a uniform grid. The area of one grid cell was treated as if it represented a single station. This representation must be treated carefully (see Section 7.I.2), but the advantage is that no one station, and its peculiar local conditions, dominates a large area. Further, it is unlikely that all stations in the grid cell will have missing observations at the same time, so that it is much easier to construct a continuous observational time series than if individual stations were being used. Finally, the consequence of the approach is that the observations appear to be on a uniform grid, which almost always makes mathematical calculations considerably easier. Later in the book we shall see that virtually all climate models use some form of regular grid.

Once a decision about the elements and their method of expression is made, the next stage in developing a classification scheme is to identify threshold values which specify an important change in the impact of the parameter. It might be decided, for example, that a temperature of 18 °C is an important threshold, since above this temperature no residential heating is required. However, for a scheme emphasising agricultural applications, this value may have little meaning. Instead, a threshold of 5 °C may be more useful, since many plants commence growth once this temperature is reached.

Once the classification scheme, with appropriate threshold values, has been developed, the climate types are automatically established. Thereafter it is conceptually simple to develop a map of the resulting climate regions. Data for the whole area of interest are examined and standard cartographic techniques used to produce the regions. There are, however, several problems inherent in producing such a map. Climate is a spatially continuous variable but observations are available only for discrete points. Hence regional boundaries must be established by interpolation. The accuracy of such interpolation depends greatly on the density of the network of observation stations or the amount of available satellite information. Some land areas have a sparse network and any boundaries developed can only be approximate. Over the oceans the problem may be even more severe since most observations are at island stations, which probably do not represent true ocean conditions. The problem of drawing accurate boundaries is also acute in mountainous terrain. Not only is there usually a paucity of observational information, but also rapid spatial changes are common, presenting challenges to cartographic representation. In many classification schemes this problem is avoided by specifying a category 'mountain climates', with a definition emphasising the spatial variations in the area. Finally, there is a problem associated with the boundaries themselves. In most cases there is not an abrupt boundary between climate types, but rather a transitional region, so that almost all boundaries that appear on maps of climatic regions should be interpreted as transitional zones.

### 7.4.1 Methods of classifying climate

From the preceding comments it is obvious that classifications are possible on all time and space scales. However, the formal development of classifications and regions has traditionally been on the global or continental scale, producing regions approximately the size of a European nation or a US state, with climatic normals being emphasised. These schemes are the ones of concern here. Given the number of choices that have to be made in developing a classification, it should not be surprising that numerous schemes, even on this restricted scale, have been proposed. For convenience the approaches can be divided into two types:

genetic – relating to the origin of the features (emphasising atmospheric dynamics);
empirical – relating to the observation of features (emphasising climatic observations).

Each has particular strengths and weaknesses.

**Genetic** classifications emphasise the role of the climate controls in creating climate and its various regional expressions. Thus they can be exceedingly useful in helping to understand climate and the nature and impacts of climatic change. Indeed, in a somewhat informal way, previous portions of this chapter have provided the basis for this type of classification. The emphasis on the dynamic nature of climate, however, makes it exceedingly difficult to produce a succinct summary of the climate of a region while retaining the dynamic flavour. It can succeed provided the user is already familiar with the way in which synoptic climatology translates into the more familiar climatic elements, such as cloud amount, temperature or precipitation. Even then, it provides no direct quantitative information about precipitation amounts or actual temperatures. Furthermore, regionalisation, in the sense of creating distinct boundaries between regions, is not appropriate for this type of classification.

**Empirical** classification schemes aim to produce a quantitatively defined series of distinct regions without regard to the causes of the climate. Our discussion of classification above implicitly concerned empirical schemes. The parameters most frequently used are temperature and precipitation, or variations on these, such as evapotranspiration or soil moisture. The result is a strictly defined scheme based on two or three variables. This type of classification has a long history and the names given to some regional types, such as 'Mediterranean climate' or 'humid continental climate', have become widely known and evocative of particular conditions. Indeed, the major strength of this approach is the succinct way in which a great amount of information can be conveyed. In addition, reasonably clear regional boundaries can be established. Traditionally, empirical classifications use summarised data, such as monthly averages or annual totals, and so do not deal with climate in its role as a collection of individual weather events. Section 7.A.2 indicates that in some cases it is possible to overcome some of these drawbacks.

The empirical and genetic approaches should be viewed as complementary. Although the regions defined by each may not coincide, when used together judiciously, they can give a quantitative indication of the average conditions for specific elements, the variations about the average likely to be encountered, and the sequence of weather events that are likely to occur. We adopt this dual approach in the subsequent chapters.

### 7.4.2 Climate classification schemes

The genetic approach, lacking rigorous quantification, does not lend itself to formal schemes, and none have been seriously proposed. For the empirical approach, however, numerous schemes, many for particular applications, have been created. At one end of the scale are simple regionalisations based on a single parameter. These are rarely thought of as true classifications, but rather as regional maps for a specific purpose. Figure 3.A.1 is a typical example. At the other end of the scale are schemes involving numerous parameters. One of the best known of these is the 'rational' classification created by Thornthwaite in 1948. He postulated that the surface water balance was the single most important characteristic of climate in any area, a contention which would certainly find support when agricultural productivity is the main concern. This water balance depends not only on precipitation and evaporation at a particular time, but also on their seasonal variability. This led to the creation of a 'moisture index' as one of the pertinent variables in the scheme. Significant threshold values were then derived using information similar to that contained in Figure 5.10. The resulting classification is somewhat complex and is not very appropriate for arid areas. Hence it has not been extensively used on a global scale. Nevertheless maps of climate types for the mid-latitude continents convey in symbolic form a tremendous amount of information pertinent to agriculture.

A group of schemes intermediate in complexity between these extremes has been based on the classification first proposed by Köppen in 1918. Originally devised as an aid to understanding world-wide vegetation distributions, it has been modified and generalised by several workers. Now it is widely accepted as the apotheosis of empirical classifications, although it still betrays its application origins in the choice of parameters used. We shall explore this scheme in a little more detail, since it forms a very convenient framework for our own division of the globe into climatic regions. In the detailed discussions of the various regions, however, we shall not be overly concerned with rigorous definitions of either the scheme or the boundaries of the resultant regions.

Monthly and annual normals of mean temperature and total precipitation are the input variables. The scheme divides these into a series of categories, the boundaries of which represent some vegetation-based threshold value (Table 7.1). Each region is categorised, in symbolic form, by a series of two or three letters. The first letter initially separates dry from moist climates and then, for the latter, divides them on the basis of temperature. A second letter then refines this to define the degree of aridity for the dry climates and the temporal distribution of precipitation for the moist ones. A final letter is used to characterise the seasonal variations for mid- and high-latitude climates.

Simply as an example of this scheme, we can choose the 'Cfa' climate. In brief, this is a mild humid climate with a hot summer but no dry season. More specifically, we can say that precipitation exceeds evaporation on an annual basis. From a botanical standpoint, this implies, to a first approximation, that there is sufficient moisture for tree growth as the natural vegetation. The 'C' climate is defined as one with an average temperature in the coldest month between −30 and 18 °C and at least one month with average temperatures above 10 °C. Thus there is a distinct summer and winter. The middle 'f' represents a climate where precipitation in the driest month exceeds 30 mm, so that there is no dry season. The final 'a' indicates that this warmest month is, in fact, above 22 °C.

Using this scheme it is possible to produce a map showing the major climate regions of the Earth

*Table 7.1* Criteria for classification of major climatic types in the modified Köppen system (based on annual and monthly means of precipitation in millimetres and temperature in degrees Celsius)

| Letter symbol | | | Explanation |
|---|---|---|---|
| 1st | 2nd | 3rd | |
| A | | | Average temperature of coolest month 18 °C or higher |
| | f | | Precipitation in driest month at least 60 mm |
| | m | | Precipitation in driest month less than 60 mm but equal to or greater than $(100 - r)/25$[a] |
| | w | | Precipitation in driest month less than $(100 - r)/25$ |
| B | | | 70% or more of annual precipitation falls in the warmer six months (April–September in the Northern Hemisphere) and $r/10$ less than $2t + 28$[a] |
| | | | 70% or more of annual precipitation falls in the cooler six months (October–March in the Northern Hemisphere) and $r/10$ less than $2t$ |
| | | | Neither half of year with more than 70% of annual precipitation and $r/10$ less than $2t + 14$ |
| | W | | $r$ less than one-half of the upper limit of applicable requirement for B |
| | S | | $r$ less than upper limit for B, but more than one-half of that amount |
| | | h | $t$ greater than 18 °C |
| | | k | $t$ less than 18 °C |
| C | | | Average temperature of warmest month greater than 10 °C and of coldest month between 18 and 0 °C |
| | s | | Precipitation in driest month of summer half of year less than 40 mm and less than one-third of the amount in wettest winter month |
| | w | | Precipitation in driest month of winter half of year less than one-tenth of the amount in wettest summer month |
| | f | | Precipitation not meeting conditions of either s or w |
| | | a | Average temperature of warmest month 22 °C or above |
| | | b | Average temperature of each of four warmest months 10 °C or above; temperature of warmest month below 22 °C |
| | | c | Average temperature of from one to three months 10 °C or above; temperature of warmest month below 22 °C |
| D | | | Average temperature of warmest month greater than 10 °C and of coldest month 0 °C or below |
| | s | | Same as under C |
| | w | | Same as under C |
| | f | | Same as under C |
| | | a | Same as under C |
| | | b | Same as under C |
| | | c | Same as under C |
| | | d | Average temperature of coldest month below –38 °C (d is then used instead of a, b or c) |
| E | | | Average temperature of warmest month below 10 °C |
| | T | | Average temperature of warmest month between 10 and 0 °C |
| | F | | Average temperature of warmest month 0 °C or below |
| H | | | Temperature requirements same as E, but due to altitude (generally above 1500 m) |

[a] In formulae $t$ is the average annual temperature in °C, and $r$ is average annual precipitation in millimetres.

(Figure 7.10). Even a cursory examination of the map indicates that patterns of climate are repeated from continent to continent, reflecting the overall control that the general circulation places on the climate. The details differ between continents, of course, depending on the particular configuration and topography of each. The map, combined with the table, gives the first feel, and much detailed information, for the type of climate to be experienced in any part of the world. In the chapters that follow we shall use the framework of the Köppen system to explore regional climates in more detail. Climatological data from the stations shown in Figure 7.11 will be used to illustrate the climate types. However, we shall combine this empirical approach with a more genetic approach both to describe and explain those regional climates.

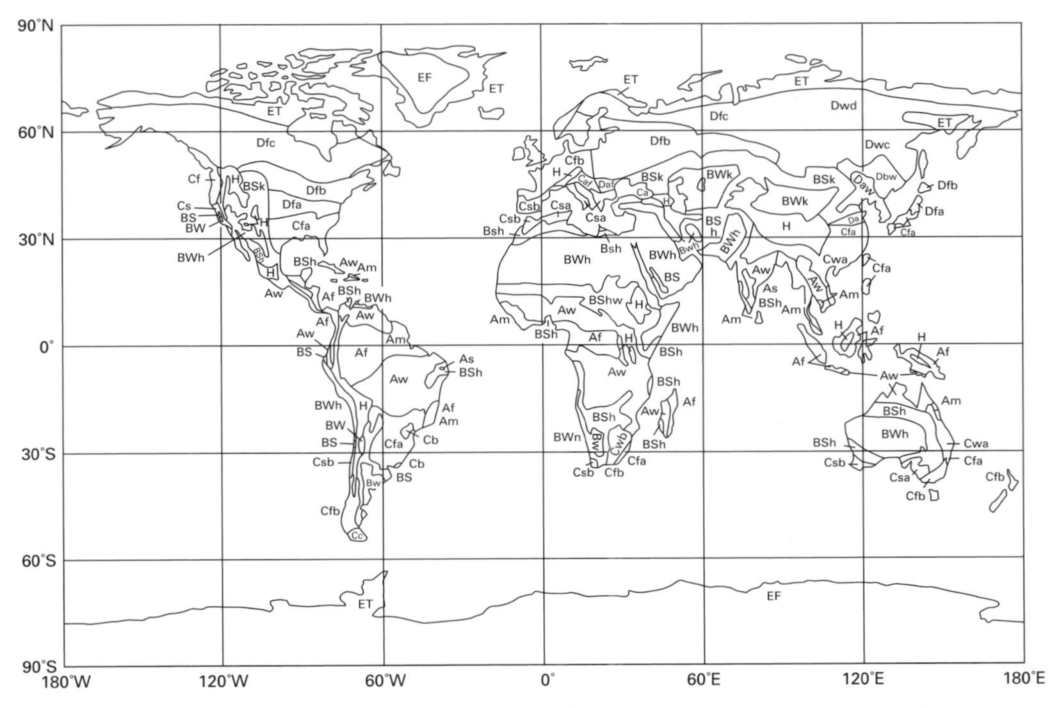


*Figure 7.10* The Köppen classification of climate. The climatic types are listed in Table 7.1. Note that the two primary forcing factors are the latitude zone and the degree of continentality of the location.

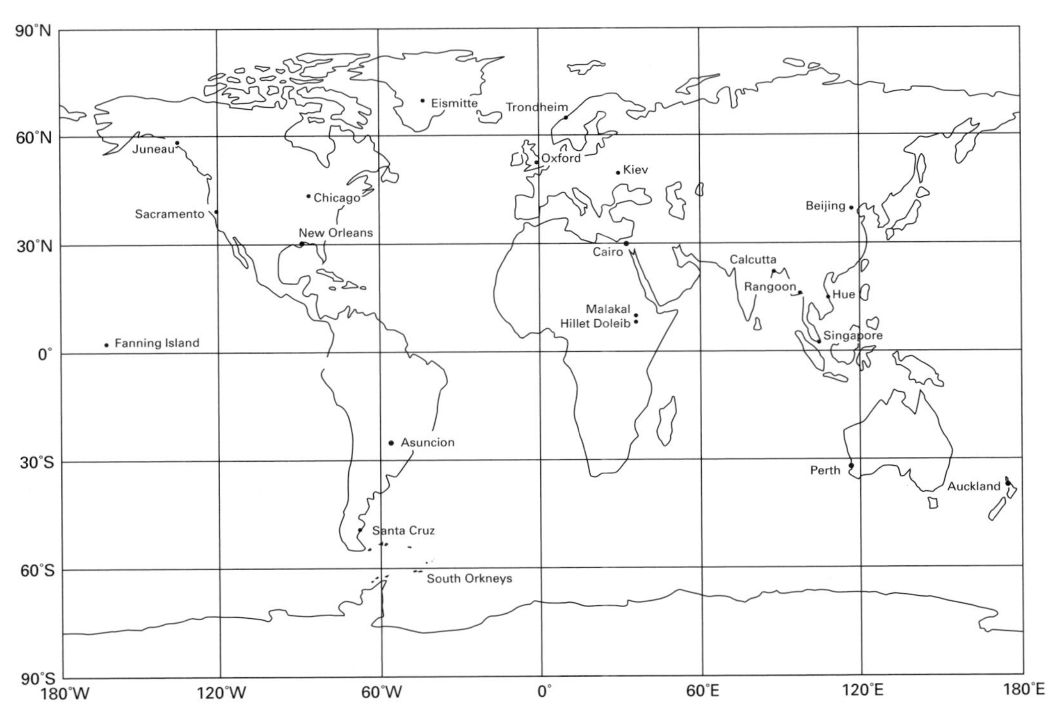


*Figure 7.11* Location maps for all stations used in subsequent chapters to illustrate the Köppen climate classes. Climatological diagrams for these stations are given in Figures 8.7, 8.8, 8.10, 8.13, 9.13, 9.14, 9.15, 9.17, 9.18, 9.20 and 9.22.

### Box 7.I Combining and comparing observations

Climatic information, as indicated above, is needed on a variety of space scales, from specific points to broad regions. Our observational instruments are to a good approximation either surface-based and point-specific or space-based giving areal averages. The two are complementary and are often used together. However, in combining them it must be borne in mind that they do not measure the same things, and that they are not interchangeable. These concerns are considered here.

*7.I.1 Satellite views and surface views*

Throughout the text we have used observational information obtained both from satellites and from ground-based sensors. Even when these both appear to be measuring the same thing, there are in fact differences which can be important. The specific case for temperature, where the actual physical entities are different, was mentioned in Box 3.I. However, the different viewpoints can give different perceptions and different values (Box 4.I). This can be illustrated by considering observations of cloud amount (Figure 7.I.1). Neither the satellite nor the surface values are inherently superior, although the surface one might be more useful if we were considering sunshine totals, or visibility, at the point, while the space-based one is probably more useful if we are concerned with global cloud amounts.

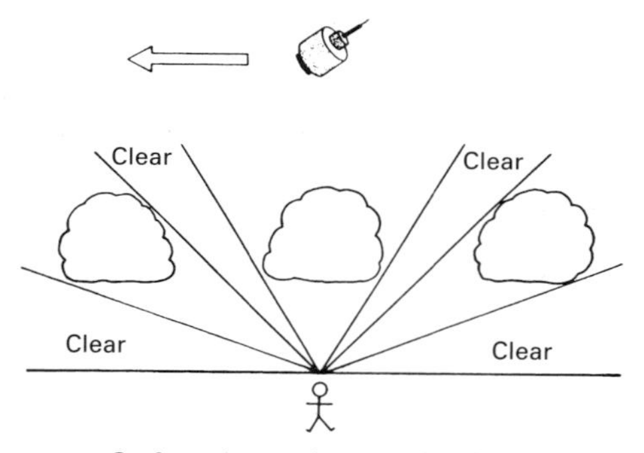


*Figure 7.I.1* Schematic illustration of the way the same cloud configuration can result in different cloud amount observations when surface and satellite retrievals are compared. The satellite, which passes rapidly over clouds obscuring about half the land surface, reports 50% cloud cover, while the surface observer, who sees the sides of the clouds and hence less than half the dome of the sky, reports 75% cloud cover.

*7.I.2 The problem of scale*

It is easy to take a set of observations from a group of surface stations and average or otherwise combine them to get an areal value, as was done for Figure 7.A.1. The result is not, however, the same as would be obtained from a single satellite measurement of the same area. The difference can be explained using a simple, rather stark and extreme example. Imagine a mountainous area having a group of surface observing stations. Most of these are likely to be in valleys, largely because the instruments must be observed and maintained by humans, who tend to live in valleys, not on mountain tops. Hence the result, if we average the observations, is certainly an areal average, but biased towards valley conditions, rather than representing the whole mountain area. The satellites are likely to take a much more even-handed view, although what is actually seen by the sensor is not always immediately apparent. Although more sophisticated and expensive surface instruments, less dependent on humans, can be deployed in rugged and remote locations, the more even network may minimise but cannot remove the differences.

The satellite observations themselves, of course, do not give a uniform coverage of the whole field of view. Certainly the satellite observation integrates information from all areas of the view, but the spatial resolution for areas near the edge is less than that at the centre. Consequently conditions near the centre of the view are given more weight and the satellite observation is thus biased towards this central area.

Overall, therefore, it is very difficult to obtain a true areal value for any climatic element. For those which do not vary too rapidly with distance, it is likely that good approximations can be made without too much trouble. This is true for temperature, which is spatially conservative even on as short a time scale as daily values. Rainfall displays the opposite situation. Short-term, say hourly or even daily, totals can vary very rapidly with space, and it is difficult to produce areal values. As the time scale increases, and we begin to consider monthly or annual precipitation totals, the spatial variability decreases, and the problem becomes somewhat simpler. Nevertheless, all areal values must be treated with caution.

# CHAPTER 8 Tropical weather and climate

Tropical climates are usually characterised as climates where there is no true temperature distinction between summer and winter. They normally occur between 30°N and 30°S. This is the area of the Hadley cell, so that much of the weather is controlled by the prevailing barotropic conditions. In some areas the weather and climate can be summarised almost completely simply by considering this cell. However, in other areas mid-latitude influences are superimposed, creating the monsoon climates. Thus it is very convenient to divide this tropical area into two distinct, but interlinked, climatic types, which are treated in separate sections below.

There are two additional features of the tropics which we consider in this chapter: hurricanes and the El Niño/Southern Oscillation (ENSO) phenomenon. Both have their origins in the tropics, both have impacts in the tropics, but both also have major consequences for the weather of the mid-latitudes. Hence, while considered as tropical conditions, they serve as a reminder that all of the features we tend to treat separately in this book are in fact interlinked to create the patterns of climate and climate variability of the planet as a whole.

## 8.1 Tropical climates – the Hadley cell

A Hadley cell, part of the general circulation (Figure 7.3), occurs in each hemisphere. The weather is very closely related to the airflow patterns (Figure 8.1). Air flows equatorward at low levels in each cell. These airflows converge at the Intertropical Convergence Zone (ITCZ) and rise to create cloud and precipitation. The latent heat released during this ascent provides much of the energy needed to continue the whole cellular circulation. The air ascends towards the tropopause where it diverges and begins to flow poleward. As it flows, it cools by long-wave radiation loss,

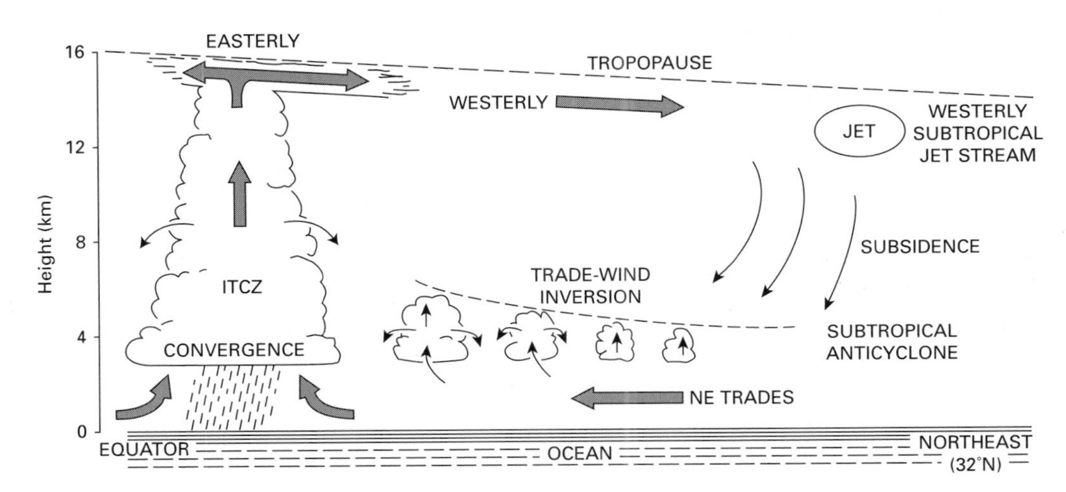


*Figure 8.1* Schematic diagram of the air flow and associated weather in the Hadley cell regime.

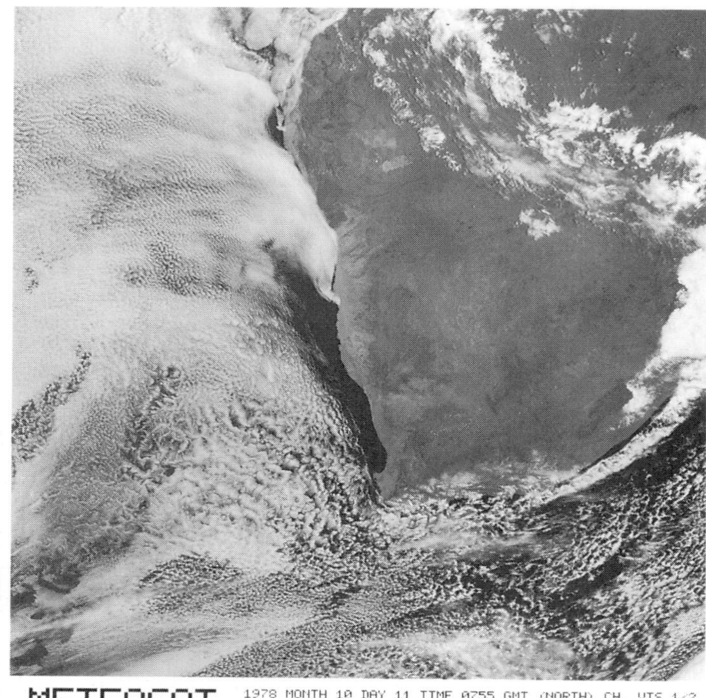


*Figure 8.2* Offshore fog and low-level stratus and stratocumulus adjacent to the Namibian coast (visible METEOSAT image for 11 October 1978). (METEOSAT image supplied by the European Space Agency.)

increases in density and descends. Although the descent covers a wide area it is particularly concentrated around 30° latitude. In the upper troposphere at these latitudes the interaction between Hadley cell air and that converging from mid-latitudes, combined with a strong Coriolis force, leads to the creation of a subtropical jet stream. This has a major influence on mid-latitude weather, and will be considered in the following chapter. Within the Hadley cell the descending air is adiabatically warmed, so that it arrives back at the surface as a dry, cloudless airstream. Once at the surface the air diverges, some flowing as the **trade winds** towards the ITCZ to complete the cell. The trade winds are, of course, influenced by the Earth's rotation, although there is not complete deflection, so that in the Northern Hemisphere the near-surface flow is the northeast trades. The airflow aloft, known as the counter trades, is from the southwest. In the Southern Hemisphere the corresponding winds are the southeast trades and the northwest counter trades.

The major area of cloud formation in the tropics is around the ITCZ, which is a region of instability. Virtually all of the rest of the tropics is subject to more or less persistent subsidence inversions in which warm descending air traps somewhat cooler air right at the surface. This feature is usually best developed, and closest to the surface, at the outer, poleward limits of the Hadley cells (see Figure 4.9). However, it persists throughout much of the trade-wind area, although rising to a higher level as the equator is approached. This feature is known as the **trade-wind inversion**. Any tendency towards instability caused by surface heating can rarely overcome this dynamically created feature. Although the surface trade winds are able to pick up a great deal of moisture when they flow over the ocean, they cannot release it, or the associated latent energy, until they reach the ITCZ. Hence high humidities but low precipitation amounts are a feature of several tropical areas.

A small amount of cooling, however, can easily bring the moisture-laden trade winds to saturation. Fogs can occur when the air flows over a cool surface. Such fogs are common, for example, when the air passes over the cool ocean currents off the coasts of Peru and northern Chile, or off the Namibian coast (Figure 8.2). This can be a frustrating sight for

*Figure 8.3* METEOSAT image (visible channel) for 7 July 1979 showing clearly the position of the ITCZ identified by the band of cloud close to the equator over Africa and the adjacent Atlantic Ocean. (METEOSAT image supplied by the European Space Agency.)

dwellers in these desert areas, although at higher elevations, as Section 4.A.2 indicates, useful cloud water may be extracted. Topographic barriers can also lead to orographic cloud and fog in the trade-wind zone. However, significant vertical development will occur only when the barrier is sufficiently large to break through the inversion. This is well illustrated by conditions on the main Hawaiian islands, clustered around 20–23°N and reaching altitudes well in excess of 1000 m. Orographic uplift over the upwind, northeast-facing slopes commonly leads to annual precipitation totals around 10 000 mm. Indeed, in Chapter 5 we noted that the station on Mount Waiale-ale can lay claim to being the wettest station in the world. At low altitudes and on lee slopes in Hawaii, in contrast, annual precipitation totals may approach two orders of magnitude less than this.

Thus the only major precipitation-forming mechanism of wide geographical extent in the Hadley cell is the ITCZ itself. The ITCZ occurs in the low pressure region circling the Earth near the equator and commonly called the **equatorial trough**. This has traditionally been characterised as a rather wide zone of cloudy, near-calm conditions, the **Doldrums**. However, it appears that gentle easterly winds, rather than calms, predominate in many areas. Further, the ITCZ itself is not a complete cloud band, but a series of fairly well developed cloud clusters separated by larger clear sky areas (Figure 8.3). Within it the cloud characteristics show great spatial and temporal variability.

The precipitation-producing clouds in the ITCZ are almost entirely convective. They are organised on two scales: small groups of convective cells with a seeming random distribution within the ITCZ, which remain as small-scale features; and larger and better-organised cloud groups exhibiting some organisation and having the potential to persist and intensify. The

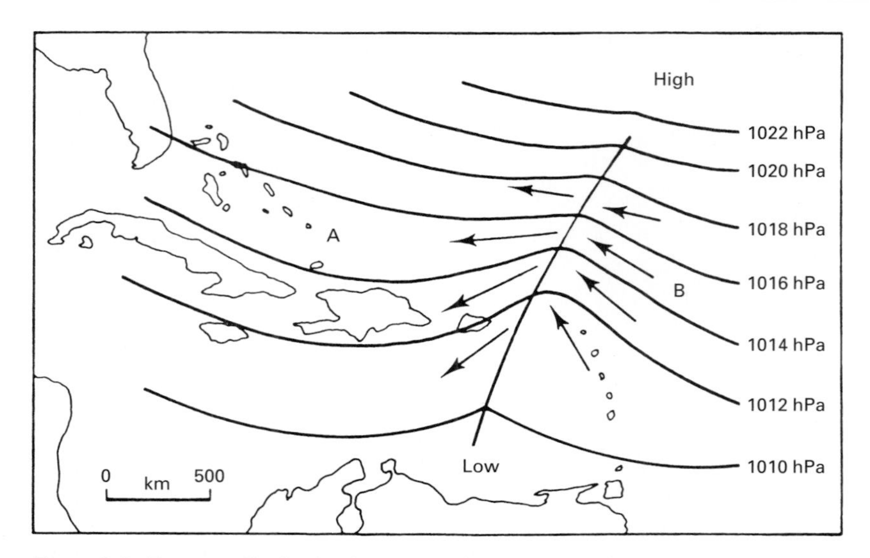


*Figure 8.4* Pressure distribution in an easterly wave in the Caribbean. Winds blowing from the direction of B converge towards the wave; they diverge in the region of A. (This feature is illustrated in the geostationary satellite image in Figure 8.18). (From Critchfield, 1983.)

major example of the latter are African **easterly waves**. These have long been observed in the western tropical Atlantic (Figure 8.4), but are known to originate frequently off the African coast. These are similar to, although by no means as well defined as, the synoptic-scale depressions of mid-latitudes (Chapter 9). They are troughs of surface low pressure which slowly move westward, being weakly defined at the surface but extending some distance into the atmosphere, and sloping away eastward with height. Behind the trough cumulonimbus clouds with intense rain and thundery showers occur. There may be about 60 of them in a typical year, each wave lasting three or four days and travelling some 2000–2500 km westward within the trade-wind flow. While dynamic processes over the tropical ocean appear to play the major direct role in easterly wave formation, upstream tropospheric conditions over Africa also have an influence. Variations in the north–south temperature gradient from the Guinea coast into the central Sahara modify the position and intensity of the low level jet which links, in a way not yet fully understood, with the wave formation process. Using climatological averages, the situation for formation is favourable during the April–October period. Not only are ocean waters warm, but also the location of the ITCZ over the Gulf of Guinea (Figure 8.5) creates a suitable north–south temperature gradient. Since these waves appear to be precursors of Atlantic hurricanes, they play a role in hurricane forecasting, a topic discussed in Section 14.2.

### 8.1.1 Seasonal movement of the ITCZ: tropical moist climates

In July the ITCZ is likely to be around 25°N over the Asian continental interior and 5–10°N over the oceans. The average January position is at about 15°S over land and close to the equator over water (Figure 8.6). This reflects the influence of continentality, including the unequal distribution of land and water between the hemispheres, on the location of maximum solar heating and moisture availability (Figure 8.5). Since the ITCZ is effectively the only rain-producing system for most of the tropics, this seasonal variation is highly significant for the description and analysis of climate types within the tropics. For stations near the equator the ITCZ is always fairly close, so its influence is felt in most months. Thus there are no dry periods. The detailed seasonal distribution may, of course, be influenced by the exact position of the ITCZ as well as by local topography. For example, the distribution at Fanning Island, a low-lying island in the Pacific Ocean, reflects the dominance of the nearly overhead ITCZ in the Northern Hemisphere

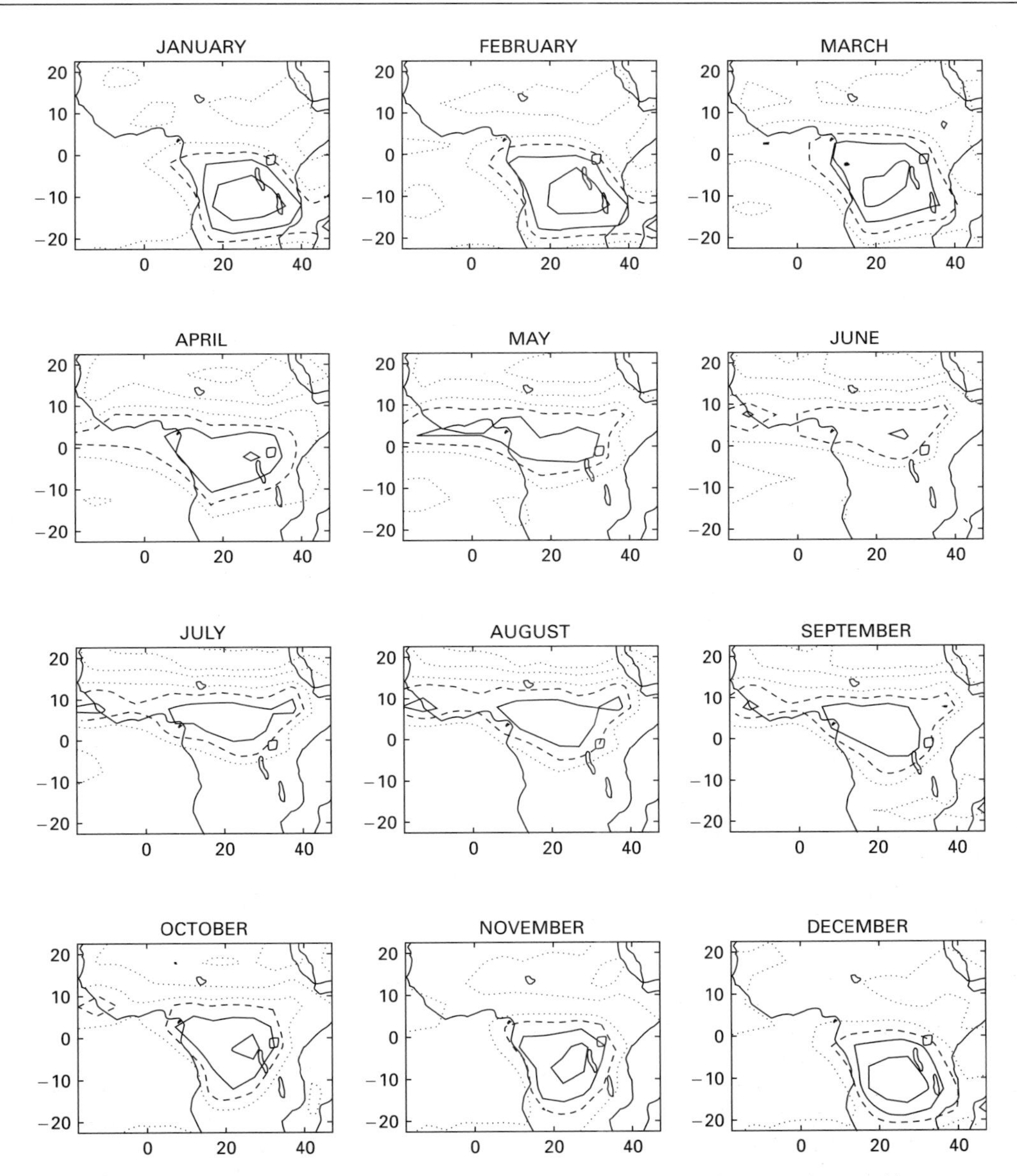


*Figure 8.5* Mean monthly position of the ITCZ over tropical Africa as indicated by the monthly average outgoing long-wave radiation for the June 1974–December 1991 period. The deep convective clouds of the ITCZ, with their cold, high tops emit small amounts of radiation compared to the surrounding cloud-free areas, where most of the radiation emanates from the warmer surface. The dashed line (235 W $m^{-2}$) represents the convection centre, while the dotted lines show 250, 265 and 280 W $m^{-2}$ in sequence. (After Moron, 1995.)

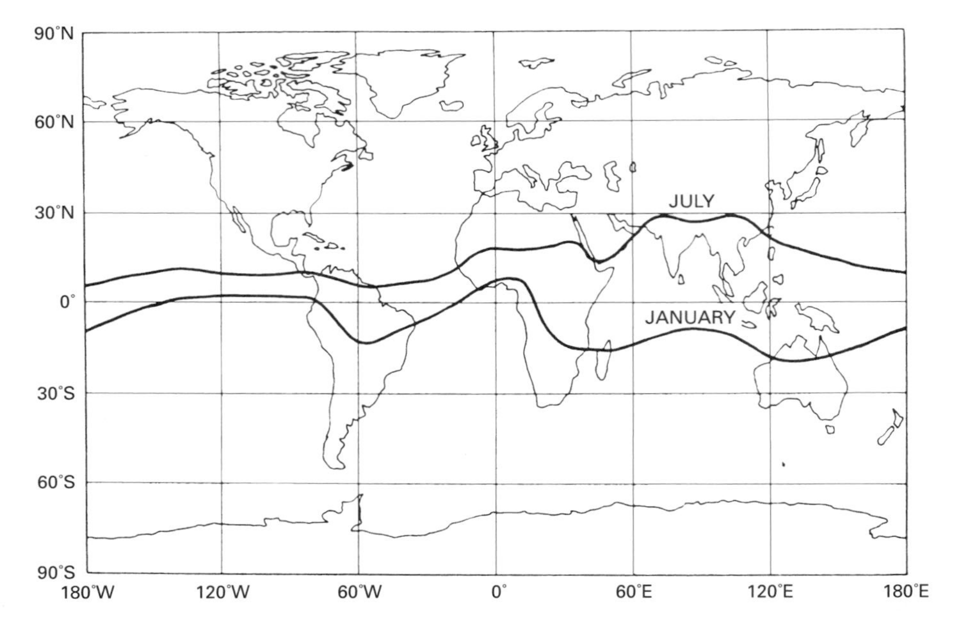


*Figure 8.6* Mean positions of the Intertropical Convergence Zone (ITCZ) in January and July. (From Critchfield, 1983.)

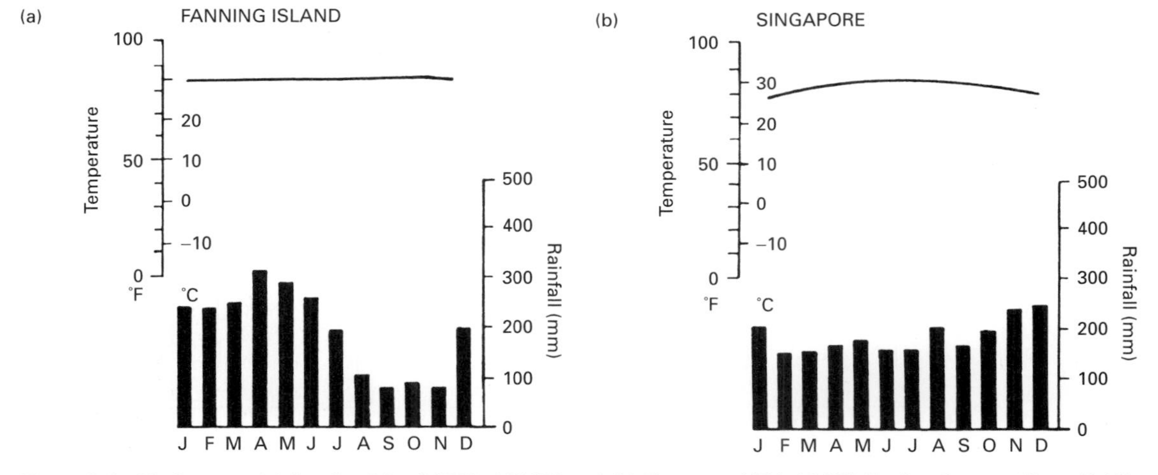


*Figure 8.7* Af climates: (a) Fanning Island (4°N, 160°W) and (b) Singapore (1°N, 104°E) (for location see Figure 7.11). (From Tanner, 1971.)

winter (Figure 8.7a). In contrast, Singapore (Figure 8.7b) has a more even precipitation distribution, partly as a result of orographic effects. At both stations temperatures remain high throughout the year, and the resulting climate type is Af.

Away from the equator, rainfall is concentrated in the summer, when the influence of the ITCZ can be felt. The further one progresses poleward, the smaller the total precipitation and the shorter the length of the rainy season. A coastal station such as Calcutta, India (Figure 8.8a), has a relatively long season and high totals partly because of the proximity to the ocean and partly because of its position relative to the ITCZ. Temperatures remain high throughout the

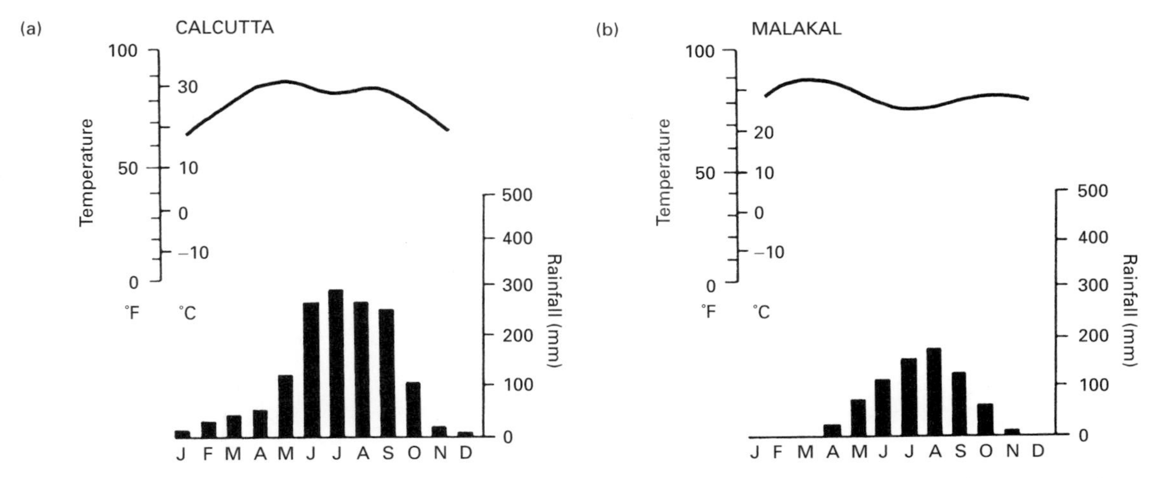


*Figure 8.8* Aw climates: (a) Calcutta (23° N, 88° E) and (b) Malakal (9° N, 31° E) (for location see Figure 7.11). (From Tanner, 1971.)

year, although Calcutta's location at 23°N leads to at least an embryonic summer and winter. Thus this is an A climate, with the distinct wet and dry seasons creating an Aw climate. Malakal, Sudan (Figure 8.8b), falls within the same group. Being closer to the equator, the seasonal temperature changes are not as marked, minimum temperatures occurring during the cloudy conditions when the ITCZ is most nearly overhead. It has a very well developed wet/dry regime. Only in the summer, with the ITCZ nearly overhead, does significant precipitation occur. During the other seasons there are no mechanisms for releasing any moisture within the trade winds, although in this particular situation the trades, having originated over the deserts to the north, are themselves rather dry. Indeed, Malakal is a station that is close to the desert margin, but it can still be regarded as an A climate since the summer precipitation is sufficient to maintain a positive annual water balance.

The difference between these two tropical climatic regimes is readily apparent when the components of the energy budget are compared (Figure 8.9). For both the Af and Aw climates the net radiation, $Q^*$, is approximately uniform throughout the year. Figure 8.9 shows a composite typical of the whole regime. If individual stations were emphasised there would be slight maxima during the times of overhead sun, perhaps compensated by a decrease resulting from increased cloud cover, leading to the specific temperature regimes indicated by Figures 8.7 and 8.8. The two climates differ markedly, however, in the disposition of this net radiation. The continuously wet regime maintains a low Bowen ratio throughout the year, evapotranspiration is continuously high and water is never limiting. The Aw climate, however, has one season when the Bowen ratio is approximately equal to that of the Af region, but the other season is one of water stress, with a high Bowen ratio. This becomes increasingly marked as we move into the true desert, B type, climates.

### 8.1.2 Desert climates

Only a few kilometres away from the Aw station at Malakal, the station at Hillet Doleib (Figure 8.10a) is classified as a desert climate, BSh. The atmospheric conditions of the two places are essentially the same; the only difference is a slight decrease in annual total precipitation at Hillet Doleib. This is sufficient to create a net annual water balance that is negative, leading to the change in classification. Certainly the outer limbs of the Hadley cell circulation, areas which are rarely influenced by the ITCZ, are desert areas. In the case of Hillet Doleib, there is some ITCZ-associated rainfall and a BS climate results. As one moves poleward this influence continues to decrease and the true desert, BW, is encountered (Figure 8.10b). The deserts created by the descending limb of the Hadley cell are generally hot deserts, BWh or BSh, and are concentrated around 30° latitude. Cold deserts, occurring farther poleward, may be spatially connected to the hot ones, but usually have a mechanism of formation that is not connected with the Hadley cell circulation. They may also lack the short wet

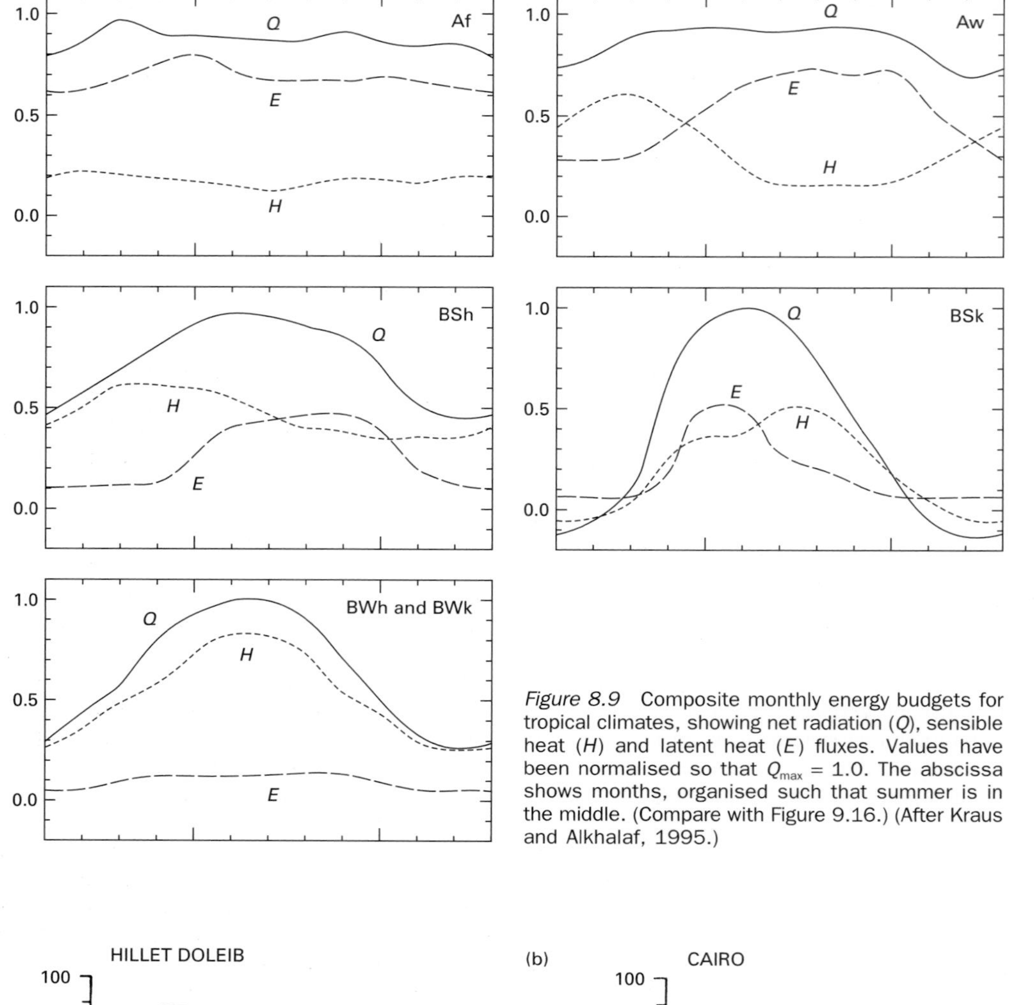


*Figure 8.9* Composite monthly energy budgets for tropical climates, showing net radiation ($Q$), sensible heat ($H$) and latent heat ($E$) fluxes. Values have been normalised so that $Q_{max}$ = 1.0. The abscissa shows months, organised such that summer is in the middle. (Compare with Figure 9.16.) (After Kraus and Alkhalaf, 1995.)

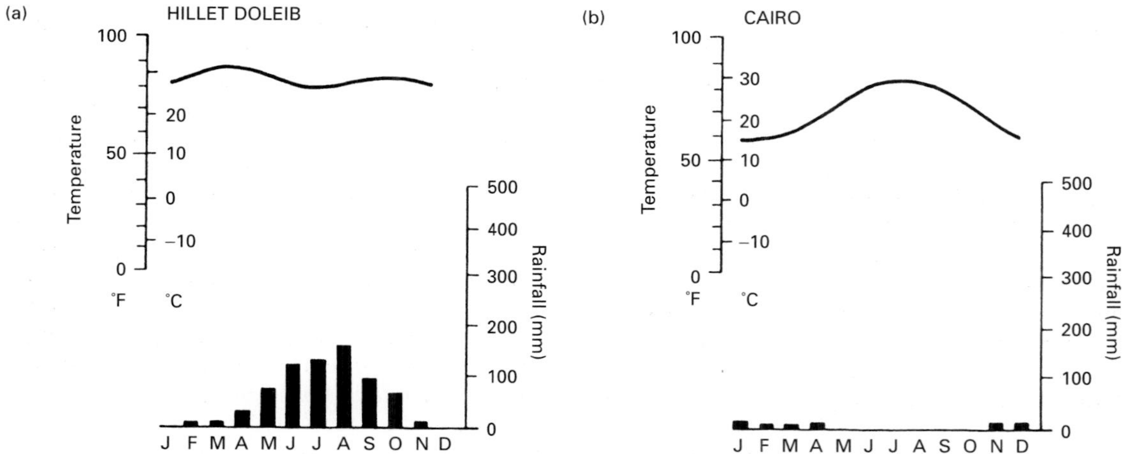


*Figure 8.10* Hot desert (BS and BW) climates: (a) Hillet Doleib (9°N, 32°E) and (b) Cairo (31°N, 31°E) (for location see Figure 7.11). (From Tanner, 1971.)

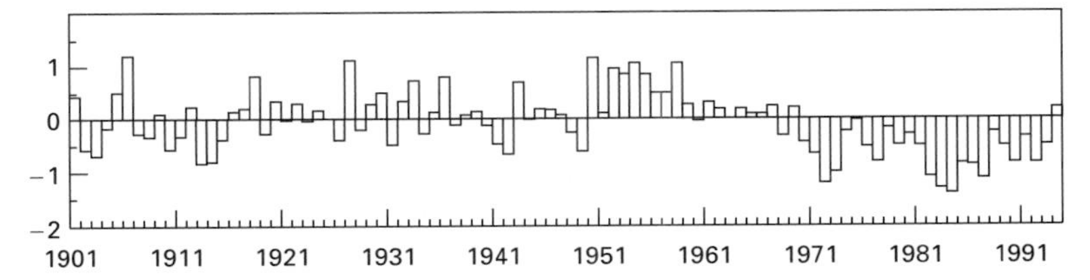


*Figure 8.11* Annual rainfall index for the Sahel region of Africa. The index is derived from observational records for a group of 22 stations, standardised to account for the average precipitation and variability at each station. The region used is indicated in Figure 8.12. (After Nicholson *et al.*, 1996.)

season characteristic of the semi-deserts associated with tropical circulations.

Within the tropics the transition from moist through semi-desert to desert climates represents a gradual increase in the annual average Bowen ratio, and thus a change in the partitioning of energy between evaporation and surface heating. Semi-desert climates, whether tropical or not, have an annual value close to unity, with individual seasons being slightly above or below this value (Figure 8.9). Indeed, the mechanisms of desert formation are such that semi-deserts in hot regions tend to have one clearly marked season where the ratio is well below unity and one where it is above. In contrast, the semi-deserts in cold regions tend to have similar ratios throughout the year, and hence no division into wet and dry seasons. In the true desert climate the Bowen ratio is high all year round whether or not it is a hot or a cold desert.

The choice of Hillet Doleib and Malakal as typical stations for specific climatic regions, together with the discussion of the Bowen ratio, serves to emphasise that fine distinctions are likely to be more a function of a classification scheme and its rigid boundaries than a real change in climate between them. This underlines our earlier caution regarding the placement of faith in any one classification scheme.

### 8.1.3 Precipitation reliability – the Sahel

As the precipitation amounts decrease as one moves away from the equator, so does the reliability of the precipitation. In the extreme case, a station such as Cairo, Egypt (Figure 8.10b), may record an average annual precipitation around 25 mm, but this is very misleading because there may be several years without rain followed by one with a few intense storms and a rainfall of 100 mm or so. Within the wet/dry tropics, receipt of precipitation depends on the movement of the ITCZ. The amount of movement poleward, and the intensity of the zone as it moves, all vary from year to year. In the Sahel region of Africa, one of the largest areas with a wet/dry tropical climate, a run of wet years followed by a sequence of dry ones has happened with some regularity in the recent past (Figure 8.11). A sequence of wet years started in the early 1950s, encouraging settlement and agriculture. However, the late 1960s and early 1970s were dry. While individual years with low rainfall may create hardship, such a sequence of drought years may bring disaster (see Box 11.I).

The causes of these annual variations are by no means clear. In our discussion so far we have simply linked them to the Hadley circulation, and implied that this dynamic system dominates over any surface energy considerations. Certainly if we consider the surface energy budget (Figure 8.9), it is clear that as the dry season replaces the wet one, the net radiation increasingly drives surface heating. Beyond this, however, the magnitudes, or even the nature, of the processes acting are unclear. Surface heating will certainly foster unstable conditions, but in the absence of moisture and in the presence of the trade-wind inversion, clouds are unlikely to form. However, if the vegetation is well established, it maintains a surface with a lower albedo than the bare desert (Table 2.1), creating more unstable conditions with higher humidity. A recent model study, of the type discussed in Section 13.3, explored the implications of foresting part of the Sahel (Figure 8.12). The higher albedo of the new model deciduous forest enhanced surface heating in June, prior to the arrival of the ITCZ cloud (Figure 8.5). This created deep dry convection, which in turn enhanced the normal eastward drift of low level air in the equatorial trough, and increased the moisture inflow centred on 15°N. In July this increased moisture stimulated the local hydrological cycle, with the vegetation itself not only fostering movement of water from the soil into the air but also

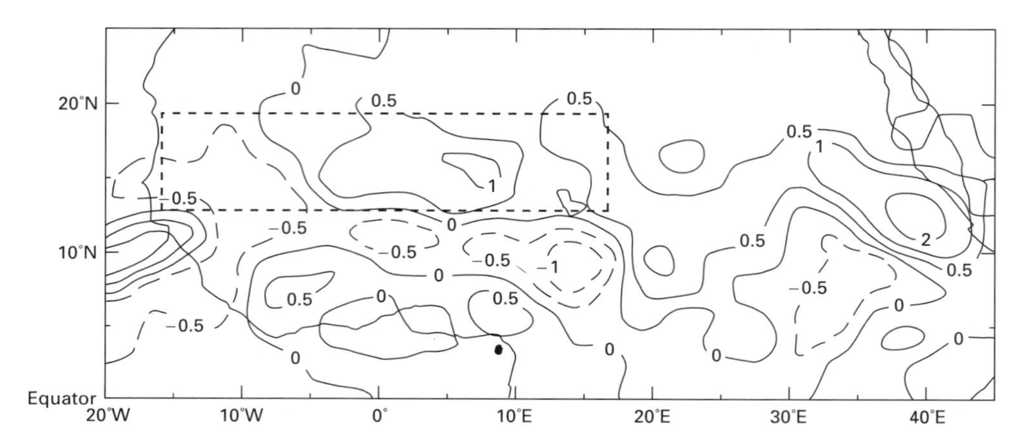


*Figure 8.12* Model-based estimates of changes in precipitation arising from a change in vegetation from the current semi-desert shrub to forest in the area indicated by the dashed box. Contours are in millimetres per day. The region within the dotted line is the area defined as the Sahel for Figure 8.11. (After Xue and Shukla, 1996.)

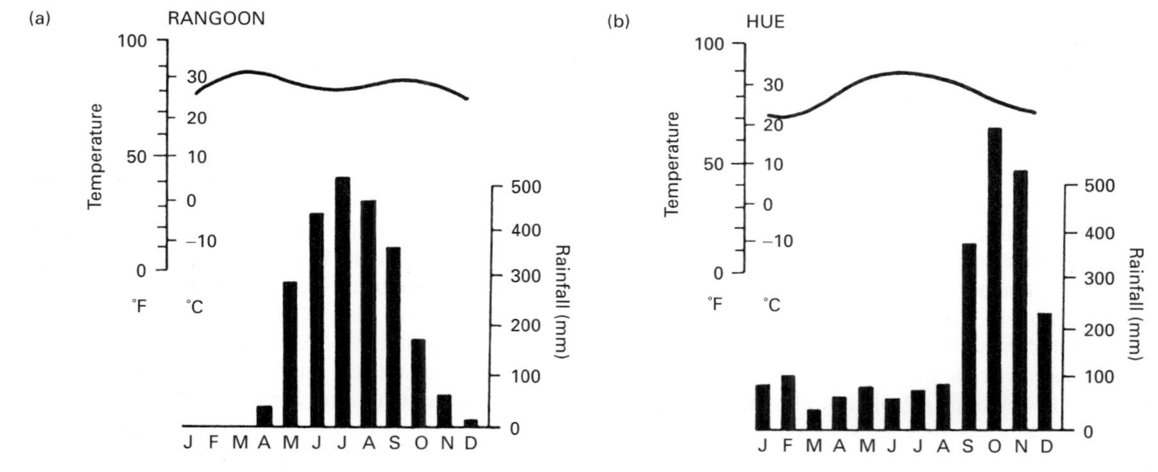


*Figure 8.13* Am climates: (a) Rangoon (16°N, 96°E) and (b) Hue (16°N, 108°E) (for location see Figure 7.11). (From Tanner, 1971.)

increasing turbulence and encouraging evaporation. The modified surface energy budget encouraged ITCZ activity somewhat stronger and more northerly than usual, and a precipitation increase followed in several areas (Figure 8.12). Thus it appears that the Sahel vegetation may serve as a catalyst in driving the seasonal ITCZ motions. However, much more investigation is needed before the process is fully understood, or before forecasts of seasonal precipitation can be made.

## 8.2 Tropical climates – the monsoon regime

Monsoon climates can be characterised as those where climate changes seasonally as a result of seasonal changes in the wind regime. In particular, the major monsoon areas exhibit twice yearly reversals in the prevailing wind direction. These lead to distinct wet and dry seasons (Figure 8.13). Although these graphs in Figure 8.13 look similar to those of Figure 8.8, the important difference is the amount of precipitation in the wet season. The very large amounts prevent any suggestion that these are dry climates, although there may be very definite dry periods. The similarity of the diagrams masks a further difference. The wet season for the Aw climates is one of convective precipitation associated with the nearby ITCZ within the general Hadley circulation. For the monsoon (Am) climates the differing wind regimes lead to changes in all elements of the climate. The wet season is one of hot, onshore humid winds and extensive cloud decks, while the dry season tends to be somewhat cooler with low humidity and offshore winds. For Rangoon, Burma (Figure 8.13a), temperatures are relatively low during the monsoon rains, largely because of the presence of clouds, and during the dry season, mainly because of the cooler winds. Temperatures are at a maximum during the intermediate seasons, when neither of these regimes is fully established. At Hue, Vietnam (Figure 8.13b), there is no clear dry season, orographic effects providing some rainfall throughout the year. Overall temperatures are lower than at Rangoon, at the same latitude, and there is a distinct seasonal temperature trend created not so much by the wind and cloud conditions as by the solar altitude variations.

### 8.2.1 Monsoon circulations

The prime cause of monsoon circulations is a thermally direct cell resulting from surface temperature differences (Figure 8.14). These differences are created by the continentality effect in conjunction with a suitable distribution of land and sea. While this favours extra-tropical regions, this type of circulation can only become well established in approximately barotropic conditions. Hence the potential regions of occurrence are limited. Thus, although a monsoon type of circulation may be developed over several areas in the relatively quiescent conditions of summer, it is rarely accompanied by the necessary wintertime reversal of wind direction required for a true monsoon regime. Most of the western hemisphere is unaffected, although there is a small area with seasonal wind reversals near the mouth of the Amazon in South America. Similarly, only small portions of the west coast of Africa display a monsoon climate. Although the whole tropical portion of the continent is potentially part of the regime, the Hadley cell normally dominates the circulation patterns. In general it is only the continent of Asia, and particularly the southern and eastern portions, that displays large areas of monsoon climate. Even here, other circulation features and surface effects may override the monsoon effect, so that by no means all of the region can be classified as monsoonal (Figure 7.10).

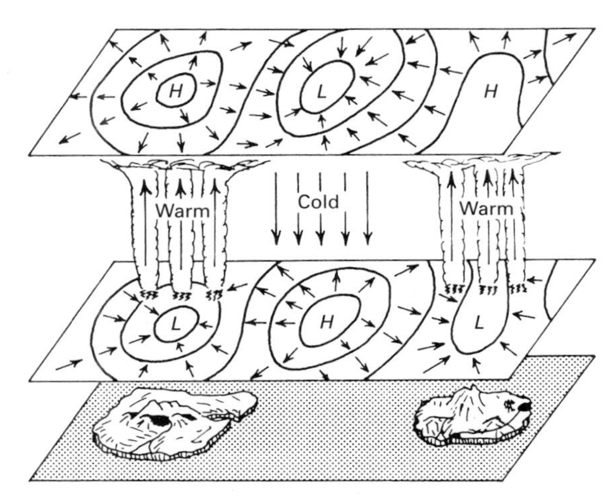


*Figure 8.14* Idealised representation of the monsoon circulations. The islands in the figure represent the tropical continents in the summer hemisphere. Solid lines represent isobars or geopotential height contours near 1000 hPa (lower plane) and 14 km or approximately 200 hPa (upper plane). Short solid arrows indicate the cross-isobaric flow. Vertical arrows indicate the sense of the vertical motions in the middle troposphere. (From Wallace and Hobbs, 1977.)

### 8.2.2 The Asiatic monsoon

The Asiatic monsoon is itself the result of a complex interaction between the distribution of land and water, topography and tropical and mid-latitude circulations. The simple model of Figure 8.14, reflecting the seasonal pressure changes over Asia (Figure 6.1), provides a good first approximation. This is particularly true for summer, where the circulation pattern is driven by low pressure centres over the northern part of the Indian subcontinent and northern Southeast Asia (Figure 8.15). Warm moist air is drawn into the thermally created low pressure areas of the continental interior where it rises, releasing both precipitation to create the wet season and latent heat to provide the energy necessary to continue the system. Over the oceans is a compensating descent of cold dry air. Once this pattern is established the onshore winds bring the monsoon rains. The moisture-laden winds are highly susceptible to orographic influences, so

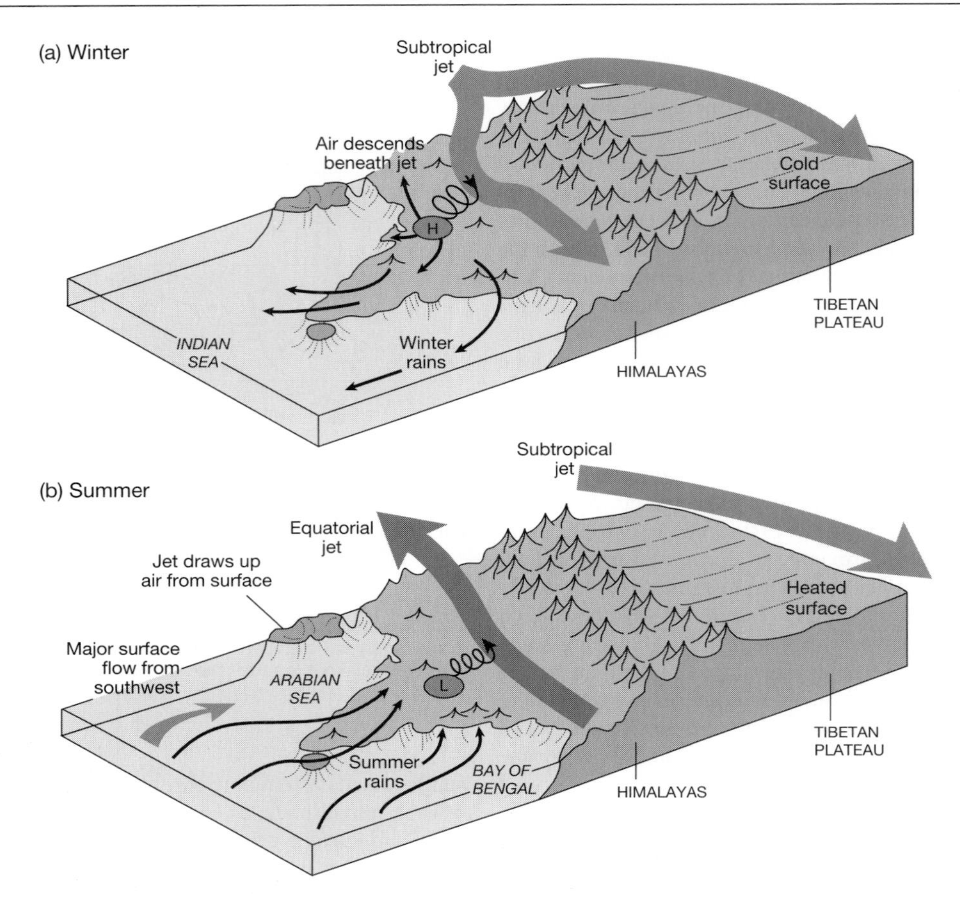


*Figure 8.15* Simplified diagram showing the airflow patterns over southern Asia in (a) winter and (b) summer. Note the change in position and direction of the jet streams, responding to seasonal changes in solar forcing and consequent changes in near-surface pressure patterns.

that the coastlands of India, for example, which are backed by mountain ranges, receive the most rainfall. The interior, of course, also receives rain since the whole area is one of rising air. The Himalayas effectively provide a barrier to the north, confining the circulation to the area south of them.

The detailed pattern of climates associated with this summer circulation is, however, much more complex. The Tibetan Plateau, at a height exceeding 4000 m above sea level, provides a high level heat source which appears to have a significant influence. Above the plateau is a high pressure region, which is part of the zonal subtropical high pressure belt. As the equatorial trough associated with the ITCZ moves to about 25°N over India at the height of the summer, a strong north to south pressure gradient is established. This is reinforced by a north to south temperature gradient and together they lead to the development of an easterly jet stream at a height of about 150 hPa (15 km). This jet extends from the South China Sea across Southeast Asia and central India into the Arabian Sea, and may reach the southeastern Sahara of Africa. The dynamic effects associated with this jet can lead to a precipitation enhancement to the north of the main axis in Southeast Asia and to a decrease to its south, over the Arabian Sea and the Horn of Africa (Figure 8.15).

With the onset of autumn, two related general changes occur. First, the thermal contrasts between land and water decrease, weakening the circulation. Second, the westerlies of mid-latitudes begin to migrate southwards. Eventually part of this westerly airstream blows to the south of the Himalayas and completely disrupts the tropical circulation. The easterly jet is replaced by strong westerly winds aloft. At the surface a north to south pressure gradient extending

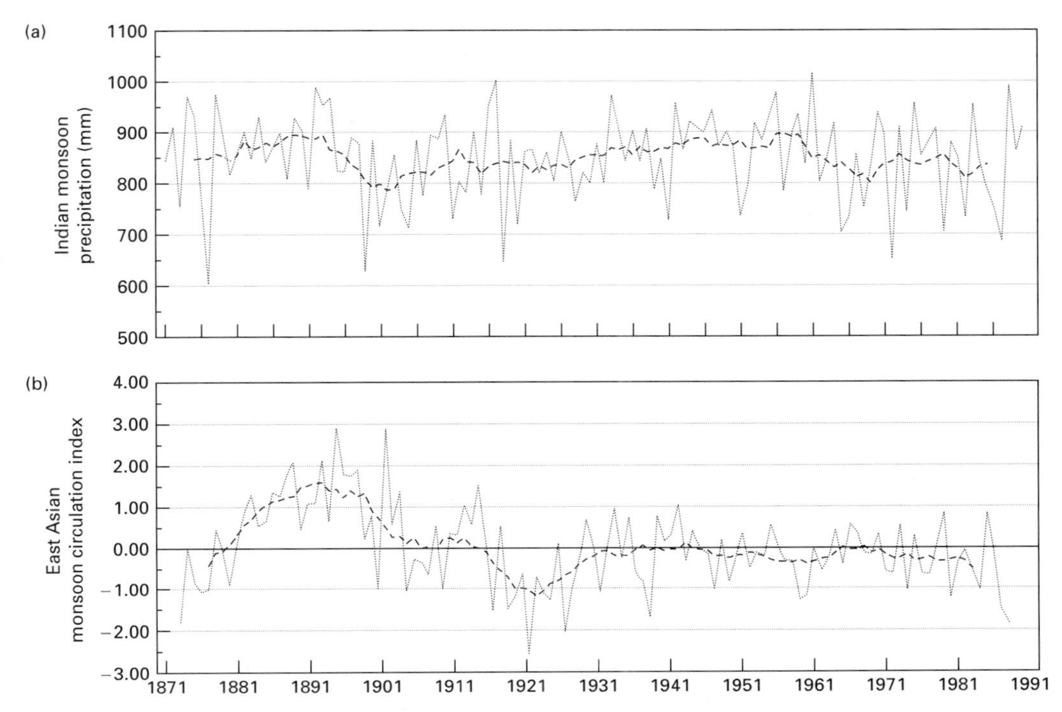


*Figure 8.16* Annual summer values of (a) average monsoon precipitation over India, based on the average of 366 observing stations, and (b) East Asia monsoon circulation index, based on pressure differences between 110 °E and 160 °E. In both cases the dotted lines indicate annual values, the dashed ones decadal running means. ((a) From Parthasarathy *et al.*, 1991; (b) from Shi and Zhu, 1996.)

from Siberia almost unbroken to the equator is established, with a high pressure centred over Siberia (Figure 6.1). The resulting airflow near the surface is from the north or northeast. This air combines with that resulting from subsidence below the westerlies to ensure that dry air covers the region, leading to the relatively cold conditions as noted for Hue. Over much of India, air has blown downslope off the Himalayas. Although this air has been warmed adiabatically, it still gives relatively cold conditions.

This type of pattern persists through the spring. As summer approaches, the conditions for the summer monsoon circulation are slowly developed. The onset of the actual circulation is delayed, however, by the continued influence of the westerlies south of the mountain barrier. As soon as the main line of the westerlies moves sufficiently far north that the flow south of the Himalayas ceases, the summer monsoon circulation starts. The very rapid establishment of warm moist airflow from the south creates the phenomenon known as the 'burst' of the monsoon. Cool, dry conditions are replaced by warm humid ones, with copious rainfall, almost overnight.

### 8.2.3 Reliability of monsoon rainfall

The time of the monsoon burst is highly dependent on conditions in mid-latitudes. In general, the later the burst occurs, the smaller is the total rainfall in the area. Thus there is great variability in the annual rainfall amount, with variability occurring on several time scales (Figure 8.16a). Spatially, the variability, as with the Sahel, is inversely related to the annual amount (Figure 8.17), so that while the rains rarely fail on the west coast, the northwest is highly susceptible to drought. Failure of the monsoon, or even its delay, significantly affects the agriculture of the region and thus the livelihood of millions of people.

Accurate prediction of the onset of the monsoon is thus of vital importance for the whole of south and east Asia. However, as suggested, this is a complex problem requiring an understanding not only of the circulation patterns of the tropics, but also consideration of mid-latitude circulations as well. Some consideration will be given later (Section 14.2) to the general possibilities of forecasting some months in the future. It is sufficient here to note that the East Asian circulation

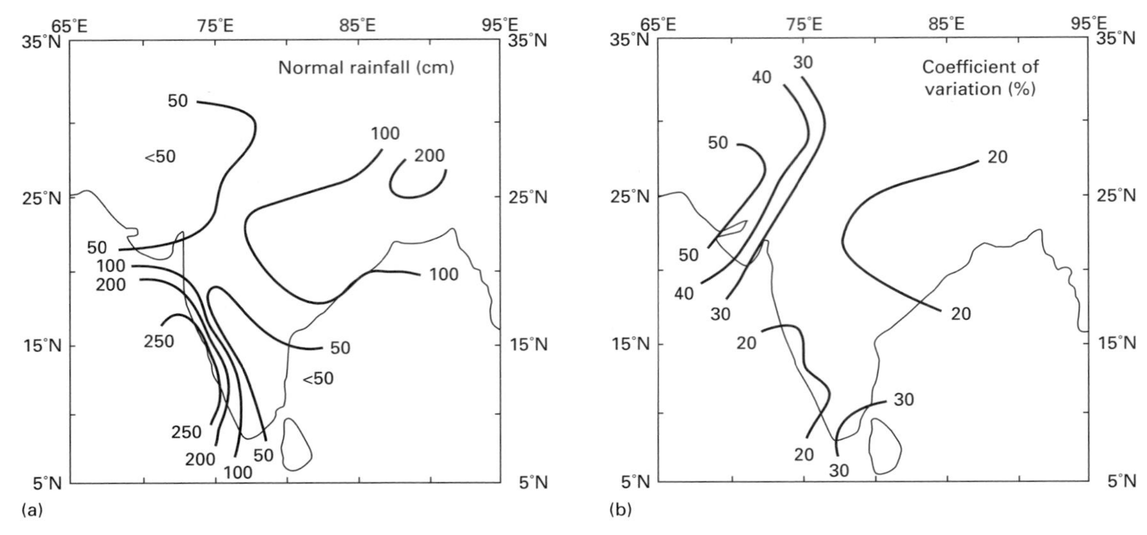


*Figure 8.17* Averages of (a) total monsoonal (June–September) precipitation over India and (b) variability of precipitation amounts (After Kripalani *et al.*, 1995.)

index (Figure 8.16b), where positive values are indicative of weak monsoonal circulations, has been related to Chinese rainfall amounts, and may be more generally applicable to the whole of the monsoon region.

## 8.3 Hurricanes

Hurricanes are intense circular vortices with winds over 34 m $s^{-1}$ (75 mph) spiralling around a low pressure centre (Figure 8.18). The centre is small, calm and cloudless. Surrounding this is a bank of convective cloud, usually twisted into a piled mass of cumulonimbus cloud through which air ascends in a spiral motion. When the air from each of these ascending regions reaches the tropopause, it moves outwards, creating a veil of cirrus which may extend several hundred kilometres radially out from the hurricane centre. These systems are known as 'hurricanes' in the Caribbean and the Gulf of Mexico regions, 'tropical cyclones' in the Australian region, 'typhoons' in the western north Pacific and 'cyclones' in the Bay of Bengal. 'Hurricane' has, however, become the accepted general name in meteorology. Similar storms, but having speeds below 34 m $s^{-1}$, are known as 'tropical storms'.

### 8.3.1 Formation and development of hurricanes

Hurricanes originate in the tropics and commonly increase in strength as they move westward and poleward. Commonly the maximum intensity is reached as they cross into mid-latitudes. Thereafter it is usual, but not universal, for the poleward component of motion to increase while the intensity decreases. Generally their energy is rapidly dissipated once they leave the tropics.

The region of origin over the tropical oceans is restricted to latitudes between about 7 and 15° (Figure 8.19). Equatorward of 7° the Coriolis force is too weak to initiate a circular motion, while poleward of 15° sea-surface temperatures are too low. Sea-surface temperatures greater than 27 °C are required to ensure sufficient evaporation to provide the latent heat release within the storm needed to maintain its energy. It is usually in the late summer and early autumn that sufficiently high temperatures are achieved, creating a distinct hurricane season.

The development of an organised cluster of tropical convective cells appears necessary for hurricane formation. Within the Atlantic Ocean, this organisation is often provided by African easterly waves (Section 8.1). It is estimated that about 60% of tropical storms and weak hurricanes, and 85% of strong hurricanes, develop in conjunction with them. There are also indications that these same waves may have a role in the formation of hurricanes in the eastern Pacific Ocean. The clouds must be organised in such a way that a distinct low pressure centre can be established over a sea surface. The low surface friction of the ocean prevents significant cross-isobaric flows, which would tend to fill in the central low pressure, but at the same time allows sufficient

*Figure 8.18* GOES East satellite visible channel image for 8 August 1980 at 1700Z showing hurricanes Allen in the Gulf of Mexico and Isis to the west of Mexico in the eastern North Pacific. The concentric cloud pattern spiralling about the central eye is clearly discernible in both hurricanes. Allen was a particularly intense hurricane with aircraft measurements of wind speeds higher than 80 m s$^{-1}$. Note also the area of activity in the eastern Caribbean which is associated with an easterly wave (see Figure 8.4). (Weather Satellite Picture Interpretation, Ministry of Defence.)

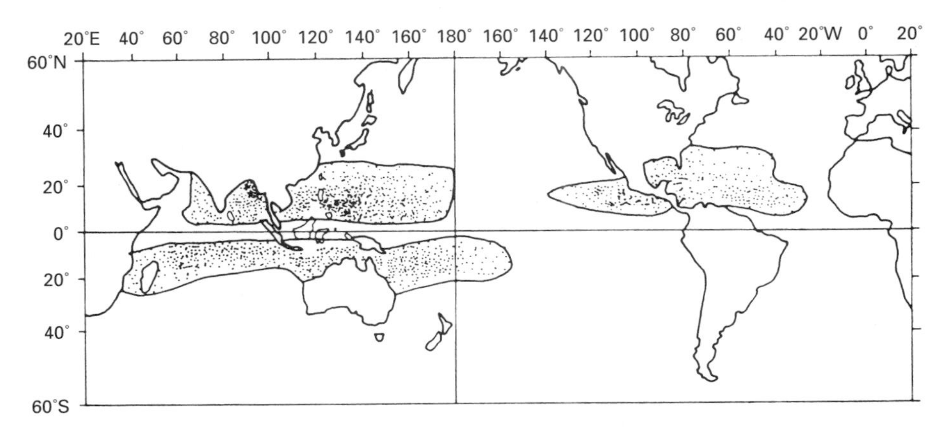


*Figure 8.19* Geographical distribution of the points of origin of hurricanes in the 20-year period 1952–1971. Each dot represents the first reported location of a storm which subsequently developed sustained winds of 20 m s$^{-1}$. About two-thirds of these storms eventually developed winds of hurricane force (>34 m s$^{-1}$). (Courtesy of NOAA.)

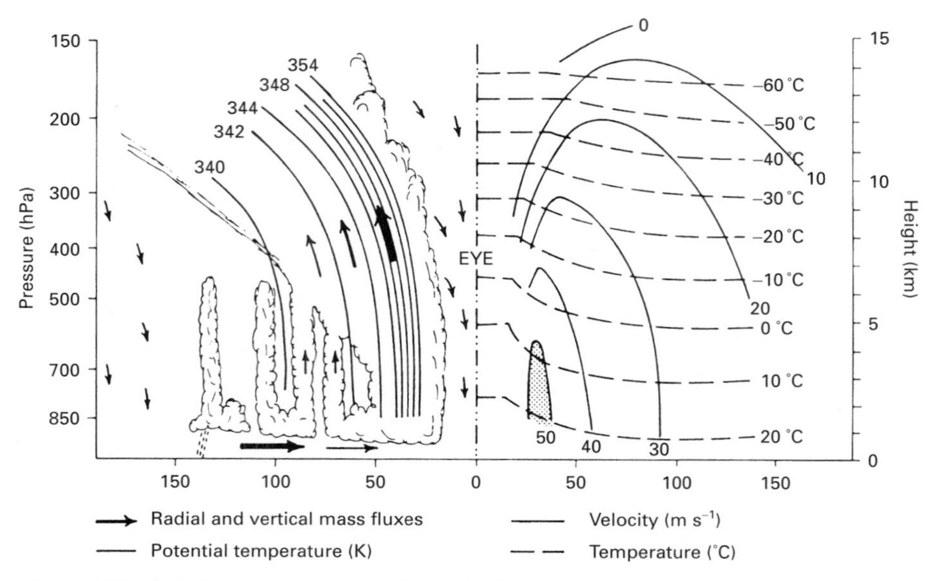


*Figure 8.20* Radial cross-section through an idealised axially symmetric hurricane. On the left the radial and vertical mass fluxes are indicated by arrows and equivalent potential temperature (K) by solid lines. On the right, isopleths of tangential velocity (m $s^{-1}$) are shown as solid lines (i.e. the speed of rotational flow around the 'eye') and temperature (°C) by dashed lines. (From Wallace and Hobbs, 1977.)

such flow to maintain the vertical motions within the storm. These vertical motions are generally sustained by **tropical upper tropospheric troughs**, located at about 200 hPa. These dynamic features, maintained near the tropopause by a balance between subsidence warming and radiative cooling, create vertical wind shear over the embryonic hurricane system, and force vertical ascent from above.

### 8.3.2 Mature hurricanes

From their source regions the storms usually start to move in a westward or northwestward direction (southwestward in the Southern Hemisphere), becoming better organised and usually stronger as they move. Eventually they have the 'classic' hurricane characteristics (Figure 8.20). In the centre is the region of descending air, giving cloudless, near-calm conditions known as the **eye**. Immediately adjacent to this is the area of upward spiralling winds, with a strong horizontal component and a smaller vertical one. Thus the convective clouds are twisted into a piled mass of cumulonimbus cloud. Some thunder and tornadic activity may occur as a result of this convection, although the frequency of this is subject to controversy. The spiralling cloud mass is densest immediately adjacent to the eye, giving the **eye-wall** cloud. Additional convective bands, also twisted into a spiral, occur further from the eye. Eventually the ascending air, some hundred kilometres from the eye, reaches the tropopause, creating a veil of cirrus.

In this mature phase, hurricanes contain sufficient energy to cut across the poleward limb of the Hadley cell and move into middle latitudes. Once there they tend to recurve under the influence of the prevailing westerlies and move towards the northeast (or southeast in the Southern Hemisphere). In theory, as they move away from their warm source waters they rapidly lose the energy source needed to sustain them. This should become particularly acute when they move over land, with its lower moisture supply and increased surface friction. In practice, although this is basically the case, hurricanes may travel a great distance, in a highly unpredictable direction, over mid-latitude areas (Figure 8.21).

*Figure 8.21* Hurricane paths over the North America region originating in the North Atlantic in the period 1886–1980. (a) The total for July (64); (b) the total for September (268). (Courtesy of J. Pelissier, NOAA, US Department of Commerce.)

(a)

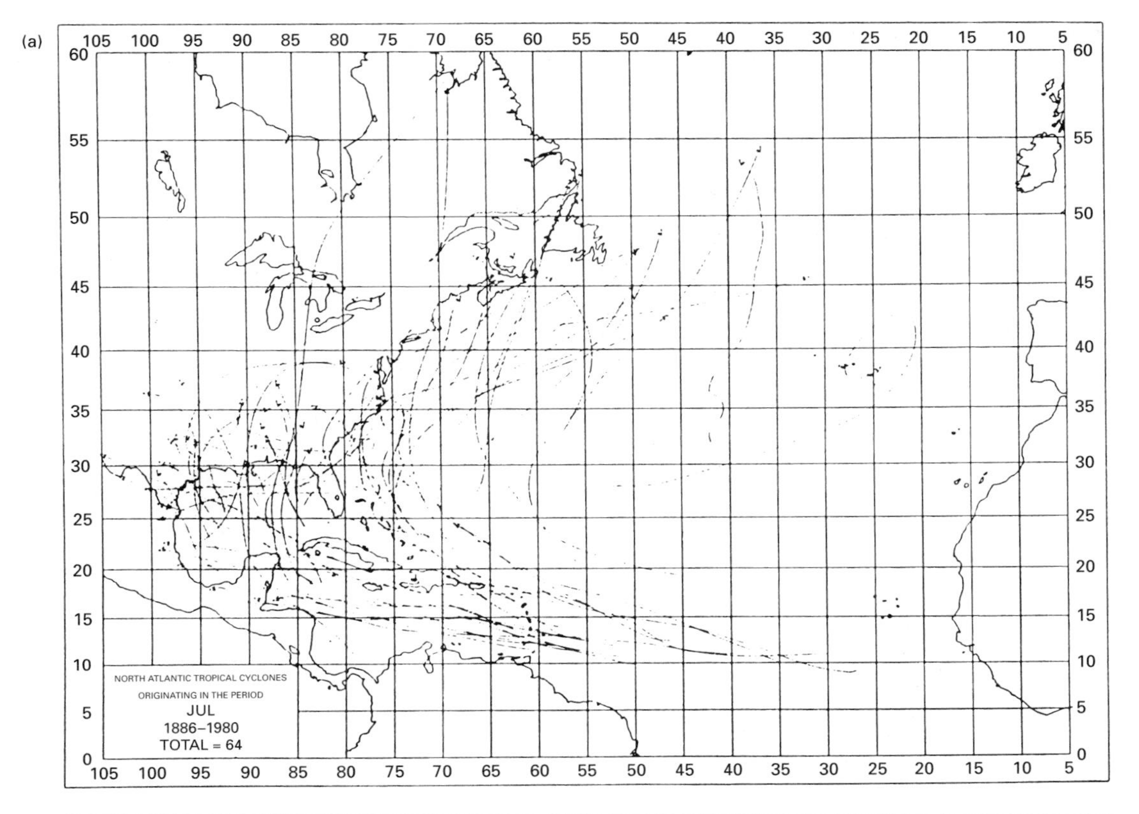

NORTH ATLANTIC TROPICAL CYCLONES
ORIGINATING IN THE PERIOD
JUL
1886–1980
TOTAL = 64
105 100 95 90 85 80 75 70 65 60 55 50 45 40 35 30 25 20 15 10 5
60 55 50 45 40 35 30 25 20 15 10 5 0


(b)

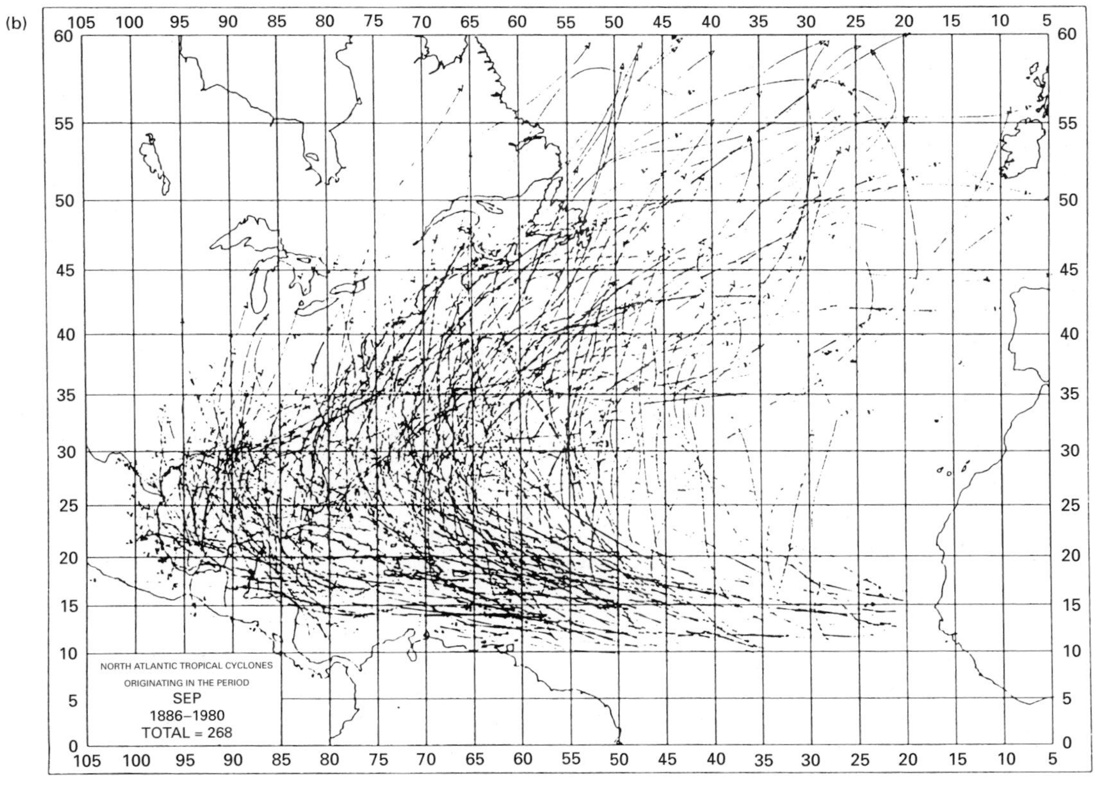

NORTH ATLANTIC TROPICAL CYCLONES
ORIGINATING IN THE PERIOD
SEP
1886–1980
TOTAL = 268
105 100 95 90 85 80 75 70 65 60 55 50 45 40 35 30 25 20 15 10 5
60 55 50 45 40 35 30 25 20 15 10 5 0

The idea of naming storms sequentially on an alphabetical basis originated in the Second World War, but it was not until 1978 that there was international agreement to use male as well as female names for these intense storms. The ocean source areas clearly differ since Atlantic hurricanes rarely get to the letter M (13 a year), whereas in the eastern Pacific the letter P (16 storms) is frequently used, and in the west Pacific the end of the alphabet is often reached.

### 8.3.3 The global role of hurricanes

We usually think of hurricanes as agents of destruction, and indeed one of the major weather modification efforts mounted soon after the Second World War was to modify either their intensity or their track. The effort, which involved the use of cloud seeding and provided a tremendous amount of information about hurricanes, was eventually abandoned, in part because of the realisation that the storms had a vital role in maintaining the general circulation and hence the present climate system. They are extremely efficient in moving excess energy away from the tropical regions towards the poles. As transient eddies they contribute a major fraction of this energy transport (Figure 7.5). Without it, there would either be a marked change in climate worldwide, or some other feature would be developed by the atmosphere to perform this transfer function. Hence it is now regarded as infinitely preferable to establish measures to ameliorate their impact (Box 8.A), rather than to attempt to remove them from the planet.

**Box 8.A Impacts: hurricanes and ENSO**

Throughout our discussion of tropical weather and climate, a major underlying theme has been the impact that the atmosphere has on human lives. In this section we look specifically at this in the context of ENSO and hurricanes.

*8.A.1 The effects of hurricanes*

Hurricanes are feared because of their destructive capabilities. The pressure gradients around the eye of the storm frequently produce sustained winds in excess of 50 m $s^{-1}$ (100 mph). This can have a direct destructive effect. Equally, or more, destructive in coastal areas are storm surges resulting from local uplift of the sea surface under the central low pressure (Figure 8.A.1). Finally, the rainfall rate associated with the rapid vertical air motions often exceeds 0.03 mm $s^{-1}$, and the total rainfall as a storm passes overhead may be in excess of 1000 mm, with the associated possibility of flash flooding in many mountainous areas.

Because of the role of hurricanes in global energy exchanges, the major human response to them must be amelioration of their impacts, not attempts to remove them. Amelioration takes two forms: providing appropriate structures in hurricane-prone areas, and providing timely warnings so that appropriate actions can be taken. The meaning of 'appropriate structure' will vary from area to area. In some cases it may imply the provision of additional strengthening (compared to non-hurricane regions) to buildings to ensure that they withstand the wind force. This may include, for example, roofs adequately clamped to walls, windows readily boarded, and walls well anchored to solid foundations. In other areas it may be best to adopt the opposite strategy, and make a minimal investment in a structure, anticipate loss during a hurricane, and rebuild afterwards. These are non-climatological decisions, but must be based partly on climatological information concerning features such as storm frequency, probability of winds exceeding certain thresholds, or expected storm surge height.

The second amelioration activity is the provision of timely warnings, allowing adequate time for the storm-proofing of buildings or the evacuation of threatened areas. This is an example, perhaps the most extreme example, of the three-stage forecasting problem considered in Section 6.A.1. Modern methods of tracking using satellites, combined with the increasing understanding of the physical processes within the storms, is allowing an increased lead time for warnings. This is still, of course, far from a completely exact science, and the farther ahead the forecast, the greater the margin of uncertainty.

*8.A.2 ENSO impacts*

It is becoming increasingly evident that ENSO is responsible for fluctuations in weather and climate, and thus for human consequences, in many parts of the globe. This is most noticeable when an El Niño event occurs. Such events have been occurring at

**Box 8.A (cont'd)**

(a)

(b)

*Figure 8.A.1* Hurricane destruction: the Roundtowner Motel in Panama City Beach, Florida (a) before and (b) after it was hit by hurricane Eloise in 1975. (Courtesy of National Weather Service, NOAA, US Department of Commerce.)

relatively regular intervals throughout the last 300 years (Figure 8.A.2), and presumably long before. The immediate impacts on the equatorial coastlands of South America, of course, led to the original identification of the phenomenon. In normal circumstances these coastlands are dry and rather barren. The cold upwelling ocean waters, however, are nutrient-rich and fish are abundant. During the early years of European colonisation the onset of El Niño was an event to be welcomed. It brought rain to the poor agricultural regions and ensured a good harvest. Fishing was not a major activity and there was little

**Box 8.A (cont'd)**

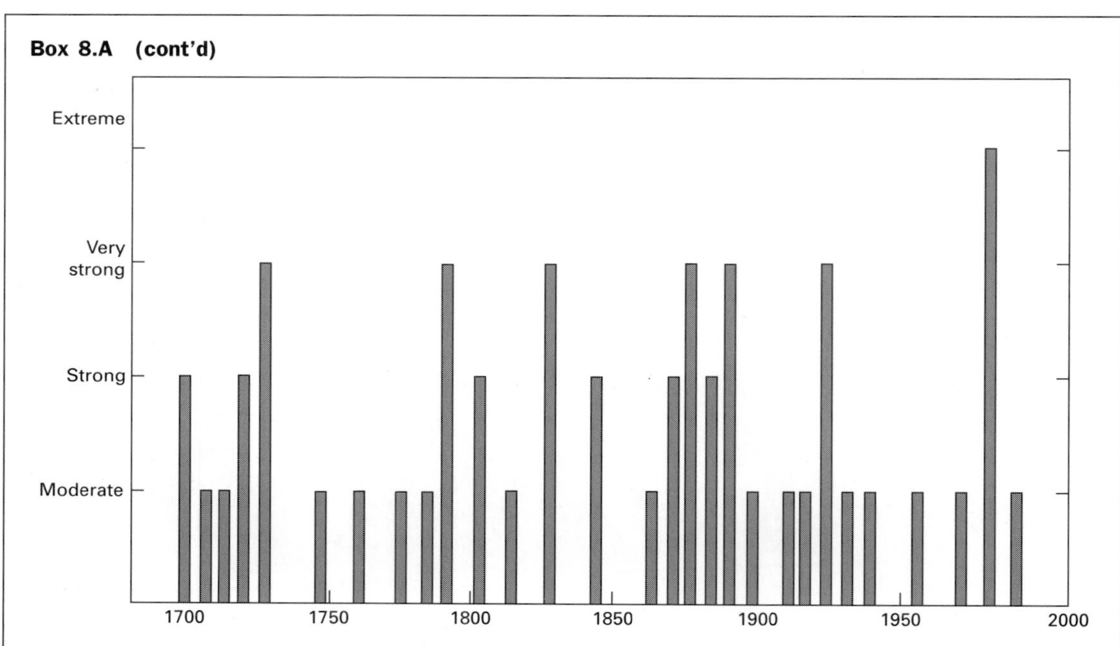


*Figure 8.A.2* Times of El Niño events. The magnitude of each event is estimated primarily from written records referring to conditions in the Peruvian coastlands. (After Glantz *et al.*, 1987.)

impact. Now the Peruvian economy in particular is highly dependent on fishing, while much of the original agricultural land has been urbanised for an increasing population. Hence, now an El Niño is a disaster, causing flooding and destruction on land, and economic collapse in the water. This disruption was particularly severe in the major events in 1982–1983 and 1997.

The 1982–1983 event was the first one where world-wide impacts were recognised and monitored (Figure 8.24). Estimates of the monetary influence were therefore possible (Table 8.A.1). However, these are very general estimates of the impacts associated only with drought (and the fires which they allowed), flooding from excess rainfall, and the estimated increase in Pacific hurricane activity. These were directly attributed to the weather created by El Niño. Indirect impacts, an example of which is given in Box 11.A, probably added much to the total monetary impact and the social consequences.

*Table 8.A.1* Economic impacts of the 1982–1983 El Niño event

| Category | Area | Impact (US$ million, 1983) |
|---|---|---|
| Floods | Bolivia | 300 |
| | Ecuador, N. Peru | 650 |
| | Cuba | 170 |
| | US Gulf States | 1270 |
| Hurricanes | Tahiti | 50 |
| | Hawaii | 230 |
| Drought/fires | Southern Africa | 1000 |
| | S. India, Sri Lanka | 150 |
| | Philippines | 450 |
| | Indonesia | 500 |
| | Australia | 2500 |
| | S. Peru, Bolivia | 240 |
| | Mexico, Central America | 600 |

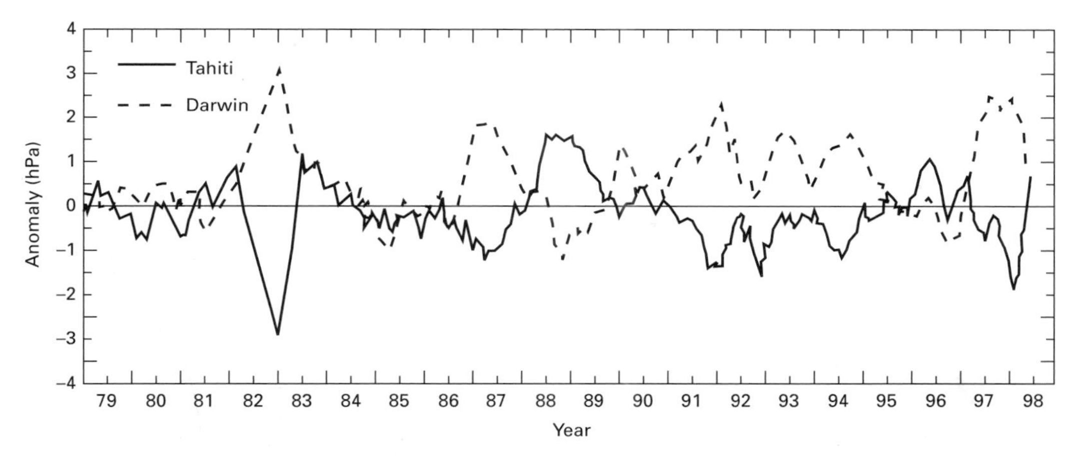


*Figure 8.22* Sea-level pressure anomalies (departures from normal) at Darwin and Tahiti (1937–1984). El Niño events commonly occur when the difference between the two is high, although not all periods of great difference lead to such events.

## 8.4 El Niño and the Southern Oscillation

One of the most exciting recent developments in climatology has been the increased understanding of the El Niño/Southern Oscillation (ENSO) phenomenon. Through efforts combining observations, notably from satellites, and theory, especially from coupled ocean–atmosphere models, we have come to understand a great deal about the processes occurring in the equatorial Pacific region. Further, through analyses of climatic events and impacts, we have identified links between the conditions in this region and those in several others, even those remote from the Pacific. Finally, we have started to use this information to create long-term climate forecasts for several months in advance. The research in the Pacific has stimulated similar activities in the Atlantic Ocean, and the first signs are emerging that there may be a comparable atmosphere–ocean system there. All of this is new and tentative, but represents one of the great areas of climatic advance in the last decade.

### 8.4.1 ENSO in the Pacific

The ENSO phenomenon represents a set of interlinked actions involving atmospheric and oceanic circulations, and creates atmospheric conditions having human consequences over numerous portions of the globe. The Southern Oscillation portion refers to the slow see-saw in pressure across the equatorial Pacific first noticed by Sir Gilbert Walker in the early years of the twentieth century. Now this is formalised as the **Southern Oscillation Index**, the pressure difference between Tahiti in the mid-Pacific and Darwin in northern Australia (Figure 8.22). These two stations represent the Southeast Pacific area of high pressure and the Indonesian low, respectively (Figure 6.1). There is thus a Hadley-like circulation, modified by ocean conditions and termed the **Walker circulation** (Figure 8.23a). Near the coast of South America the winds blow offshore and, through surface drag effects, move the ocean surface waters westward. These waters, warmed by the positive energy budget in this region, are replaced by cooler water upwelling from the deeper ocean, so that there is a temperature gradient of some 5 K across the equatorial Pacific. In the east the air is stabilised by the cold water and cannot rise and join the normal Hadley circulation as one would expect in this potential ITCZ region. Instead it flows westward, forming the southeast trade winds across the South Pacific, to the warm western Pacific, where it gains moisture and heat, and rises. Some of the air then flows eastward aloft to complete the cell.

This normal circulation is periodically reversed (Figure 8.23b). Pressure rises near Darwin and falls in the eastern Pacific. Commonly this fall is most marked around Tahiti, so that very weak easterly winds may persist over the extreme eastern Pacific, with the westerlies dominating over the rest. The changed wind regime, however, prevents the upwelling of cold water off the Peruvian coast, and traditionally the cell reversal has been signalled by a sea surface temperature increase of as much as 4 K in this region.

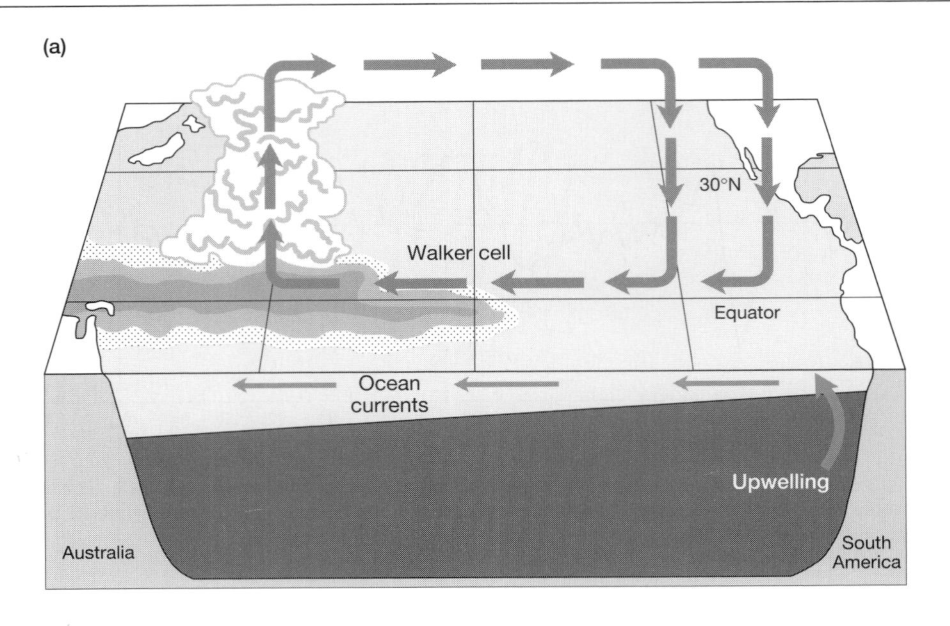


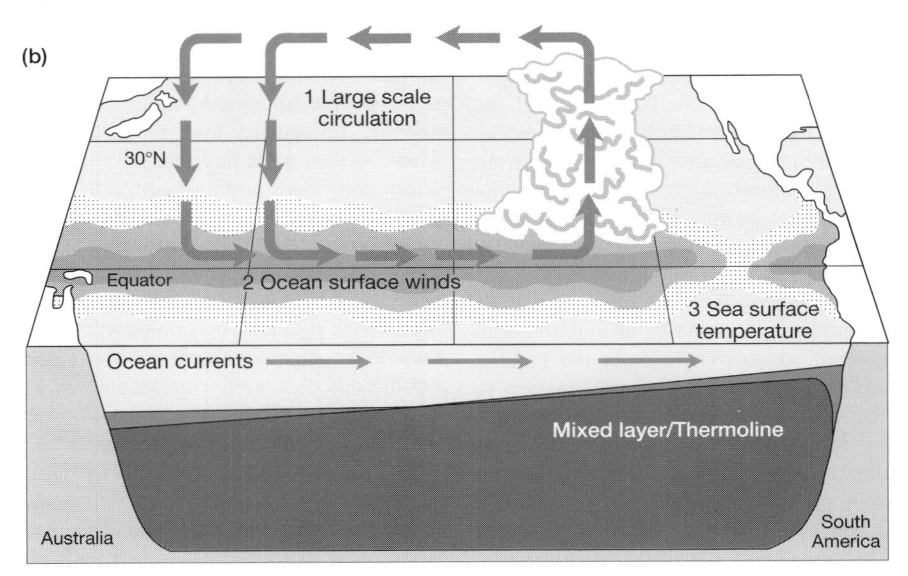


*Figure 8.23* The two modes of the Walker circulation that form part of the Southern Oscillation: (a) Walker circulation; (b) El Niño conditions. In the period of El Niño the circulation is opposite to the normal flow. (After Earth Space Research Group, University of California Santa Barbara web site.) (www.crseo.ucsb.edu/geos/13.html)

This effect, called **El Niño** ('the boy (Christ) child') because of its near coincidence with the Christmas period, occurs about every seven years, and is now known to have climatic and economic consequences in many regions of the globe.

Recent observations have shown that prior to the onset of El Niño it is usual for the pool of warm water off the northern Australian coast to expand eastward (Plate 4). It is tempting to suggest that this is a direct oceanic response to the surface wind

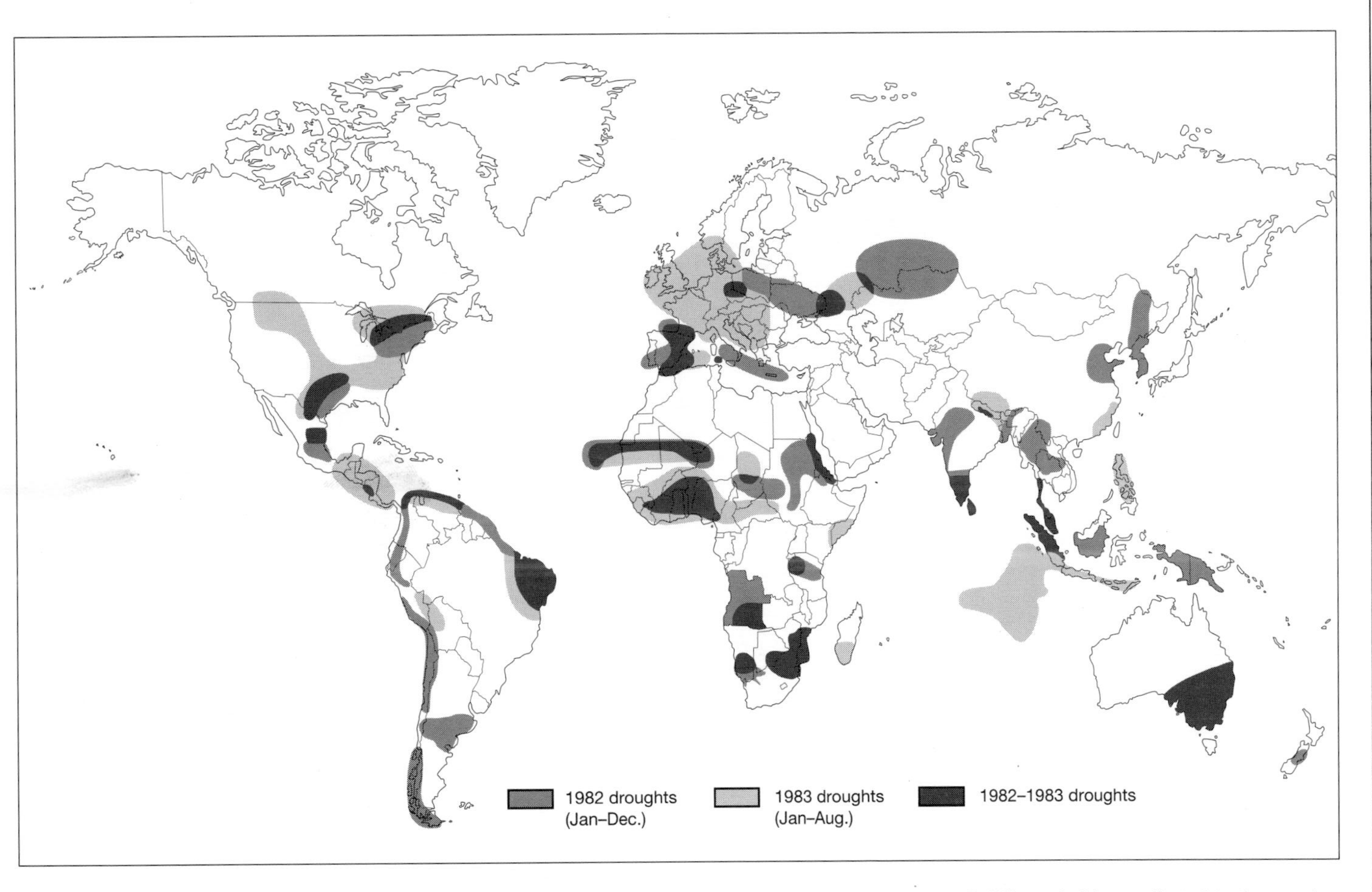


*Figure 8.24* Areas experiencing drought during or immediately after the 1982–1983 El Niño event. Although El Niño probably contributed to the onset or severity of each of these, it is not clear that it was the causal factor for all of them. (After Glantz *et al.*, 1987.)

regime created by the reversal of the Walker circulation. Certainly as the water moves eastward hydrostatic instability is enhanced and the eastern Pacific has excess precipitation (Figure 8.23b). At the same time, descending air creates dry conditions over Indonesia and Australia. However, there are also changes occurring in the oceanic circulation which interact with the overlying atmosphere. In normal conditions the easterly equatorial wind and current move water westward. As a result, not only is sea level almost 1 m higher off Australia than off South America, but also the thermocline which separates warm surface water from the colder deeper ocean, is tilted towards the west. When conditions reverse, surface water flows effectively downhill, while at the same time there is interaction across the thermocline, which itself is being modified. In addition, the interchanges with the deep ocean, along with differences in evaporation and precipitation, influence water salinity and density. This in turn generates horizontal and vertical motions, influencing surface conditions and the overlying atmosphere. Whether cause or effect, this must be associated with the meteorological changes we have described. Certainly these oceanic effects play a profound role in the rate and amount of movement of the surface waters, and thus on the establishment of the El Niño event.

### 8.4.2 The global scope of ENSO

The area of ENSO covers some 10% of the Earth's surface, so that it is not surprising that changes in the energy balance signalled by changes in sea surface temperature can have an impact far from the Pacific. Most of the direct connections that have been established so far are related to precipitation or severe storm anomalies (Figure 8.24). These frequently have a tremendous economic or social significance (Box 8.A). However, potential links between ENSO and climatic anomalies of various types around the world are being investigated. In addition to the implications for enhanced understanding of climate processes, these emerging links contain possibilities for long-term climate forecasting (Section 14.2). They also indicate the need for continuous monitoring of the ocean surface conditions (Box 8.I).

#### Box 8.I Monitoring the ocean surface

This chapter has emphasised the importance of monitoring conditions in the ocean. As with land, satellite and surface observations are complementary, and are treated separately here.

*8.I.1 In situ oceanic observations*

Until recently, oceanic climate information came only from island stations or from passing ships. Islands, irregularly scattered spots in a vast ocean, are by their very nature unrepresentative of that ocean, but their results must be used whenever a long period of record is required. They can often be supplemented by the water temperature observations made by **ships of opportunity**. Originally temperatures were measured by raising a bucket of sea water to the deck and inserting a thermometer. Now a thermometer located about 2 m below the water-line near the engine water intake is most commonly used. Neither method is entirely suitable for the measurement of absolute sea surface temperatures, but can give indications of spatial and temporal variations. These measurements, made primarily by volunteers on commercial ships, are restricted to the major shipping lanes (Figure 8.I.1).

The recognition of the importance of ocean processes has more recently stimulated the deployment of buoys as observing platforms. Fixed buoys can be placed in a regular pattern to monitor wind speed and direction, air temperature and humidity, and water temperatures at various levels (Figure 8.I.1). These are now semi-permanent features of our observing system. Drifting buoys, measuring water temperatures and ocean surface movements, are less permanent, although some may provide information for many years. Further, ocean surface height information is provided by a network of tide gauges established on many island and continental shorelines. Virtually all of these various instruments relay their observations to a central location via satellite and thus allow continuous monitoring.

*8.I.2 Satellite observations of the ocean*

In many ways the ocean is the simplest surface to which remote-sensing techniques can be applied. Albedo and emissivity vary little over large areas and topography is minimal. Thus, unlike buoy and ship data, the satellite-derived measurements of ocean surface conditions are homogeneous in space and regular in time. For sea surface temperature, upwelling radiation from three channels, with wavelength ranges 3.55–3.93, 10.3–11.3 and 11.5–12.5 μm, is measured. The latter two channels sense only thermal infrared radiation while the first senses a mixture of reflected solar radiation and emitted thermal

**Box 8.I (cont'd)**

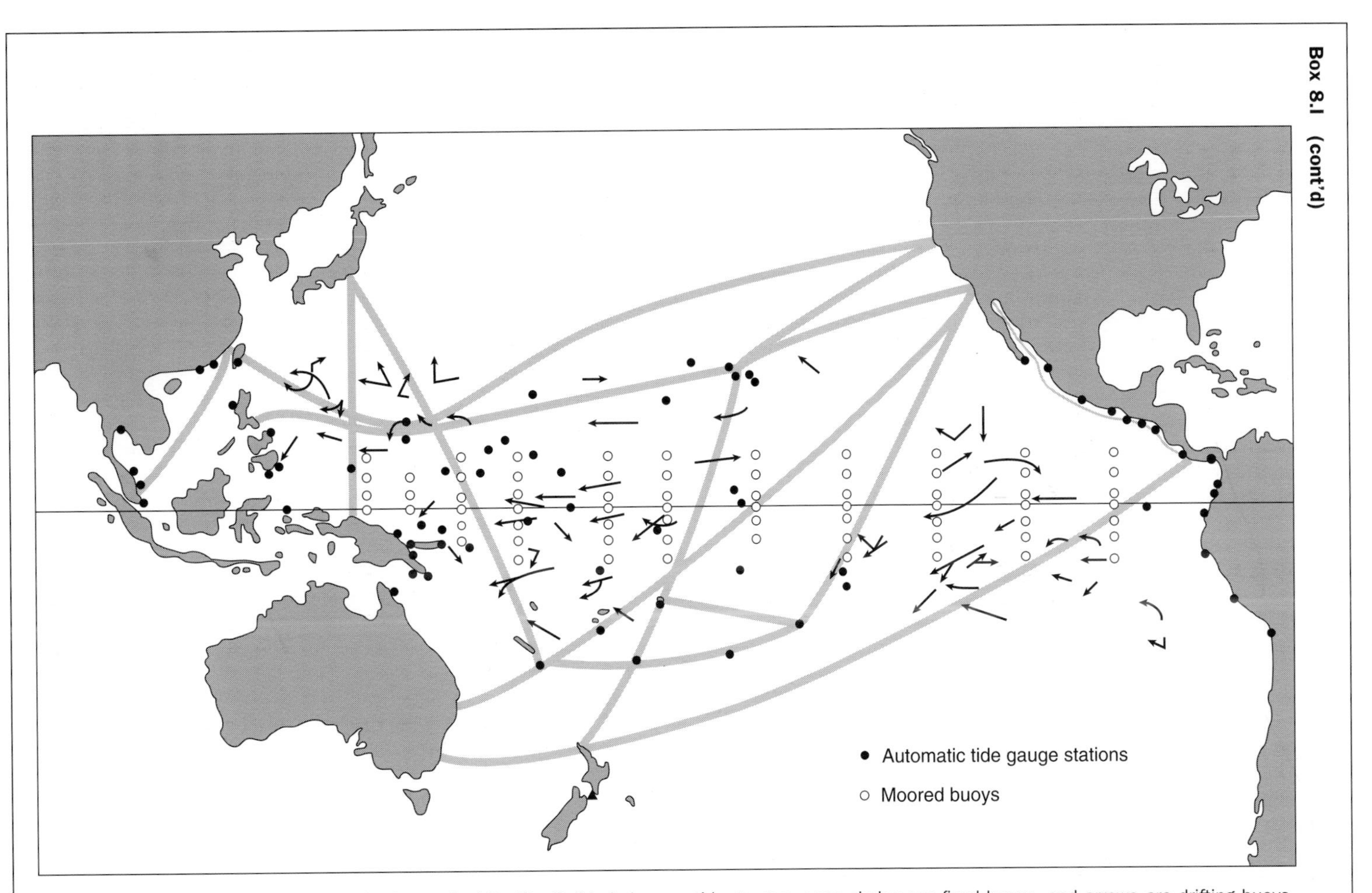


*Figure 8.I.1* Surface monitoring sites in the tropical Pacific. Solid circles are tide gauges, open circles are fixed buoys, and arrows are drifting buoys, suggesting likely directions of movement. The major areas of ship observations are shown by the grey shading. (After University Corporation for Atmospheric Research, 1994.)

**Box 8.1 (cont'd)**

infrared radiation during the day but only emitted radiation at night (see Figure 2.3). Empirical equations have been established from which sea surface temperatures can be derived. The combination of satellite, buoy and ship data has allowed the development of maps of sea surface temperature (e.g. Plate 4), which have now been routinely available for several years. Such maps have played a role in the identification and analysis of the ENSO phenomenon as well as now helping to forecast the onset of an El Niño event.

More recent advances in remote sensing, however, have begun to provide additional information. Monitoring differences in upwelling radiation across the 0.7 μm region, for example, allows estimates of photosynthetic activity (see Figure 2.8). This relates to the phytoplankton concentration in the ocean. Areas of high concentration commonly indicate areas where cold, nutrient-rich water is rising from the deep ocean. This has implications for the ENSO system, and for fish stocks, before the manifestation of any features on the ocean surface.

# CHAPTER 9 Extra-tropical weather and climate

In extra-tropical latitudes the regional climates are produced by the juxtaposition of baroclinic and barotropic conditions, by the strong influence of continentality, and by the seasonally variable latitudinal temperature gradient. Individual climatic regions may also reflect topographic influences. These various controls create not only distinct climate regimes, but also ensure that the day-to-day variability of the weather is a strong feature of the climates.

The approach we adopt in this chapter, therefore, is to consider first the role of baroclinic conditions in creating the basic atmospheric features of the region and then to describe these features. This information is then used to allow consideration of the climatic regions of mid-latitudes. Finally, the special characteristics of the polar regions are investigated.

## 9.1 The mid-latitude baroclinic zone

The basic characteristics of the extra-tropical atmosphere as part of the global circulation were established in Section 7.2. The major characteristic of mid-latitudes is that the general equator-to-pole temperature gradient is commonly concentrated in a rather narrow latitudinal band. Here baroclinic conditions are established, and Rossby waves, jet streams and polar fronts result. These in turn produce synoptic and meso-scale weather features. Seasonal changes and daily variations of the baroclinic zone create changes in short-term weather. Outside the baroclinic zone are barotropic areas with much calmer weather characterised as air mass regions. However, these may become involved in the movements associated with the baroclinic regions, and may at times be responsible for starting such movements. All of these features are interconnected, but it is convenient to start with consideration of the baroclinic zone.

### 9.1.1 Rossby waves, polar fronts and jet streams

The general flow pattern of the mid-latitude portion of the general circulation is from west to east, giving the Westerlies of Figure 7.4. This flow pattern, commonly resembling a series of horizontal wave-like oscillations, influences the whole depth of the troposphere and covers most of the mid-latitude region. When viewed from above a pole (Figure 6.2), this is seen as the circumpolar vortex, a band of relatively rapidly moving air circling the calmer region of descending air immediately above the pole. From the mid-latitude perspective we see the waves as being best developed and well defined in a relatively narrow band at restricted latitudes (Figure 9.1). These are the **Rossby waves**, also called **planetary** or **long** waves. It is in this band that the major poleward temperature gradient is established and thus that baroclinic conditions are likely to occur. As a consequence, it is in this region that the polar front and the polar front jet stream form (Figure 9.2).

A **front**, in meteorological parlance, is a region with rapid horizontal temperature change. Usually it appears on the Earth's surface as an elongated, rather narrow, region of temperature contrast separating broader areas of relatively uniform temperatures. There are various forms of front, the most familiar in mid-latitudes being those associated with depressions (Section 9.2). Here, however, we are concerned with the front that is generated by the baroclinicity associated with the Rossby waves, the **polar front** (Figure 9.2). This major temperature discontinuity, although not continuous around the hemisphere, usually covers a significant portion of it. Certainly it may extend as a single entity for several thousand kilometres, although frequently it is modified and distorted, and may be broken, by the depressions which form along

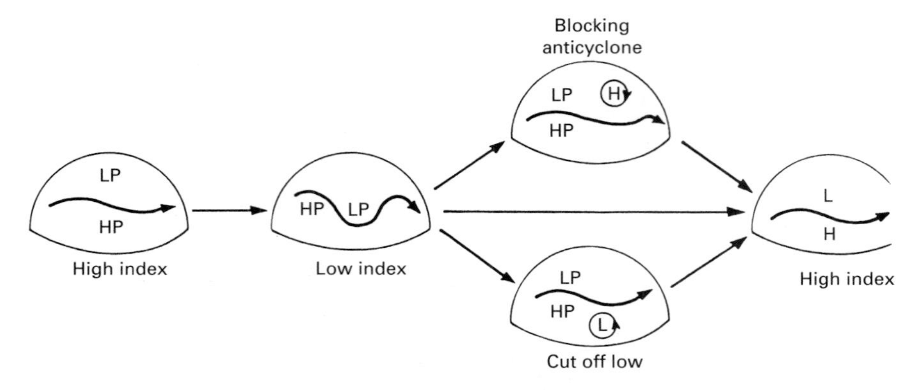


*Figure 9.1* Schematic illustration of the change in the Rossby wave flow pattern in the transition from high to low index and then from low to high index regimes.

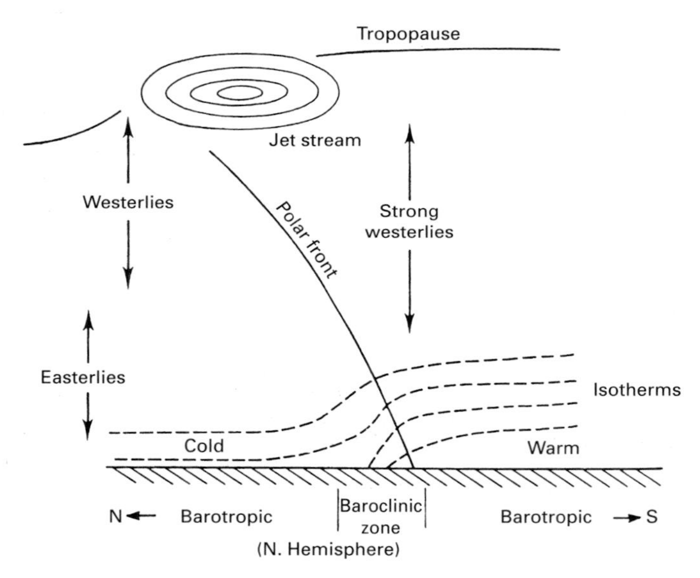


*Figure 9.2* Schematic cross-section of the mid-latitude atmosphere showing the position of the polar front, the jet stream and the baroclinic and barotropic zones.

it. The polar front also separates the warm air of the tropics from the colder, but equally barotropic, polar air. It extends upwards, as a **frontal surface**, through much of the depth of the troposphere.

The baroclinic conditions foster an increase in wind speed with height along the frontal surface. This leads to the development of a **jet stream** at or just below the tropopause. In general, a jet stream is a ribbon-like belt of rapidly moving air, a few hundred metres in depth, a few kilometres wide and hundreds, or even thousands, of kilometres long. The one considered here (Figure 9.3), strictly the **polar front jet**, is usually the longest, strongest and most persistent. However, the subtropical jet above the poleward-descending limb of the Hadley cell (Section 8.1), and the equatorial jet of the Indian subcontinent (Section 8.2), are similar.

Jets are regions of **clear air turbulence**. This, which is exactly what its name implies, is associated with the great **wind shear** around the jet. Wind shear

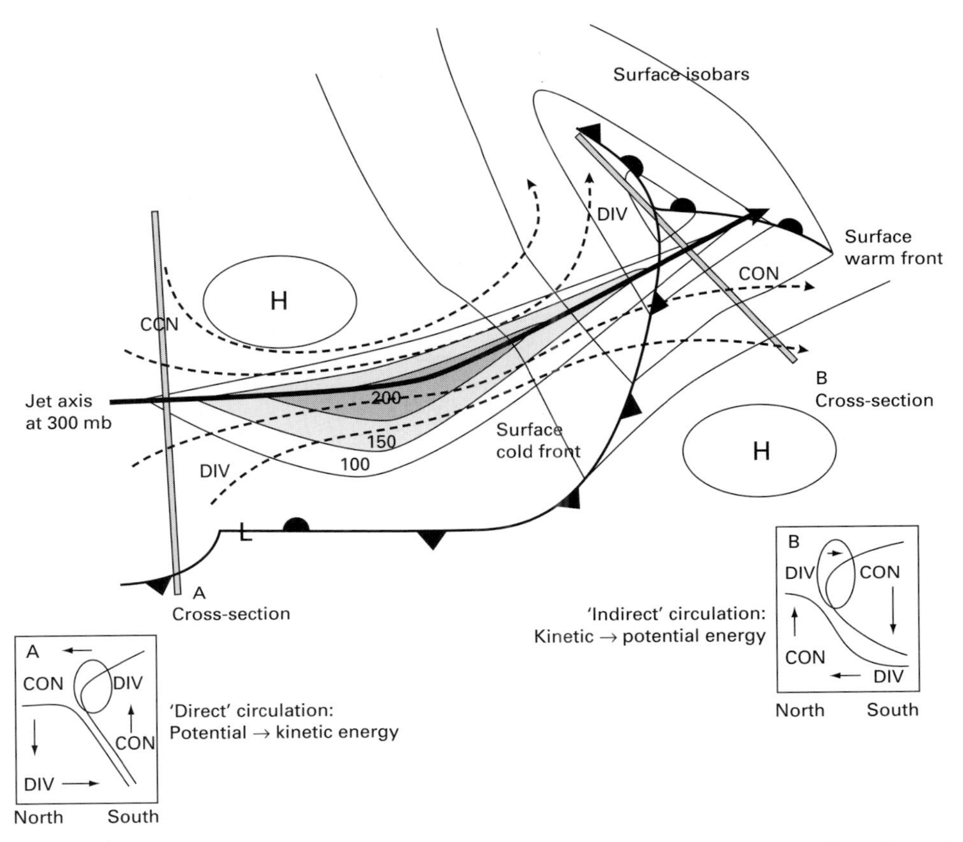


*Figure 9.3* Schematic view of a polar front jet at approximately 300 hPa. Isotachs (lines of equal wind speed) for 100, 150 and 200 km $h^{-1}$ and stream lines (the path of an individual air molecule) (dashed lines) are indicated near the jet axis (thick solid line). Surface isobars (thin solid lines) and fronts (conventional symbolism) are included. Shaded bars indicate the locations of the two cross-sectional sketches. Areas of convergence (CON) and divergence (DIV), associated vertical motions and energy conversions, are indicated. (From Reiter, 1996.)

occurs when the wind speed or direction, or both, changes rapidly with distance. The shear produces the turbulence, which is a hazard to aircraft operations. In many cases, such as around thunderstorms where updrafts and downdrafts can occur in close proximity, there is at least the visible evidence of the cloud to give a warning. With the jets, however, the turbulence does not give rise to organised large-scale motions and the air remains clear.

### 9.1.2 Changes in Rossby wave position

The location of the Rossby wave and the associated polar front and jet stream varies with time. The winter waves tend to be equatorward of, and stronger than, their summer counterparts. This is a direct response to the greater equator-to-pole temperature gradient of the cold season (Figure 7.1). They may also be influenced by the character of the underlying surface. Continentality effects, although commonly confined to near-surface conditions, alter the latitudinal pressure gradient and hence the wind field. Topographic barriers, such as the Himalayas (Section 8.2) and the Rockies (Section 9.3), play a role. Finally, in some cases the wave location appears to be constrained by the location of the polar ice margin, for reasons suggested in Section 7.3.

On a shorter time scale, the fluid mechanics of the flow dictates that changes in amplitude and configuration will occur. The Rossby waves tend to move

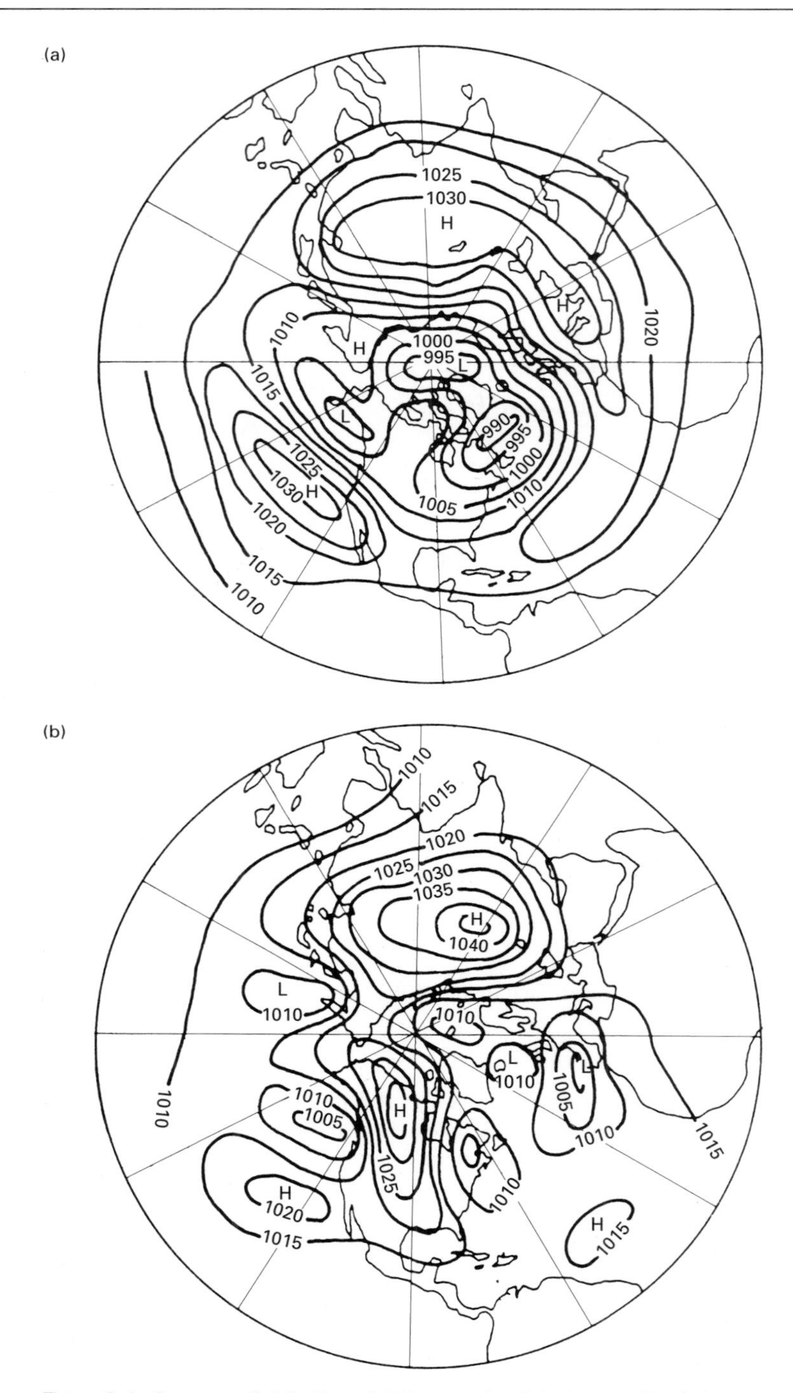


*Figure 9.4* Pressure distributions (hPa) at sea level during two 7-day periods showing (a) a high index and (b) a low index. (From Willett and Saunders, 1959.)

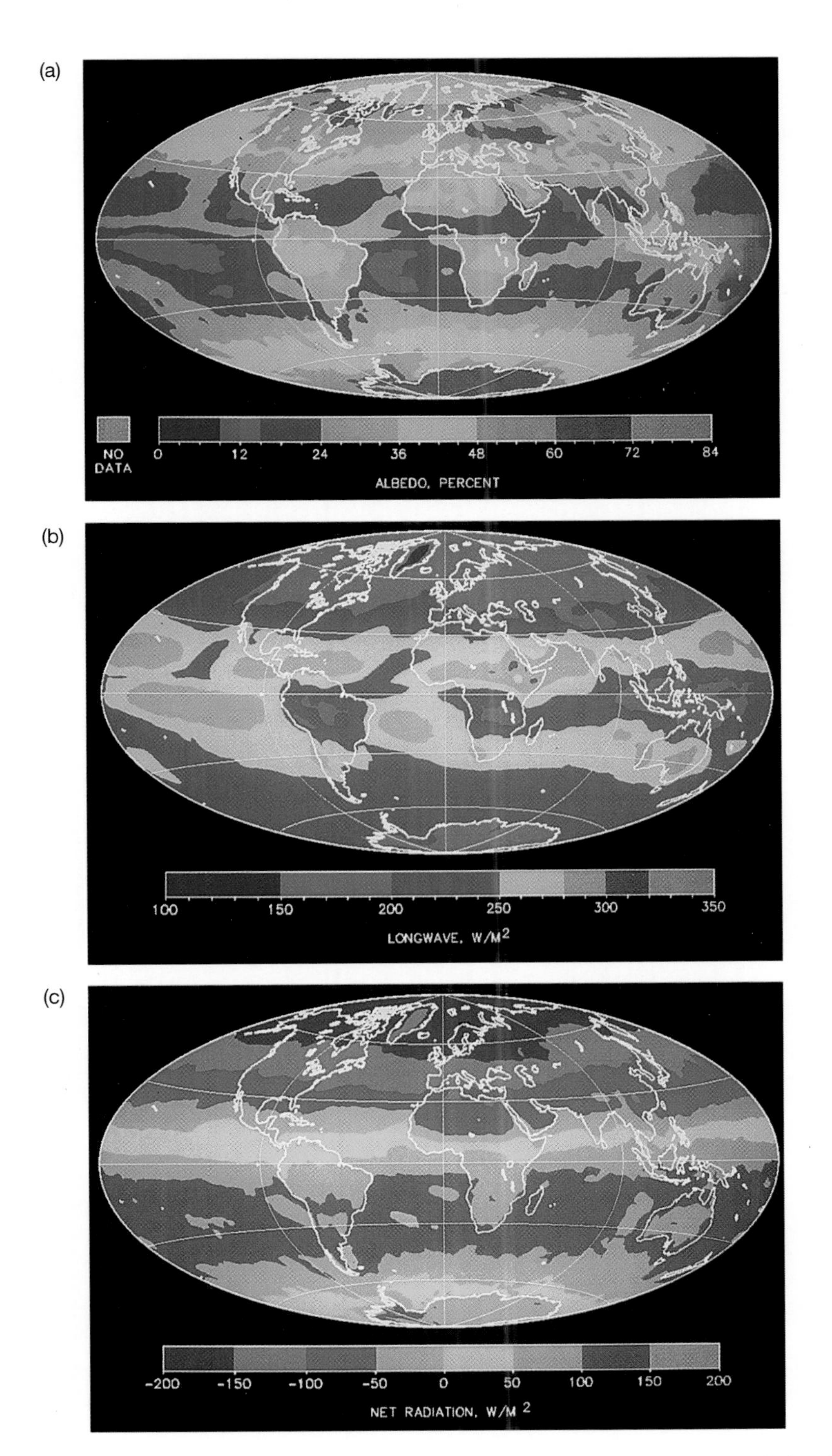


*Plate 1* Monthly average (a) albedo, (b) emitted long-wave, and (c) net radiation for January as determined by the ERBE satellite (From Harrison *et al*., 1993, reproduced by kind permission of Cambridge University Press).

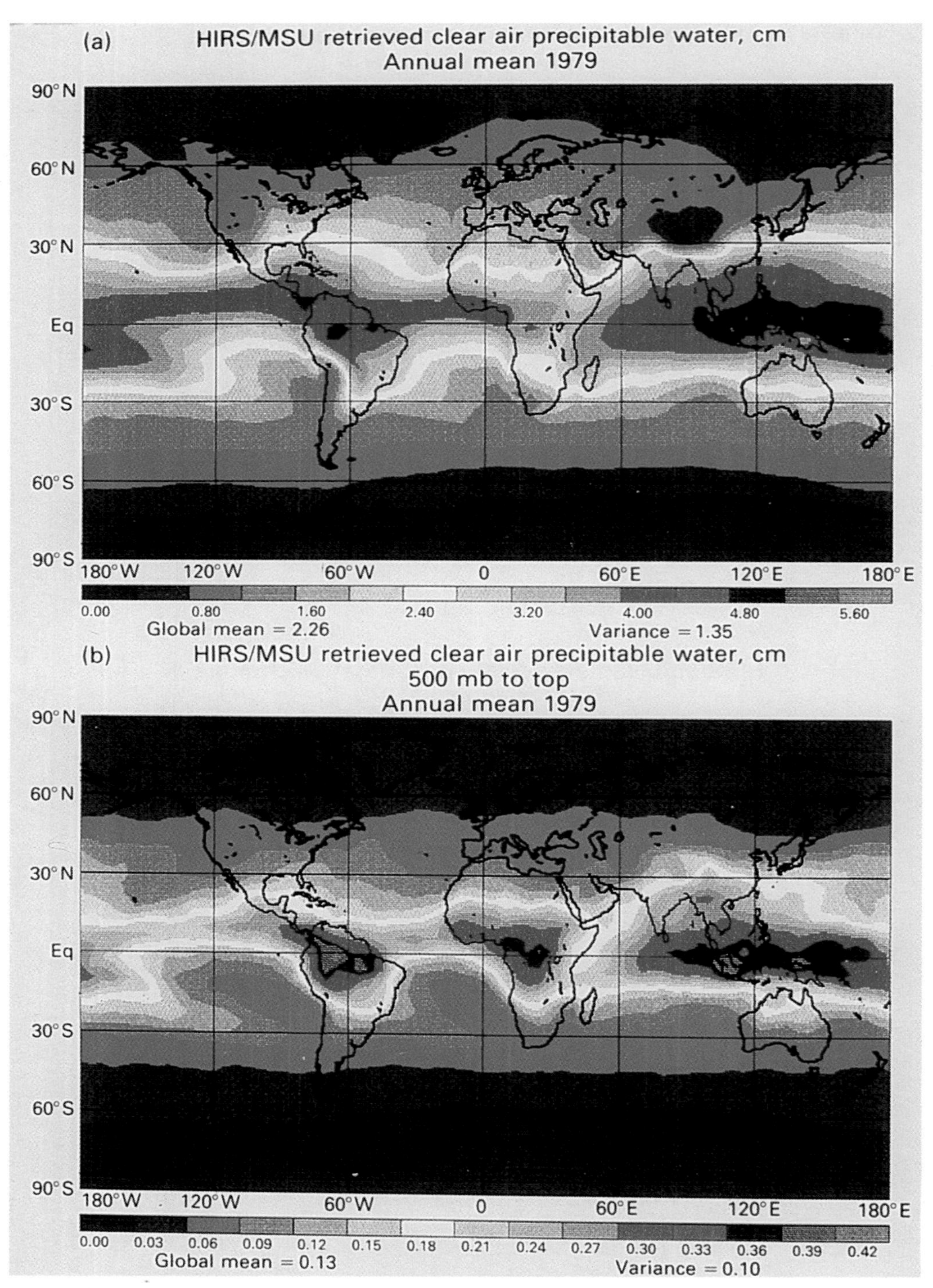


*Plate 2* Atmospheric water vapour content in centimetres of precipitable water. Values are derived from satellite observations of the humidity throughout the atmospheric column. (The PWV in kilograms is given by the PWV in metres multiplied by the density of water). (From Suskind, 1993, reproduced by kind permission of Cambridge University Press).

(a)

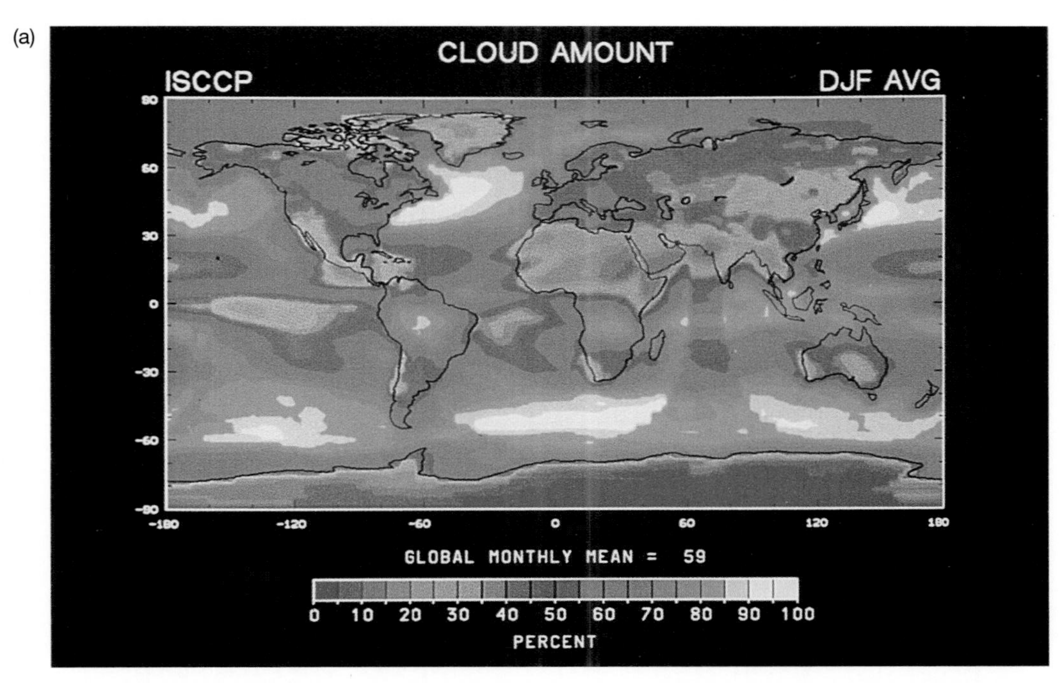


(b)

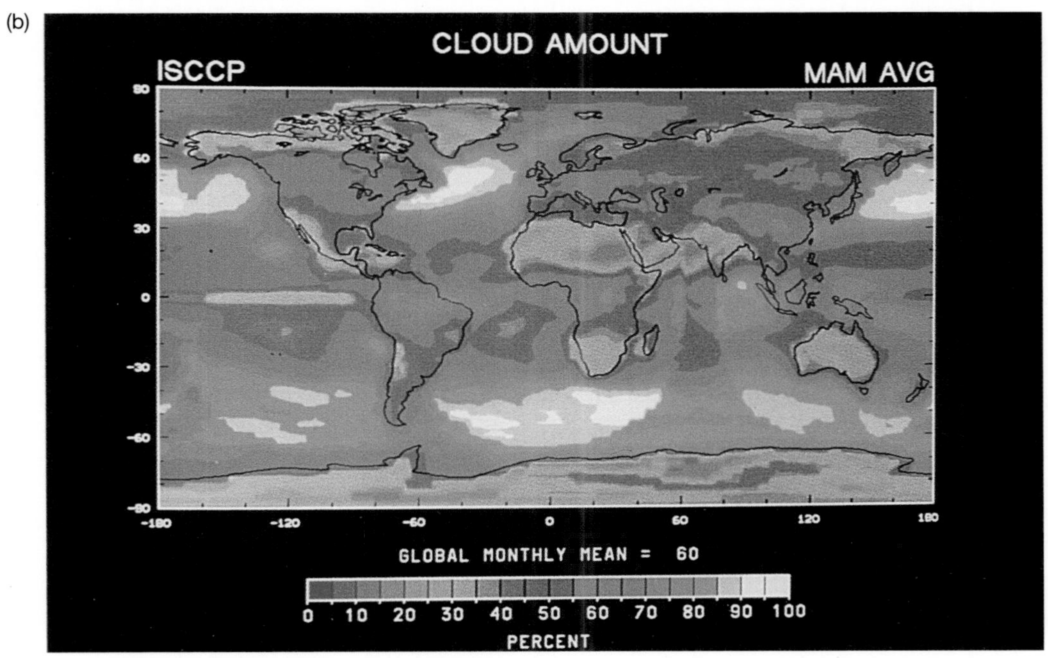


*Plate 3* continued on next page

(c)

(d)

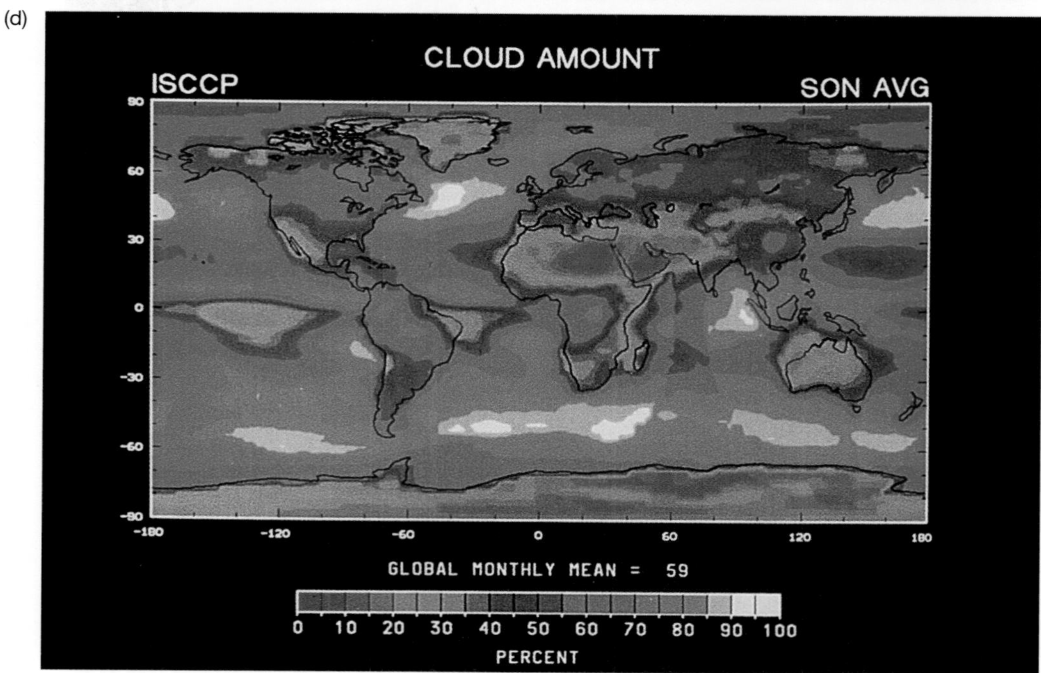


*Plate 3* Global distribution of seasonal mean cloud amount for (a) December-February, (b) March-May, (c) June-August, and (d) September-November. These long-term data are the result of the international Satellite Cloud Climatology Project, a major international effort to provide a uniform, world-wide cloud data set. Note the persistence of the mid-latitude desert areas and the regions of extensive cloud cover, particularly associated with the mid-latitude depression belts and the intertropical convergence zone. The movement of the latter is particularly apparent (From Rossow, 1993, reproduced by kind permission of Cambridge University Press).

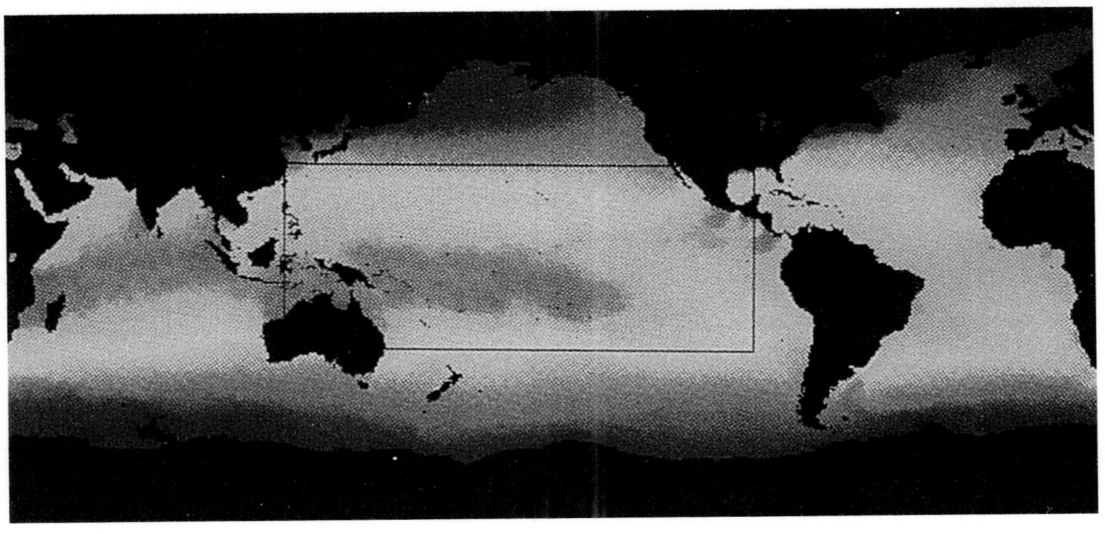

*Plate 4* The Typical Pacific sea-surface temperature pattern during (a) a normal year and (b) an El Niño event. This is a composite of many satellite observations, with the dark water colour of Australia representing ocean surface temperatures above 28°C (From Ho *et al*., 1995, reproduced by kind permission of the American Meteorological Society).

(a)



(b)

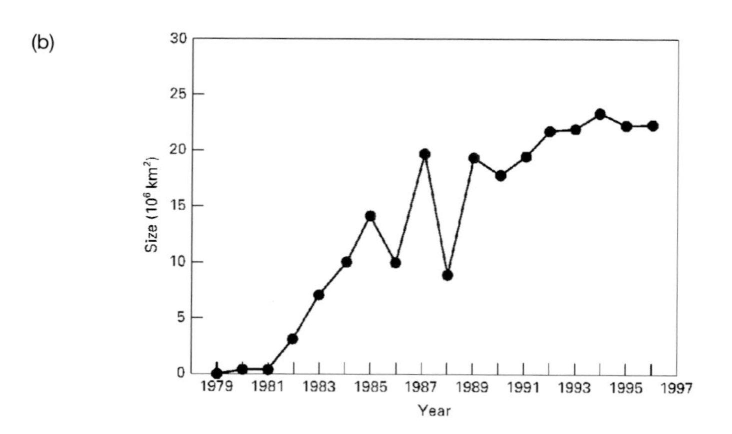


*Plate 5* (a) Distribution of stratospheric ozone over Antarctica for selected October day between 1979 and 1990, showing the low ozone values near the pole (from http://jwoky.gsfc.nasa.gov – courtesy of NASA. (b) Annual maximum extent of the area of the Antarctic stratosphere where ozone levels are less than 200 Dobson units, the value typically regarded as the threshold of the 'Ozone hole'. Note the suggestion that the rapid increase in size is slowing as a result of bans on CFC usage (based on data from NASA).

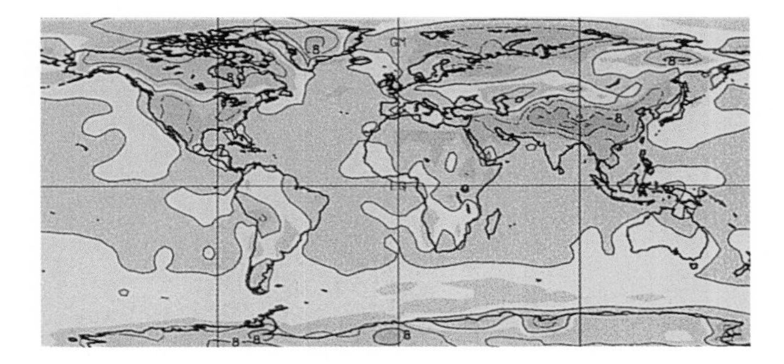

(a) Difference of model average from observed

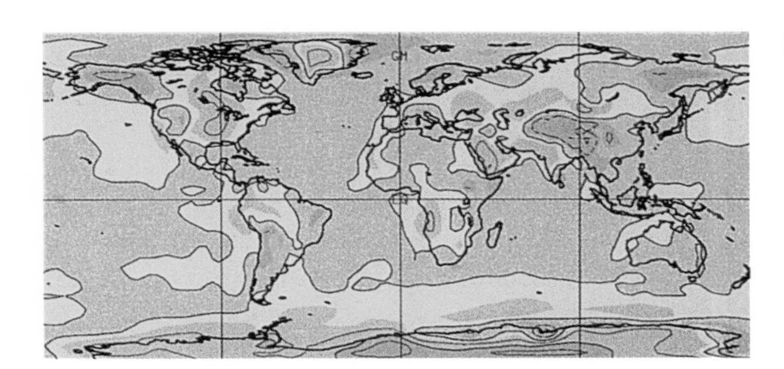

(b) Difference of model average from observed

(c)

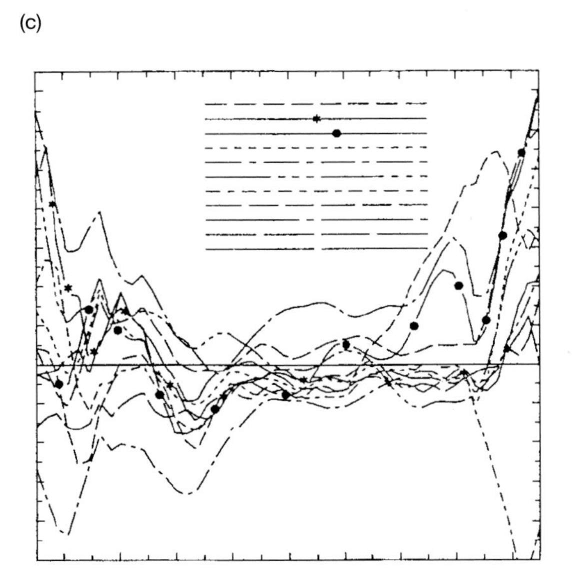

(d)

*Plate 6* Global patterns of difference between observed (Jenne, 1975) and modelled surface air temperature for (a) December-February and (b) June-August. The model results are an average of those from 11 models. The zonally averaged differences from observations for the 11 models individually are shown in (c) and (d) (From Gates *et al.*, 1996, reproduced by kind permission of Cambridge University Press).

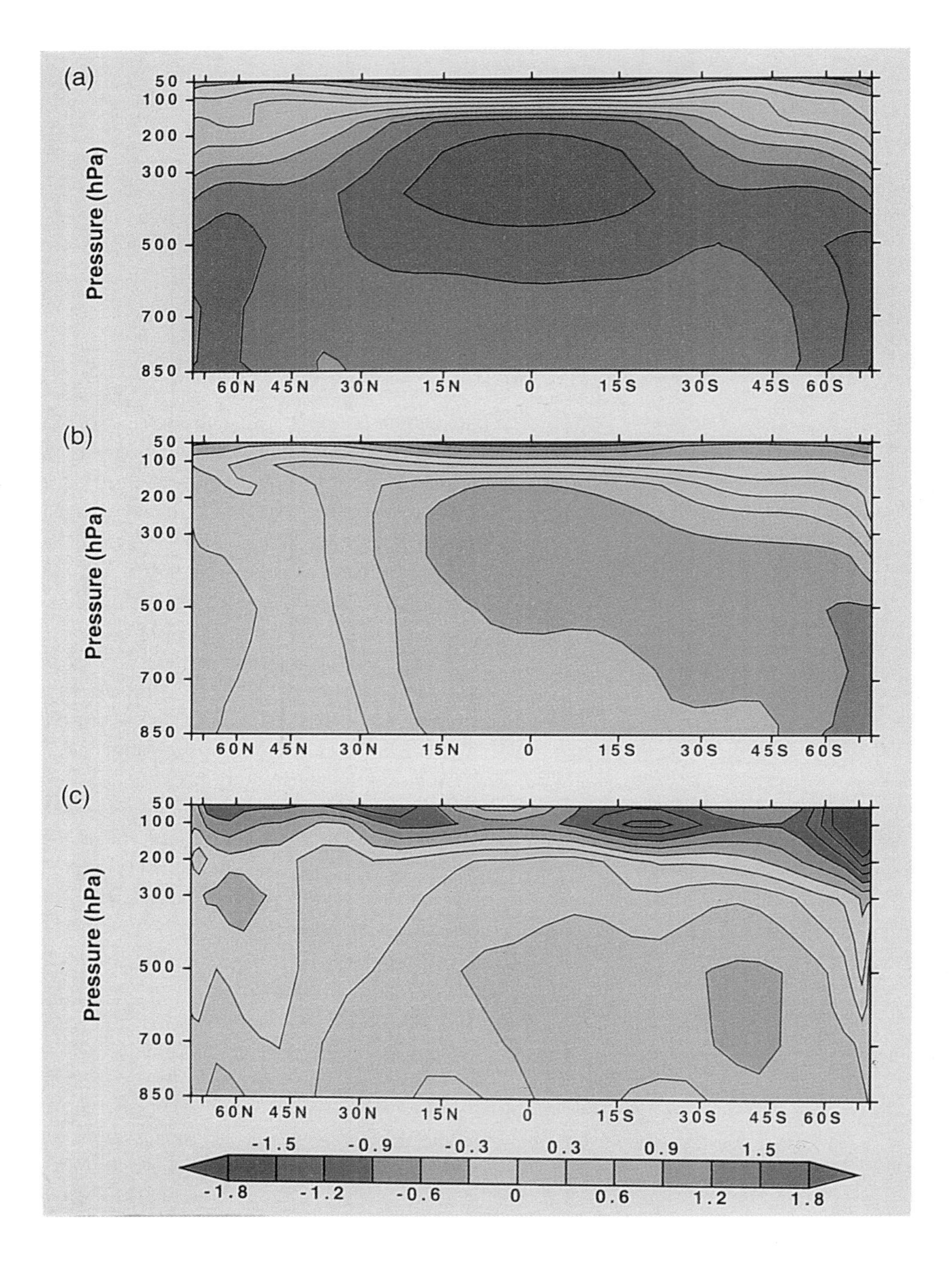


*Plate 7* Changes in the zonal mean annual average temperature as a function of height (expressed as a pressure level) between the pre-industrial and current periods, as determined by (a) a model using greenhouse gases alone, (b) a model incorporating greenhouse gases and sulphate aerosols, and (c) observations. Stratospheric cooling and tropospheric warming are apparent in all cases. The model including sulphates also shows indications of the asymmetrical nature of the response between the two hemispheres. Note that the results from observations are highly sensitive to the time period over which they are collected (From Santer *et al.*, 1996, reproduced by kind permission of Cambridge University Press).

eastward about 15° of longitude per day. They may be retarded, or even 'anchored', by a large topographic barrier such as the Rocky Mountains. In general, however, there is a tendency for a pseudo-cyclic change (Figure 9.1). This can be conveniently characterised by the zonally averaged pressure difference between two latitude circles, a difference known simply as the **index**. The latitude circles usually chosen for the Northern Hemisphere are 35 and 55°, latitudes which approximately correspond to the southern and northern limits of the main Rossby flow and which have an adequate spacing of observing stations. As the wave amplitude changes there is an **index cycle**. Starting with a strong zonal flow, there is a high index and a small number of relatively smooth Rossby waves. As the amplitude of the waves increases and the index decreases, the main axis of the strong westerlies is displaced southward and may become discontinuous. Distinct high and low pressure centres are formed (Figure 9.4). Thereafter the amplitude may slowly decrease again, or a cut-off may occur, leaving either a **blocking anticyclone** as a stationary atmospheric feature, or a **non-frontal depression** as a feature that usually moves eastwards slowly. The whole cycle may take 20 to 60 days to complete, but it is very irregular in length and in the speed with which each stage progresses. This index, though similar to some considered in Section 14.2, is not intended to have any predictive or explanatory aspect, and is simply used as a description of flow patterns.

Changes in pattern may also be produced when the jet stream, being dynamically unstable, creates meanders of its own. These propagate downwards into the general westerly flow and influence the polar front and the Rossby waves. Similarly, the polar front may change position and affect the other two features. Changes in the polar front position will also arise because of the baroclinic nature of this feature. In such baroclinic conditions vertical motions associated with convergence and vorticity will be developed. In particular, these will lead to the development of **frontal depressions**. These have cyclonic motion around a low pressure centre and probably originate near the 500 hPa flow. As they develop and extend down through the atmosphere to become surface weather features, they will not only move in response to the Rossby wave motion, but will also become sufficiently large to affect this flow themselves. They are also responsible for poleward energy transport by transient eddies, supplementing the energy transport provided by the Rossby waves (Figure 7.7). In general terms, a high index circulation will have strong zonal flows where deep, vigorous depressions are the main transporters of energy northward. With a low index the depression circulation is much weaker and the Rossby waves transport the energy directly.

## 9.2 Mid-latitude weather

Since the various features considered above all act and interact to create mid-latitude climates, a simple description and explanation of such climates is difficult. Indeed, climatic descriptions in mid-latitudes seem to reinforce the notion of climate as a 'series of weather events'. Hence we here provide a brief discussion of typical mid-latitude weather systems prior to synthesising the information to develop climatic regions. To simplify matters, we shall temporarily regard Rossby waves and jet streams as steering mechanisms and concentrate on the weather-producing features of depressions, anticyclones and air masses. To avoid an unnecessarily large number of parentheses, we describe the circulation and movement of systems in the Northern Hemisphere. The direction of circulation is reversed in the Southern Hemisphere.

### 9.2.1 Depressions (cyclones)

A depression is a synoptic feature with cyclonic (counter-clockwise) air circulation around a low pressure centre. Near the surface the air spirals inwards towards the centre, creating convergence, uplift and a tendency for cloud formation.

The classical model of depression development and characteristics was produced by the 'Bergen school' of meteorologists in the early 1920s. Although many refinements have been made, particularly in the light of upper air soundings and satellite observations, the main features of this model are still pertinent (Figure 9.5). In the model the depression originates with a **front**, traditionally the polar front. When a wave begins to develop on this front, an embryonic cyclonic circulation is initiated, with a distinct low pressure region developing at the crest of a wave. The front becomes divided into warm and cold portions. As depression development progresses, pressure falls at the crest of the wave and the trailing cold front moves more rapidly than the leading warm front. The region of warm air between the fronts, the **warm sector**, becomes progressively smaller. Eventually the cold front overtakes the warm one and **occlusion** takes place. Thereafter the depression fills in. The polar

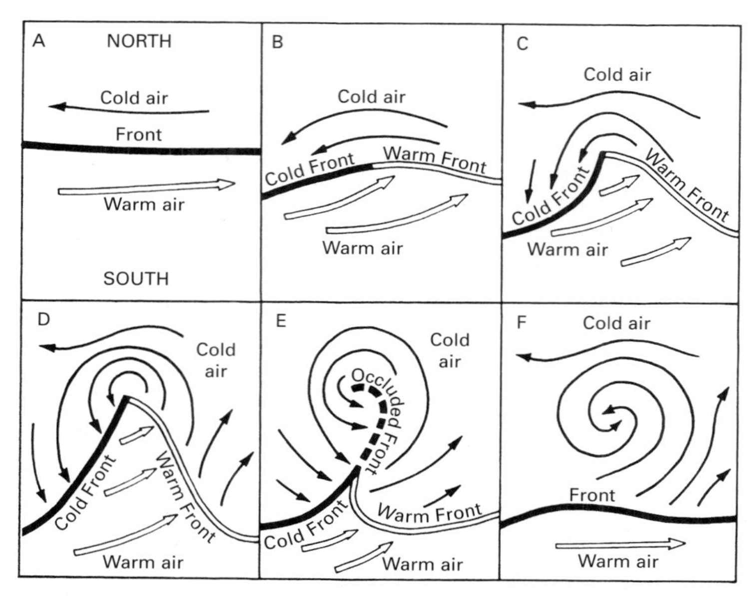


*Figure 9.5* Plan view of the six idealised stages in the development and final occlusion of a depression (or extra-tropical cyclone) along the polar front in the Northern Hemisphere. Stage D shows a well-developed depression system and stage E shows the occlusion.

front is frequently re-established somewhat farther south than it was previously. The whole sequence may take a week or so for completion, during which time the system may have moved two or three thousand kilometres in a general eastward direction. In addition, new depressions may form on the trailing cold front of an old one. Thus there can be **depression families**, usually a sequence of three or four depressions occurring at any one time, each being younger than its predecessor and occurring progressively farther south.

There is a distinct sequence of weather conditions associated with the passage of a depression. Figure 9.6 illustrates in schematic form the surface weather map conditions of an idealised 'mature' depression, together with a cross-section along line AB, showing conditions in the area with the most marked weather sequence. The warm front is a relatively gently sloping feature with a preponderance of horizontally developed cloud, while the cold front is much steeper and vertical cloud development is common. Satellite photographs show these cloud distributions very clearly (Figure 9.7).

The actual sequence of events for a particular depression system will not follow the above model exactly. Indeed, an individual weather map is unlikely to show the features as clearly as is suggested here although many are easily recognisable (e.g. Figure 9.8). However, as climatologists rather than meteorologists, we are mainly concerned with identifying the general characteristics of depressions, not the particular characteristics of any single one. Studies of numerous individual depressions in the Southern Hemisphere, especially in the southern Indian Ocean, suggest that our model may have to be modified somewhat if we are to characterise conditions in the southern mid-latitudes correctly.

### 9.2.2 Anticyclones

In mid-latitudes we can often think of **anticyclones** (Figure 9.9) as the opposite of depressions. Certainly in a general sense they are, being regions of high surface pressure and divergence, giving air spiralling clockwise (in the Northern Hemisphere) out from the centre. Descending air dominates. They are regions of generally clear skies although strong surface heating often leads to local cumulus cloud formation.

There are several classes of anticyclone, depending on their size and mode of origin. The descending limbs of the Hadley cells are large, more or less permanent anticyclones. Similarly, the high pressure

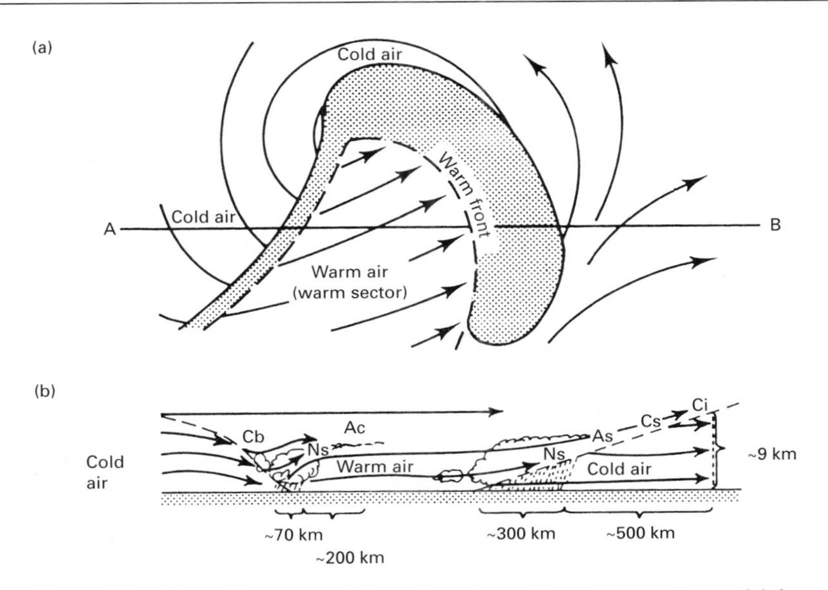


*Figure 9.6* Schematic representation of a depression system as seen (a) in plan and (b) in cross-section. The cross-section is taken along the line AB. Cloud types: Cb, cumulonimbus; As, altostratus; Ac, altocumulus; Cs, cirrostratus; Ns, nimbostratus; Ci, cirrus.

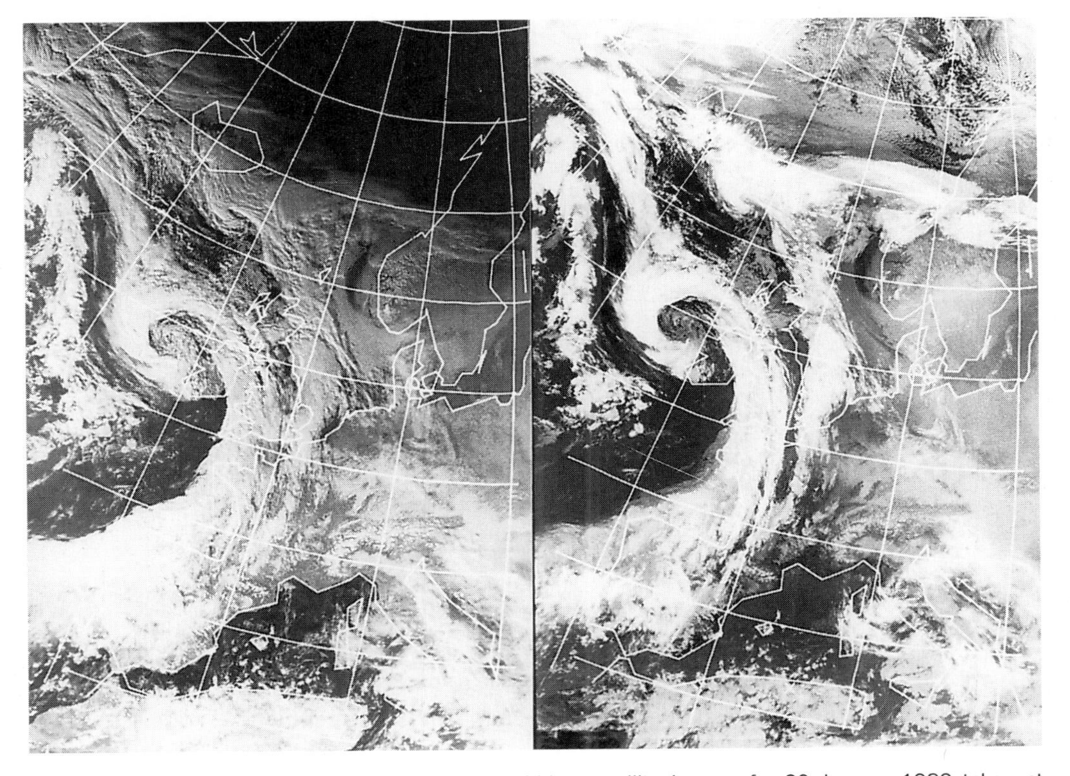

*Figure 9.7* The visible and infrared NOAA polar orbiting satellite images for 20 January 1982 taken at 1354Z, clearly showing the partially occluded depression system adjacent to the UK. Compare this image to the surface chart in Figure 9.8. (Courtesy of P. Baylis, Department of Electrical Engineering, University of Dundee.)

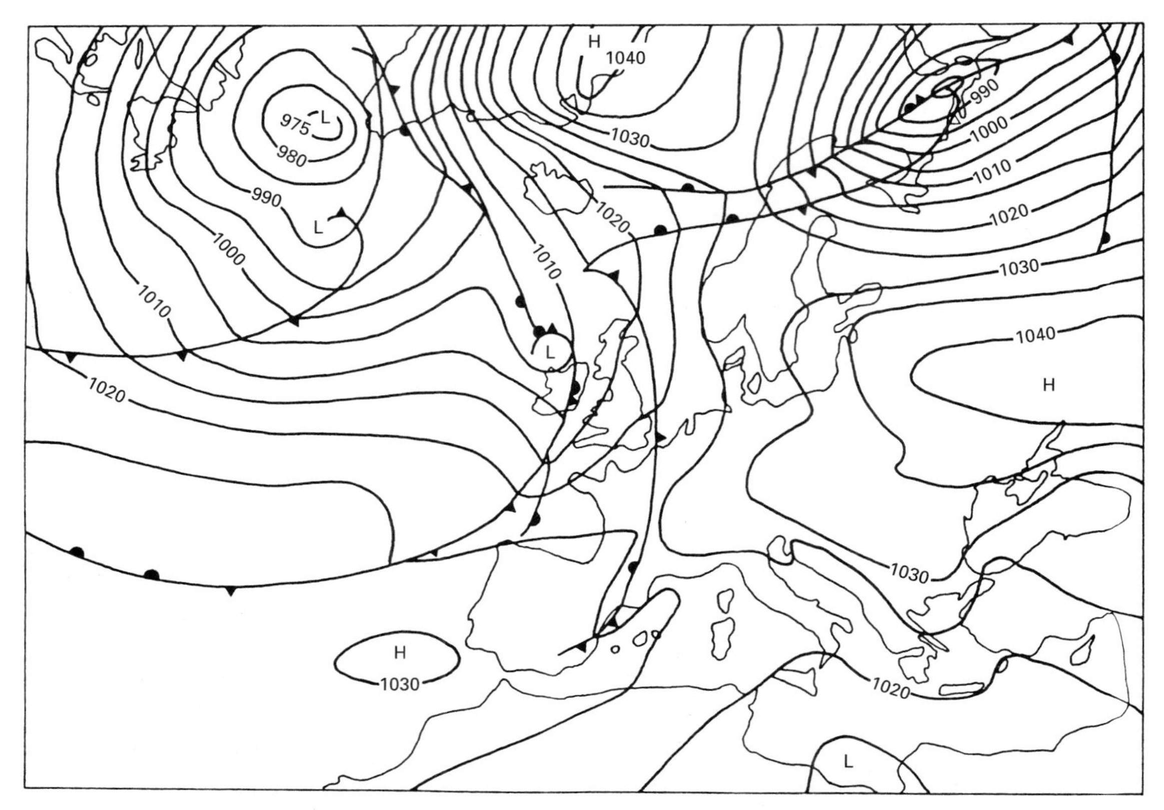


*Figure 9.8* Surface chart for 1200Z on 20 January 1982, showing a well-developed and partially occluded depression system approaching the UK with, ahead of it, a weaker depression system with a much wider warm sector. (From *European Meteorological Bulletin*, 1982.)

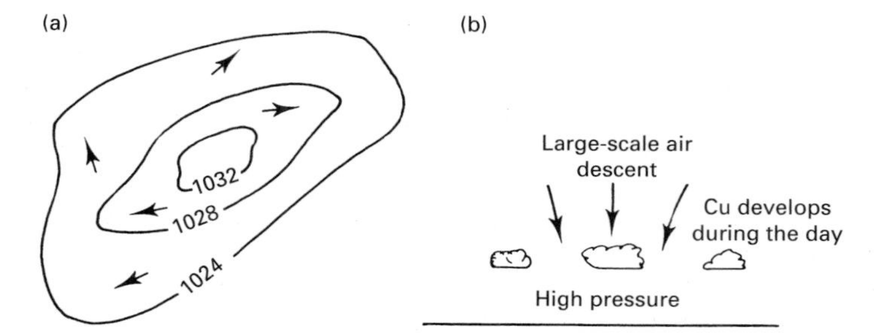


*Figure 9.9* Schematic representation of an anticyclone as seen (a) in plan (units of pressure are hPa), and (b) in cross-section. Anticyclonic conditions can cause large-scale air pollution by restricting vertical mixing. Winter nights are often very cold as clear skies allow rapid cooling, but high values of incident solar radiation can lead to cumuliform cloud development on summer days.

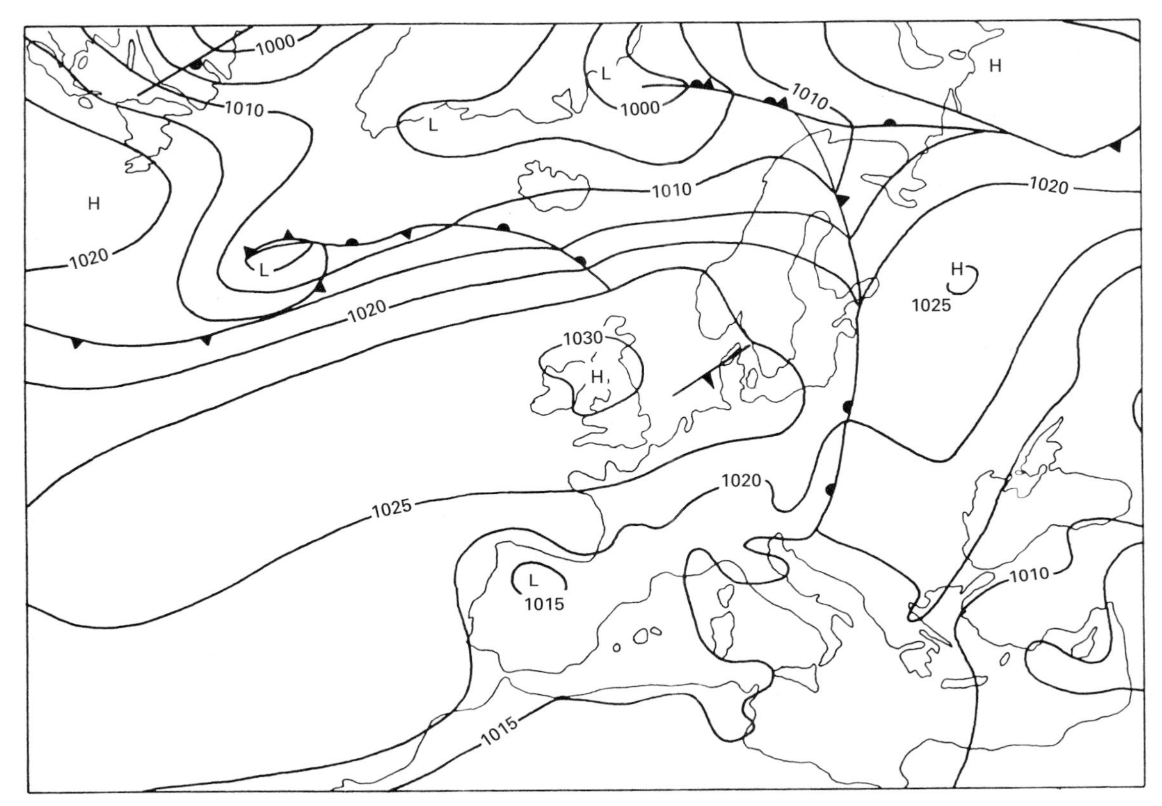


*Figure 9.10* Surface chart for 1200Z on 18 July 1982, showing a well-developed anticyclone centred on the UK which is forcing depression systems to travel far to the north of their usual path. (From *European Meteorological Bulletin*, 1982.)

region of wintertime interior Asia and the companion region over Arctic Canada are persistent seasonal phenomena. There is, however, a group of anticyclones that are small-scale synoptic features roughly comparable to depressions.

These **travelling anticyclones** occur in a number of ways. They occur in embryonic form between individual members of a depression family, giving a period of clear weather between depression passages. Commonly behind the last member of such a family there is a 'burst' of cold air of polar origin which establishes a high pressure region. Similar scale anticyclones are also established as a result of the index cycle (Figure 9.1). These usually take the form of **blocking anticyclones**. These features can persist for several weeks, travelling eastward only very slowly, if at all. Thus they can greatly influence regional weather conditions. It is in a blocking situation with the anticyclone over northwestern Europe, for example, that Britain commonly experiences warm dry airflows from the south or southeast. This gives respite from the continuous stream of depressions, which are now diverted around the blocking feature and pass to the north of the country (Figures 9.10 and 9.11).

### 9.2.3 Air masses

Away from the main line of the Rossby waves are areas where conditions are barotropic. In such regions when calm or near-calm conditions persist for several days, **air masses** develop. These are features having horizontally uniform temperatures and humidities, the values being dictated by energy and moisture exchanges with the underlying surface, and generally calm and cloud-free weather. Eventually an air mass will move, either because it has become sufficiently

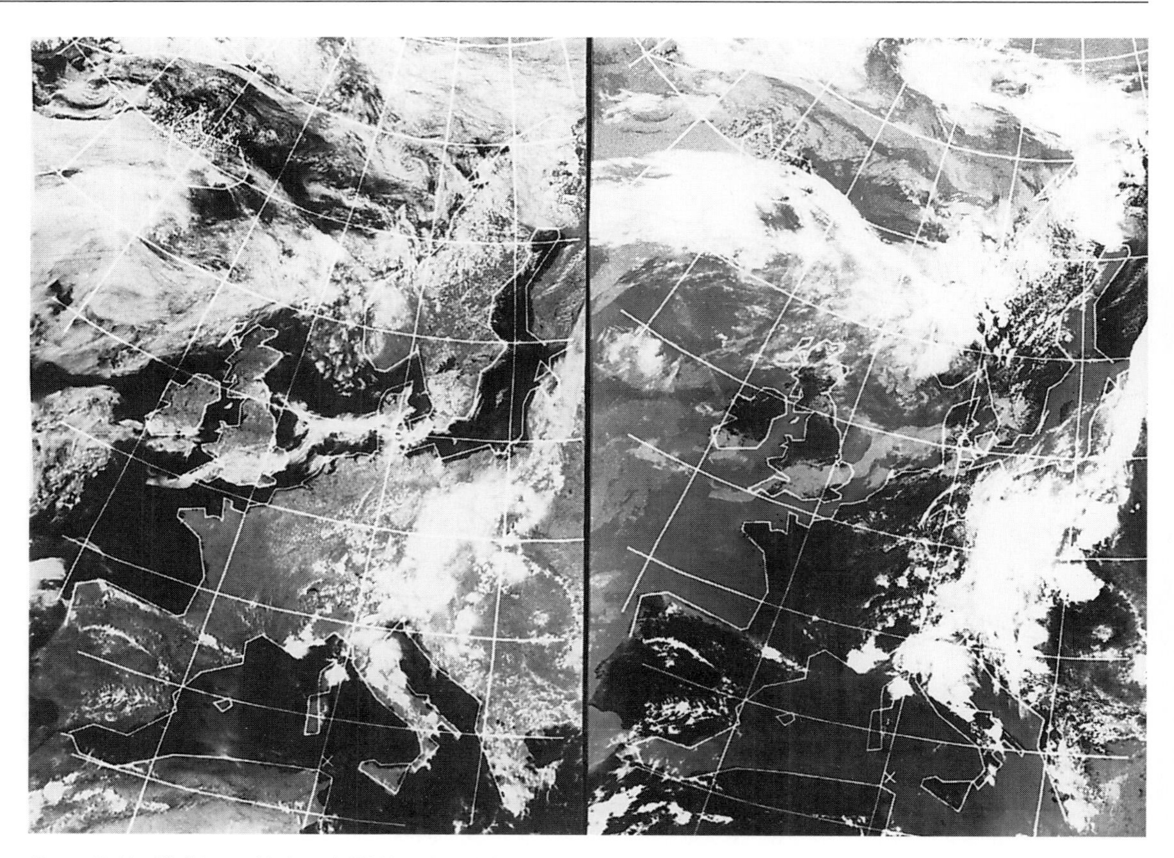

*Figure 9.11* Visible and infrared NOAA polar orbiting satellite images for 1348Z on 18 July 1982. The position of the weak depression system being forced northward (cf. Figure 9.10) is clearly seen. In addition, the use of combined information from two wavelength regions for identification of cloud height is clearly illustrated by the low-lying stratus cloud over southern Ireland and southern England. This cloud band appears bright in the visible image but is relatively dark in the infrared image, suggesting that it is low level cloud with temperatures not very different from those of the surface. (Courtesy of P. Baylis, Department of Electrical Engineering, University of Dundee.)

strong to influence the general circulation in mid-latitudes, or because changes in the wave circulation force it to move into a new area. In either case it will take its characteristics with it into the new area and thus affect the weather there.

Theoretically any area on Earth could become an air mass **source region**, but in order for the characteristics to develop sufficiently to become a significant feature of the circulation, an area of the order of a million square kilometres or more is needed. Thus in practice there are only a few common source regions (Figure 9.12). Most notable are the centres of the semi-permanent high pressure areas at the outer limits of the Hadley cell and, in winter, the thermally induced high pressure regions over the poleward continental margins. The other common source regions are the semi-permanent low pressure regions in the high mid-latitudes, especially those over the oceans. Since the air remains for several days over a source region, it takes on the thermal and moisture characteristics of the region. Its stability is also influenced by the underlying surface; air which comes into a cold source region will lose heat to the underlying surface and will tend to become a stable air mass, while air moving into a warm source region will tend to become hydrostatically unstable. The main characteristics of the common types of air masses are summarised in Table 9.1. The variation in stability

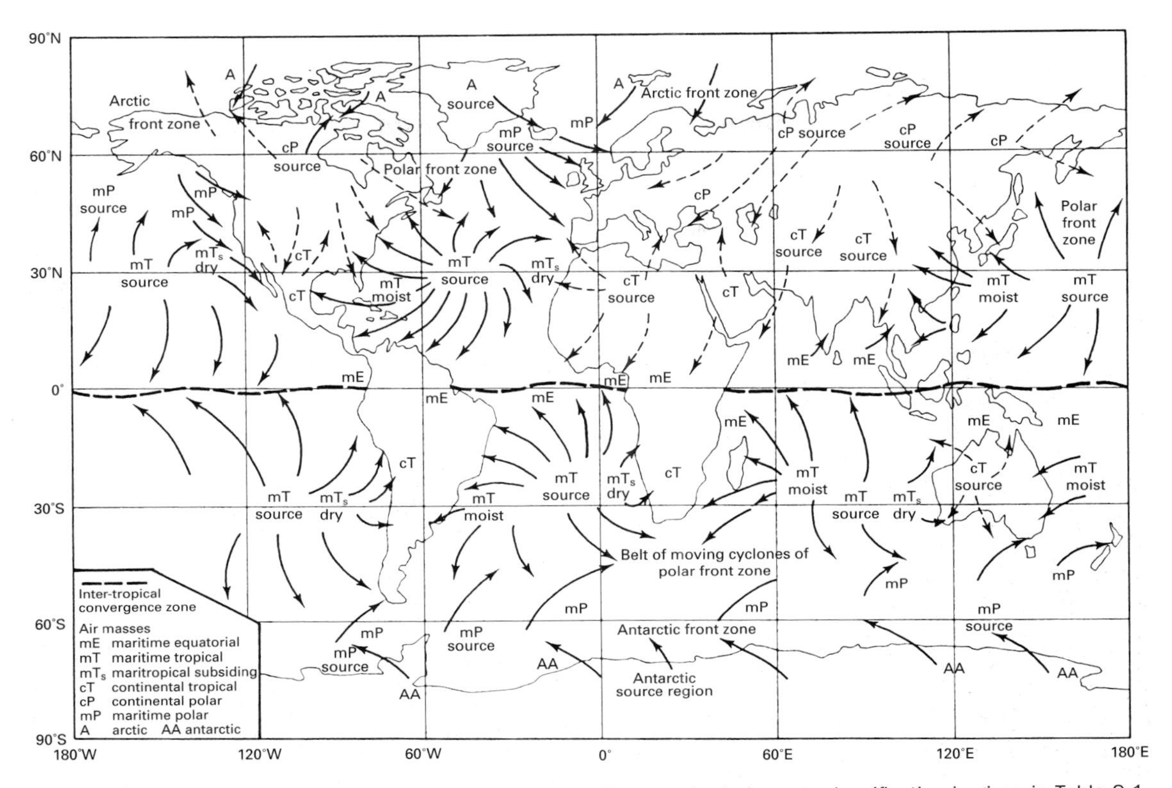


*Figure 9.12* Global distribution of air mass source regions. The standard air mass classification is given in Table 9.1.

*Table 9.1* Classification of air masses

| Major group | Subgroup | Source regions | Properties at source |
|---|---|---|---|
| Polar (P) | Maritime polar (mP) | Oceans poleward of approx. 50° | Cool, rather damp, unstable |
| | Continental polar (cP) | Continents in vicinity of Arctic Circle | Cold and dry, very stable |
| | Arctic (A) or Antarctic (AA) | Polar regions | Cold, dry, stable |
| Tropical (T) | Maritime tropical (mT) | Trade-wind belt and subtropical oceans | Moist and warm, stability variable: stable on east side of oceans, rather unstable on west |
| | Continental tropical (cT) | Low latitude deserts, chiefly Sahara and Australian deserts | Hot and very dry, unstable |
| | Maritime equatorial (mE) | Equatorial oceans | Warm, moist, generally slightly stable |

indicated for the maritime tropical air mass arises because of the effects of ocean currents on the underlying temperature. The basic condition is one of near-neutral stability, but the warm waters of the west side of the ocean convert this to an unstable tendency, while cold currents on the east side lead to stability. It was fashionable at one time to provide numerous subclasses of air mass types but the resulting variations were of degree, rather than nature. Hence the simple scheme used here, using six basic types, is adequate for all practical purposes.

Changes in the pattern of the general circulation will cause these air masses to move from their source regions. In general, polar air masses will move over increasingly warm surfaces. When such movement takes place the air mass is symbolised as mPK or cPK, the K indicating that the air mass is colder than the surface over which it is moving. These air masses will be warmed from below and will decrease in stability as they move. Thus some cloud formation may become associated with them. However, they do tend to move as a coherent body of air, frequently taking on anticyclonic characteristics as they move into the main line of the westerlies. They can thus themselves influence the route of the Rossby waves. The tropical air masses, when they move, will become more stable, since they are warmer than the surfaces over which they pass. The appropriate symbols are therefore cTW and mTW. These, however, are only general guides. Regional variations can easily occur. mT air that affects the southeastern United States, for example, has its source in the subtropical high pressure region of the North Atlantic Ocean. It is likely to be mTW in winter when the continent is cold, but mTK in summer when the continent is considerably warmer than the ocean.

Much mid-latitude weather forecasting was developed using air mass movements as the basic guide, an approach termed **air mass analysis**. Modern meteorology no longer uses this approach extensively; however, from a climatological viewpoint it provides a useful concept. Since air masses, anticyclones and depressions are closely connected, and each in turn influences, and is influenced by, the Rossby waves and the jet stream, movements in any one feature lead to movements in them all. Furthermore, we can think of the polar front and depressions as precipitation-controlling features, and air masses and anticyclones as temperature-controlling features. Thus air mass analysis combines most mid-latitude circulation features and provides a relatively simple means of characterising mid-latitude climates. Indeed, this approach has often been used as a framework for a genetic classification of climate.

## 9.3 Mid-latitude climate regions

The analysis of the climate regions of mid-latitudes can be approached by considering the frequency and seasonality of the influence of each of the mobile features on particular areas. In general, the Rossby waves, because of their great variability in position from day to day, allow penetration of polar air masses equatorward at certain times, and tropical air masses poleward at others. Further, since the Rossby waves are likely to be closer to the equator in winter than in summer, it follows that polar air can more easily penetrate farther equatorward in winter than in summer, thus enhancing the seasonal contrasts anticipated from global energy considerations. Since the precipitation-bearing depressions are guided by the Rossby waves, we can also anticipate a seasonal precipitation distribution in keeping with the seasonal changes in wave position. Further, since depressions are guided from west to east, it is easy to speculate that they are likely to yield the most precipitation on the west coasts, with a gradual decrease in amounts inland. However, it is likely that the truly dry climates in mid-latitudes will be confined to continental interiors, since on east coasts moisture can be advected over the continents both as the result of air mass movements, particularly moisture-laden mT air, and monsoon-like motions associated with continentality effects.

It is difficult to go beyond these general statements for most regions because superimposed on these general effects are a host of local features. These include the mesoscale atmospheric systems, notably the thunderstorms considered in Section 5.3, and the influence of local surface effects to be analysed in Chapter 10. In addition, the details for each continent are very much influenced by topography. However, some statements applicable to all continents can be made for a group of climates which might be called 'west coast climates', which in the Köppen system have the symbols Cs, Cf and Df as one moves successively poleward. These provide our starting point.

### 9.3.1 West coast climates

The **Mediterranean** climate, Cs, is one of warm to hot dry summers and cool, wet winters (Figure 9.13).

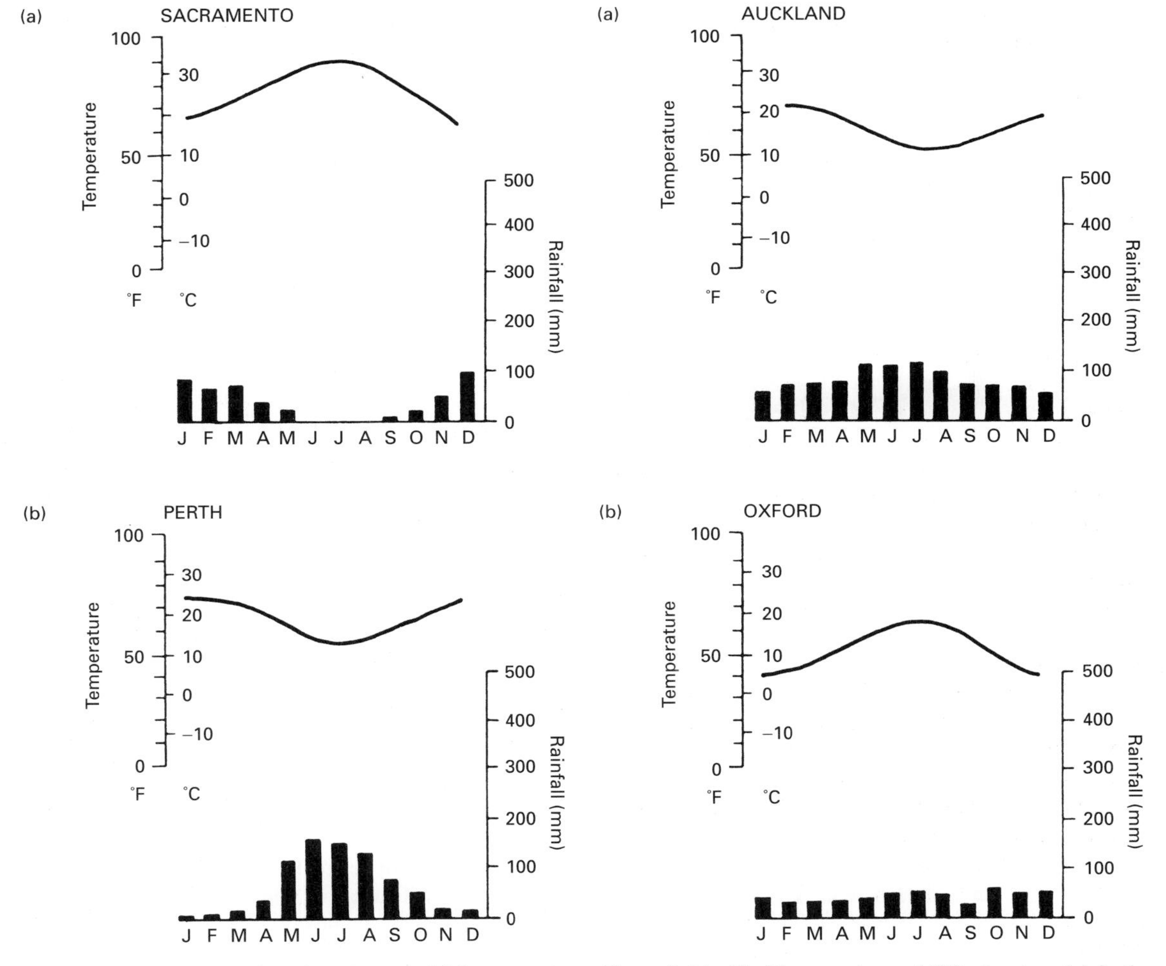


*Figure 9.13* Mediterranean (Cs) climates: (a) Sacramento (39°N, 22°W) and (b) Perth (32°S, 116°E) (for location see Figure 7.11). (From Tanner, 1971.)

*Figure 9.14* Maritime west coast (Cf) climates: (a) Auckland (37°S, 174°E) and (b) Oxford (52°N, 2°W) (for location see Figure 7.11). (From Tanner, 1971.)

During the summer the areas are influenced by the poleward limb of the Hadley cell, with clear, cloudless conditions. The Rossby waves, with their depressions, are poleward of the areas, and only occasionally does a trailing cold front bring summer rain to the regions. In winter, however, these depression tracks are more nearly overhead and bring precipitation to these areas. Their west coast position makes it relatively rare for cP air to move off the continent against and across the planetary waves, so that very low temperatures are rarely encountered. Cold spells are usually associated with mP air, which is considerably less cold than cP air and is likely to give cool, cloudy conditions.

Poleward of the Mediterranean climatic region are the Cf climates, frequently called **marine west coast** climates, which are influenced by depressions all year round and so have rain at all seasons (Figure 9.14). On average, of course, they are cooler than their equatorward neighbours. This is most marked in the summer, where frequent cloudy periods ensure relatively low monthly temperatures. Again, however, winter cP is a relatively rare air mass, although not as uncommon as farther equatorward.

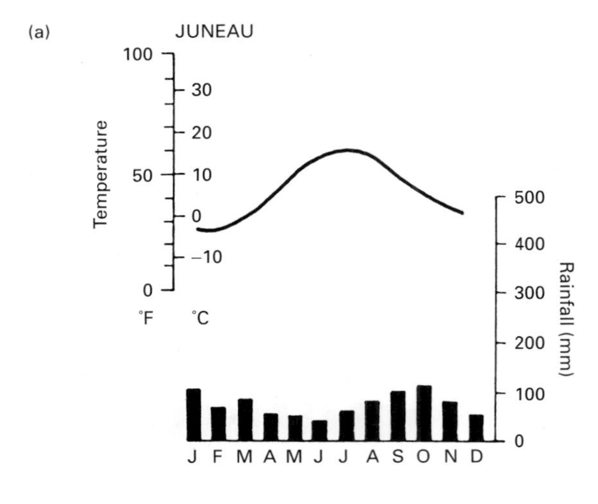


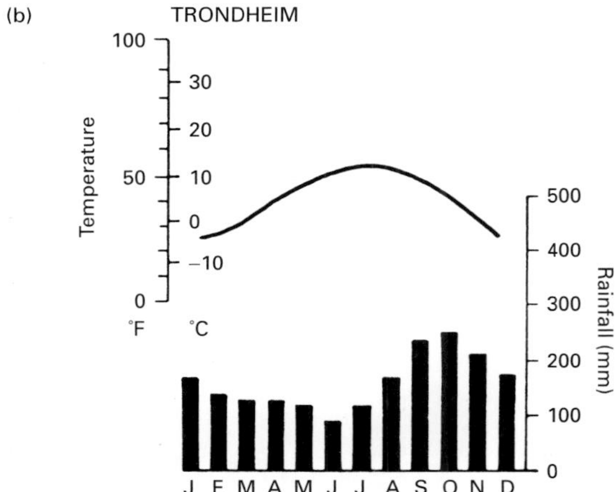


*Figure 9.15* Cold west coast (Df) climates: (a) Juneau (58°N, 134°W) and (b) Trondheim (63°N, 11°E) (for location see Figure 7.11). (From Tanner, 1971.)

Proceeding poleward to the Df climate, which may be called the **cold west coast** climate, the major change is a lowering of temperatures throughout the year and a resultant much shorter growing season (Figure 9.15).

The summer minimum in precipitation so characteristic of the Mediterranean climate is generally carried through to the other west coast climates. The major reason is that the winter depressions tend to be much more vigorous than the summer ones. The trend is certainly very well developed in North America and New Zealand, but is less clear in Europe, being completely obscured at Oxford, England (Figure 9.14b). This difference is largely a result of the different topography of the continents and of the stations chosen. In Europe there is no mountain barrier oriented north to south and the west coast climates penetrate a great distance inland. The maritime influence can be felt a thousand kilometres from the sea, the major changes inland being a slow decrease of precipitation amounts and an increase in temperature range as the continentality effects are enhanced. The Alps serve mainly to provide a distinct separation of the Mediterranean climates of the south from the more temperate ones to the north. In North America the situation is almost the reverse. The Rocky Mountains provide a barrier to the penetration of west coast climates inland, but there is a more gradual transition as one moves north along the west coast. Only in northern Europe is there a comparable mountain barrier, but even here the differences are great. The Rocky Mountains steer depressions into the Gulf of Alaska and tend to ensure that they remain there, whereas the mountains of Scandinavia force the depressions to detour to the north, but allow them to continue their eastward progress. The west coast of South America is on its equatorial side akin to North America, but as one moves poleward, strongly resembles Scandinavia and has a similar climate. West coast climates also occur in restricted areas of southern Africa and Australasia.

There are also marked differences in energy regimes between the three west coast climates (Figure 9.16). The impact of the summer dryness is the most obvious feature of the Mediterranean type. Evapotranspiration is close to zero and sensible heat is responsible for removing the surplus of net radiation. In many ways this Cf summer resembles a hot desert climate (Figure 8.9). However, as winter approaches, the sensible heat flux decreases towards zero and virtually all of the net radiation drives the latent heat. Indeed, until the onset of the dry season the Bowen ratio remains around 0.5, despite the rapid increase in the amount of energy involved. The marine west coast climate displays a relatively small change in Bowen ratio throughout the year, with winter being close to 0.5 and summer nearer 0.8. However, it is clear that water is never a limiting factor. For the colder Df climate, however, net radiation is negative in the winter, when the ground is likely to be frozen and with an high albedo snow cover. The sensible heat flow is from the air towards the surface, and the small amount of evaporation will occur during periods of midday thawing. Outside this cold period there is likely to be an excess of precipitation

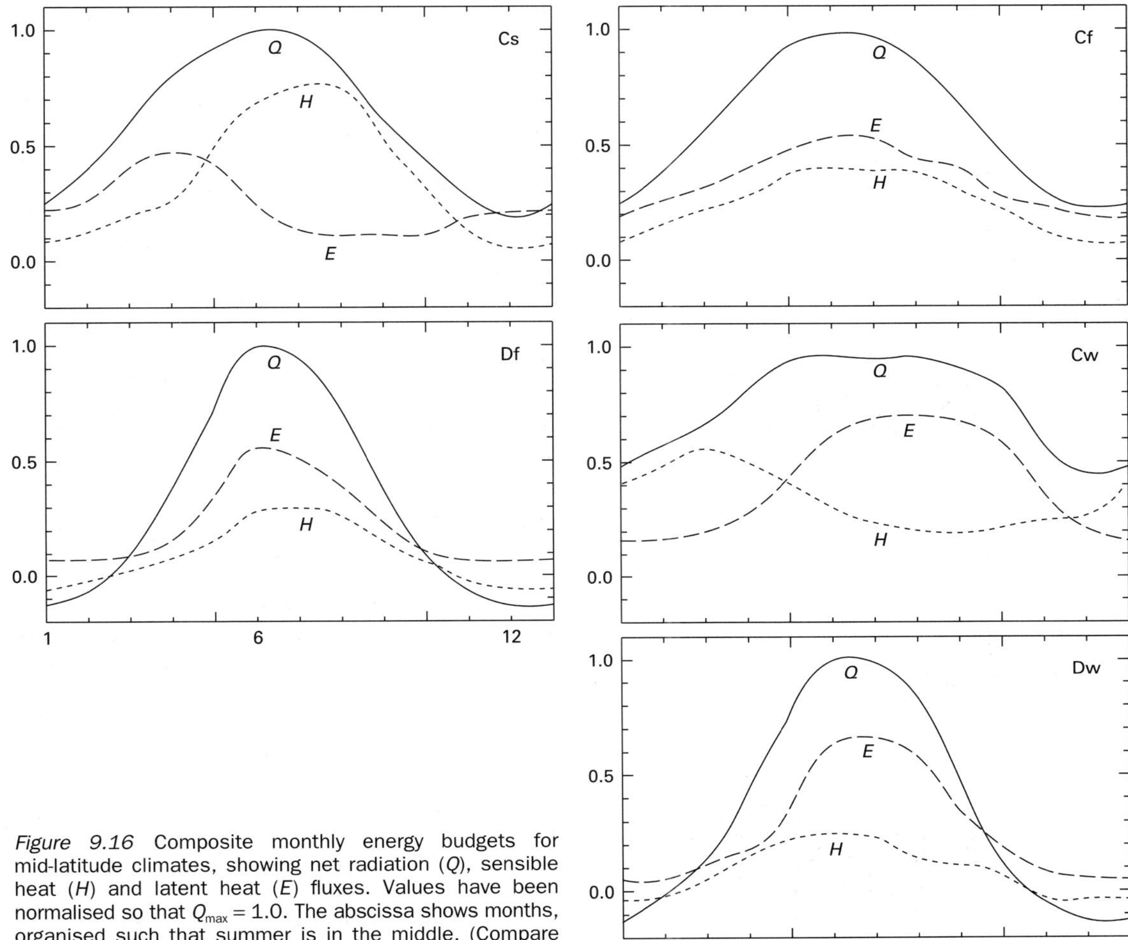


*Figure 9.16* Composite monthly energy budgets for mid-latitude climates, showing net radiation ($Q$), sensible heat ($H$) and latent heat ($E$) fluxes. Values have been normalised so that $Q_{max} = 1.0$. The abscissa shows months, organised such that summer is in the middle. (Compare with Figure 8.9.) (After Kraus and Alkhalaf, 1995.)

over evaporation and a moist surface. Consequently the latent heat flux considerably exceeds the sensible heat one.

### 9.3.2 Continental interior and east coast climates

Away from the west coasts a slow decrease in precipitation amounts and an equally slow increase in the annual temperature range would be anticipated, and would be expected to continue until, relatively close to an eastern coast, a maritime influence was again encountered. Although this is generally the case, each continent has topographic differences which, to varying degrees modify the pattern.

For Eurasia the west coast influence can easily penetrate inland and the **interior desert** climates are not reached for thousands of kilometres. Further, they tend to be cold deserts, BSk and BWk, occurring on the northern periphery of the east–west oriented Alps–Himalayan mountain system. This mountain system acts as a barrier to the penetration of warm air from the tropics. Further, the regions themselves are commonly high plateaux, not only fostering low temperatures but also serving to steer to the north any depressions which penetrate this far. Indeed, the northern regions of Eurasia are Df climates in the west (Figure 9.17b), expressing the inland penetration of the west coast regime, and Dw in the centre and east (Figure 9.17a), responding in part to the greater

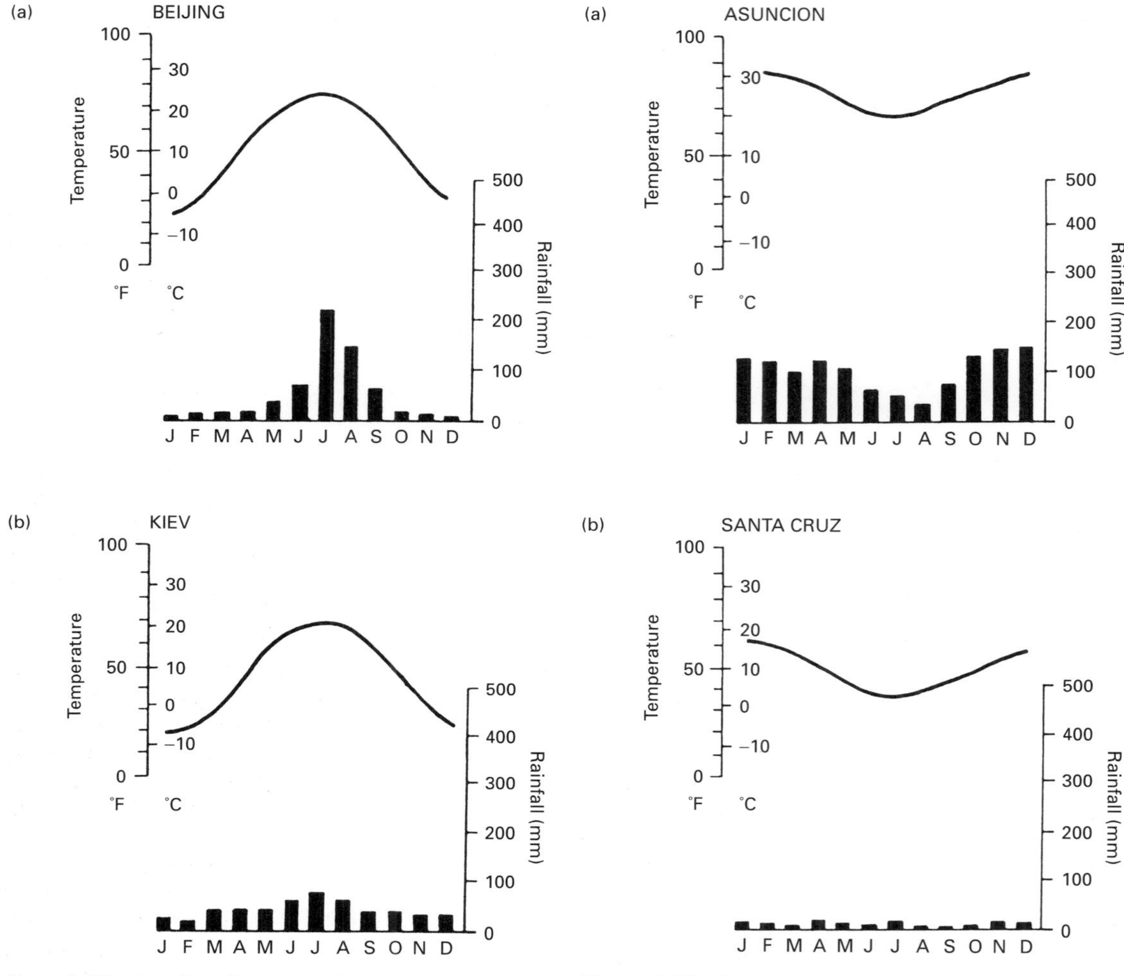


*Figure 9.17* Eurasian climates: (a) Beijing (40°N, 116°E), Dwa climate, and (b) Kiev (51°N, 31°E), Dfb climate (for location see Figure 7.11). (From Tanner, 1971.)

*Figure 9.18* Southern Hemisphere east coast interior climates: (a) Asuncion (25°S, 58°W), Cf climate, and (b) Santa Cruz (50°S, 69°W), BWk climate (for location see Figure 7.11). (From Tanner, 1971.)

frequency of depressions in summer than in winter. The two have similar energy regimes, the major difference being that the time of the major evaporation period is during the snowmelt period of spring in the Df climate, and during the summer wet season for the Dw region (Figure 9.16). The eastern coastal region of Asia is also, of course, part of the monsoon region and certainly the effects of this regime, particularly the precipitation distribution, can be seen far away from the area usually regarded as the true monsoon area.

The relatively small amount of land in the Southern Hemisphere precludes extensive development of continental interior and east coast climates. They are best developed in South America, where there is an area of east coast Cf climate. Asuncion, Paraguay (Figure 9.18a), displays the seasonal temperature variations and the distinct summer maximum in precipitation typical of such climates. The topographic situation and climate is similar to that of the moist southeast United States. South of this moist region is a much drier area. Although this corresponds in latitudinal and topographic setting to the southwestern deserts of the United States, its extent is restricted so that only the cold desert, symbolised by BWk, occurs (Figure 9.18b).

Over the other southern continents the arid areas are hot, BWh, deserts. They occur primarily as a result of the descending limb of the Hadley cell and thus are a feature of the tropical circulations. The south and west margins of these continents, however, do have regions influenced by the westerlies, with the resultant moist west coast climates clearly indicated in Figures 9.13 and 9.14. These are all warm, C type climates. Only at the southern tip of South America do any of these southern continents penetrate far enough poleward to experience a cold, Köppen type D, climate.

### 9.3.3 Dynamic climatology of North America

For all North American climates the Rocky Mountains provide a real barrier. In the case of climates away from the west coast, 'barrier' is a highly appropriate word. The desert areas are in the gigantic rain shadow of these mountains. Here rain shadow is used in both its traditional sense and in the sense that the Rocky Mountains influence the position of the Rossby waves. Commonly these waves have a crest at, or close to, the northern Rocky Mountains. They thus commonly swing south over the western interior, reach a trough over the Mississippi River valley and then move off the continent in a northeastward direction (Figure 9.19). Since there is no appreciable east–west oriented mountain barrier, these waves are capable of moving cP air deep into the interior and very far south. They can equally easily move mT air of Gulf of Mexico origin to the margins of the Arctic. Thus short period temperature fluctuations can be very marked.

Expressing these ideas in a complementary way, the polar front can vary in position over a very wide area (Figure 9.19). Thus, as far as temperatures are concerned, the extreme southeast of the continent is dominated by mT air all the year round. As one moves northwestward the time of mT dominance is slowly decreased until somewhat north of the US/Canadian border cP air dominates throughout the year. This effect is combined with the continentality effect to produce not only lower mean annual temperatures in the interior, but also a much higher annual temperature range (Figure 9.20).

The cP air mass is cold and dry. As it moves south it may become slightly unstable and produce some cloud, but rarely does it contain enough moisture to provide significant precipitation. The warm and humid mT air mass on the other hand, has its source region over the tropical Atlantic or the Gulf of Mexico. The major moisture source for the North American continent east of the Rocky Mountains is Atlantic air. Without the action of depressions there would be a very marked decrease in precipitation from the east coast inland. One type of depression forms on the polar front west of the Mississippi. Frequently the warm sector consists of mT air, which has had a long trajectory over land. Hence precipitation is rather sparse. However, once air from the Gulf of Mexico is incorporated directly, usually in areas east of the Mississippi, the vigour is enhanced and precipitation amounts increase. A second type of depression is the storm that manages to cross the mountain barrier from the west. Usually such a storm loses much of its moisture but maintains its energy while crossing. Consequently it can rapidly replenish itself as it moves over a moist surface, or can release its remaining moisture over the interior. These storms are the main source of precipitation in the interior of the northern United States and southern Canada.

The warm and humid mT air gives rise to convective precipitation, especially in summer when its instability is increased as it moves northwestward over a hot land surface. Consequently the eastern US has a high frequency of thunderstorms (Figure 5.6), and in places more than 75% of the summer precipitation is from this source. In addition, when humid, unstable near-surface mT air moving northwestward is overlain by eastward moving dry air which has been adiabatically warmed by descent after crossing the Rockies, conditions are ideal for tornado formation. A relatively small trigger, such as a squall line (Figure 5.8), can release the latent instability and initiate tornado development. The resulting distribution of tornado incidence clearly reflects the dual air flows required (Figure 9.21). Although tornadoes can occur wherever severe thunderstorms are possible, only in North America do topography and airflow patterns combine to create frequent major tornado outbreaks. Hence only in this region are they regarded as a significant part of the regional climatology.

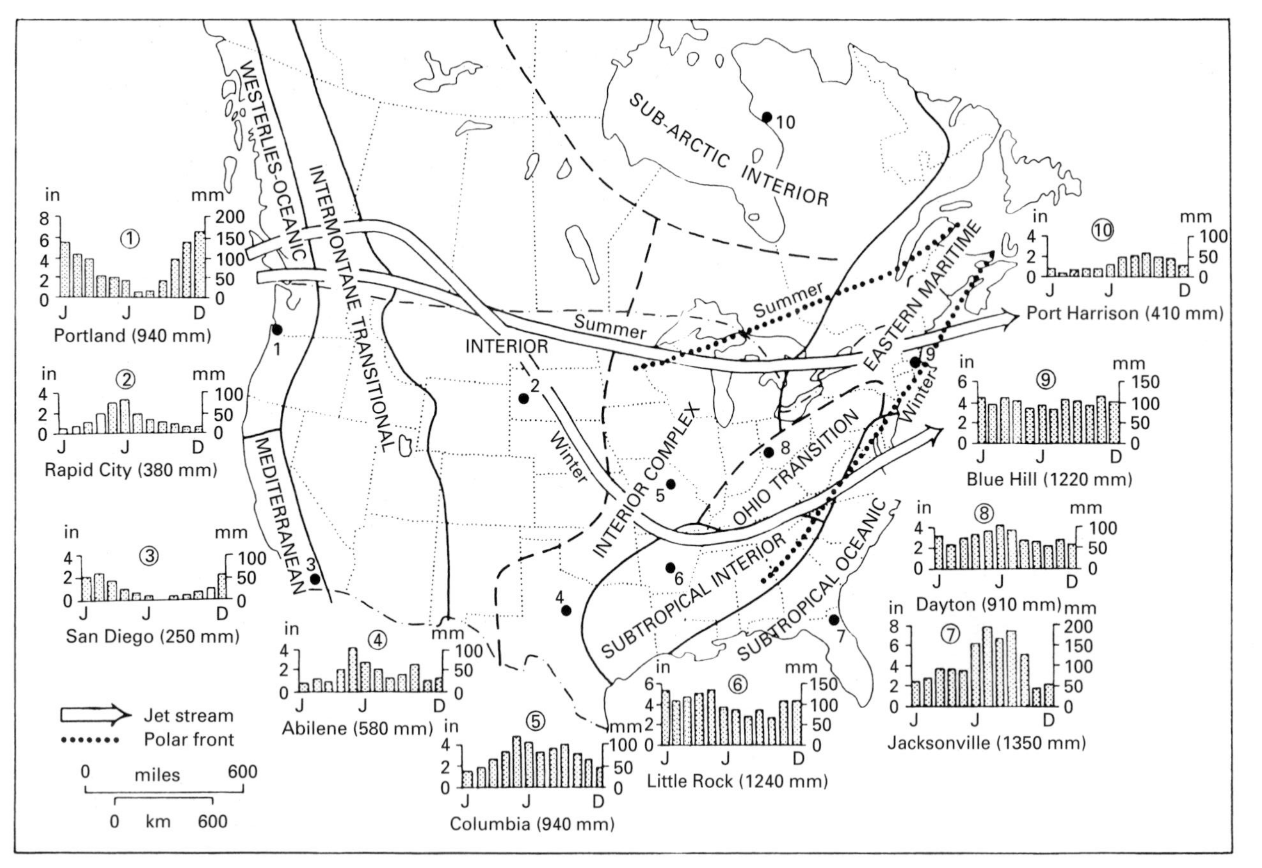


*Figure 9.19* Dynamic climatology of North America showing the summer and winter positions of the jet stream and the polar front. Note the orographic anchoring of the jet stream by the Rockies. Rainfall regimes are shown on the inserted histograms (months of January, June and December are indicated) together with the total annual rainfall in parentheses for the 10 numbered stations.

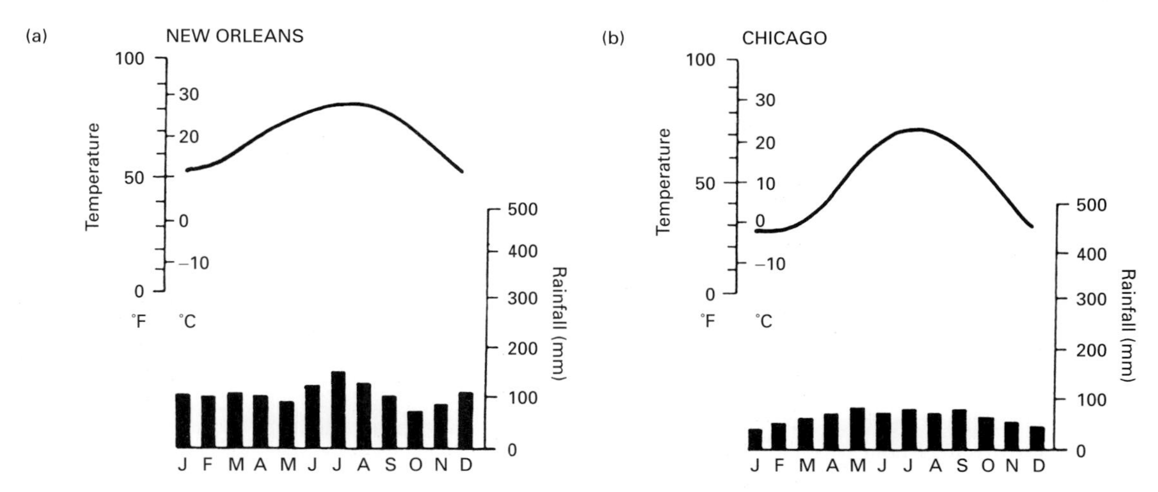


*Figure 9.20* Continental interior climates of North America: (a) New Orleans (30°N, 90°W), Cfa climate, and (b) Chicago (42°N, 88°W) Dfa climate (for location see Figure 7.11). (From Tanner, 1971.)

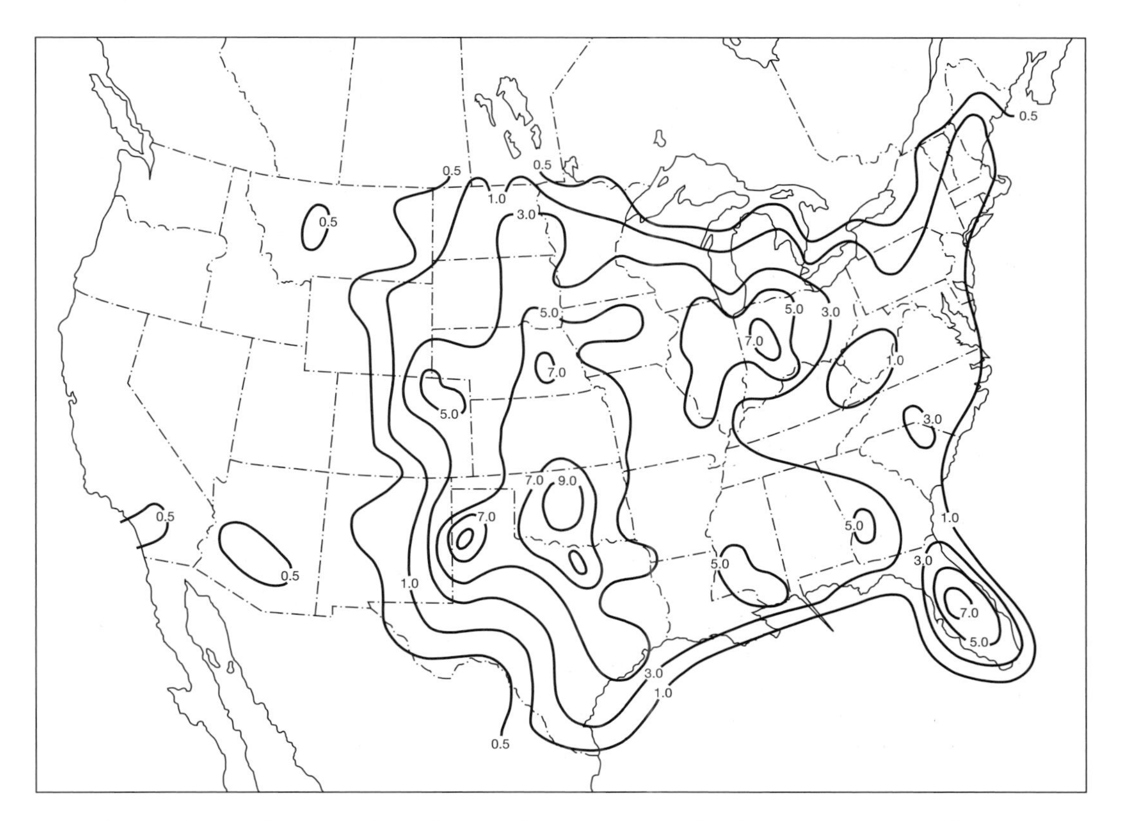


*Figure 9.21* Average annual tornado incidence in the continental United States. (Courtesy of NOAA.)

## 9.4 Polar climates

Polar climates are, of course, cold climates. In the Köppen system they are defined as those climates where the average temperature of the warmest month is not above 10 °C. Even with this constraint, however, there are variations in the climate from place to place. In particular, continentality effects occur and annual temperature ranges can be large in inland areas (Figure 9.22a). Diurnal temperature changes are not as marked as in mid-latitudes, mainly because solar zenith angle changes are small during the day. Indeed, right at the poles there is no change in zenith angle in any given 24-hour period. Instead there is a 'day' six months in length, followed by an equally long 'night'. As a result, freezing days, days when the temperature never rises above 0 °C, are relatively uncommon in maritime areas even in winter, although in inland areas periods of many days continuously below freezing are common.

In areas where the mean annual temperature is at or below −9 °C, **permafrost**, permanently frozen ground, can exist. The ground may be completely frozen in winter, but a thin surface layer may thaw if air temperatures above freezing occur during summer. This promotes growth of some sparse vegetation, which acts as an insulating layer for the lower levels, which therefore remain frozen. Although there is this annual near-surface freeze–thaw cycle, the depth of permafrost must be, at least partially, a function of palaeoclimatic conditions since deep layers are insulated from current surface conditions and temperature changes take a long time to penetrate. In Canada, where over 50% of the land area is permanently frozen, the depth is about 2 m at the southern edge of the permafrost zone, and as great as 300 m in the north.

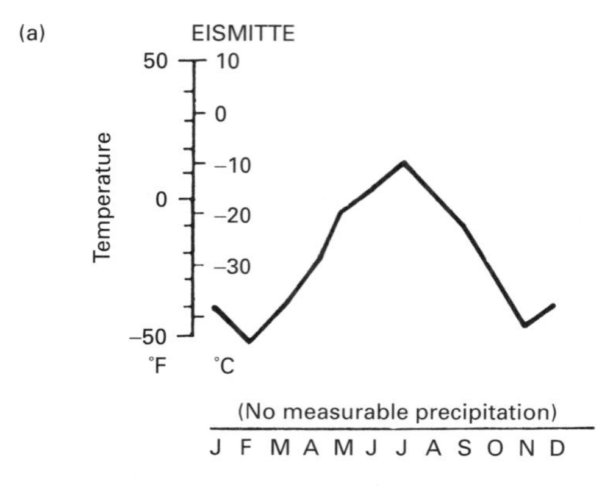


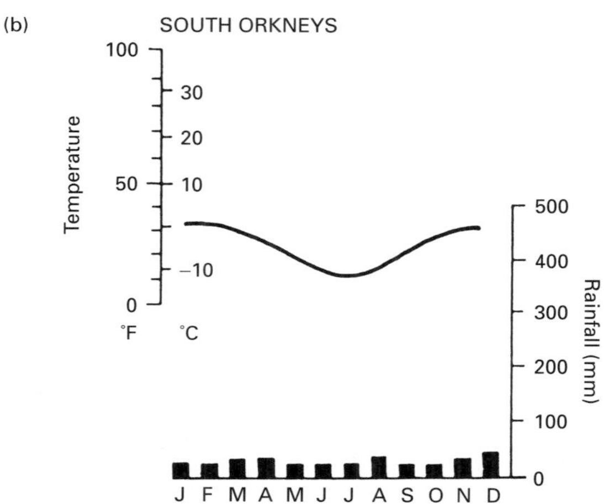


*Figure 9.22* Polar (EF and ET) climates: (a) Eismitte (71°N, 41°W) and (b) South Orkneys (61°S, 45°W) (for location see Figure 7.11). (From Tanner, 1971.)

### 9.4.1 Aridity and windiness at high latitudes

Low precipitation (e.g. Figure 9.22a) is the second major characteristic of climates of the polar cell. While this cell is relatively weak and certainly not a major feature like the Hadley cell, the basic tendency is towards stable conditions and relatively little vertical motion. This stable tendency is enhanced near the surface in winter, when a radiatively caused inversion is common in the lowest 1 km of the atmosphere. Furthermore, the cold air is not capable of holding a great deal of moisture and so, however vigorous the release mechanisms, the resulting precipitation amounts must be small. Although precipitation amounts are low, in summer conditions are often favourable for slow, gentle uplift and low stratus cloud is common. Cloudiness may be around 40% in winter, but may rise to over 70% in summer.

Since there is little energy available for evapotranspiration, most of the precipitation that does fall remains on the ground. Hence in summer, when the ground is unfrozen, the surface is likely to be wet. This effect is exaggerated in the permafrost region where infiltration is negligibly small even in the summer. There is an additional feature associated with moisture in the winter time. When temperatures fall below around −30 °C, atmospheric moisture freezes rapidly into tiny ice crystals in the air close to the

(a)

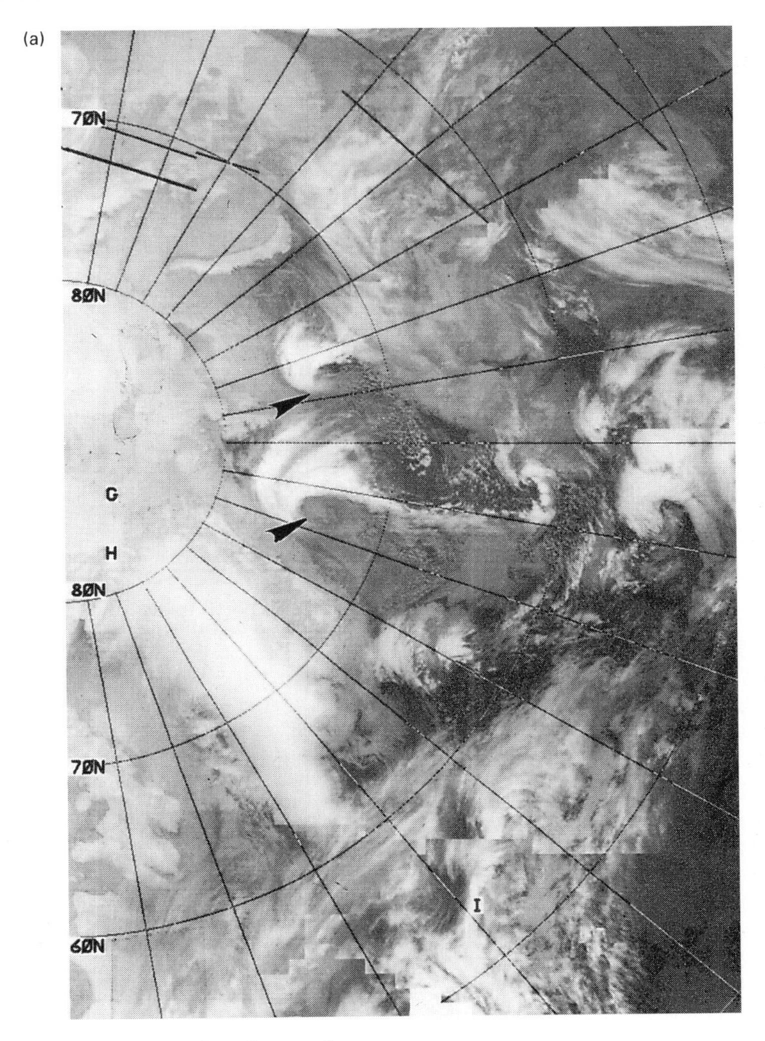


*Figure 9.23* continued on next page

ground, forming an **ice fog**. This can be a serious hazard for people and animals, since the ice crystals can block nasal passages and cause suffocation.

Wind conditions associated with the polar climates of the Antarctic continent are very different from those of the Arctic Basin. The interior of the latter is frequently a region of calms. The Arctic margins can also be calm when the polar cell dominates. At other times this marginal area is strongly influenced by mid-latitude conditions. Depressions are often steered for thousands of kilometres along the boundary between sea ice and open water (Figure 9.23a) and can bring strong winds as well as precipitation to the Arctic. While similar steering occurs along the margins of Antarctica, the high plateaux and mountains in the interior of this continent themselves generate strong winds. Cold dense air originating in the interior sweeps down the slopes creating almost continuous windy conditions. Port Martin (66°55′S, 141°24′E), believed to be the windiest place on Earth, has an average annual wind speed of 64 km $h^{-1}$, while individual 24-hour averages may exceed 105 km $h^{-1}$.

(b)

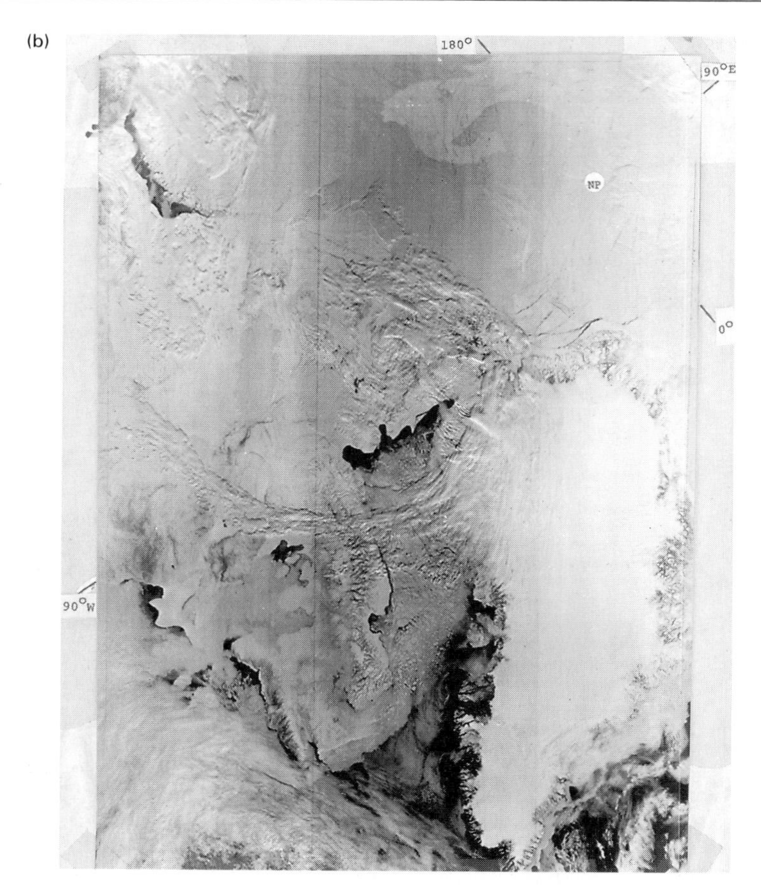


*Figure 9.23* (a) Defense Meteorological Satellite Program (DMSP) infrared image for 12 January 1979 showing two cyclones (arrowed) travelling close to the sea-ice edge in the Greenland and Norwegian Seas. (b) DMSP image of Greenland and the North Pole (dark spot) for 20 May 1979 showing polynyi (dark areas) and cloud shadowing on the bright surface. (Courtesy of D. Robinson, Rutgers University.)

### 9.4.2 Sea ice

The hostile nature of the polar climates means there is a rather sparse network of surface-based observation stations. However, fluctuations in the cryospheric extent, vital for understanding climatic variations, can be monitored by satellite. There appear to be some relatively small but significant long-term fluctuations in sea-ice extent. Seasonal fluctuations are such that in the Northern Hemisphere the sea-ice extent grows from around $7.0 \times 10^6$ to $14.1 \times 10^6$ km$^2$ from summer to winter. In contrast, the seasonal change in the Southern Hemisphere is from $2.5 \times 10^6$ to $20.0 \times 10^6$ km$^2$, reflecting the different continental configurations of the two polar areas. It has also been observed that less than usual ice cover in the Barents and Kara Sea area is associated with more than usual ice in the Chukchi Sea. Similarly, the extent of ice off Alaska in August appears to be associated with the amount off Greenland in the previous June and July. Temporal variations of this type must be reflected in the surface climates of the polar regions, and must influence mid-latitude weather and climate (Figure 9.23a).

The presence of **polynyi**, open water areas, within the sea ice is also easier to study using satellite observations (Figure 9.23b). These are the only places where the relatively warm waters below the sea ice can interact directly with the air. Hence they are of great importance in establishing the energy balance of the area and understanding the processes which create polar climates.

Despite our lack of understanding of polar climates, or even an adequate description of them, it is clear that the polar regions, through their interaction with mid-latitudes, have a profound effect on the climate of the whole globe. It is also clear that the polar regions can be modified relatively easily by anthropogenic influences. Atmospheric pollutants have been observed in the polar regions. Since these pollutants typically have a much lower albedo than the polar surface, marked changes in the surface energy balance can be anticipated. Although there have been few proposals to modify the polar surface directly, water storage and irrigation schemes on the polar margins have the potential to modify the cryospheric extent and the salinity of the Arctic Ocean. Since all segments of the Earth's atmosphere are linked, this in turn could have an impact on the climate of mid-latitude or even tropical areas.

**Box 9.A Applications old and new**

*9.A.1 Housing design*

One of the oldest applications of climate information is for housing design. This is so well established and ubiquitous that we commonly fail to recognise it as an application. Although design cannot be controlled by climate alone, since the availability of suitable building materials is an obvious constraint, it is generally true that a house correctly designed for its climate optimises energy use and human comfort.

In many instances an area's 'traditional' building style incorporates the climate information. In many mid-latitude climates the annual average temperature may approach 18–20 °C, which is usually regarded as the optimum for human comfort. While seasonal temperatures may depart markedly from this, theoretical heat transfer considerations and practical observations of well water temperatures indicate that a relatively constant temperature, close to the annual mean, occurs not too far below ground. Although a hole in the ground is not necessarily a desirable residence, it is probably no coincidence that many early humans dwelt in caves – horizontal holes in the ground. Lacking suitable topography for a cave, a thick-walled house with small openings is a reasonable approximation. In desert regions, where the diurnal rather than the annual temperature cycle is paramount, such houses are common. The design also has the advantage of keeping out any sand-laden desert wind, while a flat roof can catch any of the rare precipitation.

Protection against precipitation may be the major concern for people in wetter climes. A sloping roof provides this. In cold regions one with a sufficiently steep pitch also prevents snow accumulation and building stress. Snow, however, is an insulator, so that an igloo-like structure minimises heating needs. Certainly, thick insulating walls are required in any cold region. In hot regions shade is the vital concern. In hot deserts the cave is again ideal, but in humid conditions it is unbearable. Here as much breeze as possible is needed, so few walls but much shade is required. The porch with overhanging roof was the solution adopted in the US south. This allowed solar radiation to penetrate into and warm the house in the cool, low sun winter period, but prevented direct radiation during the hot summer. High albedo white paint provided further cooling.

This brief discussion indicates some of the connections between housing and climate. In many regions the various climatic elements, or the seasonal variations, suggest contradictory strategies. The United Nations has formulated a series of steps with associated look-up tables, commonly called the Mahoney Tables, to show the best strategy in an area (Table 9.A.1). The guidelines foster the concept of minimum energy use for maximum human comfort. In many industrialised nations where energy for space heating or cooling has become relatively cheap, these climatically driven design criteria have been replaced by more fashionable, if less efficient, styles. If energy becomes more scarce and expensive, or even if fashions change, a return to more climatically appropriate styles can be anticipated.

*9.A.2 Synoptic climatology*

Most climate classifications and descriptions emphasise one or possibly two aspects of the climate, although it is recognised that the climate involves many elements. It is also recognised that these various elements commonly occur in specific combinations. A north wind may often be associated with sunny, cloudless conditions with low temperatures and low humidity. The specification of the nature, frequency and causes of these combinations is the main focus of **synoptic climatology**. As such, one aim is to provide a succinct means of characterising the common weather associated with the climate in a particular area. Our discussion of the dynamic climatology of

**Box 9.A (cont'd)**

*Table 9.A.1* Basic information and steps needed for establishing a climate-sensitive design for a residential dwelling

1. *Obtain monthly climate information needed*
   A. Mean maximum/minimum temperature:
      calculate annual average maximum/minimum and determine day/night comfort zone (look-up table)
      calculate monthly anomalies (for day/night stress)
   B. Mean maximum/minimum relative humidity (early morning/early afternoon observations): assign to high/moderate/low humidity classes
   C. Total precipitation: liquid equivalent
   D. Wind (prevailing direction and secondary direction): 16 point compass directions only

2. *Calculate monthly comfort response indicator*
   Yes/no for each month for each of following six indicators:
   A. Air movement essential: high temperature and high humidity; *or* high temperature with moderate humidity but small diurnal temperature range
   B. Air movement desirable: moderate (in comfort range) temperature, but high humidity
   C. Rain penetration protection needed (N.B. not just shelter): generally needed if precipitation >200 mm
   D. Thermal storage in walls: large diurnal range and low or moderate humidity
   E. Outdoor sleeping desirable: high night temperature with low humidity
   F. Cool season problems: low daytime temperatures

3. *Add total number of months with each indicator*
   Weight for design depends on frequency of occurrence

4. *Follow design criteria table*
   Resolves conflicts between contradictory indicators
   Provides guidelines, not complete designs, for the following:
      layout (best orientation, including courtyard)
      spacing (between buildings)
      air movement (room juxtaposition)
      openings (windows/doors)
      walls (thickness)
      roof (thickness)
      outdoor sleeping (provision)
      rain protection (need)

North America contained several concepts derived from synoptic climatology.

As the name implies, the study is based on the kind of information contained in the analysis of day-to-day weather patterns. Indeed, until recently the development of a synoptic climatology involved reviewing a long sequence of daily weather maps and getting a subjective feel for the common kinds of weather in an area. Then, equally subjectively, the maps were divided into a manageable set of characteristic types. Terms such as 'cold front passage' or 'unstable maritime tropical air mass', usually self-explanatory, were used. Once the categories were identified, each day could be placed, with greater or lesser confidence into an appropriate class and frequencies developed. A major drawback was that it was time-consuming, with any scheme restricted to a rather small area. Moreover, subjectivity meant that it was not repeatable, and different analysts would come to different conclusions.

Now more objective, computer-based, schemes are used. The frequency of occurrence of the various combination of elements is determined and the most common modes identified. These provide the synoptic types directly. Then daily frequencies can be calculated, giving the same kind of result as before. Although objective, the initial elements to include, the number of classes to use, the combinations to investigate, and even the exact method used to match combinations, must be pre-specified, and all are somewhat subjective. Once the decisions have been made, however, the results should be independent of the

**Box 9.A (cont'd)**

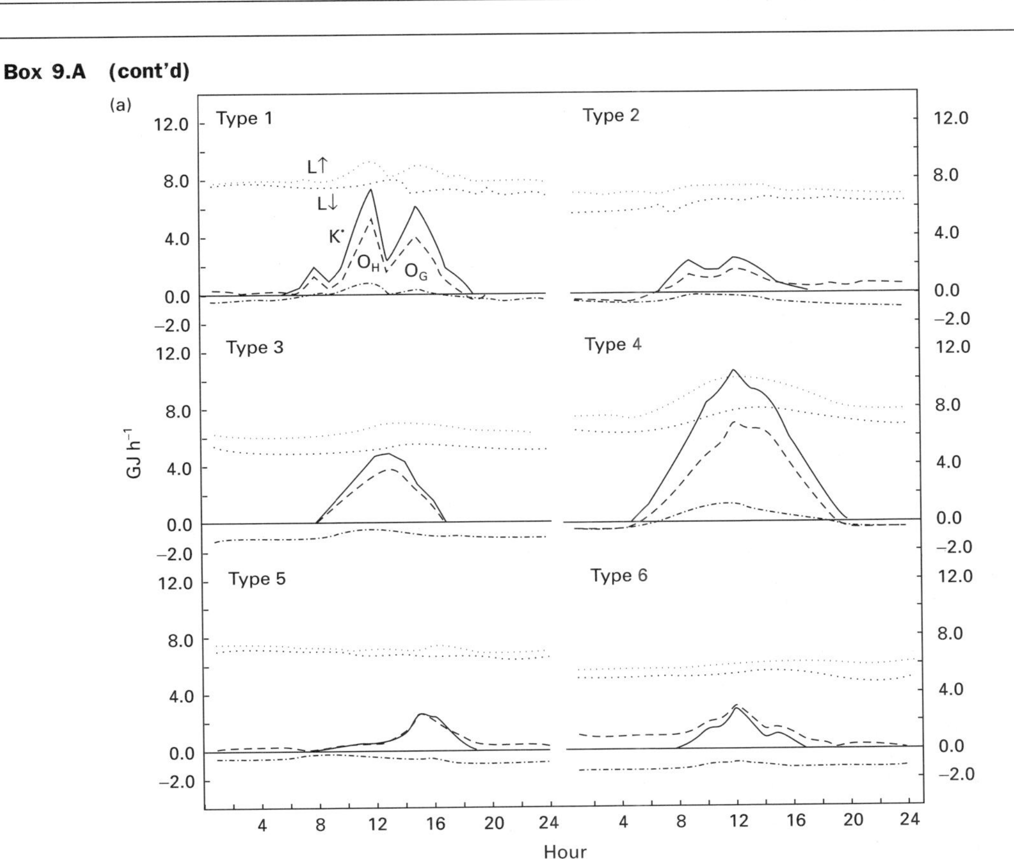


*Figure 9.A.1* The results of a principal component analysis of daily temperature, pressure, humidity and wind speed, followed by a cluster analysis, to show the weather types influencing the energy budget of urban canyons in Boston, USA, showing (a) the typical daily energy budgets of the six types, and (b) the frequency of occurrence of each by month. Symbols are as used throughout this text except that $Q_H$ and $Q_G$ show heat transfer by conduction and convection respectively. Positive values indicate fluxes from the surface. (After Todhunter, 1989.)

person undertaking the analysis, while the same method can readily be used for different times and places. Thus, for example, the analysis for Boston, USA (Figure 9.A.1) could be undertaken equally well for London, England, or Sydney, Australia.

When using many different elements and combinations in analyses such as these, care must be taken to ensure that each is given the correct importance or 'weight'. In particular, many climatological elements have both spatial and temporal **autocorrelation**, where one observation provides a good estimate of the next. As a simple example, a temperature observation at one hour provides a very good estimate of the temperature both for the next hour and at another location a few kilometres away. An observation of hourly total precipitation would allow much poorer comparable precipitation estimates. Identifying the various degrees of autocorrelation, and compensating for them, is rarely as obvious as this simple example indicates. From our consideration of air masses, for example, temperature, humidity and cloud amount may be autocorrelated. It is common to use a principal component analysis (Box 7.A) to remove as much autocorrelation as possible and to reduce the data to manageable proportions. Then a **cluster analysis** is used to determine the synoptic categories. This is an objective scheme to evaluate the numerical 'distance' between the components for each day (or location in a spatial analysis) and create clusters of the days which are closest together. This two-stage numerical approach was used, for example, in the development of Figure 9.A.1.

**Box 9.A (cont'd)**

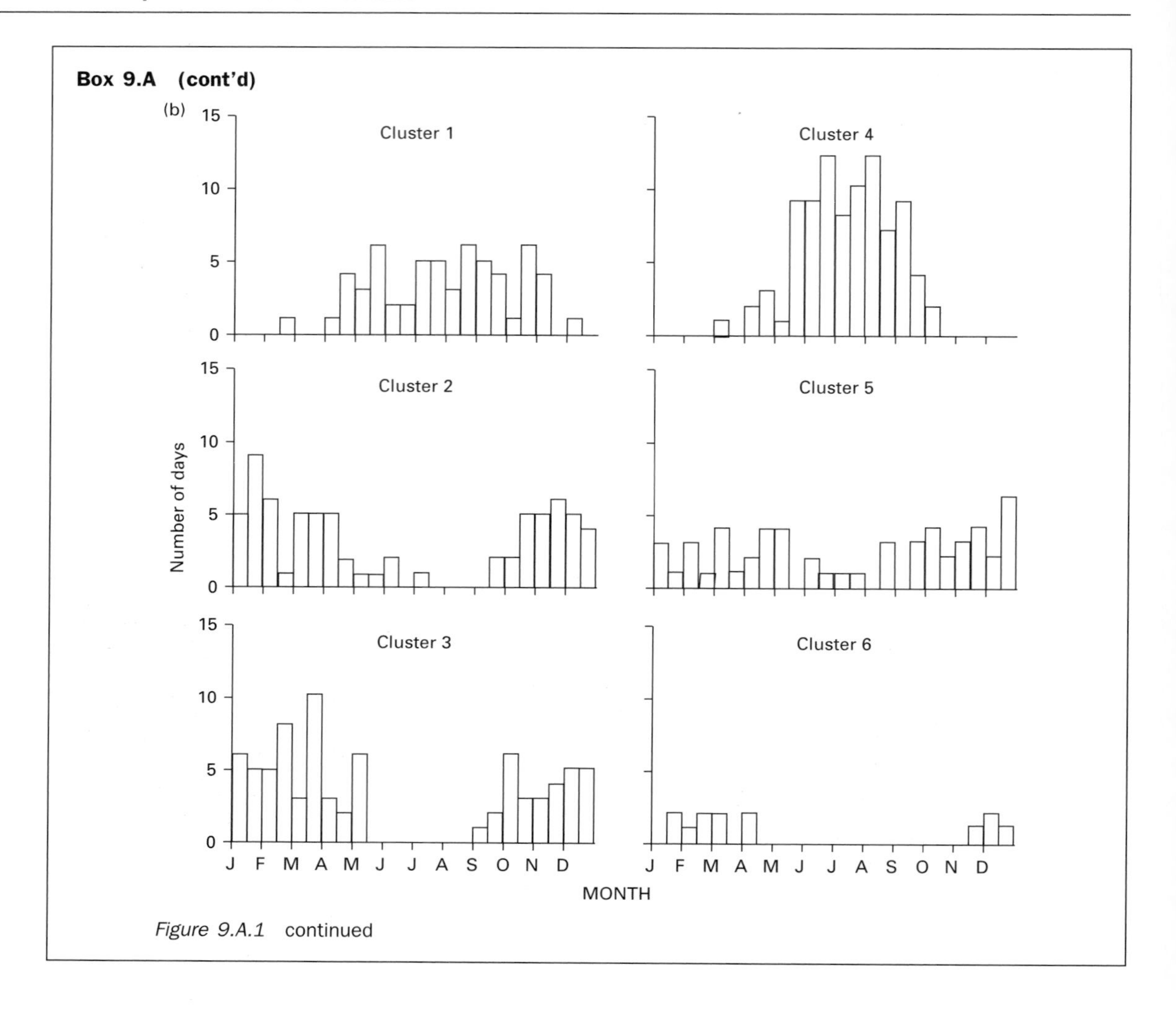


*Figure 9.A.1* continued

**Box 9.I Observational networks**

In our previous consideration of instruments, we discussed various aspects of observational networks. However, two facets of particular concern for the present chapter concern the density of observations needed to define a climatic region, and the difficulty of monitoring in harsh, especially polar, environments.

*9.I.1 Observational network density*

Although it may be climatologically desirable to 'measure everything everywhere' this is a practical and financial impossibility. Instead, many networks will be devised for specific purposes. The aim will be to devise the optimal network density to meet the specific needs. Here the density includes both the spatial distribution of the instruments and the time frequency with which they make observations. This will depend both on the need and its perceived economic or social value, and on the degree of autocorrelation of the element of concern. Comparing the need for precipitation information between western Britain and western North Carolina, USA, for example, the former has a much greater need to monitor rainfall closely because of its greater role in the national water supply. Hence it is likely to be economically more beneficial to maintain a denser network. Climatologically, it seems likely that the spatial variation of precipitation in the mountains of North Carolina is greater than that in western Britain,

**Box 9.I (cont'd)**

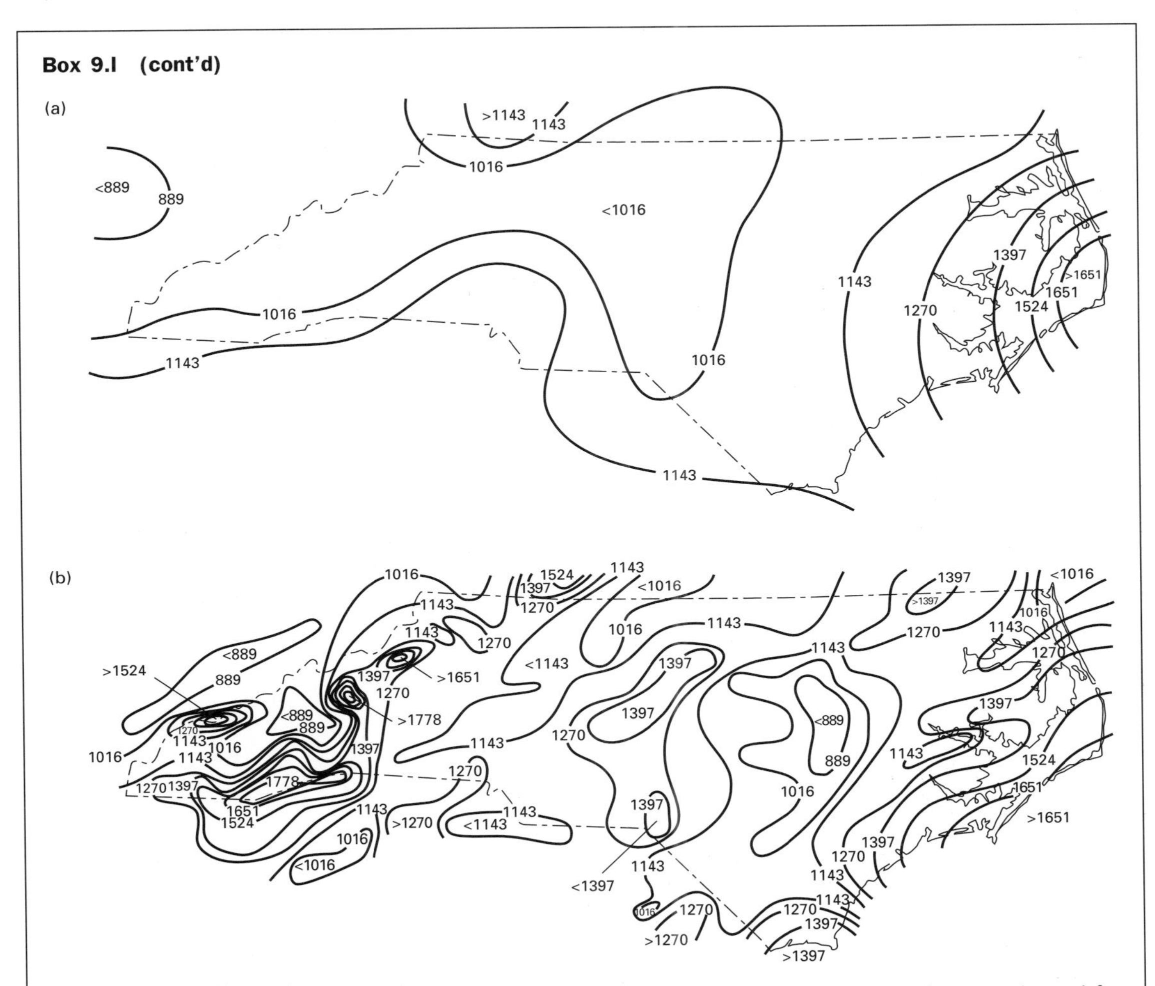


*Figure 9.I.1* Annual total precipitation (mm) for North Carolina in 1985 as deduced from the network used for (a) synoptic forecasting and (b) climate monitoring. In this year there was a tendency for the former to underestimate over most of the state, but this is not always the case. (Courtesy of Grant W. Goodge, National Climatic Data Center/NEDIS/NOAA.)

given the major convective component of the former. Certainly in many cases need seems to be related to population density, and indeed in most regions of the world there is a close relationship between network density and population density.

Our surface climatological information comes from networks maintained for a variety of specific purposes. Most national meteorological services maintain a network of observing sites specifically for their own weather forecasting purposes. This network must have a density sufficient to identify synoptic situations and a reporting speed sufficient to allow short-term forecasts. Other networks, with greater density but less need for speed, may support purposes such as water resource assessment or soil moisture monitoring. Care must be taken when using observations that the purposes for which they are used are congruent with the purpose for which they were made. For example, the synoptic network of the US National Weather Service can be used for rapid assessment of precipitation amounts during a specific storm. However, using only that network when estimating monthly or annual precipitation totals could be very misleading (Figure 9.I.1). A much denser network, designed primarily to assess precipitation totals for water supply purposes, is available. It gives a very different, and more pertinent, annual total.

**Box 9.I (cont'd)**

*9.I.2 Problems in harsh environments*

Much of our discussion of observations has assumed a rather benign environment for the network, the instruments and the observer. This may not be the case. In almost any location there are likely to be times when the environment is 'harsh'. It is much easier to make observational errors, for example, when forced to read the instruments in the middle of a blizzard rather than in the middle of a balmy spring day. However, some environments, notably polar and mountainous ones, are perpetually harsh.

In such environments, special care must be taken for any network. In most cases not only is there a harsh environment, but also the population density is probably low and support personnel or supplies are lacking. The instruments themselves may be operating close to their physical limits and thus may not behave in an entirely predictable fashion. Should a failure occur, speedy replacement is unlikely. Indeed, routine maintenance is probably less frequent than in more favourable conditions, so that failure is more likely. In addition, the network density is sparse, so that comparisons between nearby stations are not reliable as indicators of instrument problems. These can be problems even for instruments that are either visited regularly or which telemeter their observations via radio or satellite relay to a central location. In some cases instruments, such as storage gauges for precipitation, are read only at the end of a winter snow season, so that problems may go unheeded for months.

Despite all of these potential problems we do have a great deal of data from arctic and alpine environments. Perhaps the major constraint is that is it necessary to pay attention to any and all information concerning the quality of the observations and the observing techniques. Theoretically this is true for all data. In these regions it is vital.

# CHAPTER 10 Local climates

The smallest scale of climate variations can conveniently be termed local climates. The spatial range for these varies from a few square centimetres, for example the conditions around a growing plant, to a few square kilometres, for example the climate of a city. Although this, as Figure 1.8 indicates, spans the range of what is conventionally both micro- and meso-climate, they are treated together here since this is the scale where humans interact directly with the climate. It is the local climate that we experience every day, that dictates what crops we grow, that determines our home heating bills and influences our city drainage system design. It is also on the local scale that we can deliberately modify the climate to help meet our need for comfort and also the scale upon which the major inadvertent modifications have already taken place.

The local climate depends for its general characteristics upon the regional climate and ultimately upon the global climate system. It is therefore useful to bear in mind that the local climate of a particular place is a variation on the regional climate. Indeed, the mechanisms acting to create a local climate are essentially the same as those creating the global climate. The major differences are those of emphasis. In particular, the character of the surface and how it varies spatially and interacts with the overlying atmosphere are the most vital considerations.

The character of the surface includes virtually everything we see when we look at a landscape: the type of surface, whether it be grass, forest, concrete or water; the nature and size of upstanding objects such as fences, trees or tall buildings; the general topography of the area; and its overall altitude. They all influence the surface character and thus the local climate. Summarising these effects, we can say first that the surface characteristics and their variations are the major determinant of local differences in the energy balance and thus local temperature variations. These are likely to lead to air density and pressure differences, which create local winds when the regional atmospheric conditions are favourable. These regional conditions themselves can be influenced by the surface characteristics, producing local cloud and precipitation regimes. These not only influence the local water balance, but also the local energy balance, thus bringing us full circle. Thus this chapter relies heavily on the energy and water budget concepts of Chapters 3 and 5. There, however, the emphasis was on vertical exchanges. As a result of the discussion of winds and regional climates in Chapters 6–9, we can now add the horizontal dimension to give the full flavour of local climates.

## 10.1 Factors controlling local climates

Consideration of the factors controlling local climates can be approached through the energy and water budgets discussed earlier. Only selected pertinent aspects are recapitulated here.

The governing equation for the energy budget (equation (3.2)) is:

$$\Delta E = Q^* - (H + LE + G) \qquad (10.1)$$

where $\Delta E$ is the local energy change of the surface, $Q^*$ is the net radiation, $H$ and $LE$ are the sensible and latent heat fluxes to the atmosphere, and $G$ is the heat flux into the underlying medium. $Q^*$ can be expanded to emphasise the surface itself (equation (2.13)):

$$Q^* = (1 - A)K\downarrow + L\downarrow - \varepsilon\sigma T_s^4 \qquad (10.2)$$

where $K\downarrow$ and $L\downarrow$ are the incoming fluxes of solar and terrestrial radiation, respectively; and $T_s$, $A$ and $\varepsilon$ are the surface temperature, albedo and emissivity, respectively. Variations in $A$ and $\varepsilon$ with surface type have already been considered (Table 2.1). Here we need to pay particular attention to $H$, $LE$ and $G$. For the first two, we have already considered heat conduction and convection on the macro-scale (Sections 3.1 and 4.1). However, for local climates it is necessary to consider these effects on the micro-scale. We do this prior to considering the heat flow into the ground.

### 10.1.1 Heat and moisture in the surface boundary layer

Near the surface the effects of friction on the airflow pattern are marked. Adjacent to the surface is the **laminar boundary layer**, a layer never more than a few millimetres thick, where there is no turbulent mixing and all heat transfer is by conduction. Above this is the **turbulent boundary layer**, where the air is mixed as a result of both hydrostatic and dynamic instability. The thickness of this layer depends on the nature of the underlying surface, but can be visualised as extending, with various modifications, throughout the friction layer of the atmosphere up to the point where the geostrophic approximation becomes valid (Section 6.2). For most of this chapter, however, we shall be concerned only with the lowest few tens of metres of this turbulent boundary layer.

In the laminar boundary layer the rate of energy transfer is dictated by molecular processes. Thus the sensible and latent heat fluxes are given by:

$$H = -\rho c_p K_h(\partial T/\partial z), \quad LE = -\rho L d(\partial q/\partial z) \quad (10.3)$$

where $q$ is the specific humidity; $K_h$ and $d$ are, respectively, the thermal diffusivity of air (0.16–0.24 $cm^2\ s^{-1}$) and the diffusivity of water vapour in air (0.20–0.29 $cm^2\ s^{-1}$). In conditions of rapid surface heating or evaporation the vertical gradients necessary to maintain fluxes can become very great. It is not unusual, for example, to find a temperature gradient of 30 °C $mm^{-1}$ in this very thin layer in the afternoon over a desert surface.

These equations, and the ones that follow immediately, indicate that the transport of any gaseous entity is proportional to its gradient. They can be applied, suitably modified, in a variety of circumstances. For example, the flux of carbon dioxide to and from the surface of the Earth, with or without vegetation, and in various concentrations, can be analysed with an eye to agricultural productivity and the impacts of climate change. More generally, the paths and concentrations of trace gases, of natural or human origin, malignant or benign, passing between the Earth's surface and the atmosphere, can be examined. They underpin, for example, our consideration of air pollution in Section 11.2. Furthermore, similar equations also relate the way in which frictional stress or drag controls the flow of momentum from the atmosphere to the surface. This helps create swaying trees or grass over land, and waves on water. Despite all these implications, here relatively simple exposition will emphasise the direct links with the surface energy and water budgets.

Before extending our discussion to the turbulent boundary layer, we need to consider the variation of wind speed with height, the **wind profile**. A typical set of wind profiles is presented in Figure 10.1, in which it should be noted that the vertical axis (height) has a logarithmic scale. When conditions are near neutral (i.e. at 0600 and 1900 hours) the wind speed increases almost logarithmically with height (drawn as a straight line on these axes). The general equation for these neutral curves can be shown to be:

$$u = (u^*/k) \ln (z/z_o) \quad (10.4)$$

where $u^*$ is the friction velocity, which depends on the wind speed; $z_o$ is the roughness length and $k$ is a constant, the **von Karman constant**, with a value approximately equal to 0.4. $k/u^*$, which corresponds to the slope of the line of the profile. Although this equation holds in all stability conditions at heights below about 1 m, above this height stability influences the results. In stable conditions (e.g. 2200 hours local time) the wind speed increases more rapidly with height than in neutral conditions, largely because there is no hydrostatic instability to aid in turbulent mixing. In unstable conditions (e.g. 1400 hours) this hydrostatic instability combines with any dynamic instability present to enhance mixing and thus smooth out the profile.

The **roughness length** is the height above the surface at which the wind speed would go to zero if the turbulent layer extended completely to the ground, and if there were no upstanding elements to obstruct the wind flow. The values for a particular surface are independent of wind speed or atmospheric stability, provided, of course, the surface elements do not bend in the wind. Typical values are given in Table 10.1.

Within the turbulent boundary layer the energy transfer rate is controlled by a process analogous to that for the laminar layer, but with the molecular processes replaced by turbulent ones, which are more efficient and thus more rapid. The amount of turbulent mixing is largely controlled by the wind profile. Several approaches can be used to characterise this turbulence. For our purposes a full appreciation of the techniques of deriving them, and their respective

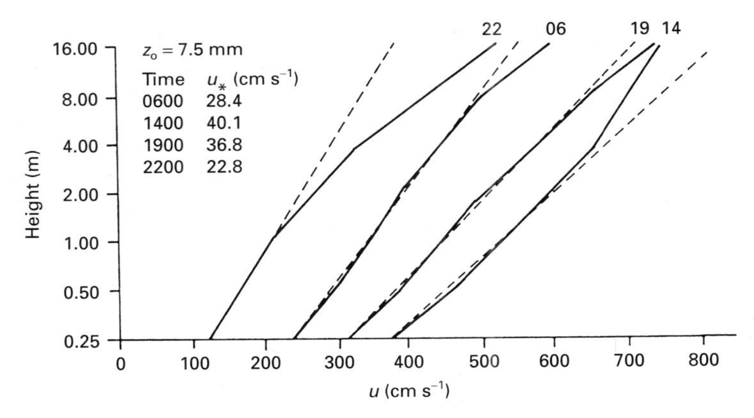


*Figure 10.1* The average variation of wind speed with height and time of day in the layer near the ground. The observations (solid lines) were taken over short grass at O'Neill, Nebraska, during the summer. The dashed line is a best-fit straight line using the three lowest observed values. (From Sellers, 1965.)

*Table 10.1* Roughness lengths ($z_0$) for various surfaces

| Type of surface | Height of stand (cm) | $z_0$ (cm) |
|---|---|---|
| Fir forest | 555 | 283 |
| Citrus orchard | 335 | 198 |
| Large city (Tokyo) | | 165 |
| Corn | 300 | |
| $u^{a}_{5.2}$ = 35 cm s$^{-1}$ | | 127 |
| $u_{5.2}$ = 198 cm s$^{-1}$ | | 71.5 |
| Corn | 220 | |
| $u_{4.0}$ = 29 cm s$^{-1}$ | | 84.5 |
| $u_{4.0}$ = 212 cm s$^{-1}$ | | 74.2 |
| Wheat | 60 | |
| $u_{1.7}$ = 190 cm s$^{-1}$ | | 23.3 |
| $u_{1.7}$ = 384 cm s$^{-1}$ | | 22.0 |
| Grass | 60–70 | |
| $u_{2.0}$ = 148 cm s$^{-1}$ | | 15.4 |
| $u_{2.0}$ = 343 cm s$^{-1}$ | | 11.4 |
| $u_{2.0}$ = 622 cm s$^{-1}$ | | 8.0 |
| Alfalfa brome | 15.2 | |
| $u_{2.2}$ = 260 cm s$^{-1}$ | | 2.72 |
| $u_{2.2}$ = 625 cm s$^{-1}$ | | 2.45 |
| Grass | 5–6 | 0.75 |
| | 4 | 0.14 |
| | 2–3 | 0.32 |
| Smooth desert | | 0.03 |
| Dry lake bed | | 0.003 |
| Tarmac | | 0.002 |
| Smooth mudflats | | 0.001 |

[a] The subscript gives the height (in metres) above the ground at which the wind speed, $u$, is measured.

strengths and weaknesses, is not required. Instead we can quote one treatment, the aerodynamic method, as an example. In this the fluxes are calculated by replacing the (molecular) thermal diffusivity in equation (10.3) by the turbulent diffusion coefficient given by $ku^*z$. These equations can then be integrated over a height $z_1$ to $z_2$. This gives, by use of the logarithmic wind profile (equation (10.4)), values for the fluxes $H$ and $LE$ as:

$$H = -\rho c_p k^2(\Delta T \Delta u)/(\ln z_2/z_1)^2$$
$$LE = -\rho L k^2(\Delta q \Delta u)/(\ln z_2/z_1)^2 \qquad (10.5)$$

where $\Delta u$, $\Delta T$ and $\Delta q$ are the differences in wind speed, temperature and specific humidity between heights $z_2$ and $z_1$, respectively. Since the wind speed at any height depends on $z_o$, values of the fluxes in equation (10.5) depend on the surface roughness and thus on the type of surface.

An analogy has often been drawn between this bulk aerodynamic approach and Ohm's law (which says that the electrical resistance is equal to the potential divided by the current). Replacing the potential by the concentration and the current by the flux leads to a climatic analogue of the electrical resistance, called the surface resistance. This permits relatively easy consideration of the separate roles played by plants and soil in producing the moisture flux. Thus, for example, the bulk stomatal resistance of vegetation can be calculated and used as part of the above method to determine the role of transpiration in the latent heat flux.

### 10.1.2 Heat flow to and from the underlying surface

The flow of heat, $G$, into the underlying medium is the final component of the energy balance to be considered here. The contrast between land, which allows only conduction, and water, which allows both conduction and convection, has been treated in Chapters 3 and 5. For local climates we are essentially concerned only with land surfaces. In these conditions $G$ is given by:

$$G = -K\Delta T/\Delta z \qquad (10.6)$$

where $K$ is the thermal conductivity of the medium. The thermal conductivity and the related conductive capacity vary between surfaces (Table 3.1), and this leads to variations in the rate of heat flow, the range of surface temperatures and the depth of penetration of the heat. Typical annual temperature variations at various depths within a clay soil have been shown in Figure 3.1. These results, a decrease in amplitude of the temperature wave with depth and an increasing delay in the time of the maximum with depth, are typical of all surfaces. Generalising the results, the lower the conductive capacity the greater the near-surface temperature change, while the greater the thermal conductivity, the greater the depth of penetration of the change.

The surface composition also influences the rate of water movement through the underlying medium, and hence the evaporation rate. A sandy soil allows rapid drainage of precipitation, minimising the water available for evaporation. Hence the Bowen ratio is often high and the soil surface is relatively warm. A clay soil, on the other hand, tends to retain water near the surface, have a lower Bowen ratio and cooler conditions. Extreme examples of this difference, the oasis effect and urban climates, will be treated in detail later. We introduce it here to emphasise the close link on this local scale between the energy and water budgets.

## 10.2 Interactions between surfaces

Having considered the way surface type influences the various fluxes, with the consequent implications for temperature and humidity over the individual surfaces, we can add a level of complexity by allowing interaction between surface types. In this section we consider the situation when a wind blows over an area of relatively flat terrain. The surface type change can then be one of roughness or moisture status, or both.

Whenever the wind blows from one surface type to another there is almost always a change in surface roughness. This leads to changes in the wind profile and thus a change in the fluxes. This is likely to be seen most clearly when tree and grass surfaces are juxtaposed (Figure 10.2). When the wind blows from

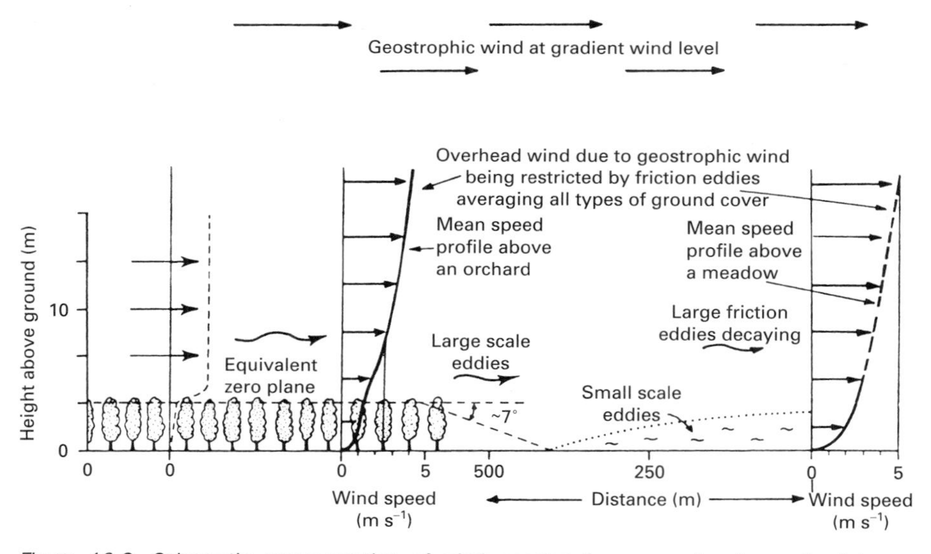


*Figure 10.2* Schematic representation of wind passing from an extensive orchard to an open field. The effect of the orchard dominates for a short distance near the surface before an internal boundary layer, dominated by small-scale eddies, is established. Thereafter an increase in wind speed occurs at all levels in the friction layer. (After Munn, 1966.)

the orchard onto the open field there will be a rapid decrease in the roughness length. The near-surface wind speed will increase rapidly with height, enhancing the rate of the sensible and latent heat transfers in the lowest layers. The opposite effect will occur when the wind is blowing in the opposite direction, or when it moves from an open field into a region of taller irrigated crops.

### 10.2.1 The oasis effect

The second type of surface change, that of moisture availability, can be demonstrated by postulating a stark surface contrast between a desert and an oasis (Figure 10.3). Indeed, the process is frequently called the **oasis effect**. From the energy budget perspective, it is akin to moving from the conditions at El Mirage (Figure 3.2c) to Tempe (Figure 3.2b). We can assume that the wind approaches the oasis having blown for some distance over a hot dry desert. The air is therefore in equilibrium with the desert surface, being itself warm and dry. Upon reaching the oasis edge evaporation rapidly increases, with sensible heat being extracted from the air to supplement the radiant energy to establish the high rates. The temperature thus decreases somewhat as the air begins to come into equilibrium with the new surface. Indeed, a new **internal boundary layer** is created, increasing in depth as more of the oasis is crossed and adjustment continues. This adjustment is a continuous process, however, since the humidity gradient is decreasing as more water vapour is evaporated into the air. Eventually, provided the oasis is large enough, complete adjustment will be achieved and the air will be in equilibrium with the new surface. Thereafter the energy balance components and the temperatures will not change downwind. Few situations are as stark as that of an oasis, but the adjustment process and the development of internal boundary layers will occur whenever the surface type changes. This may involve wind blowing over a city surface into an urban park, blowing from irrigated to unirrigated land, or simply from a wet to a dry surface.

### 10.2.2 Sea breezes

When regional winds are light, the conditions are opposite to those needed for the oasis effect. In this case the differences in surface temperature, translated into pressure differences, themselves create the flow. This is most likely, and perhaps only likely, to occur as the result of differential heating of land and water along a coastline (Figure 10.4). It is a small-scale analogue of the monsoon circulation considered in Section 8.2. The more rapid heating of the land during the daytime (Figure 10.4a) results in the development of a temperature gradient across the coast. This leads to ascent over the land and descent over the sea and hence a pressure gradient which initiates an airflow from sea to land, the **sea breeze**. At the same time there is a compensating return flow aloft. The flow develops through the day and by the middle of the afternoon may extend several tens of kilometres inland. The uplift over the land may lead to cumulus cloud formation (Figure 10.4b), but rarely is this sufficiently developed to give increased precipitation. At night the situation is reversed and the flow is from the colder land to the warmer sea, as a **land breeze** (Figure 10.4c). This is generally less well formed than the sea breeze since surface temperature contrasts tend to be smaller.

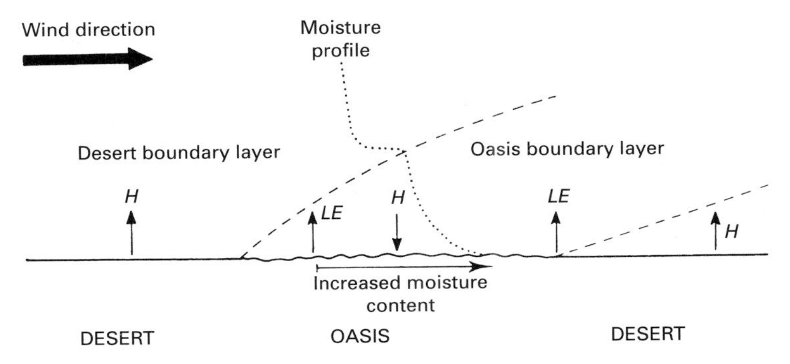


*Figure 10.3* Schematic representation of the oasis effect. Note that the direction of the sensible heat flux, *H*, changes. This is because sensible heat is removed from the air to achieve maximum evaporation from the oasis. *LE* is the latent heat flux.

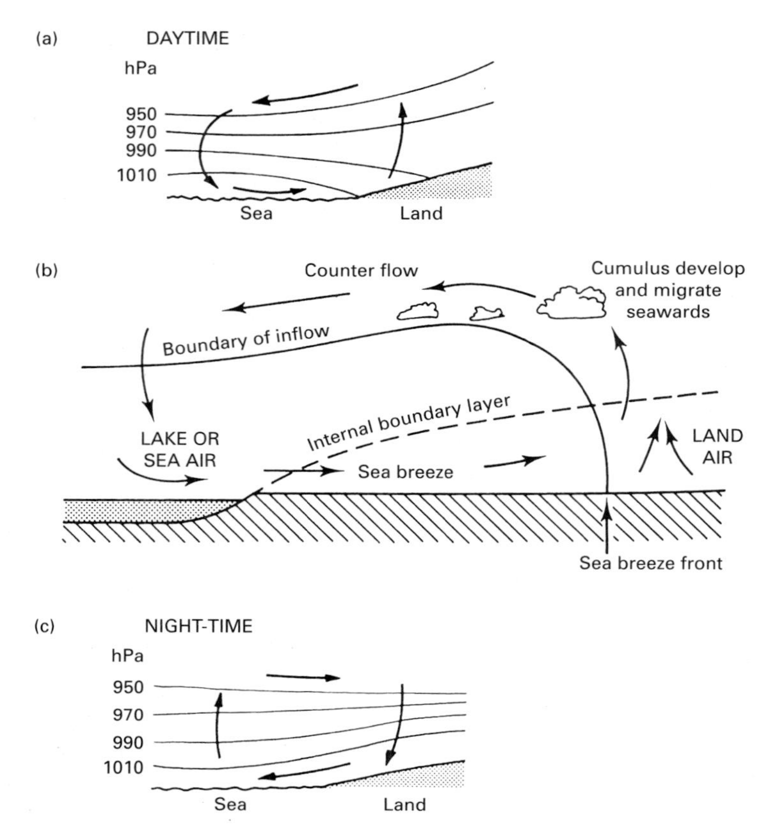


*Figure 10.4* Schematic representation of (a) and (b) daytime and (c) night-time land/sea breezes.

## 10.3 The urban climate

Probably the most marked land-surface contrast occurs between a city and its rural surroundings. Differences in surface materials, drainage characteristics, sources of heat, configuration of surfaces and pollutant loading act to change all aspects of the local climate of a city. The major changes are summarised in Table 10.2.

### 10.3.1 The surface energy budget in urban areas

All components of the energy budget at a city surface are influenced by the presence of the city. Incoming solar radiation is depleted, and incoming long-wave radiation enhanced, by pollution in the urban atmosphere. Frequently there is increased cloud amount, which has the same effects. While rural and urban surface materials have similar albedos, the surface geometry associated with urban canyons creates multiple reflections, with some absorption at each interaction, leading to an overall lower albedo for a city. Although this enhances short-wave absorption, the effect is partially offset by the smaller amount of radiation actually reaching an urban surface. For long-wave emission, the surface geometry ensures that less energy is lost from a city than from a rural area at the same temperature.

Thus net radiation and surface temperature is commonly somewhat higher in a city than in a surrounding rural area. The enhancement is modified by the alteration in the other surface energy fluxes. Both the latent and sensible heat fluxes are influenced by the increase in surface roughness as one moves from the relatively smooth countryside through the suburbs into the city centre. The increase in turbulence associated with this tends to increase the magnitude of these fluxes.

The major differences in the latent heat flux between rural and urban areas, however, occur because

*Table 10.2* Comparison of urban and rural climatic parameters

| Element | Parameter | Urban compared with rural (– less; + more) |
|---|---|---|
| Incoming radiation | On horizontal surface | –15% |
| | Ultraviolet | –30% (winter); –5% (summer) |
| Temperature | Annual mean | +0.7 °C |
| | Winter maximum | +1.5 °C |
| | Length of freeze-free season | +2 to 3 weeks (possible) |
| Wind speed | Annual mean | –20 to –30% |
| | Extreme gusts | –10 to –20% |
| | Frequency of calms | +5 to +20% |
| Relative humidity | Annual mean | –6% |
| | Seasonal mean | –2% (winter); –8% (summer) |
| Cloudiness | Cloud frequency and amount | +5 to +10% |
| | Fogs | +100% (winter); +30% (summer) |
| Precipitation | Amounts | +5 to +10% |
| | Days (with <5 mm) | +10% |
| | Snow days | –14% |

of the differences in the surface moisture content. For many regions on Earth a rural surface is a moist surface and the latent heat flux is a significant contributor to the surface energy balance. Most city surfaces, however, are designed to ensure rapid removal of rainwater and only during or immediately after a rainstorm does a city have a significant latent heat flux. Otherwise the flux is confined to vegetated areas, which usually cover only a small fraction of the total city surface. Thus for much of the time the city acts in the opposite way to an oasis, having flux characteristics akin to the desert conditions of Yuma (Figure 3.3) and El Mirage (Figure 3.2), with energy exchanges restricted to the sensible and ground heat fluxes. These, while acting in the same way for city and country, must remove a greater amount of energy in the city in order to attain an equilibrium temperature. This requires an increased temperature gradient, which in turn requires a higher surface temperature. Further, the conductive capacity for city materials is generally lower than for moist soil and a greater subsurface temperature gradient is needed to maintain the same ground heat flux. Finally, the air above a city is frequently directly heated by industrial processes, exhausts from automobiles, and heat rejected by buildings. This also requires the temperature of

*Table 10.3* Hypothesised causes of the urban heat island

1. Increased counter radiation ($L\downarrow$) due to absorption of outgoing long-wave radiation and re-emission by polluted urban atmosphere.
2. Decreased net long-wave radiation loss ($L^* = L\downarrow - L\uparrow$) from canyons (tall buildings and narrow sidewalks) due to a reduction in their sky view factor by buildings.
3. Greater short-wave radiation absorption ($K^* = K\downarrow - K\uparrow$) due to the effect of canyon geometry on the albedo.
4. Greater daytime heat storage due to the thermal properties of urban materials and its nocturnal release.
5. Anthropogenic heat from building sides.
6. Decreased evaporation due to the removal of vegetation and the surface 'waterproofing' of the city.
7. Decreased loss of sensible heat due to the reduction of wind speed in the canopy.

the city surface to increase in order to maintain the temperature gradient necessary to remove heat from the surface by the sensible heat flux.

### 10.3.2 Urban heat islands

The net result of the energy exchanges discussed above and summarised in Table 10.3, is to create an **urban heat island**, making the city a few degrees warmer than its surroundings. The intensity of the heat island is largely a function of building density and the amount of incorporated vegetation, as would be expected from energy considerations. There is a general relationship between city size and heat island intensity:

$$T(\text{urban} - \text{rural}) = P \log p \qquad (10.7)$$

where $p$ is the population. The constant of proportionality, $P$, appears to differ between European and North American cities, probably reflecting their different urban characteristics. Nevertheless, cities of all sizes display an urban heat island (Figure 10.5).

In human terms, the negative aspects of this increase in urban temperature include aggravation of heat stress during prolonged heat waves. On the other hand, in cold climates, increased temperatures are a benefit since they increase comfort and tend to reduce power demand. Some cities, such as Stuttgart, Germany, have deliberately incorporated vegetation into their city planning to mitigate the heat island effect. Others, such as Dayton, Ohio, USA, have experimented with 'green parking lots', car parks whose surface is an intermixture of tarmac and grass, with the same objective. Such strategies, unless adopted city-wide, have only a local effect, since we must consider a city as being composed of a series

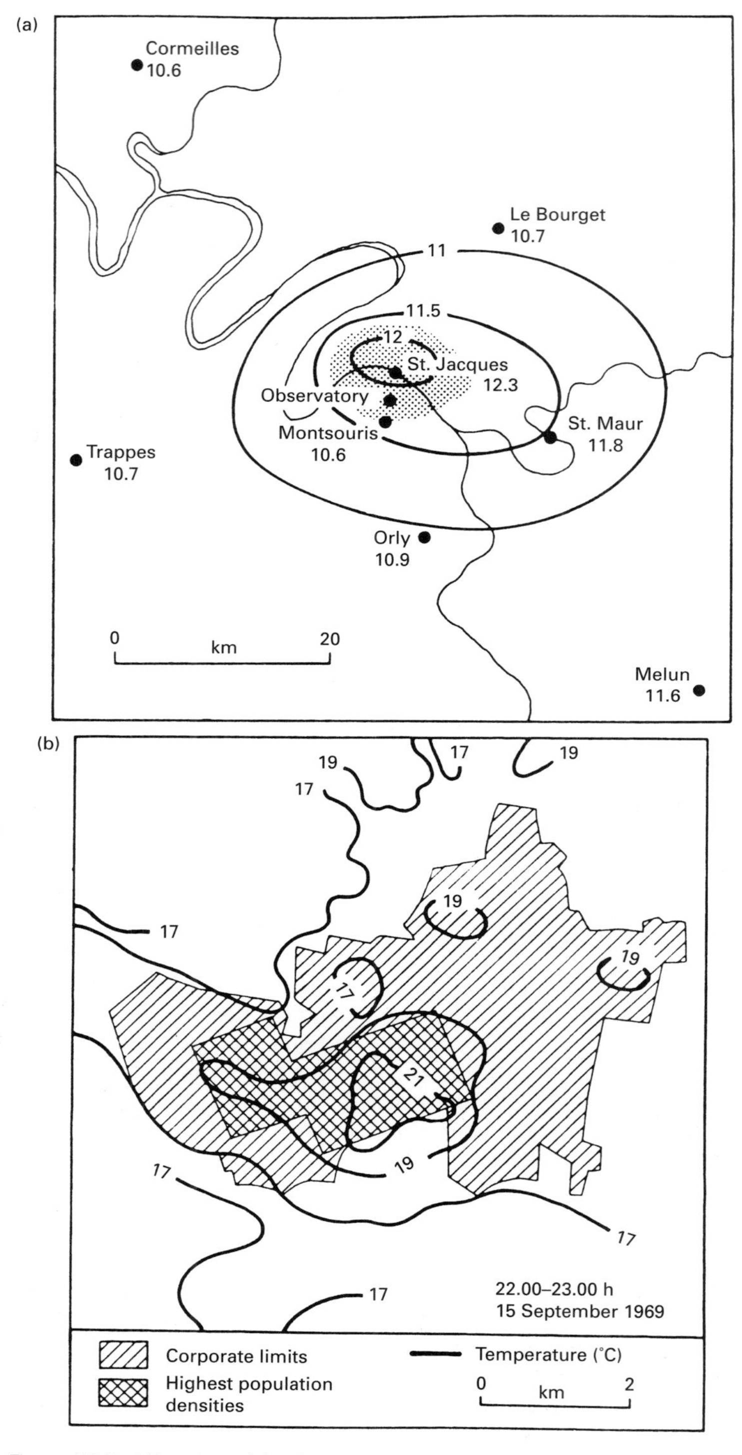


*Figure 10.5* Urban heat island maps at (a) Paris, France, and (b) Chapel Hill, North Carolina, USA. (After (a) Dettwiller, 1970; (b) Kopec, 1970.)

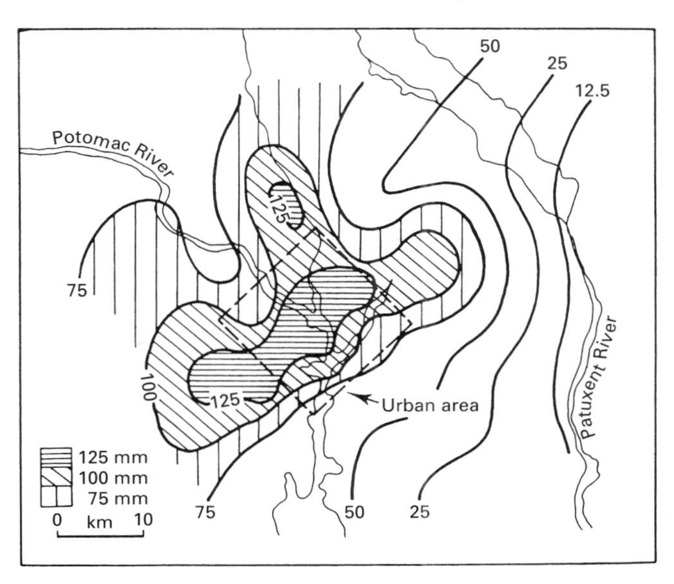


*Figure 10.6* Distribution of rainfall during a storm over the Washington DC area on 9 July 1970 showing the precipitation enhancement due to the presence of the conurbation. The wind was from the northeast. (From Hess, 1974.)

of interlinked microclimates. Certainly in calm conditions this is obvious since the temperature variations within a city can be clearly sensed by any urban dweller. For example, during a summer afternoon in Columbia, Maryland, USA, the temperature over a lawn was found to be 31 °C, whereas a nearby car park had a temperature of 44 °C.

The urban heat island itself influences other elements of the climate. By providing a hot spot, convection can be enhanced, leading to an increase in cloudiness over and immediately downwind of the city. In addition, the increased surface roughness usually associated with a city may cause it to act as a barrier to the regional windflow, forcing the air to rise, further increasing the possibility of increased cloud amount. In some cases these effects are suspected of being the agents responsible for increased precipitation downwind of the city (Figure 10.6). However, the role of the regional climate in such enhancement must be stressed. Regional airflows must contain the necessary moisture and must be susceptible to the vertical motions induced by the city. Estimates of precipitation enhancement by the city are difficult to obtain since it is difficult to extract the urban effects from the usual temporal and spatial precipitation variations. Most of our information, therefore, comes from cities in areas of relatively homogeneous terrain with a well marked prevailing wind direction. The influence of cities in other situations is not as clearly demonstrable.

### 10.3.3 Urban air circulation

In calm or near-calm conditions the city may generate its own circulation (Figure 10.7). This can be particularly marked at night, when warm air rising from the city surface is replaced by cool rural air. If the rising air encounters an elevated inversion, a not uncommon situation, it is forced to spread laterally. As it spreads it is cooled radiatively, its density increases and it descends over the rural areas. It is then 'sucked back' into the city to complete the circulation. It is a microscale version of the Hadley circulation, but truly thermally driven. An urban dome is thus created, which can often be seen when pollution sources are present in the city. This pollution is retained within the urban circulation and concentrations increase throughout the night. Pollution concentration will not decrease until the increase in incoming solar radiation changes the regime as it breaks the overlying inversion in the morning.

In windy conditions the city acts as a much rougher surface than does the rural area, leading to a decrease in wind speed within the city. On one scale this leads to spatial variations in wind of concern to urban planners (Section 10.A.2). On another it leads to the

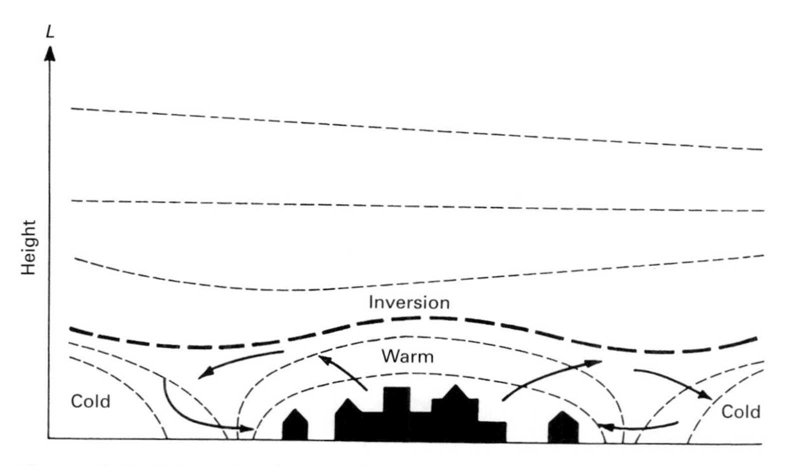


*Figure 10.7* Schematic diagram of typical urban circulation patterns when regional winds are light. The dashed lines are isotherms. The temperature inversion rises over the 'urban dome'.

**Box 10.A Airflow around obstacles: applications and impacts**

In regions of high wind speed it is frequently advantageous to apply our knowledge of the interaction between wind and obstacles to reduce the speed of the airflow. It is also advantageous, especially in an urban setting, to use that same knowledge to minimise the impact of wind speed variations. This section addresses these complementary concerns in turn.

*10.A.1 Shelter belts and snow fences*

The reduction in wind speed resulting from an increase in surface roughness can be used in the construction of **shelter belts**. In any rural area subject to high winds from a specific direction, a belt of trees planted upwind of the area to be protected can reduce damage or danger (Figure 10.A.1). Such belts provide a semi-permeable barrier to wind flow. The wind profile begins to adjust to the new surface, but, since the obstacle is thin, it never reaches complete adjustment. Instead it begins to return to the upwind profile. The effect, however, is that some low-level wind passes through the gaps between the trunks with markedly reduced speed. At higher levels the more densely packed crowns provide a more substantial barrier and force the air to flow over them. Any subsequent wake that is created by this wall of tree crowns is concentrated above and just downwind of the belt, being prevented from reaching the ground by the presence of the low-level airflow. Thus near-surface wind speeds are reduced until the wind profile readjusts, which is at a distance equivalent to several tens of tree heights downwind. Such belts are frequently developed to protect high value and sensitive crops (Figure 10.A.2), or to provide shelter for human dwellings.

A similar phenomenon is produced by **snow fences**, which are intended to prevent snow accumulation in particularly sensitive areas, such as roads. A slatted fence is installed some distance away from the road on its upwind side. As the airstream containing blowing snow encounters the barrier, its speed is reduced in much the same way as that of the shelter belt. This speed reduction reduces the snow-carrying ability of the wind and the snow is deposited just downwind of the fence. Beyond that region the wind speed will again increase, but now its snow load has been reduced and it sweeps across a clear roadway without impeding visibility or traction.

*10.A.2 The planning implications of urban airflow*

A city surface is a rough surface and so serves to slow the wind. However, that can be a misleading statement. A city will have spots which, for many wind directions, remain virtually calm, while nearby streets may channel the wind and produce high speeds. The calm spots may foster high pollution levels, while high winds may make city streets difficult for walking. The turbulent wind flow may also

**Box 10.A (cont'd)**

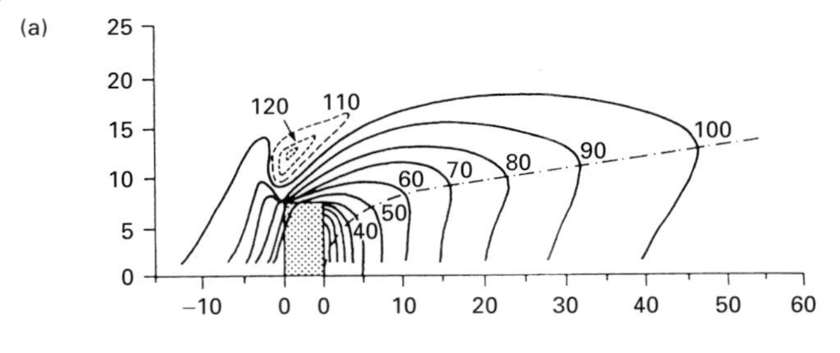


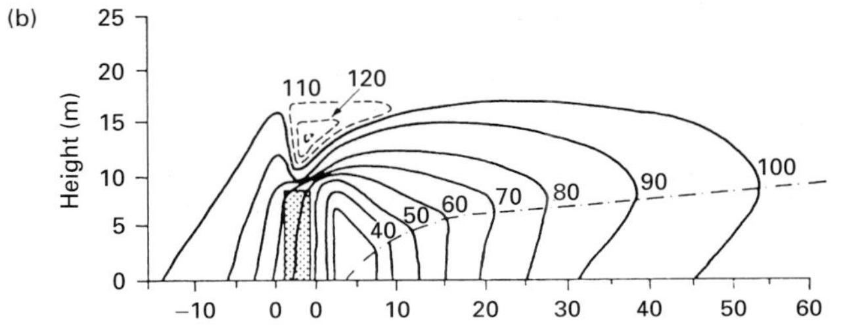


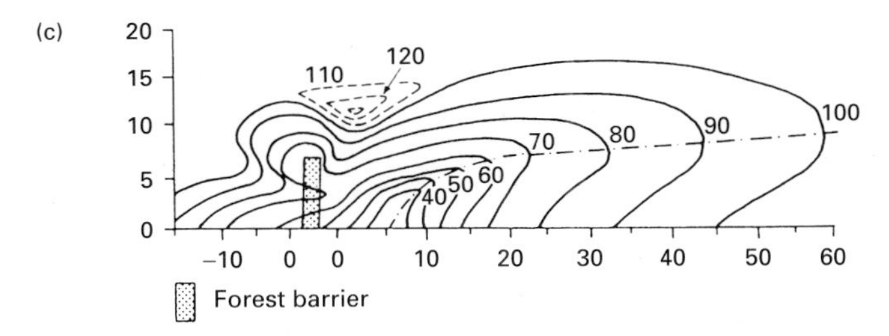


*Figure 10.A.1* Percentage of undisturbed wind speed at different heights experienced as a result of (a) thick, (b) moderately wooded and (c) thin forest belt as measured at Kirov, Russia. The abscissae are multiples of the forest belt height. There is a great reduction in speed immediately downwind of the thick belt, but the effects of the thinner belts persist for a greater distance downwind. (From Munn, 1966.)

produce tremendous stresses on buildings which act as obstructions at the end of channelling streets. Identification of these potential trouble spots prior to the construction of new buildings depends on a knowledge of the airflow around obstacles and is of great concern to city planners.

The airflow around an obstacle depends on the characteristics of the flow and the geometry of the obstacle and is difficult to predict even for a single isolated building. From observation in wind tunnels and in the ambient atmosphere, and from model simulations, some generalisations are possible. When air encounters an obstacle the flow separates and accelerates (Figure 10.A.3). In turbulent flow this will increase the velocity with which eddies impact the upwind face of the building, increasing the stress on that building. However, once past the obstacle, the nature of the airflow depends greatly on its speed. In light steady winds the air will decelerate and return to its upwind conditions (Figure 10.A.3a). At higher speeds airflow separation will occur and a wake vortex will develop. This will not be a continuous phenomenon, but will intensify the turbulence inherent in the wind, creating highly turbulent and variable conditions in the building wake. Frequently, vortex shedding will occur at more or less regular

**Box 10.A (cont'd)**

*Figure 10.A.2* Cypress trees used as wind breaks near Les Beaux, Provence, France. (Courtesy of A. M. Harvey.)

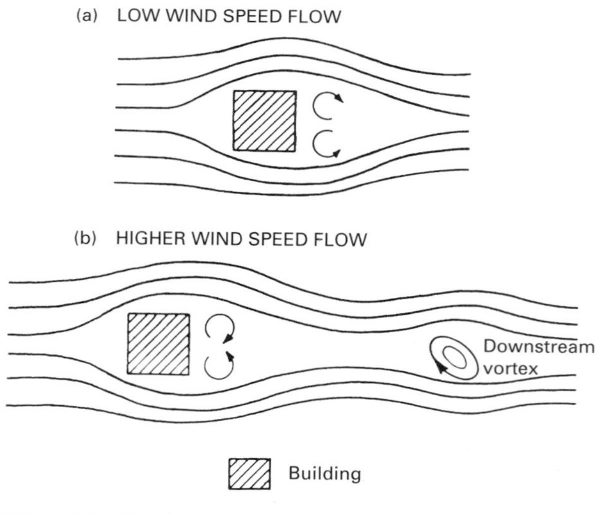


*Figure 10.A.3* (a) Schematic representation of airflow around an obstacle. At the higher wind speed (b), downstream vortices can be created. (From Munn, 1966.)

**Box 10.A (cont'd)**

(a) VENTURI EFFECT

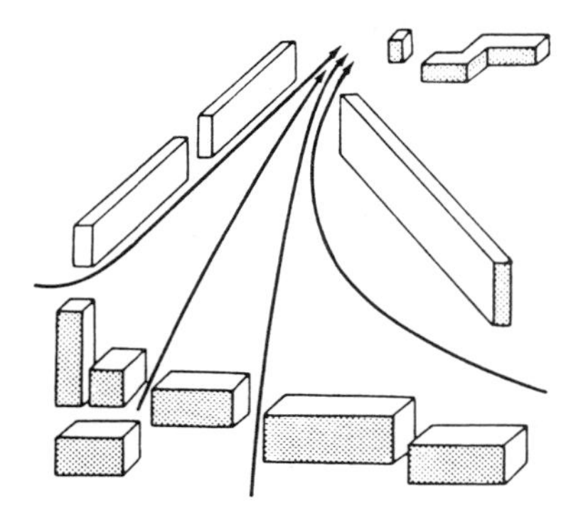

(b) TRANSVERSE CURRENTS

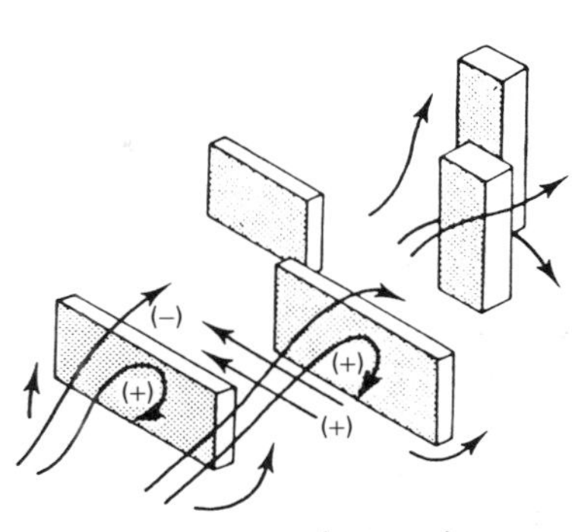


*Figure 10.A.4* Two typical airflow patterns around buildings, illustrating flow resulting in (a) the Venturi effect and (b) transverse currents. (After Thurow, 1983.)

intervals; these vortices travelling downstream, rotating alternately clockwise and anticlockwise, at a speed less than that of the mean wind (Figure 10.A.3b). This stream of vortices is called a **Karman vortex street**. When the wake vortex leaves the building there is a distinct pressure drop. In a well-sealed building this may set up a pressure difference across the building wall, which not infrequently causes the blowing out of windows. Furthermore, interaction between simultaneous upwind and downwind vortices may occur, forming rapidly varying stresses which may lead to building collapse.

The common situation in most cities, of course, is not for a set of isolated structures, but for a collection of buildings which interact with the wind field in complex ways. Studies of various 'typical' situations have been undertaken. Two examples are given here. When there are two converging buildings, air is funnelled between them, causing increased wind speed as a result of the **Venturi effect** (Figure 10.A.4a). The magnitude of the effect depends on the building geometries, but can create significant problems for walking or opening doors, and may put severe stress on the buildings. As the second example, whenever buildings are approximately perpendicular to the wind (Figure 10.A.4b), pressure differences between the upwind and downwind faces of the buildings can be established, leading to unexpected transverse currents.

*Table 10.A.1* Hill slopes and wind speeds which demand the same muscular power

| Hill slope | 1/20 | 1/10 | 1/7 | 1/5 | 1/4 | 1/3 |
|---|---|---|---|---|---|---|
| Wind speed ($m\ s^{-1}$) | 9 | 13 | 15.5 | 18.5 | 21 | 24 |

In addition to the danger posed by the stresses placed on the buildings as a result of the turbulent airflow, changes in both the wind speed and its turbulence have an impact on pedestrian comfort and safety. The effect of a steady wind can be compared to the effect of walking uphill (Table 10.A.1). When turbulence is incorporated, experiments have suggested that wind speeds above 5 $m\ s^{-1}$ are annoying, above 10 $m\ s^{-1}$ disagreeable, and above 20 $m\ s^{-1}$ they can be dangerous. However, specification of the influence of a planned building on airflow patterns cannot be predicted from theoretical considerations. At present, only by the use of wind tunnel simulations is it possible to suggest what impact an urban development will have on the wind regime and hence on the liveability of the area.

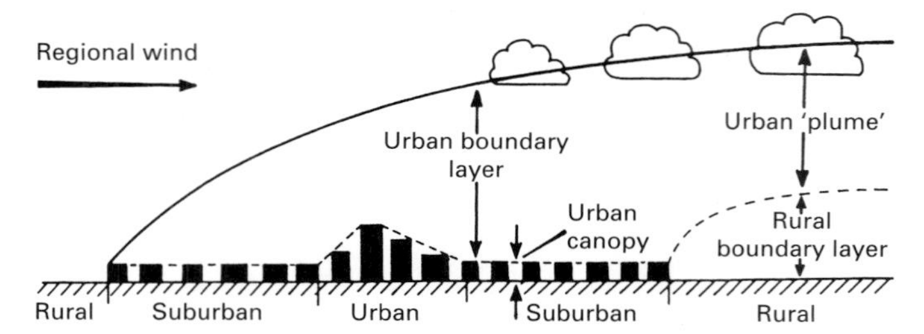


*Figure 10.8* Schematic representation of the urban atmosphere illustrating the 'urban plume' (for a modelled representation, see Figure 11.10).

establishment of an urban boundary layer in much the same way as was discussed above for rural surfaces. However, the roughness elements are typically much bigger than for rural areas and tend to act as obstacles to the windflow. The result is an increase in turbulence within the city and the development of a distinct internal boundary layer which forces the regional wind to rise as it flows over the city. The result is often called the **urban plume** (Figure 10.8). At the upper edge of such a plume, where the regional winds are forced to rise, enhanced cloud formation commonly occurs and precipitation may be enhanced.

## 10.4 The influence of topography

So far we have tacitly assumed that the landscape we are considering is flat. This, of course, is rarely the case, and topography influences the local climate as much as the surface character does. Any landscape can be considered to be composed of a series of slopes, of varying angle to both the horizontal and the north–south axis, which influence the energy and water budgets. In addition, topography also implies altitude changes, which change air density and thus also have an impact on local climates.

### 10.4.1 The energy and water budgets of slopes

A fundamental factor influencing the climate of a non-planar landscape is the solar radiation input. This is dictated by astronomical relationships (Section 2.2), and can lead to great differences between slopes, differences which vary with time of day and with season (Figure 10.9). It is difficult to summarise the results, but for the Northern Hemisphere the maximum radiation is received on south-facing slopes, while slopes inclined at an angle approximately equal to the latitude have maximum annual total radiation. The actual amount received on a particular slope during a particular time period will, of course, depend on the atmospheric conditions above the slope. These may themselves be influenced by topography; as when the slope leads to the formation of orographic clouds.

The variable distribution of solar radiation is the major driving force for variations in the rest of the energy budget, thus leading to variations in the local climate in the ways we considered earlier. This differentiation is reinforced by the variation in the water content of slopes. The rate at which water drains from slopes under the action of gravity is dictated by the angle of the slope and the vegetation and soil conditions on it. Generally, the more gentle the slope, the thicker the soil and the denser the vegetation, the slower the drainage and the greater the retention of moisture that becomes available for evapotranspiration. A typical consequence of clearing a forested slope is given in Figure 10.10.

A major result of energy and water balance differences between slopes is that the rate of temperature decrease up a slope depends on the angle and orientation of the slope. Although the basic rate of temperature change will be the environmental lapse rate, this will be modified for the slope surface by the local energy balance characteristics of that surface. In general, in the Northern Hemisphere, a south-facing slope is likely to be warmer than a north-facing one. The persistence of this difference is frequently reflected in the vegetation distribution up the slope, notably

(a)

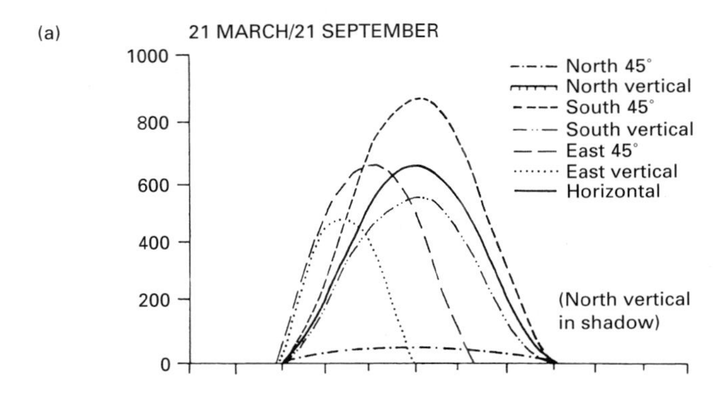


(b)

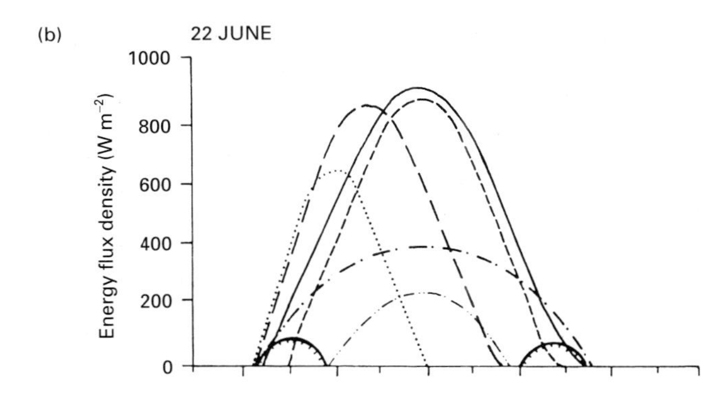


(c)

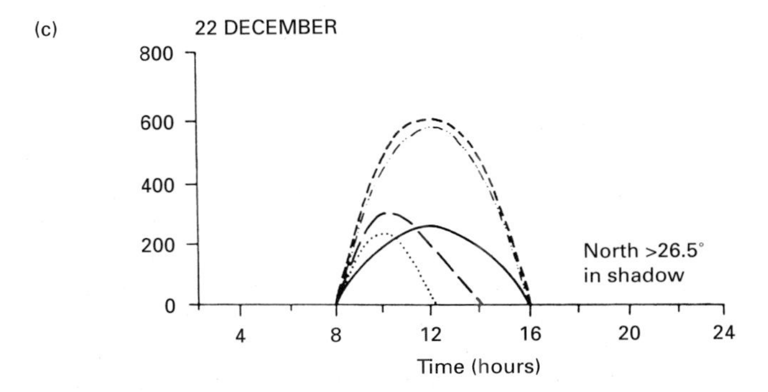


*Figure 10.9* The diurnal variation of direct beam solar irradiance upon surfaces with different angles of slope and aspect at latitude 40°N for (a) the equinoxes – 21 March and 21 September; (b) summer solstice – 22 June; and (c) winter solstice – 22 December. (From Gates, 1965.)

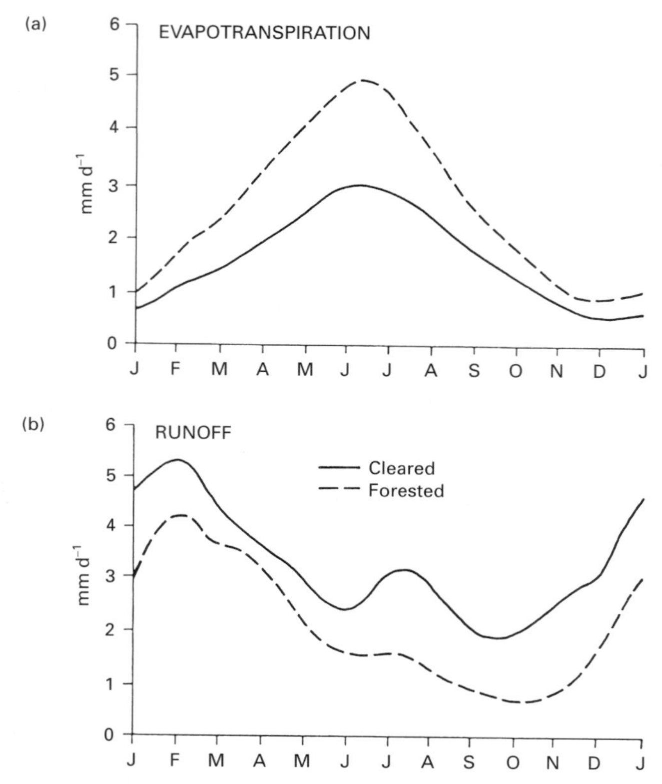


*Figure 10.10* (a) Estimated evapotranspiration and (b) estimated (cleared) and observed (forested) runoff for cleared and forested catchments near Coweeta, North Carolina. (From Sellers, 1965.)

the higher tree line on the south-facing slope (Figure 10.11).

### 10.4.2 Wind flow in non-planar topography

Local differences in the energy balance of slopes can also lead to the creation of local valley breezes. Considering the valley as consisting of three slope elements – valley floor, valley sides and mountain ridge – there will be different temperature regimes for each element. In near-calm regional wind conditions, these differences will lead to the development of a valley wind system. On clear nights long-wave radiation loss from the mountain ridge will cool that surface and its overlying air. Cooling of the valley sides and floor will be much less marked because of radiation exchanges between the two valley walls and the floor. Consequently the cooler, denser air at the top will sink to the floor, moving down as a slow, generally smooth flow. This is the **katabatic wind** (Figure 10.12a). Although the speed of the flow depends on the angle of slope and the roughness of the surface, it is found to be approximately proportional to the square root of the temperature difference between the top and bottom of the valley, the speed increasing approximately linearly with the distance from the top. However, the speed of the katabatic wind rarely exceeds a few metres per second. The temperature difference between the top and the bottom, $\Delta T$, depends on the net long-wave radiation loss, $L^*$:

$$\Delta T \propto L^{*2/3} \qquad (10.8)$$

The cold air that flows down the valley sides will collect at the valley bottom. It will then flow out of the valley unless there is an obstruction to the flow. Any obstruction will act as a 'dam' and lead to the build-up of a cold air pocket. Such a feature is known as a **frost hollow**. The obstruction could be a

(a)

(b)

*Figure 10.11* Photographs of valley treelines in the Rocky Mountain National Park in Colorado, USA. (Courtesy of A. M. Harvey.)

variety of things, either natural, such as a constriction in the valley or a glacial ridge, or artificial, such as a road or railway embankment. Frost hollows are by no means confined to valley situations. Any topographic depression is capable of producing such a feature when there is a net loss of long-wave radiation in near-calm conditions. The Gstettneralm sinkhole, near Lunz, Austria, is one of the coldest places in central Europe. It is not unusual for the base of the sinkhole, approximately 150 m below the general ground level, to record temperatures 20 °C below the surrounding area (Figure 10.13). On occasional nights temperatures below –50 °C have been measured.

The distribution of a low night-time temperature associated with cold air drainage can have important consequences for agriculture. The effect of this can clearly be seen in the distribution of vineyards in many wine-producing areas. High quality wines require that the grapes be exposed to some climatic stress, necessitating their location in regions susceptible to frost. However, frost itself is detrimental to the vine. Whenever possible, therefore, the vines will be

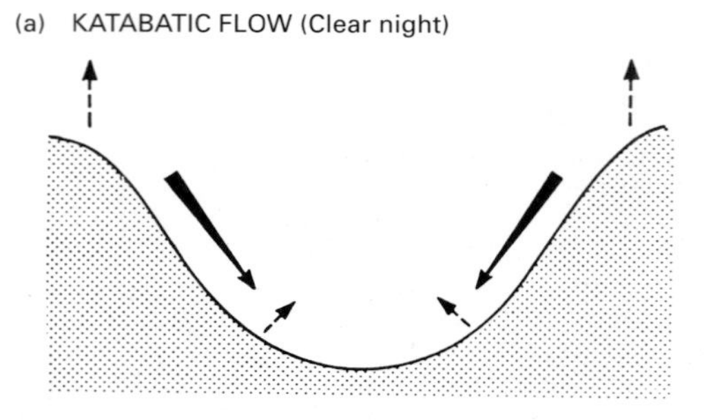


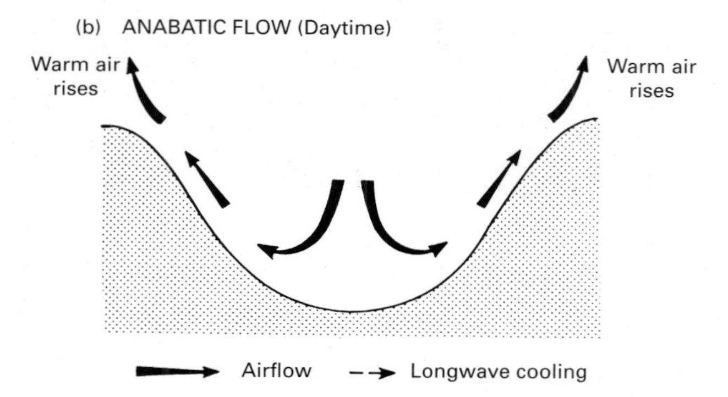


*Figure 10.12* Schematic representation of (a) katabatic flow on a clear night and (b) anabatic flow during the day.

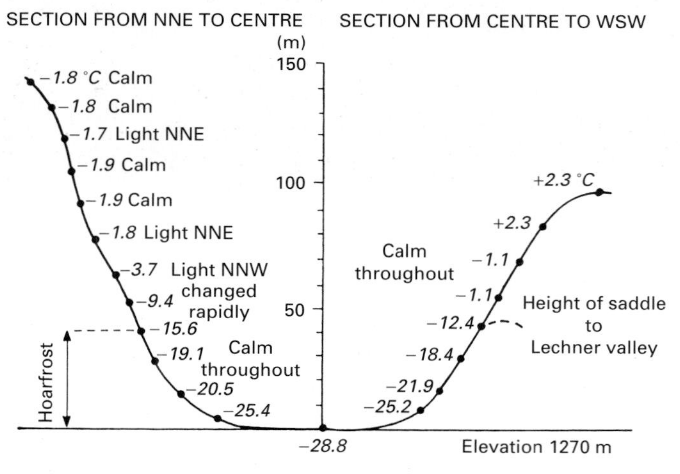


*Figure 10.13* Gstettneralm sinkhole on 21 January 1930 showing night temperatures in two sections. (After Geiger, 1965.)

(a)

(b)

*Figure 10.14* Hillslope vineyards in the Appennine Mountains in Italy.

planted on the hillsides, avoiding the colder hilltops and valley bottoms (Figure 10.14). This is a passive form of frost protection. More active forms are discussed in Box 3.A.

The **anabatic wind** is the daytime equivalent of the katabatic wind we have been considering. Again, this is most often developed when regional winds are light and conditions are cloudless. Now the valley tops receive the maximum incoming solar radiation, and the maximum heating, in the system. The air just above the valley top becomes warmer than the air at the same level over the valley itself. This warm air rises. Air then flows up the valley sides, as the anabatic wind, to replace this rising air (Figure 10.12b). Descending, adiabatically warmed air from the centre of the valley replaces the rising air, which may lead to a dynamically caused inversion, with its implication for pollution levels. However, this daytime circulation is rarely as well developed as its night-time counterpart. Only immediately after dawn are conditions favourable for its formation. Daytime regional winds are usually stronger and more unstable than their night-time counterparts, and tend to swamp any local valley circulation.

(a)

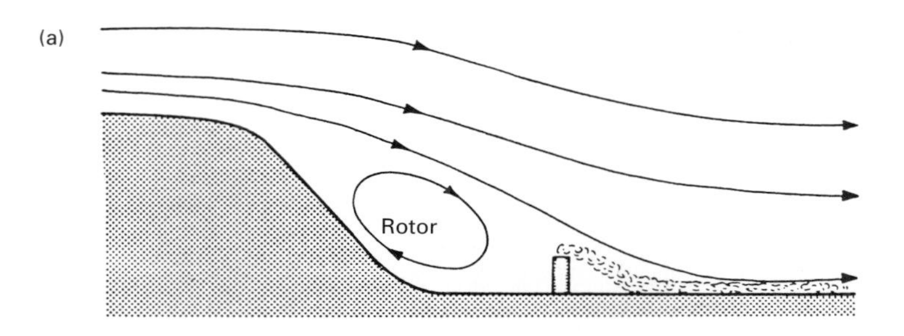


(b)

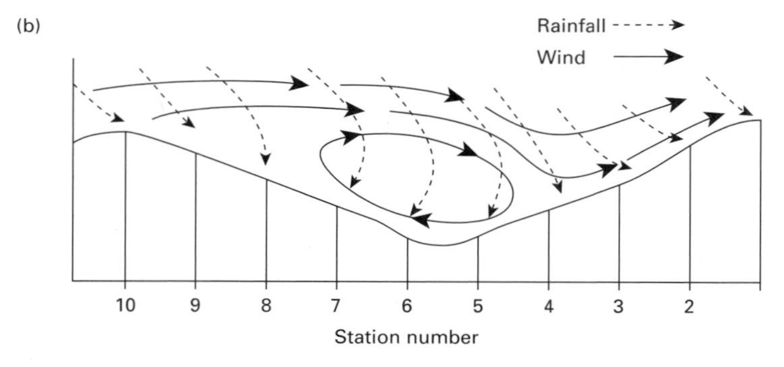


*Figure 10.15* A slope or valley orthogonal to the regional wind flow can create a rotor and downdraft which can (a) influence air pollution concentrations locally or (b) modify the direction of rainfall (shown by the dotted lines) with potential impact on soil erosion and hydrological processes ((a) After Munn, 1966; (b) after Sharon and Arazi, 1997.)

With stronger regional winds, hills and valleys can become small-scale obstacles to the wind flow in much the same way as the building considered in Box 10.A. Valleys, for example, can generate flows at right angles to the regional wind. On a somewhat larger scale, downdraughts created in the lee of a valley slope can develop rotors (Figure 10.15). These can have detrimental effects for pollution concentration in the valley and for aircraft flying through them. As the scale of the topography increases, the mountains act as airflow barriers, induce thermodynamic and eventually orographic effects, and produce the larger-scale regional climates we have already considered.

## 10.5 The influence of larger-scale atmospheric features

Local surface type and topography have a continuous, pervasive influence on local climates. However, large-scale atmospheric motions frequently have local expressions, particularly through the creation of local winds.

### 10.5.1 Local winds

The small-scale winds we have considered in previous sections have been driven by surface conditions. The local winds we consider here depend upon the interaction of the surface with the overlying atmosphere. Although all are in one sense local and unique, it is possible to divide them into three general classes. The first group is associated with the descent of air from a mountain range or high plateau (Table 10.4). The descending air is adiabatically warmed and arrives at the lower elevations as a hot, dry wind. The Föhn of Europe and the Chinook of North America are examples. Such winds will occur only when the regional winds are in the correct direction to facilitate the downslope flow. The descending air will often emulate the downwash flow of Figure 10.15, so that the wind is likely to be very gusty. Further, it will affect only a small area to the lee of the mountains unless the synoptic situation is such that it will aid in

*Table 10.4* Some local winds of the world

| Name | Location | Characteristics | Season |
|---|---|---|---|
| Bora (Latin, *boreas* = north) | Adriatic Coast | Cold, gusty northeasterly wind. Frequency at Trieste, 360 days in 10 years. Mean wind speed 14 m $s^{-1}$, summer 10 m $s^{-1}$ | |
| Chinook (from Chinook Indian Territory) | Eastern slopes of Rockies | A warm wind that may, at times, result in a sudden and drastic rise in temperature. May attain 15 or 20 °C in spring with a relative humidity of 10% | Most violent in winter (may reach 100 km $h^{-1}$) |
| Etesian (Greek, *etesiai* = annual) | Eastern Mediterranean | Cool, dry, northeasterly wind that recurs annually | Summer and early autumn |
| Föhn (German, possibly from Latin, *favonium* = growth, i.e. favouring wind) | Alpine lands | Similar to Chinook. Characterised by warmth and dryness | Most frequent in early spring |
| Haboob (Arabic) | Southern margins of Sahara (Sudan) | Hot, damp wind often containing sand. Of relatively short duration (3 h), average frequency of 24 per year | Early summer |
| Harmattan (Arabic) | West Africa | Hot, dry wind characteristically dust-laden | All year, but most effective in low-Sun season |
| Khamsin (Arabic, Khamsin = 50) | North Africa and Arabia | Hot, dry southeasterly wind. Regularly blows for a 50-day period. Temperatures often 40–50 °C. Same winds with adiabatic modifications include Ghibli (Libya), Sirocco (Mediterranean) and Leveche (Spain) | Late winter, early spring |
| Levanter (from Levant, eastern Mediterranean) | Western Mediterranean | Strong, easterly wind, often felt in Straits of Gibraltar and Spain. Damp, moist, sometimes giving foggy weather for perhaps two days | Autumn, early winter to late winter, spring |
| Mistral (maestrale of Italy = master wind) | Rhône valley south of Valence | Strong, cold wind channelled down Rhône Valley. May reach 28 m $s^{-1}$ in north. Can cause sudden chilling in coastal regions. (Note also the Bise, an equivalent cold north wind in other parts of France) | Most frequent in winter |
| Norther | Texas, Gulf of Mexico to West Caribbean | Cold, strong, northerly wind whose rapid onset may suddenly and drastically lower temperatures (also Tehuantepecer of Central America) | Winter |
| Pampero | Pampas of South America | Southern Hemisphere equivalent of the Norther | Winter |
| Zonda | Argentina | A warm, dry wind, on lee of the Andes. Can attain 33 m $s^{-1}$. Comparable to Chinook and Föhn. In dry weather carries much dust | Winter |

the removal of the cold air ahead of the Föhn. Then the wind can penetrate several hundred kilometres away from the base of the mountains. Although this penetration could occur at any season of the year, these winds tend to be most strongly developed in spring, when depression systems moving over the mountains provide ideal conditions.

A second group of winds included in Table 10.4 is associated almost entirely with changes in the synoptic situation, but changes that occur frequently enough, and produce weather changes noticeable enough, to be impressed on the local consciousness. The Mistral is an example. In winter, cold air collects in the valleys and on the plateaux of the Alps. This air will remain stagnant, with continued cooling, until it is forced to move, usually when a depression passes through the Mediterranean. It descends into the Rhône Valley and, even though somewhat adiabatically warmed, is experienced as a cold wind moving to the Mediterranean shore. A similar situation occurs in the production of the Santa Anna of southern California. Here, however, the source is the hot dry deserts of the interior, giving a hot dry wind over the Californian coast.

Seasonal changes in the general circulation pattern create the third group of local winds (Table 10.4). Here an example is the Harmattan. As the subtropical high pressure region drifts southward in winter, the humid tropical air over west Africa is replaced by dry, dusty air of Saharan origin. The return northward of this high pressure belt in spring produces the Sirocco. This blows from the Sahara across the Mediterranean basin. It arrives at the European coast as a hot wind and, although it may have picked up a little moisture over the sea, it is still dry.

**Box 10.I Defining and measuring local climates**

In many cases where climatological information is to be used for practical problems, the local scale is the area of concern. Indeed, many of the problems we have presented earlier have been essentially local. In most cases the information is needed for a site well removed from weather stations. Indeed, when considered on a local scale the surface observation network is not very dense, while the satellite observations, which give greater areal coverage, do not always provide appropriate information. There are thus two options open: take measurements at the site of interest, or use the existing observations and a knowledge of local-scale processes to estimate the required conditions.

*10.I.1 Measurement programmes for micro- to meso-scale climatology*

There are numerous instruments available for monitoring various aspects of the meso- and micro-climate. Most are highly specialised and beyond the scope of this book. It is sufficient to note that an observational programme requires time, and that instrumental cost and accuracy are often related, so it may be as cost-effective to estimate as to measure. Thus the advisability of initiating a measurement programme depends on the nature of the problem and the time and financial resources available. In many cases measurements must be undertaken. Nuclear power plants in the United States must legally maintain meteorological observations on-site to monitor the likely effects of any emissions. For some other problems, however, it may be inappropriate to initiate measurements. If, for example, a crop yield model is to be produced, a long period of record is needed in order to establish the correlations between the yield and the climatic parameters. It is impractical to wait for decades for the results. Furthermore, the results will depend on a great number of non-meteorological factors, not all of which can be measured, so that the meteorological data required need not be of exceptionally high accuracy. Data from a nearby site, with a long period of record, may therefore provide the best information.

One instrument which promises to be of great use on the meso-scale is Doppler radar (Box 5.I). This has the ability to determine conditions inside a cloud, and to determine rainfall rates on a kilometre by kilometre and minute by minute basis. It is currently used primarily as a tool for short-term prediction of severe weather, but it has an unexplored potential to provide local climatological information.

*10.I.2 Transferability of local-scale climatological data*

In transferring data from one site to another, all aspects of the local climates of the two sites must be taken into consideration, and the various effects introduced above must be incorporated. For some parameters quantitative models are available to assist in the transfer, and the increasing sophistication of spatial statistics and the power of Geographic Information Systems suggests that model-based approaches will continue to increase. Nevertheless, in many cases

**Box 10.I (cont'd)**

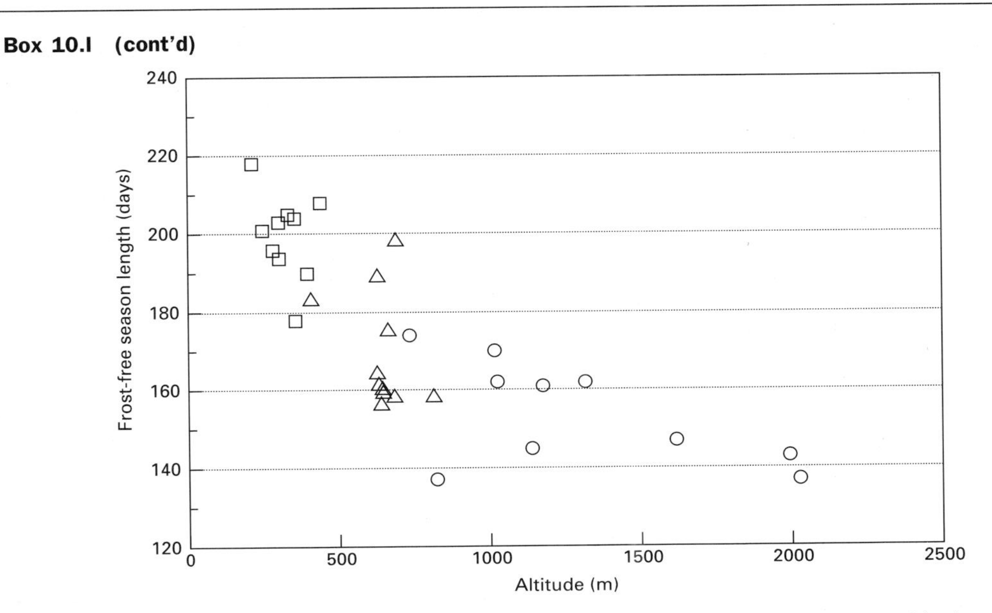


*Figure 10.I.1* Relation between average length of frost-free season and station altitude for selected stations in western North Carolina. The stations are divided into three classes on the basis of topography: □, Piedmont; △ French Broad; ○, Blue Ridge. (After Robinson, 1997.)

we must still rely on rather simple conceptual models based on the basic principles of the controls of local climates to incorporate all relevant effects.

We can illustrate some of the ideas and pitfalls involved in such data transfers using the length of the frost-free season as an example (Figure 10.I.1). Our concern is to determine the value for a particular site at an altitude of 600 m in a mountainous region in North Carolina, USA. We suspect that establishing a linear relationship between altitude and season length using the available stations within about 50 km will allow the best estimate. Initial examination of nearby stations with a range of altitudes provides a wide scatter of points. However, separation by topographical setting provides three regimes. One, the Piedmont, has sites generally on well-drained slopes and tending to face southeast. There seems to be some linear relationship in this region. The second region, the French Broad river valley, has little relief. The stations here seem to have only minor topographical differences: some are at the water's edge; some on gentle east-facing slopes; some on slight, rather dry, rises above the flat valley floor. These minor differences have a major climatic impact. Finally, the stations in the Blue Ridge region tend to be on or near hill summits, and have a reasonably linear relationship. The one clearly exceptional station, at about 850 m but having the same season length as the station at 2050 m, is in a steep-sided valley and must represent a frost hollow situation. This brief review indicates that while there is an overall response to the regional climate, and a general relationship between season and altitude, each facet of the local climate must be considered at each station before it is possible to make a realistic estimate for a particular site.

# CHAPTER 11 Human interaction with climate

So far throughout this book we have emphasised climate as a physical system governed by a set of physical laws, varying on several time and space scales, and seen on the surface of the Earth as a series of climate regions. Some relations between climate and humans have been considered, but as an adjunct to the physical emphasis. In this chapter the emphasis is reversed and we explore (1) the impact of climate and its fluctuations on human lifestyles and activities, and (2) the impact of human lifestyles and activities on the climate. Some of our earlier considerations of the applications of climate information, of course, are a direct response to the first of these two links.

This chapter also represents a change in the level of development of climatology itself. Earlier chapters presented the results of a long investigative tradition. Even when discussing major new concepts and discoveries such as ENSO, the basic physical processes are known with a great deal of confidence, and uncertainty is usually concentrated on the magnitudes of the various processes, not on their nature. From this point forward, however, we are discussing phenomena where the processes are often poorly understood, or even unknown, and their magnitudes are highly uncertain. All are the subject of investigation, debate, speculation and modification.

## 11.1 The impacts of climate on humans

The most obvious impacts of climate on human activities are those where there is a direct relationship between the two. A prolonged period without rain may lead to a water supply crisis, or conversely, an excess of precipitation may lead to flooding and building destruction. The results of such close links are called, somewhat unsurprisingly, **direct impacts**. Direct impacts may themselves create other impacts, which are **indirect**. Sometimes the terms **first-order**, **second-order** and so on are used for the various levels of impact, with the link with climate getting more and more tenuous as the order increases. The range of impacts is vast and we can only consider a few, mainly in a general way, in this section. A more detailed set of examples in one sector, electricity production, is given in Box 11.A.

In general, direct impacts are well known and clearly defined. In addition, they are often known quantitatively. In contrast, the indirect impacts are usually known only qualitatively, and are often poorly understood. The connections whereby a direct climatic impact leads to an indirect one, which in turn creates another, may be nebulous or speculative. Indeed, since non-climatic factors must almost always be incorporated, care must be taken to ensure that the links are truly associated with climate and, if so, that climate plays a significant role in the impact.

### 11.1.1 Climate variability and its impact

In most regions of the globe the local economy and lifestyle is attuned to the local climate, with an often unspoken assumption that this climate shows short-term variability about a long-term stable mean. Successful activities are likely to be adjusted to the direct climatic impacts, reflecting both local knowledge accumulated over many years' experience with the climate and, it is hoped, the more recent application of information to deal with it. Spatially, for example, frost hollows are avoided for agriculture, flood plains are avoided for settlement. Indeed, many of the local considerations introduced in Chapter 10 find a practical expression in such site selection. Temporally, activities must be sensitive to short-term climatic variability if they are to survive. Water supply reservoirs

**Box 11.A Climate impacts on an electric utility**

*11.A.1 Direct impacts*

It has been estimated that some 50% of European and 30% of North American energy usage is to overcome the effect of climate. Much of this is for space heating and cooling, which is a direct response to ambient temperature variations (see Box 3.A). This is an impact which occurs every day, although the occurrence of occasional heat waves or exceptionally cold spells defines the maximum energy demand likely to be placed on the utility. However, it is just one of a range of impacts of weather and climate on electric generating utilities (Table 11.A.1). It is unlikely that all will occur for a single utility in a single year, and some may never be relevant for a particular utility in the current climate. In most cases a utility can prepare for and respond to those common in its service area, using past experience as a guide, as with the development of the energy/temperature relationships (Figure 3.A.2). In some cases the development of the long-lead climate forecasts (Section 14.2) is helping to minimise the adverse impacts of some of these features.

If there is a significant climate change in a region, past experience may not be a sufficient response guide since the suite of impacts for a utility may change. Ice storms, for example, may appear in new areas, or vanish from old ones, influencing the way that transmission line repair crews are deployed. Even if the type does not change, the magnitude will change, and not always in ways that are immediately obvious. We can postulate a regional temperature increase. If this is evenly spread through the year in a cold region, electricity demand will decrease. If it is a hot region, demand will increase. If the change is mainly in the summer, it may have a large impact on the warm region utility and a small one in the cold region. Similarly, the elevated temperatures may increase evaporation from hydroelectric storage reservoirs if they occur in the summer, but have little impact in the winter. Hence it is clear that any estimate of future climate need be rather detailed if it is to be of use to a utility.

*Table 11.A.1* Some direct impacts of weather and climate on utilities

| Time/space scale | Atmospheric condition | Area of effect |
|---|---|---|
| Short-term/ local area | Thunderstorms | Power outages |
| ↓ | High winds | Transmission line problems; structural damage |
| | Ice storms | Transmission line problems |
| | Heavy rain | Facilities damage; flood control problems |
| | Snow storms | Maintenance crew activity; fuel transportation problems; stored hydroelectric water |
| | Tornadoes | General mayhem |
| | Hurricanes | Plant integrity |
| | Air stagnation | Air quality compliance |
| | Heat waves | Peak demands; plant operating efficiency; transportation problems |
| | Cold spells | High demand; fuel transportation |
| | Drought | Hydroelectric generation shortfall; little cooling water; river transport problems |
| | Seasonal temperature | Fuel usage |
| | Seasonal clouds | Solar energy sources |
| | Seasonal winds | Wind power possibilities |
| | Seasonal precipitation | Water availability |
| | Annual means | Rate changes |
| Long-term/ large area | Long-term means | Facilities planning |

*11.A.2 Indirect impacts*

In the previous section we emphasised single direct impacts, but many from Table 11.A.1 are linked in various ways. If summer temperatures rise, for example, not only will electricity demand rise, but the availability of water for hydroelectric generation will decrease. This will increase the potential for a drought, which will not only deplete reservoirs but also make river-borne fuel delivery difficult. Thus we can quickly widen the circle of impacts from those strictly associated with energy generation to those in the economic sectors which support it or depend on it.

The circle can go even wider, and climatic events having an impact seemingly entirely outside the energy industry will have repercussions inside

**Box 11.A (cont'd)**

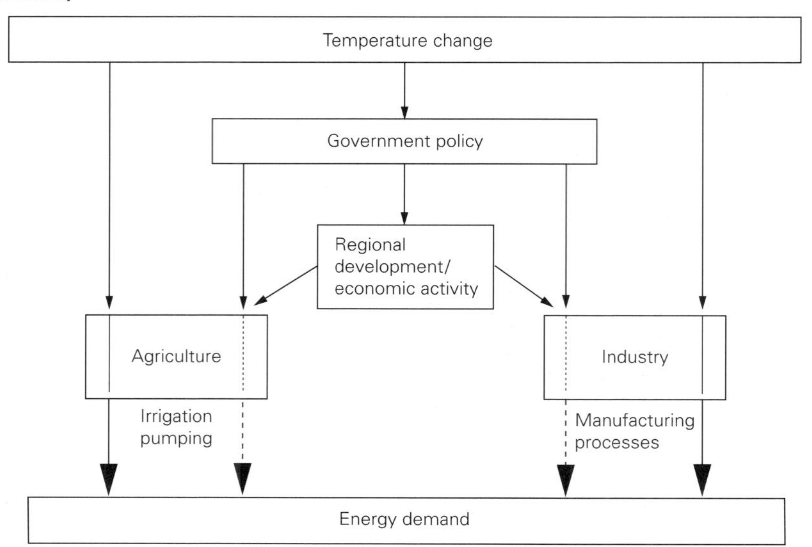


*Figure 11.A.1* Schematic of some potential direct (solid arrows) and indirect (dashed arrows) impacts as a result of a temperature change. There are some direct relationships between the temperature change and energy demand through, for example, agriculture and industry. Temperature changes may also change the level of industrial or agricultural activity, and hence the amount of regional development, while government policy, which may be a response to temperature change, can influence them all.

(Figure 11.A.1). During an agricultural drought, irrigation water may have to be pumped from deep wells to maintain productivity. That increases energy usage. If, as a result of a long period of drought, agriculture abandons an area, that also has an impact on energy. A government may establish a policy to prevent such abandonment, subsidising agriculture in some form, so that energy usage is influenced by government action to mitigate the impacts of the climate in an entirely different sector. Thus, we have an ever-widening circle of impacts, with ever more tenuous links to climate.

must be sized to ensure that they do not run dry except in the rarest dry spell, but at the same time they must not be excessively large and costly. Food storage facilities must be adequate to ensure that surplus stocks from favourable agricultural years can be retained so that a community can survive during less benign times. If the climate is such that harvests vary little from year to year, the needed facilities may be small, but in a more variable climate they must be larger, or problems will ensue. Reference to the material in Chapters 8 and 9 provides indications of regions where variability may cause problems in various activities.

However well-adjusted a particular society is to the climate and its variability, anomalous events are likely to cause problems of various severity. In any year numerous regions of the world will demonstrate impacts of climatic events. As an example, 1996 (Figure 11.1) was a year without a major El Niño event, so none of the impacts illustrated are akin to those of Figure 8.24. Indeed, some areas indicate the opposite tendency. Further, the figure generally shows the climatic anomaly rather than the impact. For direct impacts this is readily inferred. Although we usually think of anomalies leading to adverse impacts, this is not always the case. The Sahel region, for example, had

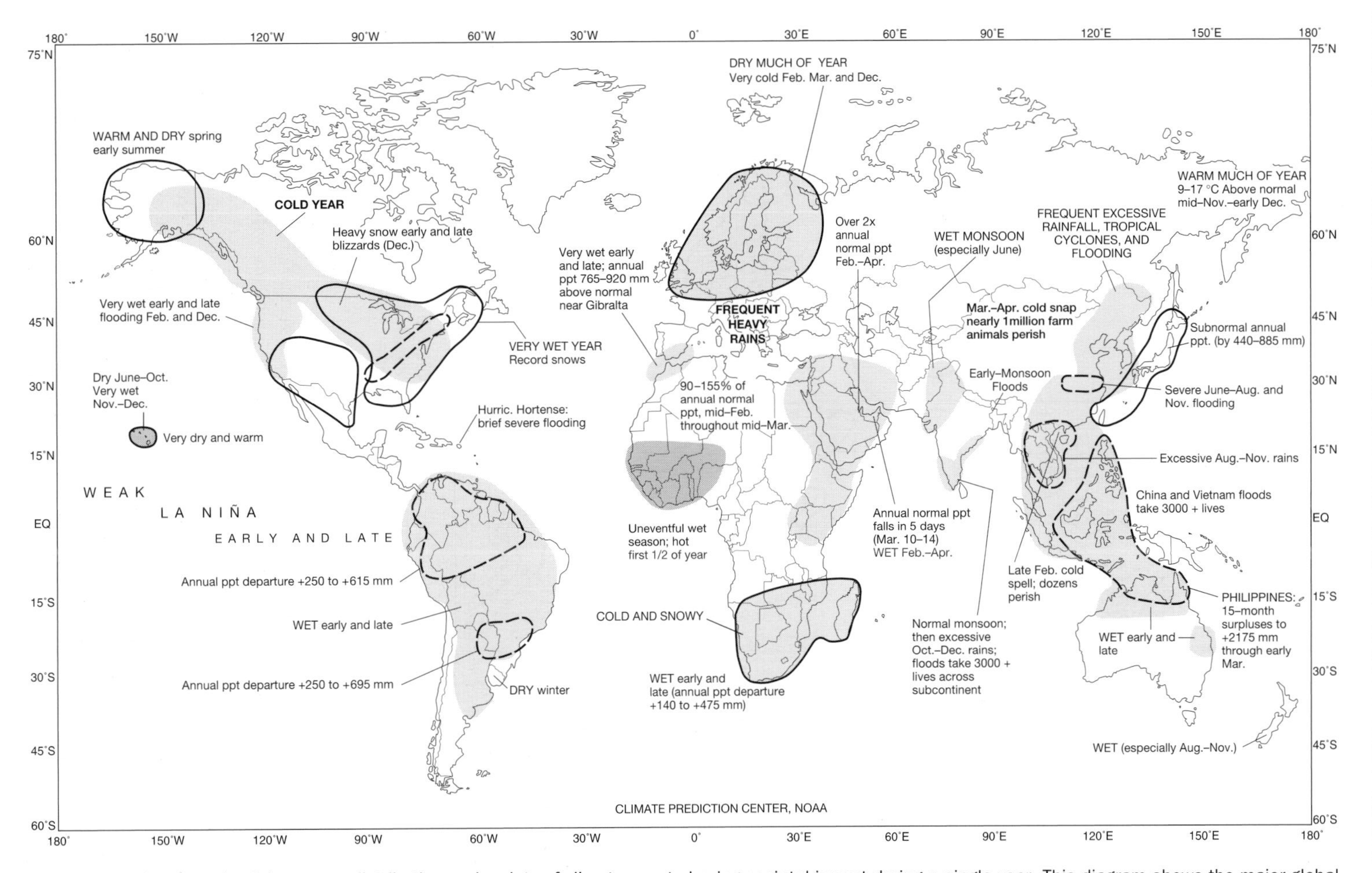


*Figure 11.1* An example of the range, distribution and variety of climate events having societal impact during a single year. This diagram shows the major global climate anomalies and episodic events in 1996. (After Tinker, 1997.)

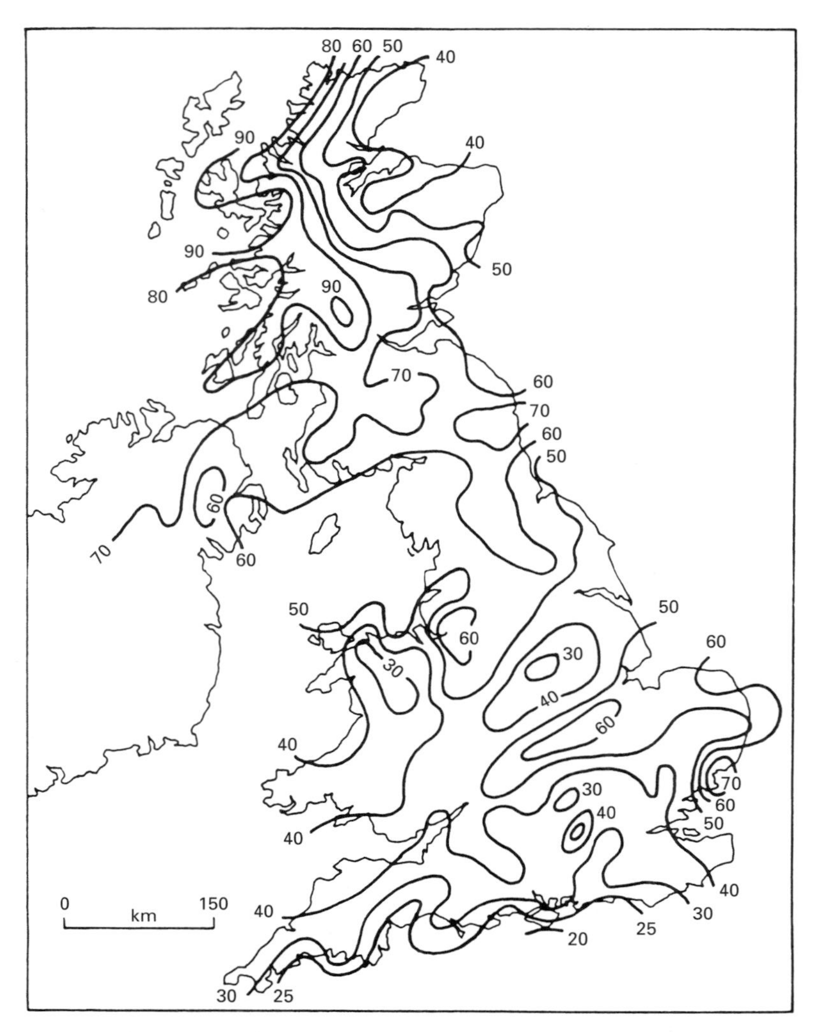


*Figure 11.2* Total rainfall in the period 1 April 1976 to 31 August 1976 expressed as the percentage of the 1916–1950 average rainfall over the UK for the months of April to August inclusive. (From Roy *et al.*, 1978.)

an 'uneventful' wet season, with adequate rainfall for vegetation growth and thus few agricultural problems.

A single climatic anomaly may have a series of impacts, the consequences of which are by no means invariably adverse. The summer of 1976 in Britain can be used as an example. This summer was exceptionally dry, qualifying as a drought with an estimated return period of around 100–200 years (Figure 11.2). The lack of rainfall certainly retarded crop growth, but had an even greater effect on crop-destroying pests. The result was a harvest that was not too much below normal (Table 11.1). The unusually dry soil caused problems with the foundations of many buildings; a problem to the owners, but a boon to the construction trade. Similarly, resort areas in Britain reported an exceptionally profitable year. Conversely, resort operators in the Mediterranean region, having the climatically normal sunny and dry summer, reported fewer British visitors. So, most climatic events will have many and various impacts on the national, and possibly the international, economy, with some sectors being favoured, some not.

*Table 11.1* Annual estimates of average (UK) crop yields in tonnes per hectare

| | 1974 | 1975 | 1976 | Percentage decrease in 1976 from 1974 yields |
|---|---|---|---|---|
| *England and Wales* | | | | |
| Wheat | 4.94 | 4.30 | 3.85 | 22 |
| Barley | 3.95 | 3.40 | 3.46 | 12 |
| Oats | 3.88 | 3.45 | 3.42 | 12 |
| Potatoes: early | 18.8 | 14.1 | 16.3 | 13 |
| maincrop | 33.9 | 22.1 | 20.4 | 40 |
| *Scotland* | | | | |
| Wheat | 5.75 | 5.58 | 4.98 | 13 |
| Barley | 4.99 | 4.79 | 4.10 | 18 |
| Oats | 3.66 | 3.45 | 3.58 | 2 |
| Potatoes: early | 24.1 | 19.3 | 19.4 | 20 |
| maincrop | 30.9 | 26.6 | 26.2 | 15 |

On a much broader scale, the series of droughts in the interior of the United States in the early 1930s, commonly known as the **Dust Bowl**, had direct impacts for agriculture which led to a long series of indirect impacts. The establishment of national and local policies for both public water supply and soil conservation, as well as a major shift in the distribution of the US population, have been attributed to this event. Indeed, some historians would credit the impact of the Dust Bowl with providing the impetus for the establishment of the whole of the current social security system. The climatically comparable long series of drought years in the Sahel (Figure 8.11) had a different impact. The agricultural failures and human suffering stimulated an international relief effort. In that sense, the impact of climate is a global impact.

### 11.1.2 The impact of a changing climate

As climatologists, we think of climatic changes as being represented by changes in the long-term mean values of a particular climatic parameter. Superimposed on this mean value will be decadal fluctuations, year-to-year variations, or short-term anomalies. The changes in mean values in the last few decades can be detected only by careful analysis of instrument records and, on the time scale of human impacts, are likely to be sufficiently slow and small as to be almost imperceptible. Much more noticeable will be changes in the variability, expressed in terms such as 'a run of exceptionally wet years', 'there seem to have been more hail storms in recent years', or even, the human memory being what it is, 'we don't get the long summers that we used to'. Any human response to an actual or perceived climate change must be heavily influenced by such a perception, whether conscious or unconscious. Furthermore, the change must involve a climatic element that is significant for a particular activity. Consequently 'climate change' *per se* will not always lead to a climatic impact.

In general the magnitude of the impact of a climate change on a human society, and how direct it is, depends on the degree to which the social and economic system is dependent on agriculture. Thus, since ecosystems respond to a climatic change in a direct way, for nomadic and semi-nomadic peoples there will be a direct and immediate response. They will move if a climatic variation causes an unacceptable deterioration in their environment provided, of course, that an alternative, more acceptable area lies within their reach. A successful nomadic society will have become adapted to an environment whose variability makes it marginal for their particular mode of existence.

Generalising this idea, it is in marginal agricultural areas that the links between climatic change and human response can be postulated to be direct and strong. Such areas represent the outer limits of a particular climatic and agricultural region. In the centre of the region agricultural practices will be well adapted to that particular climate and year-to-year variations will pose little threat. As the margins are approached, however, variability will become more significant. Usually overall production will be low, so that little surplus can be stored against the poor years that climate variability will inevitably bring. If a climate change occurs which alters the frequency of the poor years, some human response must follow.

The definition of marginality depends upon the climatic regime and agricultural practices of an area. Three such areas using different climatic indices for marginality are shown in Figure 11.3. Northern Europe is divided into agriculturally marginal and sub-marginal areas. The limits are as follows:

(a) Marginal: either less than 5 months >10 °C and no increase in precipitation deficit July–September; or less than 3 months >10 °C.
(b) Sub-marginal: either less than 5 months >10 °C and 0–50 mm increase in precipitation deficit July–September; or 5–6 months >10 °C and no increase in precipitation deficit July–September.

(a)

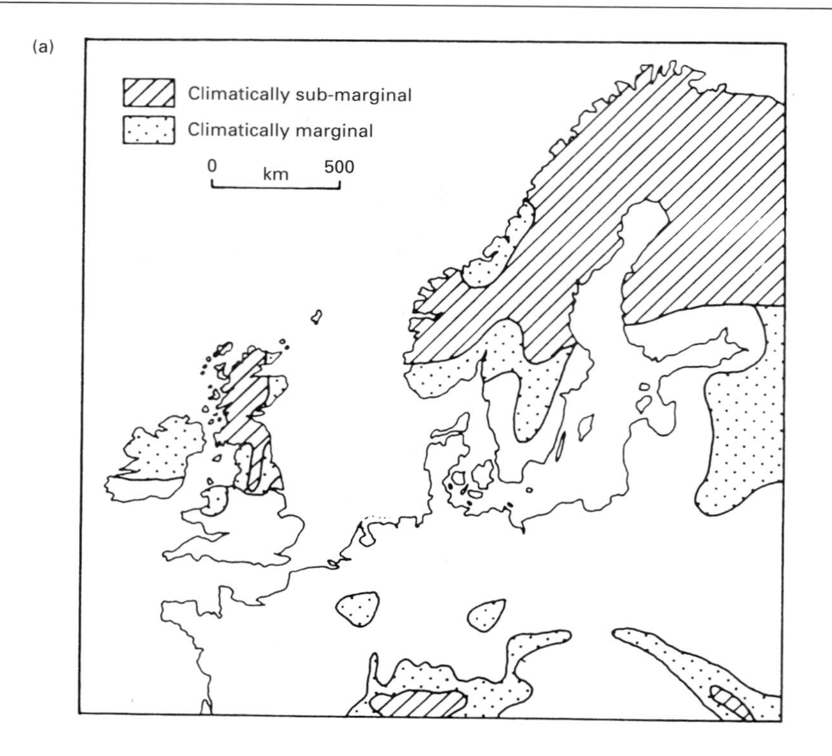


(b)

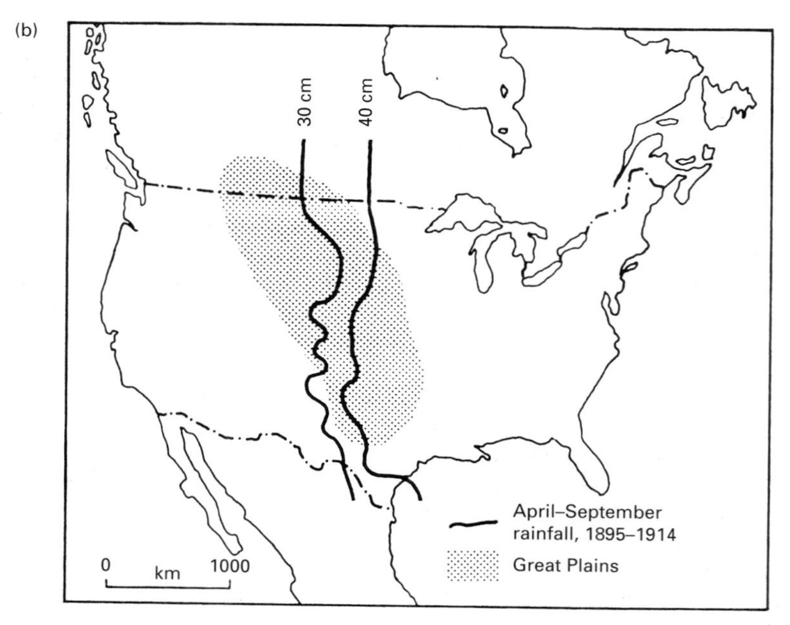


*Figure 11.3* Climatologically marginal land in (a) northern Europe, (b) the Great Plains of the USA and (c) eastern Australia. In (c) the shifts in climatic belts between 1881–1910 and 1911–1940 are shown. (From (a), (b) Parry, 1978; (c) Hobbs, 1980.)

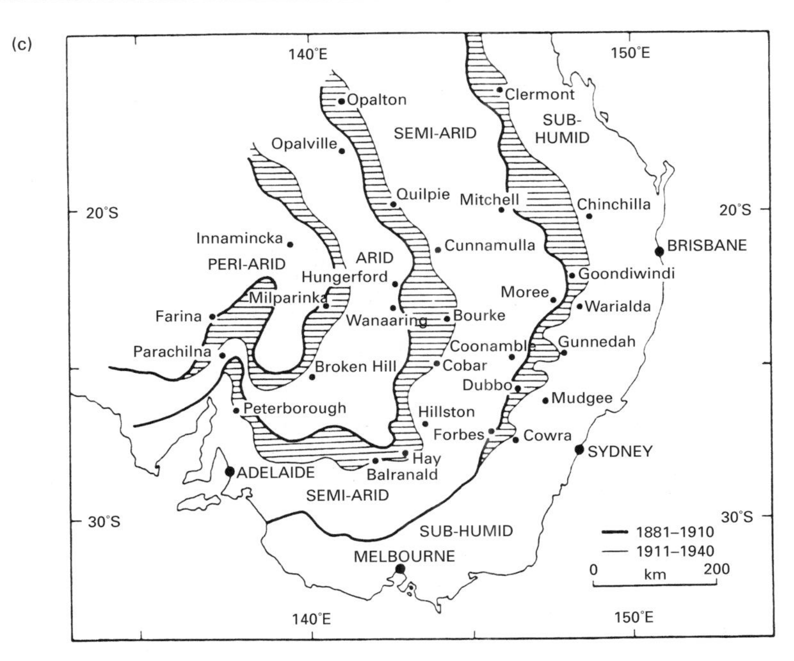


*Figure 11.3* continued

The region of marginal cultivation identified for the United States is based upon total rainfall in the period April–September rather than upon the combination of temperature and rainfall used for northern Europe. For Australia, three zones of marginality, for different climatic regimes and agricultural enterprises, are shown in Figure 11.3(c). The limits are based on temperature and precipitation values and ranges. The changes of these limits with time indicate the eastward encroachment of aridity and the establishment of new marginal areas.

A similar effect is seen, on the local scale, for the Lammermuir Hills in Scotland (Figure 11.4). As is usually the case, the higher land is more marginal. Here it has been possible to suggest that the combined isopleths of 1050 degree days plus 60 mm potential water surplus (at the end of the summer) represented the approximate cultivation limit in the area during the Middle Ages. The temperature decline during the course of the Little Ice Age (roughly AD 1400–1800) thus reduced the area suitable for cultivation. At this same time land and settlements were abandoned. Although this strongly suggests a causal relationship, an indirect impact is implied, and thus the connection must be treated with caution.

For marginal areas such as these the introduction of any factor affecting agricultural success, such as a new crop type, new grain storage technology or financial aid policies, will disrupt the close connection between the simplified climatic limit to cultivation and the settlement illustrated in Figure 11.4. Even when clearly connected to climate, the change in cultivation may be a response to seemingly minor climatic effects, such as the change in frequency of late frosts. Despite these problems, studies of this type have shown the considerable importance of the climate for crop production and the possibility of direct human response to climatic change under some circumstances.

### 11.1.3 Drought and its possible impacts

Drought is intuitively known as a period of low precipitation which has immediate adverse consequences for agriculture and longer-term ones for water supply. Beyond this, it can be extremely difficult to define (see Box 11.I). Here we are much more concerned with the impacts, starting with the very simple intuitive one that crops begin to fail in a drought. The Dust Bowl was mentioned above suggesting that

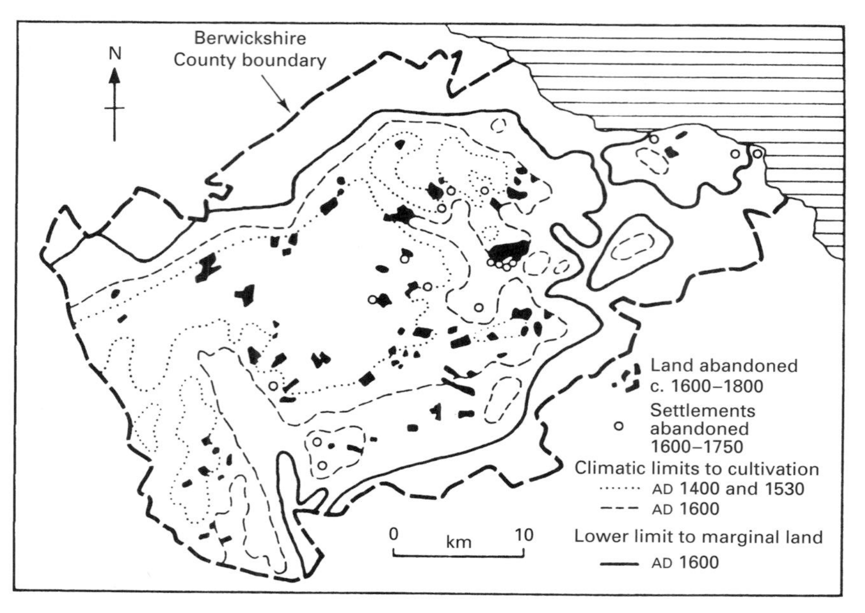


*Figure 11.4* Abandoned farmland and lowered climatic limits to cultivation in southeastern Scotland, 1600–1750. (From Parry, 1978.)

**Box 11.I Derived variables for practical use**

In previous 'Instrument' boxes we have discussed the measurement of specific climatic elements. In many cases, however, it is not a single element, but a combination, that is important. In this section we explore two such measures, both associated with the impact of climate on human activities. The first, considering human comfort, emphasises the combination of various instrumental observations; the second, dealing with drought, is more concerned with linking definitions and observations so that impacts can be assessed.

*11.I.1 Comfort indices*

Human beings generally become adapted to the 'normal' atmospheric conditions of the area where they live. Some of this adaption stems from physiological development in the early years of life and some is physiological acclimatisation, but a lot is cultural adaptation through the adoption of appropriate lifestyles, building styles, and heating and air-conditioning practices. Humans have built-in resilience for some variation about the local average climate, and only wide deviations from it cause severe problems. When heat waves occur in Britain, for example, the temperatures and humidities may be extreme for the inhabitants, but are frequently not unusual or particularly uncomfortable for someone born and raised in the tropics. This visitor may have a greater problem because houses are designed to retain heat rather than remove it, while the lifestyle offers little opportunity for a siesta during the heat of the day.

Even when the cultural influences are removed, the sensation of comfort depends on temperature, humidity, wind speed and the radiation environment, in addition to the amount of clothing and the activity being undertaken by the body. In many ways it is convenient to think of the heat balance of the human body as similar to the surface heat balance, except that the body is an active participant in the energy exchanges. Internal metabolism is an additional heat source and respiration and sweating promote active latent heat fluxes. In hot conditions heat stress occurs when internal temperatures rise as a result of a body's inability to remove excess heat. This is most likely to occur when ambient conditions are hot enough to minimise sensible heat loss, moist enough to inhibit latent heat flow, and where there are no air currents to enhance either. Large amounts of incoming solar radiation may compound the problem.

**Box 11.I (cont'd)**

*Table 11.I.1* Apparent temperatures as a function of relative humidity (RH)

| Air temp. (°C) | Apparent temperature (°C) | | | | | | | | | | | | | | | | | | | | |
|---|---|---|---|---|---|---|---|---|---|---|---|---|---|---|---|---|---|---|---|---|---|
| 60 | 52 | | | | | | | | | | | | | | | | | | | | |
| 57 | 49 | 53 | | | | | | | | | | | | | | | | | | | |
| 54 | 47 | 50 | 55 | | | | | | | | | | | | | | | | | | |
| 52 | 44 | 47 | 51 | 55 | 61 | | | | | | | | | | | | | | | | |
| 49 | 42 | 44 | 47 | 51 | 54 | 59 | 64 | | | | | | | | | | | | | | |
| 46 | 39 | 42 | 44 | 46 | 49 | 53 | 57 | 62 | 66 | | | | | | | | | | | | |
| 43 | 37 | 39 | 41 | 42 | 44 | 47 | 51 | 54 | 58 | 62 | 66 | | | | | | | | | | |
| 41 | 35 | 36 | 38 | 39 | 41 | 43 | 45 | 48 | 51 | 54 | 57 | 61 | 65 | | | | | | | | |
| 38 | 33 | 34 | 35 | 36 | 37 | 38 | 40 | 42 | 43 | 46 | 49 | 52 | 56 | 59 | 62 | | | | | | |
| 35 | 31 | 31 | 32 | 33 | 34 | 34 | 36 | 37 | 38 | 40 | 42 | 43 | 46 | 48 | 51 | 54 | 58 | | | | |
| 32 | 28 | 29 | 29 | 30 | 31 | 31 | 32 | 33 | 34 | 35 | 36 | 37 | 38 | 39 | 41 | 43 | 45 | 47 | 50 | | |
| 29 | 26 | 26 | 27 | 27 | 28 | 28 | 29 | 29 | 30 | 31 | 31 | 32 | 32 | 33 | 34 | 35 | 36 | 37 | 39 | 41 | 42 |
| 27 | 23 | 23 | 24 | 24 | 25 | 25 | 26 | 26 | 26 | 27 | 27 | 27 | 28 | 28 | 29 | 30 | 31 | 31 | 32 | 32 | 33 |
| 24 | 21 | 21 | 21 | 22 | 22 | 22 | 23 | 23 | 23 | 23 | 24 | 24 | 24 | 24 | 25 | 25 | 26 | 26 | 26 | 26 | 27 |
| 21 | 18 | 18 | 18 | 18 | 19 | 19 | 19 | 19 | 20 | 20 | 21 | 21 | 21 | 21 | 21 | 21 | 22 | 22 | 22 | 22 | 22 |
| RH | 0 | 5 | 10 | 15 | 20 | 25 | 30 | 35 | 40 | 45 | 50 | 55 | 60 | 65 | 70 | 75 | 80 | 85 | 90 | 95 | 100 |

| Heat stress category | Average temperature |
|---|---|
| Caution | 27–32 |
| Extreme caution | 32–41 |
| Danger | 41–54 |
| Extreme danger | >54 |

Equivalent and apparent temperatures are not real, measurable values. They are subjective impressions of temperature, exaggerated either because of wind or relative humidity. There are other formulae which give different results.

Although several ingenious instruments have been made to mimic and quantify the human internal temperature, none can replicate the partly subjective response of a human being to the ambient conditions and the overall concept of 'comfort'. No universally accepted quantitative measure of comfort has been developed. Rather, several indices have been proposed, most by exposing a group of volunteers to various environmental situations and assessing their responses on a qualitative scale. Most volunteers have been drawn from the ranks of the military, so the resulting indices have tended to reflect the reactions of young, fit males, rather than a true cross-section of the community. Although the environment has been specified using conventional climatological measurements, it has proved difficult to produce a single measure which uses routine meteorological observations while incorporating realistic heat balance concepts and human perception. The most commonly quoted index for hot conditions is the **temperature-humidity index** (THI) (Table 11.I.1), given by:

$$\mathrm{THI} = 0.4(T_a + T_w) + 4.8 \qquad (11.I.1)$$

where $T_a$ and $T_w$, the dry and wet bulb temperatures, are in degrees Celsius.

In cold conditions the problem is to maintain the core body temperature. The presence of wind increases the energy fluxes from the body. This may be beneficial at high temperatures, but is dangerous at low ones. Thus wind replaces humidity as the variable of concern at low temperatures. The most common formulation for the apparent temperature at low values is the **wind chill equivalent temperature** (Table 11.I.2).

### *11.I.2 Drought indices*

Although drought and its impacts were considered in Section 11.1, no effort was made there to define drought. This is of practical importance since in many countries govenment aid becomes available once an area enters a clearly specificed drought situation. The simplest definition of drought would be one where any dry period lasting longer than a predefined threshold was termed a drought. A few moments' consideration of regional climatology, however, indicates that no single world-wide definition would be appropriate. In areas with marked seasonality even a single local

**Box 11.I (cont'd)**

*Table 11.I.2* Equivalent temperatures including the effects of wind chill

| Estimated wind speed (m s$^{-1}$) | Actual thermometer reading (°C) | | | | | | | | | | | |
|---|---|---|---|---|---|---|---|---|---|---|---|---|
| | 10 | 4 | –1 | –7 | –12 | –18 | –23 | –29 | –34 | *–40* | *–46* | *–51* |
| | Equivalent temperature (°C) | | | | | | | | | | | |
| 0 | 10 | 4 | –1 | –7 | –12 | –18 | –23 | –29 | *–34* | *–40* | *–46* | *–51* |
| 2.0 | 9 | 3 | –3 | –9 | –14 | –21 | –26 | *–32* | *–38* | *–44* | *–49* | *–56* |
| 4.5 | 4 | –2 | –9 | –16 | –23 | –29 | *–36* | *–43* | *–50* | *–57* | **–64** | **–71** |
| 7.0 | 2 | –6 | –13 | –21 | –28 | *–38* | *–43* | *–50* | *–58* | **–65** | **–73** | **–80** |
| 9.0 | 0 | –8 | –16 | –23 | *–32* | *–39* | *–47* | *–55* | **–63** | **–71** | **–79** | **–87** |
| 11.0 | –1 | –9 | –18 | –26 | *–34* | *–42* | *–51* | **–59** | **–67** | **–76** | **–83** | **–92** |
| 13.5 | –2 | –11 | –19 | –28 | *–36* | *–44* | *–53* | **–62** | **–70** | **–78** | **–87** | **–96** |
| 15.5 | –3 | –12 | –20 | –29 | *–37* | *–45* | *–55* | **–63** | **–72** | **–81** | **–89** | **–98** |
| 18.0 | –3 | –12 | –21 | –29 | *–38* | *–47* | *–56* | **–65** | **–73** | **–82** | **–91** | **–100** |
| >18 – little additional effect | Little danger for properly clothed people | | | | *Increasing danger* | | | Great danger<br>Danger from freezing of exposed flesh | | | | |

definition might not be adequate. After a few further moments of reflection, a second set of concerns would be pertinent: what is the impact and on whom?

Although there are numerous impacts, experience has suggested that drought can be divided into two major classes: agricultural drought and hydrological drought. For the former, the moisture deficit is confined to the soil layer from which crop roots extract water. The prime concern is with the growing season, which is often climatically constrained by a cold or a dry season, and an agricultural drought may not persist from one growing season to the next. A hydrological drought, in contrast, is a much more deep-seated factor. Precipitation percolating below the root zone collects in deep rock layers, eventually to be removed by evaporation, horizontal flow to streams, or extraction for human consumption. While the first removal method suggests some seasonal control, the others do not. Thus hydrological drought exhibits much less seasonal constraint, is slower to respond to climatic fluctuations, and can last much longer than agricultural drought.

With such a range of climates and impacts, it is not surprising that at least eight indices have been proposed and used. The Australian agricultural approach, noted earlier, is based on the frequency of recurrence of low precipitation averaged over a specified time period. To obtain this, the historical record is ranked from driest to wettest. The current conditions are compared to this ranking, and if they fall within the lowest 20%, it is deemed that precipitation is 'much below normal'. Although this specifies precipitation conditions, not drought *per se*, the result is treated as an index of drought (Figure 11.I.1). In Britain, the current *Meteorological Glossary* defines drought simply as 'dryness due to lack of rainfall' and emphasises that it is a relative, impact-oriented, concept. Indeed, earlier editions had more specific definitions, which proved to be misleading when response plans were being considered. A similar reassessment is also taking place in the USA. The traditional **Palmer Drought Severity Index** (PDSI), an early scheme linking soil conditions, spatial precipitation variability and potential impacts, has been found to have problems for certain time scales, locations and impacts. For climatological and practical reasons, therefore, the attempt to produce a single value covering all eventualities is being replaced by a search for a more flexible set of simpler measures. In part, this is possible not only because of increased understanding of drought and its impacts, but also because more sophisticated computers allow more, or more refined, calculations. At present, therefore, the **Standardised Precipitation Index**, assessing the departure of current precipitation amounts from the norm for periods of specific length, is increasingly being tested and used. Nevertheless, many US jurisdictions currently have drought response plans which are triggered when the PDSI falls below a certain threshold. Clearly, we are still searching for the ideal approach to the characterisation of the climatological aspects of drought.

**Box 11.I (cont'd)**

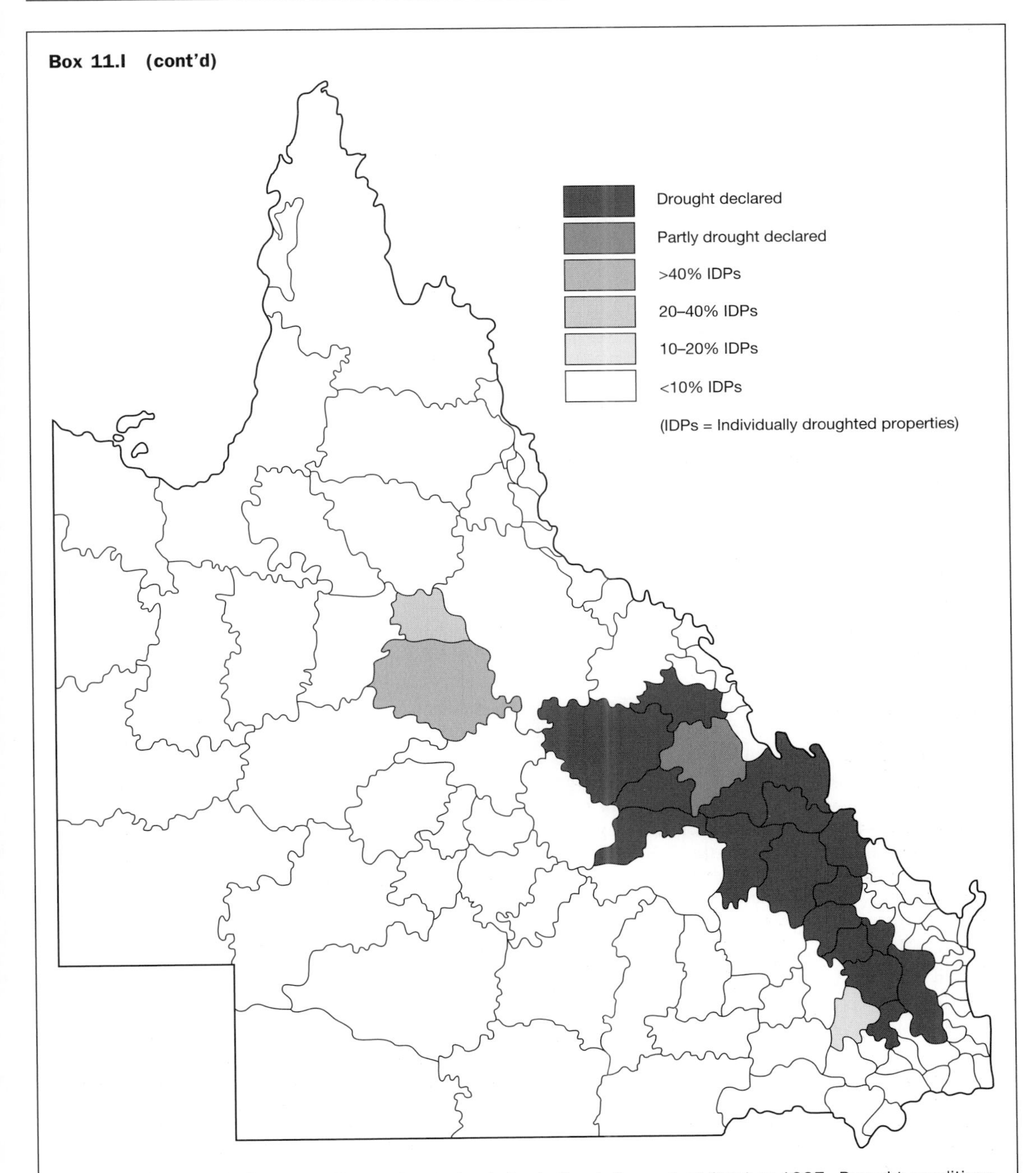


*Figure 11.I.1* The drought situation in Queensland, Australia at the end of October 1997. Drought conditions were not uniform over the state, even within the southeastern portion where drought had been declared. Along the coast northwestward from this there was a gradual decrease in the number of local areas experiencing drought. Inland areas had little drought at this time. (Redrawn from web site http://www.dnr.qld.gov.au/longpdk/carrier/drought/1997 10.gi)

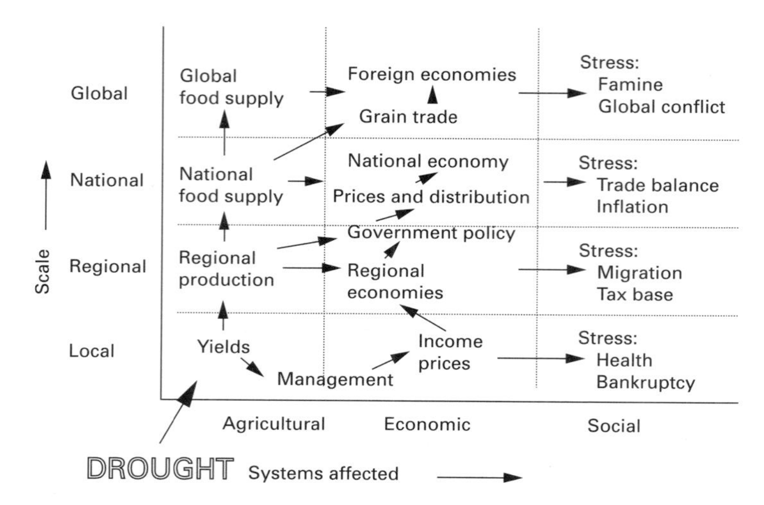


*Figure 11.5* A series of links between a drought at a specific location and economic and social impacts on a variety of scales. Although each individual link is theoretically realistic, the chained connection between them all has not been demonstrated. (After Warrick and Bowden, 1981.)

many indirect impacts follow from a crop failure. One suggested pathway leads from drought to global conflict (Figure 11.5). Each of the links in the web of this figure has validity, and indeed many have been demonstrated. Nevertheless it cannot be inferred that a particular drought will lead to a global conflict. Indeed, it seems that drought has frequently engendered international goodwill, with a sharing of resources with afflicted nations rather than conquest. This was certainly the case for the Sahel region in the 1980s. These responses, however, have been associated with variability in the current climate, not with possible consequences of global climate change. Increasing, or increasingly frequent, drought, particularly if it is in the major grain-producing regions of the world, could certainly strain international relations and the 'worst case' scenario of Figure 11.5 could become a real possibility. Indeed, this possibility was explicitly recognised in 1988, when the Canadian government hosted a conference, with climatologists, impact assessors and diplomats sharing concerns and ideas, with the eventual objective of ensuring that this worst case did not happen. This event was one of many which eventually led to the establishment of the 'Intergovernmental Panel on Climate Change', which has played a major role in studies of future climates, and which will be considered in more detail in later chapters.

If global conflict arising from drought seems a remote possibility, the human suffering resulting from present droughts is all too real. Since we cannot directly forecast droughts with any confidence, we must adopt an approach similar to that used for hurricanes (see Box 8.A). Monitoring of the evolving climatic conditions to provide early warning is vital. This must be combined with a response plan if it is to aid the amelioration of human suffering. Several governments have developed plans whereby specific actions are triggered by specific climatic conditions. In Australia, for example, government assistance to farmers commences once it is certified that for a period of 12 months the precipitation has been sufficiently low that such conditions would normally be expected only once in 20 years or more.

## 11.2 Human modifications of the atmosphere

The obvious converse of the impact of climate on humans that we have just considered is the impact of humans on the climate. This can be both deliberate and inadvertent, and can arise from changes to the atmosphere and to the surface. The former is of concern here; the latter is considered in the next section.

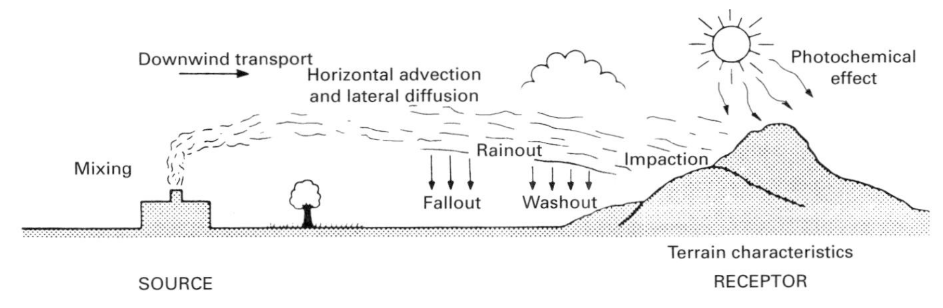


*Figure 11.6* Schematic representation of the transport and removal of air pollutants.

Alteration of the atmosphere for the deliberate modification of the climate is limited to cloud seeding (Section 5.1). This is a small-scale activity. The large energies involved in atmospheric circulations at the larger scales preclude human modification with the current energy generation technologies. This does not mean that local scale changes could not be used as 'triggers' for regional ones. However, no-one has yet seriously suggested that humankind embark on a deliberate global experiment of this nature. The complexity of the interactions and feedbacks within the climate system, which are poorly understood, would make it a very risky undertaking. Nevertheless, it can be suggested that we are inadvertently undertaking such an experiment as we emit pollutants into the atmosphere.

The gases, particles and energy emitted into the atmosphere by human activity can all be regarded as pollutants with an actual or potential impact on climate. A major concern is that this impact is growing in severity as a result of the combined increase in human population and general rise in the standard of living. Although this is increasingly creating a set of more general 'environmental problems', a full analysis of many of them requires information well beyond the scope of this book. We will focus on the physical climatological aspects. For our purposes, we will adopt a broad definition of atmospheric pollution as any and all material (gas, liquid or solid, and including energy) in the air which does not occur naturally or is present in quantities outside the range that occurred in pre-industrial times. This definition excludes the particles of natural origin, such as desert dust, pollen and sea spray which are involved in cloud and precipitation production, unless their concentration becomes excessive. With this definition it is possible to provide a single schematic diagram of the process involved in atmospheric modification (Figure 11.6). Differences occur because of the nature and size of the pollutants themselves, their residence times and dispersal characteristics, and because of the level, location and force with which they are injected into the atmosphere. Although we divide the problems into three separate subsections, most of the theoretical discussion is concentrated on the near-surface air pollution.

### 11.2.1 Air pollution

There are many sources for pollution: industry, agriculture, settlements and transport. Intensive cultivation of arable land and deforestation both enhance the natural emplacement of material in the atmosphere. This is a low-level source of wide areal extent. In contrast, most industrial sources are relatively high-level point sources, although they may be grouped together to give the effect of an areal source. In our discussion we shall assume that we are dealing with an elevated point source of industrial pollution, although extensions can be made for ground level and area sources.

The point where the pollution enters the atmosphere can be characterised by the actual **stack height** or, more often, by the **effective stack height** (Figure 11.7a). The effective stack height depends on the height of the stack itself and the exit velocity of the material. If the material is a gas, the effective height will also depend on its buoyancy and the atmospheric stability, since the gas coming from the stack acts in many ways like a parcel of air displaced from the surface. Under normal conditions, the effective stack height will be at least the height of the stack itself, and usually considerably higher. Once at this height, however, the plume will begin to level off and be influenced almost solely by the atmospheric conditions. Of great importance is the relationship between the effective stack height and the local stability conditions

(a)

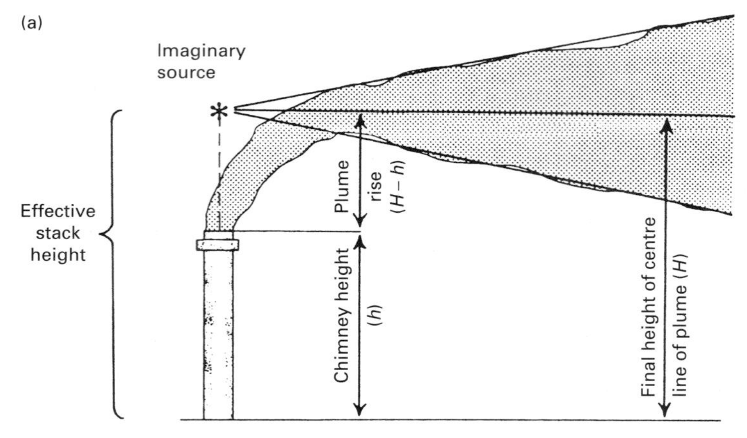


(b)

*Figure 11.7* (a) Schematic diagram showing the effective stack height, which is compounded from the actual chimney height and the buoyant plume rise. (b) Buoyant plumes for the cooling towers of an electricity generating station penetrating a low-level early morning inversion. (Courtesy of A. Abbott.)

(Figures 11.7(b) and 11.8). Any of these conditions may persist for a long period of time, or may change diurnally in the way indicated in Figure 4.10. However, the marked effects of inversions are clearly seen, emphasising that it is desirable to ensure that the effective stack height is above the local surface inversion in order to minimise ground level concentrations.

As the plume moves horizontally away from the stack under the influence of the regional wind, turbulence and diffusion within it will cause it to expand. These are pseudo-random processes. For relatively short distances, such as are commonly associated with air pollution, they are well described by the **Gaussian plume model**. This may be written in various forms for various boundary conditions, but for illustrative purposes we present it for a continuous stack emission (i.e. a point source) of strength $S$. In this case the mean concentration, $C$, is given, at a point in space $(x, y, z)$ where the origin of the axes is situated at the stack top, by:

$$C = (S/4\pi rk)\left[\exp\{-(u/4Kx)[y^2 + (z - H)^2]\} + \exp\{-(u/4Kx)[y^2 + (z + H)^2]\}\right] \quad (11.1)$$

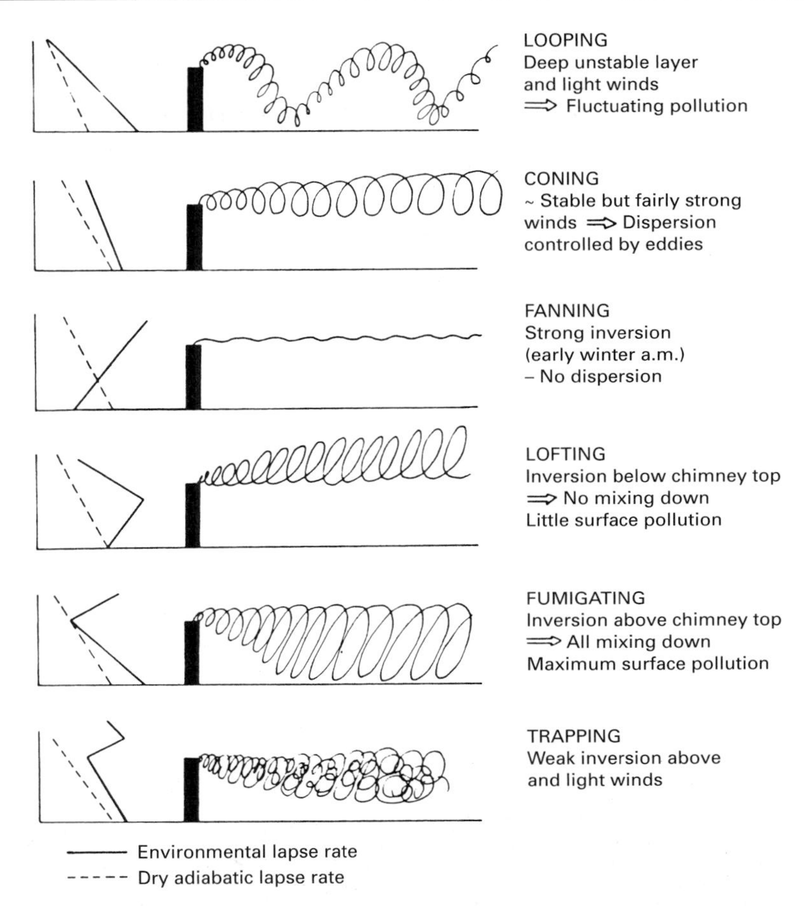


*Figure 11.8* Chimney plume behaviour as a function of the environmental lapse rate (solid line) and wind speed. Note that dispersion is also a function of turbulence, which is often locally induced (e.g. Figures 10.15 and 10.A.3). (From Sellers, 1965.)

where $H$ is the effective chimney height, $u$ the mean wind speed, $r^2 = x^2 + y^2 + (z - H)^2$ and $K$ is a diffusion coefficient. Here, for simplicity, we assume that this coefficient is the same in all three directions and does not vary with elevation. Consideration of Figure 11.8 indicates that in the vertical the coefficient will depend on the stability, and that in inversion conditions it will vary strongly with elevation.

An obvious extension, for particulate emissions, is to include the rate at which they fall under gravity. It is possible to derive expressions for the maximum height reached by these particles, $z_{max}$, their residence time in the atmosphere, $t$, and the distance of travel, $x_{max}$. These are shown in Figure 11.9 for a range of particle terminal velocities. In the figure, $m$ is the fraction of the initially injected particles that remain airborne. Those particles that are removed from the plume by gravitational settling, a process simply known as **fallout**, arrive at the surface as **dry deposition**. The particles may also be removed by **impaction** as part of the plume moves over the ground surface and the particles are collected on objects such as plant leaves. The values given in Figure 11.9 must remain as general estimates which will vary with stability and wind speed. Nevertheless, they indicate clearly that the residence time, travel distance, and height of the largest particles are several orders of magnitude smaller than for the smallest particles.

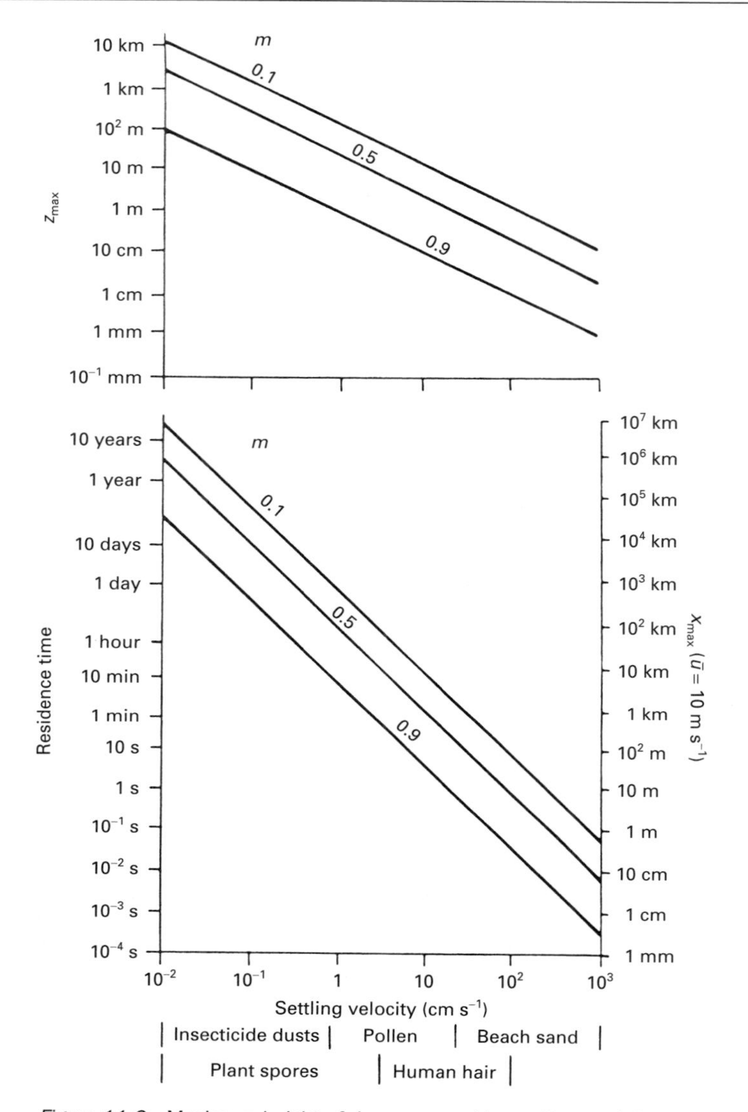


*Figure 11.9* Maximum height of rise, $z_{max}$, residence time and distance of travel, $x_{max}$, for various fractions, *m*, of particles of given settling velocity (cm s$^{-1}$) released from rest into an atmosphere with a (constant) mean wind speed of 10 m s$^{-1}$ and eddy diffusivity of 10$^4$ cm$^2$ s$^{-1}$. (From Sellers, 1965.)

This relatively simple plume model can be applied only over short distances, since the wind field is assumed to be homogeneous. Thus Figure 11.9 implies that it is most applicable for relatively large particles and short travel distances. Further, it is only appropriate for horizontally uniform terrain. In many cases this is sufficient to model and predict air pollution episodes over urban areas where the concern is for high concentrations of particulate matter close to the source. In these situations, one of the major concerns is to forecast inversions and their height, because of the relationships indicated by Figure 11.8. Indeed, one of the major consequences of the concern for air pollution, and the understanding of the processes as sketched here, has been to build ever taller stacks. These minimise the risk of being below an inversion and thus decrease the frequency of high local concentrations. However, they foster a more widespread, regional air pollution and have implications for the production of acid precipitation.

Estimates of pollution concentrations from simple plume models also assume uniform and horizontal terrain. This, of course, is rarely the case and airflow around obstacles of all sizes (see Section 10.4) can have a major influence on the plume trajectory and ground concentration. Often of particular concern in calm or near-calm conditions are frost hollow situations where katabatic winds (Figure 10.12) can lead to strong inversions in valley bottoms. In many older industrial areas, valley bottom sites were attractive since transportation was easy. Nevertheless, they have been responsible for some of the more notorious air pollution episodes. Certainly the episode at Donora, Pennsylvania, USA, in October 1948, when 5910 people became ill and 20 deaths were directly attributable to air quality, is infamous. Even the London episode of December 1952, the last of the great 'pea soup' smogs caused primarily by smoke from domestic coal fires, which contributed to some 4000 excess deaths, was influenced greatly by the valley topography.

When dispersion over a wider, and non-uniform, region is considered, the simple Gaussian plume and fallout models must be considerably modified. Indeed, there is an increasing emphasis on the use of regional atmospheric models to analyse the sources and sinks of pollutants. These bear a family resemblance to the models discussed in Section 13.3, but are primarily concerned with following parcels of air as they move from a postulated source, allowing the terrain to modify the trajectory, the stability to influence the size and concentrations, and natural evolution within the parcel itself. These models find uses in, for example, assessment of the potential impact of new industries, identification of meteorological situations likely to create high concentrations (Figure 11.10), or ensuring that regions comply with local air quality standards.

At this regional scale in particular, the particulate matter may be removed by precipitation in addition to fallout and impaction. Particles may act as condensation nuclei in the precipitation formation process and reach the ground by **rainout**. The particles may also be swept up by falling drops, incorporated into them, and fall as **washout** (Figure 11.6). The efficiency of this type of removal depends primarily on the types of processes discussed in Chapters 4 and 5, but generally the average residence time of a particle decreases as the rainfall rate increases.

So far we have mainly been concerned with solid particulates. Gaseous matter can be treated in a similar way. Indeed, it is treated as such in the following section, where we consider the broader space scale associated with acid precipitation. However, the gases also enter into chemical transformations and **photochemical effects**. These are of concern on the local scale, particularly in urban areas.

The major detrimental feature caused by gaseous air pollution at a local scale is the photochemical formation of **peroxyacetyl nitrate** or **PAN**. In the atmosphere the nitric oxide emitted by industrial processes is oxidised to nitrogen dioxide. In the presence of adequate ultraviolet radiation, this nitrogen dioxide dissociates to nitric oxide and free oxygen, the latter reacting with molecular oxygen to form ozone. Further reactions, several of which require ultraviolet radiation, occur between the nitrogen oxides, oxygen species and the hydrocarbons (originating from unburned fuel). The result is a wide range of organic substances including PAN, which together form **photochemical smogs**. Since such smogs require both bright sunshine and gaseous air pollutants for their formation, they tend to be confined to large conurbations equatorward of about 50° (for example, Los Angeles and Sydney). However, many mid-latitude cities have experienced short-term pollution incidents, when blocking anticyclonic conditions in the summer have been responsible for increased levels of insolation.

### 11.2.2 Acid precipitation

Some gases react with rain to form acid solutions. The creation of carbonic acid ($H_2CO_3$) by the solution of carbon dioxide is universal and produces natural precipitation that is slightly acid. Similarly, nitric acid is commonly formed during thunderstorm activity and leads entirely naturally to acidic precipitation. However, the major concern of **acid rain** is with the acidity produced as a result of the various oxides of sulphur and nitrogen that are placed in the atmosphere as effluent from industrial processes, especially from electricity generating plants and from traffic. Sulphur dioxide, once oxidised,

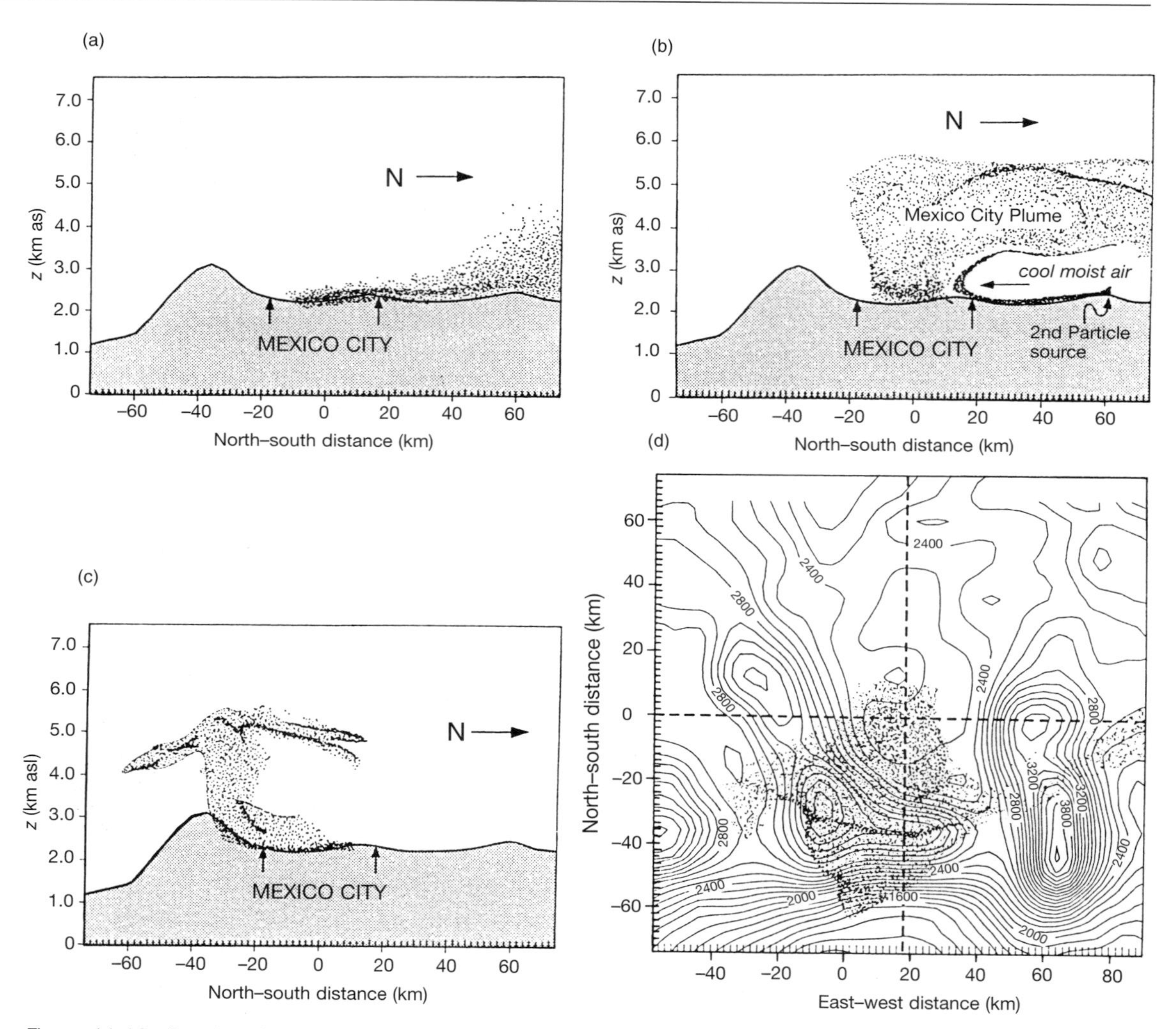


*Figure 11.10* Results of an investigation based on the use of the Regional Atmospheric Modelling System (RAMS) showing particle concentrations for three days in the Mexico City Metropolitan Area. In (a) the synoptic scale flow was strong and the particulates were rapidly removed from the city, while in (b) relatively light regional winds and thermally induced local flows produced a distinct urban plume (Figure 10.8). (c) Shows conditions when local afternoon upslope flow dominates (Figure 10.12), and (d) indicates the horizontal distribution of the particles for case (c). (From Bossert, 1997.)

may be dissolved to give $H_2SO_4$ (sulphuric acid). The acidity is often enhanced by the presence of $HNO_3$ formed as the result of large emissions of various of the oxides of nitrogen ($NO_x$) The consequence of this is that the acidity of the precipitation is increased. Acidity is measured in terms of the pH, the concentration of free hydrogen ions, and neutral solutions having a value of 7.0, and acidity increasing as the pH decreases. Relatively few unambiguous measurements of the pH of natural precipitation are available, but observations at sites apparently remote from pollution sources or trajectories suggest values around 5.6 (Figure 11.11). Recent measurements have suggested that some storms in the USA or northwest Europe may have pH values as low as 3.0.

Since the gases that are the cause of acid precipitation can travel great distances, acid rain tends to be

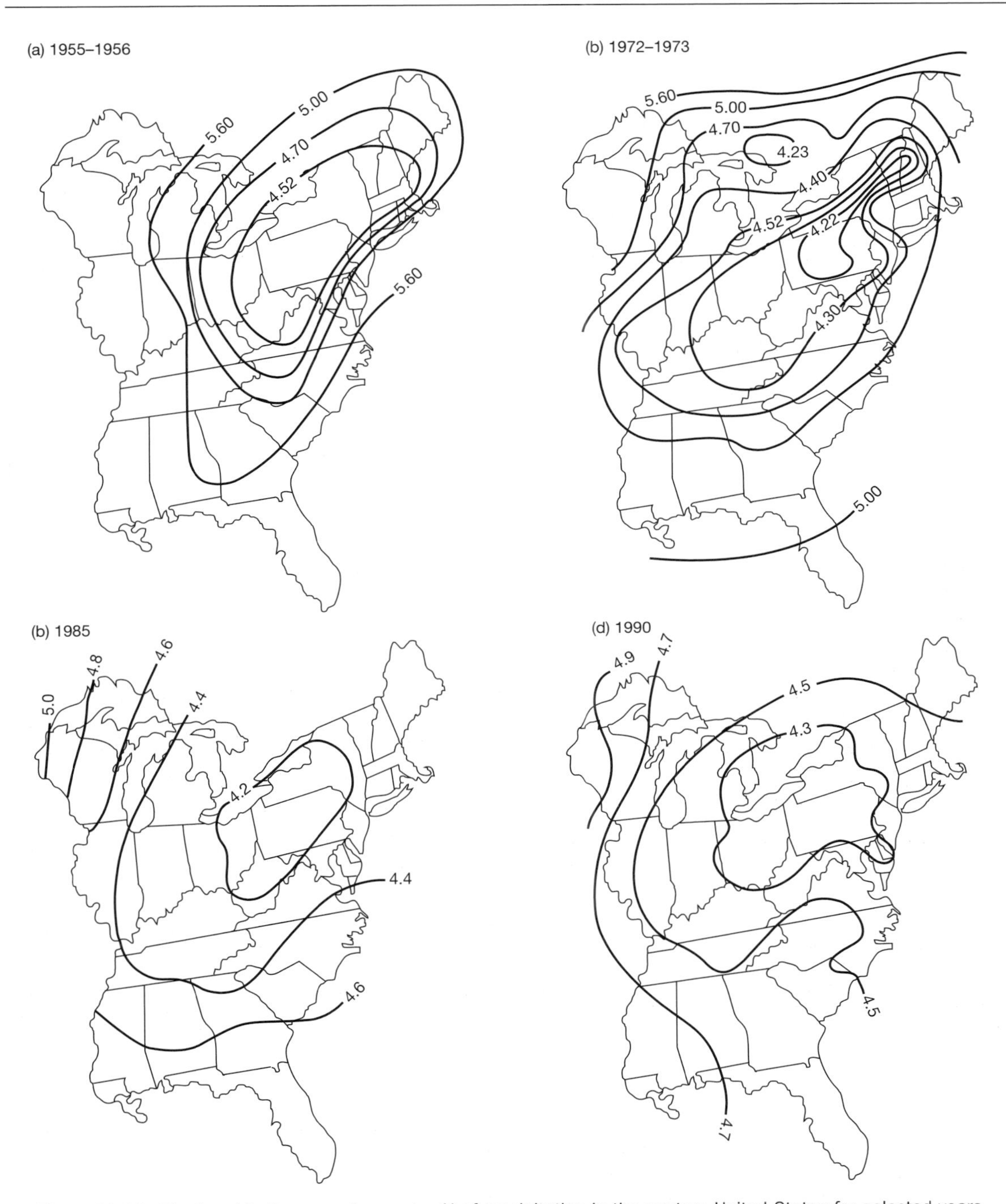


*Figure 11.11* The trend in the annual average pH of precipitation in the eastern United States for selected years between 1955 and 1990. The downward trend to 1985 is clearly seen. Later legislation to control emissions resulted in a partial reversal of that trend. (After Laws, 1993.)

the result of local processes but with regional consequences. Although 100% of the material may be deposited within a few kilometres of the site during a rainstorm, generally only approximately 10–20% of the deposition occurs within 50 km of the source. The fate of the remaining material depends greatly on the atmospheric conditions. Although dispersion commonly follows the Gaussian plume model, exceptions occur. For example, effluent from the 381 m stack at the nickel smelter at Sudbury, Ontario, Canada, which was originally constructed to rise above the common inversion level and thus minimise local air pollution, has been tracked as far as Florida, USA, a distance well over 2000 km. Although few situations are as stark as this, climatological conditions indicate that most pollutants will travel, on average, in a preferred direction. Hence the major industrial regions of the world are suspected of creating acid rain in nearby countries. Effluents from the UK and West Germany have been tracked to Scandinavia. Emissions from the industrial Midwest of the USA have been detected in northeast USA and in eastern Canada (Figure 11.11).

Wherever there is acid precipitation there will be an impact on the local ecosystem. However, this can be especially concentrated in selected mountainous areas. Much of the chemical activity that creates acid precipitation must occur in clouds, and observations suggest that the cloud drops near the cloud base are likely to be the most highly acidic. This has direct consequences for highland areas downwind of industrial sources, which may be bathed in cloud water much of the time. Here the mountain top vegetation is particularly stressed and the impact of acid precipitation most readily seen.

### 11.2.3 Ozone depletion

The very lightest gases placed into the atmosphere by human actions may be highly buoyant, rise into the stratosphere, have residence times (in the absence of chemical reactions) of several years, and be capable of travelling several times around the Earth. They disperse both by their natural diffusion and by the action of the winds in those regions. The major gases of concern are the various species of chlorofluorocarbons (CFCs), manufactured and used on the surface of the Earth for a variety of purposes. This entirely artificial gas reacts with stratospheric ozone, $O_3$. In normal circumstances this is a relatively slow process, but the net result is a gradual decrease in concentration in the ozone layer some 12 km above the surface. Since $O_3$ normally absorbs radiation with wavelengths around 0.15 μm, more ultraviolet radiation penetrates to the surface. This short-wave, high energy radiation interacts with the human body to enhance the likelihood of melanoma, a form of skin cancer. Note that the absolute energy amount at these wavelengths is small (Figure 2.3) so that this effect has little impact on the surface energy budget directly. However, CFC itself is a greenhouse gas, and so plays a role in the greenhouse effect (Section 2.5).

The rate of reaction between CFCs and ozone is highly temperature dependent. In the lower stratosphere over Antarctica this chemistry, combined with the normal meteorology, produces what has been called an **ozone hole** (Plate 5a) This is an exceptional decrease in ozone concentration during the Austral spring. The circumpolar vortex ensures that all CFC reaching the polar regions in winter is retained there. The low winter temperatures inhibit reactions. However, with the temperature rise during spring the high concentration of CFC molecules allows rapid removal of the $O_3$ molecules and hence the concentration decreases. As summer commences, an equilibrium situation, akin to that in the rest of the atmosphere, is established.

In the situation involving the ozone hole, there is a clear connection between CFC emissions, the atmospheric effect, and the potential human health consequences. As a result, an international agreement, the Montreal Protocol and its later modifications, was signed to minimise the emissions, and thus the impact (Plate 5b). This was the first international agreement involving the impact of the atmosphere and climate on society, and addressing potential impacts on a global scale. As such, it set the precedent for consideration of much more complex issues where the connections are much less well defined. We shall consider these issues as they involve global warming in the next few chapters.

The study of atmospheric chemistry has tended to be neglected by climatologists perhaps because the time scales of chemical reactions are so much shorter than those typically of importance in climatology. It is becoming increasingly obvious, however, that the chemistry of the atmosphere has a great bearing on radiative exchanges and on the transfer rates of atmospheric properties, and therefore must be treated as an integral part of the climate system.

## 11.3 Changes in the character of the surface

The role of the surface in creating and maintaining climate on the local, regional and global scales has already been considered in a variety of contexts. Any surface change we make will have some influence on the climate. Again, the changes may be both deliberate and inadvertent. We have already considered many deliberate changes, ranging from frost and wind protection to housing and irrigation, and have considered inadvertent ones particularly in association with urban climates. Most changes have been on a small scale, but it is pertinent to enquire whether consistent local modifications taking place at a large number of sites could combine to produce coherent regional, or even global, alterations in climate. Satellite observations of energy fluxes over urban areas suggest the potential for regional impacts in isolated cases. However, both modelling studies and observations suggest that the major human activities on the surface likely to lead to regional or global climatic, as well as practical consequences, are deforestation and desertification. These two are treated separately below.

### 11.3.1 Deforestation

Whenever land is cleared for agricultural purposes there is a change in surface character. This change can be especially marked when forests are replaced by cropland. At present, the amount of forest land, particularly in the tropics, is rapidly being reduced by agricultural expansion and so the surface characteristics of large areas are being changed. One area that is undergoing deforestation is the Amazon Basin in Brazil, and the links between this activity and the climate are being extensively studied. In general, forested areas have greater potential evapotranspiration than does open ground (e.g. Figure 10.10). This difference influences the whole of the water balance, and eventually the whole of the regional climate. In the case of the Amazon, moisture is fed into the tropical Walker circulation by the rainforest and the resultant uplift and condensation provides both energy to maintain the circulation and precipitation to maintain the forest (Figure 11.12a). With lower evaporation rates, the circulation may become weaker (Figure 11.12b). This simple analysis ignores other responses to the change in surface character. The albedo, the emissivity, the wind profile, the nature of the soil and hence the ground heat flux, will all change. Many of these effects will interact, so that a simple analytic solution is not possible. Instead a modelling approach is needed. Models will be considered in some detail in Chapter 13, but here we can note that both a general circulation model (Section 13.2) and a regional scale model emphasising surface conditions (e.g. Figure 13.7) are needed. Several such model combinations have been used (Figure 11.12c). One group of results suggests a relatively small change, the other set a much larger one, indicative of the uncertainty associated with any model simulation. However, almost all indicate a warming, and a decrease in precipitation and evaporation. In most cases the change in evapotranspiration is less than that in precipitation, indicating that the atmospheric effect is responsible for the anticipated surface drying.

### 11.3.2 Desertification

Changes in surface type may also have important climatic consequences in arid and semi-arid areas. The sparse vegetation natural to these areas can easily be removed. When the vegetation is removed and bare soil exposed there is a decrease in soil water storage because of increased runoff, and an increase in albedo. These changes would tend to affect surface temperatures in opposite ways. With less moisture available the decreased latent heat flux would lead to an increase in surface temperature, while an increased albedo would produce lower temperatures. Model calculations indicate that, unlike the case for the Amazon, the latter would dominate. It can then be hypothesised that the increased cooling would lead to large-scale subsidence. In this descending air, cloud and precipitation formation would be impossible and aridity would increase. The validity of this hypothesis cannot be tested by actual observations in the semi-arid regions since any surface albedo change is almost certainly a direct response to vegetation growth in the wet season, and the energy fluxes respond to the enhanced moisture (Figure 8.9). Model simulations, however, support the hypothesis (Figure 8.12).

The removal of vegetation which leads to these changes can occur either as a result of relatively minor fluctuations in the climate, or by over-grazing by livestock. The latter was held responsible for the Sahelian drought, which reached a peak in human

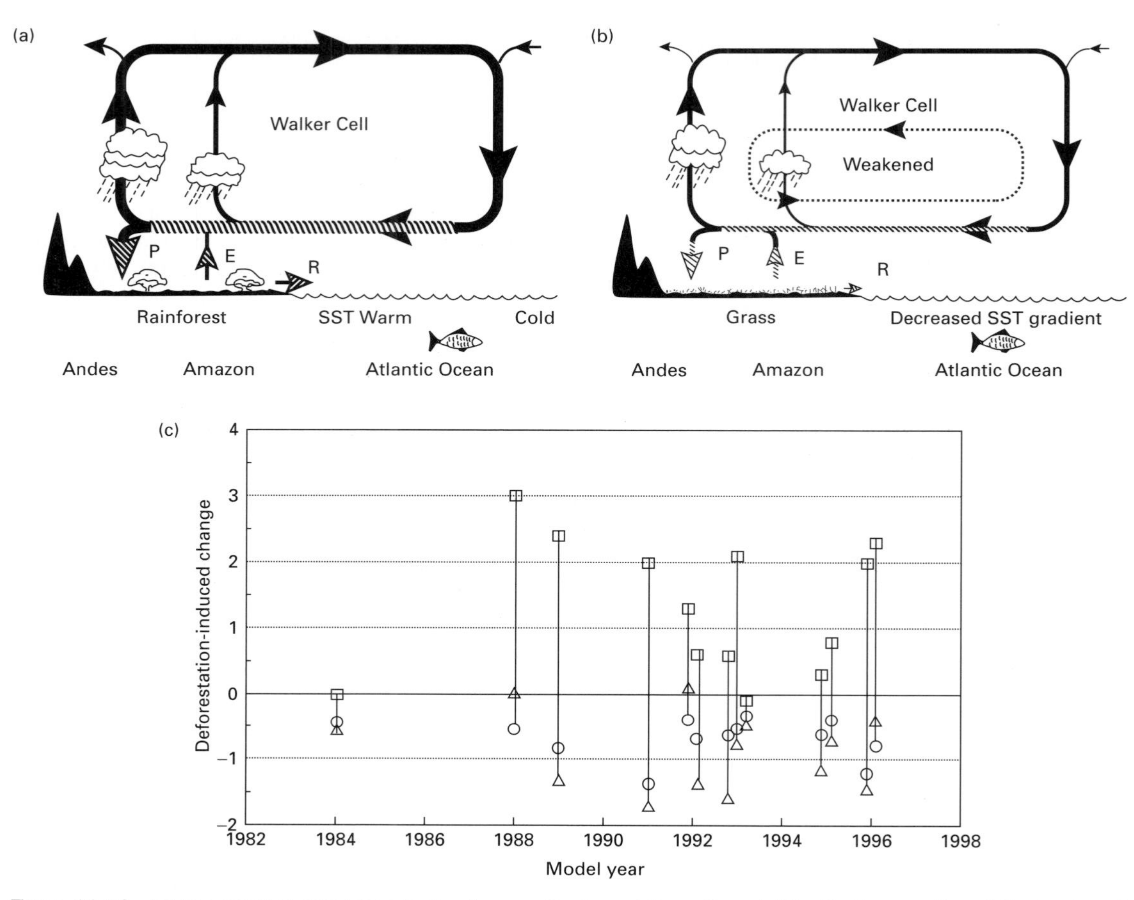


*Figure 11.12* (a) The tropical circulation in the Amazon Basin and the adjacent Atlantic Ocean with a fully developed rainforest and a vigorous Walker circulation. This behaves in much the same way as the classic Hadley Cell described in Chapter 8. (b) The postulated circulation as a response to conversion of the Amazon Basin surface to grassland. Note the weaker but more complex circulation that is created. (c) Model estimates of the change in precipitation (mm day$^{-1}$) (△), evapotranspiration (mm day$^{-1}$) (○) and surface temperature (°C) (□) as a result of deforestation, as a function of the year of publication of the results. (After Zeng *et al.*, 1996.)

suffering and global concern in the 1968–1973 period. As a consequence, the United Nations Conference on Desertification, the first major conference devoted to climate and its impacts, was convened to develop response strategies. However, this short-period drought was part of a gradual decline in precipitation for the whole of sub-Saharan Africa from the anomalously wet period of the early 1950s (Figure 8.11), and which continued into the 1990s (Figure 11.13). Hence this decline could be part of the natural climatic fluctuation, possibly with human activity acting as an additional impetus, or even as a trigger for the whole process.

The two examples considered here suggest that regional-scale climate changes can occur and that they may not necessarily be associated with global changes. However, they do require modifications in surface type that cover large areas. Consequently, although minor fluctuations are probably occurring continuously from entirely natural causes, and although the possibility of significant anthropogenic changes exists, the impact of human activity is at present small. Certainly most human modifications are on a local scale and do not as yet have the potential to alter the present distribution of regional climates.

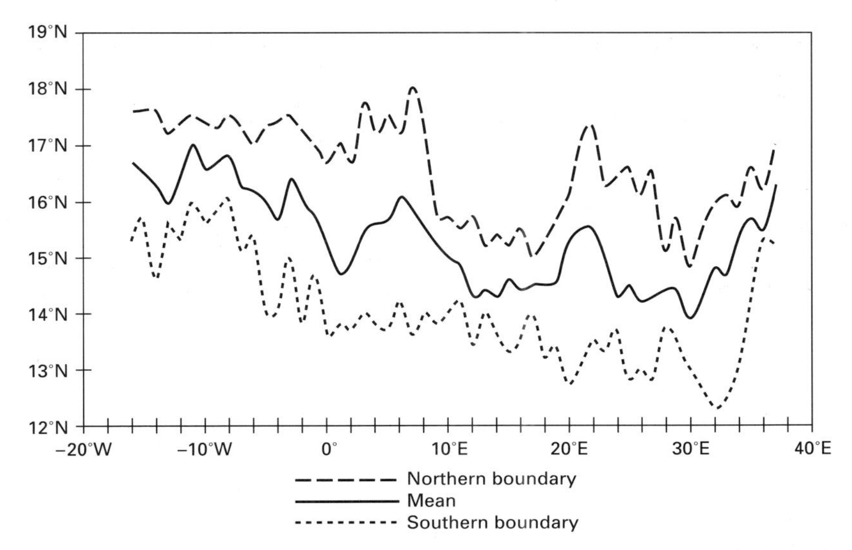


*Figure 11.13* Location of the mean (solid) and most northerly and southerly positions of the 200 mm year$^{-1}$ annual isohyet for the 1980–1990 period, by 0.5° longitude across the African continent. (After Tucker *et al.*, 1991.)

# CHAPTER 12 Climates of the past

Atmospheric conditions vary on all time scales from day-to-day weather fluctuations to climatic changes spanning millennia. The various fluctuations and changes are superimposed upon each other to give the complex time sequence of past climates. The field of **palaeoclimatology** is the study of these sequences, and is dedicated to defining and explaining the evolution of climate. While there is an intrinsic interest in this, and while knowledge of past climates can have an important bearing on other branches of knowledge, we are particularly concerned here with those aspects which permit us to understand present atmospheric processes, and which can be used to test models and provide clues to future conditions. Palaeoclimatology requires a highly interdisciplinary approach. Geologists, oceanographers, glaciologists, botanists, archaeologists and historians, among others, are all involved with climatologists in unravelling the patterns of past climates. The techniques used are equally varied. Consequently this chapter can be only an overview, and is divided into three main portions: a description of methods for revealing past climates; a review of results from these methods; and an examination of causes of climatic change.

Several initial generalisations are desirable. As we recede into the past from our present sophisticated observational network, data become increasingly sparse in time and space. While individual observations may allow a climatic snapshot at a particular time and place, estimates of past conditions must cover increasingly broad time and space scales, and local detail gives way to global smoothing. Assumptions and uncertainty increase. So, while we have in the last few decades deduced a tremendous amount about our climatic past, much is still speculation. It is another area where there is a large amount of ongoing research, so that new information is constantly being revealed.

## 12.1 Palaeoclimatic observations

Once we go back beyond the instrumental record of the last 100 years or so, we are forced to use **proxy** evidence to estimate climate. Anything which itself varies in some way with climate, and which can be measured and dated, is a candidate for use as proxy evidence.

Whatever the source of that data, the steps in converting it to climate information are essentially the same. First a link between the proxy and the climate must be established. This may be a fundamental relationship, developed through laboratory or natural experiments, such as that relating the ratio of oxygen isotopes in the air to air temperature. In these situations the links that are established may be very well defined, and refer to specific climatic elements. Others may be much more tenuous, perhaps semi-quantitative, and may refer to a combination of elements. Typically a time series of observations is used. Thus, for example, crop yield depends to some extent on growing season temperature (Figure 12.1). Yield will also be influenced by other climatic elements, such as precipitation, as well as by many non-climatic factors. Nevertheless, records of crop yield, suitably interpreted, give some indication of temperature. Once any relationship has been established, it should be tested (Figure 12.2). It is desirable, whenever possible, to test it through comparisons between the proxy and a set of observations of the present climate different from those used in the development of the relationship. Thereafter it becomes a matter of quantifying the proxy at a known time and location in the past. Commonly it is a highly skilled exercise to extract the climatic signal from the noise created by the many extraneous influences on the proxy.

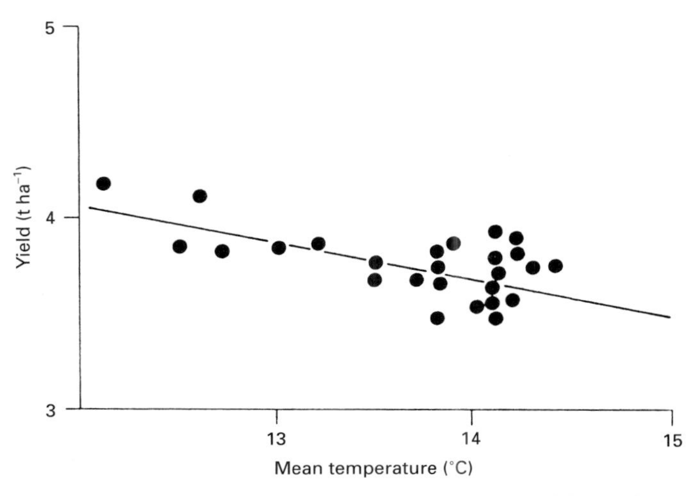


*Figure 12.1* Wheat yield (tonnes per hectare) in parts of England as a function of the mean temperature for May, June and July for the period 1941–1970. (From Monteith, 1981.)

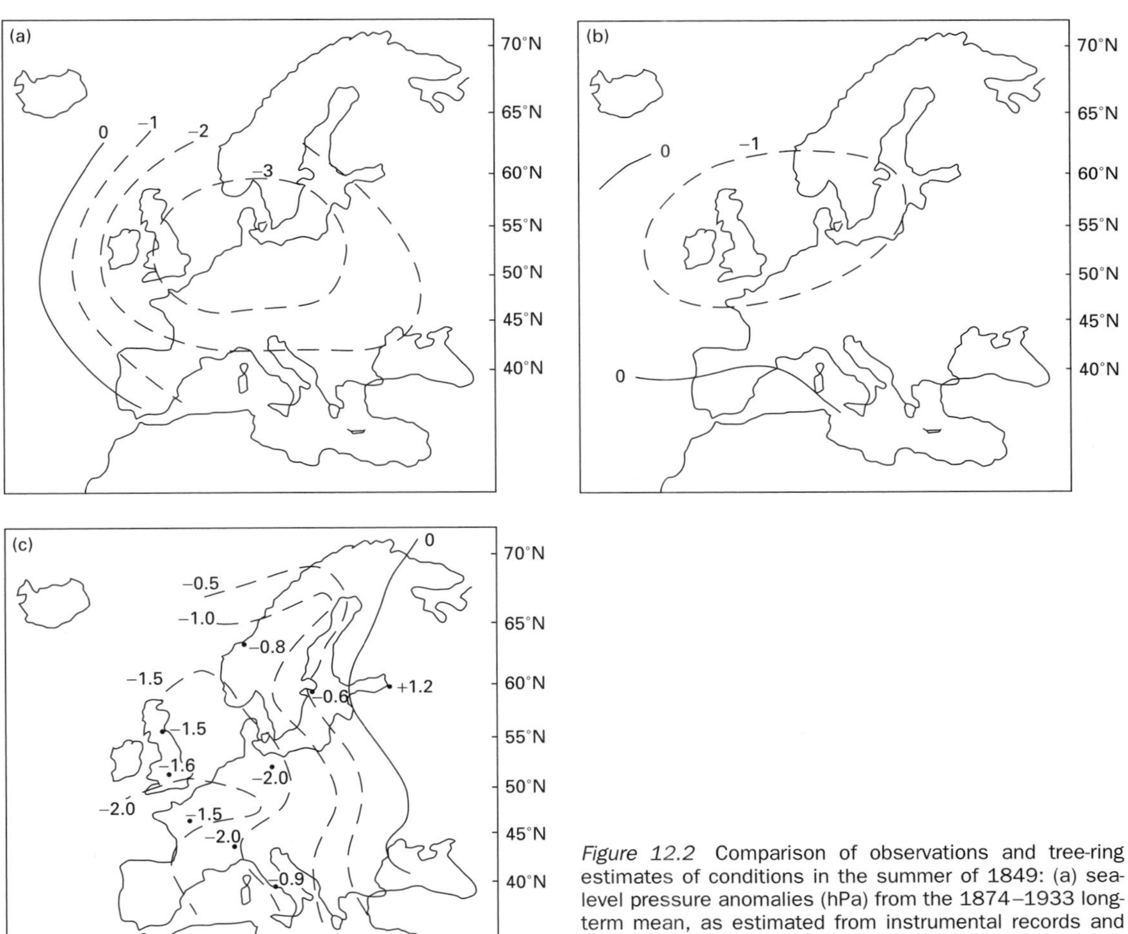


*Figure 12.2* Comparison of observations and tree-ring estimates of conditions in the summer of 1849: (a) sea-level pressure anomalies (hPa) from the 1874–1933 long-term mean, as estimated from instrumental records and (b) as determined from 75 tree-ring chronologies, and (c) tree-ring (contours) and instrumental (point value) estimates of temperature anomalies (°C) from the 1876–1975 mean. (From Hughes, 1991.)

*Table 12.1* Sources of palaeoclimatic information

| Archive | Best temporal resolution | Temporal range (years) | Information derived[a] |
|---|---|---|---|
| Historical records | day/hour | $10^3$ | T, H, B, V, M, L, S |
| Tree rings | season/year | $10^4$ | T, H, $C_a$, B, V, M, S |
| Lake sediments | year to 20 years | $10^4$–$10^6$ | T, H, $C_w$, B, V, M |
| Ice cores | year | $10^5$ | T, H, $C_a$, B, V, M, S |
| Pollen | 100 years | $10^5$ | T, H, B |
| Loess | 100 years | $10^6$ | H, B, M |
| Ocean cores | 1000 years | $10^7$ | T, $C_w$, B, M |
| Corals | year | $10^4$ | $C_w$, L |
| Palaeosols | 100 years | $10^5$ | T, H, $C_s$, V |
| Geomorphological features | 100 years | $10^7$ | T, H, V, L |
| Sedimentary rocks | year | $10^7$ | H, $C_s$, V, M, L |

[a] T, temperature; H, precipitation/humidity; C, chemical composition (air/water/soil); B, biomass; V, volcanic eruptions; M, geomagnetic field variations; L, sea level; S, solar activity.

Although the sources of suitable proxy evidence are many and varied, they can be divided into a manageable series of types (Table 12.1), each with specific, although generally very broad, time and space scales. These types are reviewed briefly below.

### 12.1.1 Palaeoclimatic techniques

A detailed review of all of the various methods of creating climate information from proxy evidence is well beyond the scope of this book. We shall examine some in detail later, but here we provide a brief introduction to the main techniques.

**Historical records** Written records from medieval sources in Europe and dynastic records in east Asia. The former frequently include monastic records indicating grain yields (often with weather-related explanations for poor yield years) or the quality of grape harvests and the subsequent vintage. The latter are famous for the long record of the time of particular events related to vegetation development. The timing of China's cherry blossom is probably the best documented, and **phrenological** studies are linking this to climate fluctuations.

**Tree rings (*dendrochronology*)** Each year a tree adds a new growth ring. The width of this ring partly depends on the environmental conditions. Especially in marginal areas, the width becomes very sensitive to the limiting factor. In dry climates ring width is often related to precipitation amount, while in cold climates it may be related to summer temperatures.

**Lake sediments** In almost any climate with a distinct wet/dry cycle, sediment will be deposited in a lake bed during the period when the rivers are most active. The deposits will be in annual layers and will contain material such as vegetation debris, or seeds. The analysis of the relative abundance of these will provide the proxy information.

**Ice cores** The seasonal cycle will provide periods when snow accumulates and periods when it either melts or ages. Provided some accumulates each year, the cycle will provide a series of datable layers, while the chemical composition of the ice and trapped air bubbles will provide the climatic information.

**Pollen** Similar to lake sediments, but not confined to lakes. Any area of accumulation, provided there is a dating mechanism, may be suitable. The material is often dated through use of radiocarbon or similar techniques, which produce the ratio of radioactive to inert forms of carbon, the ratio being predictable from the moment the sample dies. Archaeological sites, where dates can be obtained by alternative methods, are especially fruitful.

**Loess** Blown sand, primarily from dry, or especially desert surfaces, which collects either in calmer regions or where the surface roughness forces the wind to drop its load. Surface roughness is usually associated with the presence of vegetation, and thus with humid areas. The orientation of this deposited material allows estimates of the magnetic orientation.

**Ocean cores** Deep ocean cores, often extracted from depths >5000 m, provide chemical, mineralogical and biological information. For example, foraminifera (forams), simple near-surface organisms, use calcium and carbon from the near-surface waters to form the calcium carbonate of their shells. The isotopic ratio of the oxygen in the carbonate depends on the water temperature and on salinity. When these creatures die and sink, they take temperature and global ice volume information with them, ready to be extracted (see Figure 12.4).

**Corals** The many species of these marine organisms are extremely sensitive to the oceanic environment, in general growing well in warm, salty water, and dying as it gets colder. The reefs they create preserve the climatological record.

**Palaeosols** Distinct soils form under particular types of climate, and are highly dependent on the seasonal variation in the water budget. Soil, however, is slow to form and slow to change. Some soils therefore contain indications of climate conditions prior to the present era. In addition, soils can be buried (under loess for instance) and hence they and their climatic record are effectively fossilised.

**Geomorphological features** The classic set of geomorphological features is the glacial landscape of the past and present Arctic margins, with its finger lakes, poor drainage, and overdeepened and straight river valleys containing small streams. All indicate previous glaciation, and, used in conjunction with other techniques, the times of advance and retreat can be dated (see Figure 12.6).

**Sedimentary rocks** The examination of rock types and the fossil record can provide rather direct information of conditions over a span of many millennia. We use this explicitly in the next section.

## 12.2 Past climates

These various approaches, and the differing amounts and sophistication of information they yield, make it convenient to divide our discussion of past climates into three rough divisions: geological, covering the entire history of the Earth; intermediate, dealing primarily with the latest great ice ages; and historical, where more or less direct evidence of climatic conditions is available.

### 12.2.1 Geological record of climate

Over the longest possible time scale, the evolutionary history of the Earth itself, the climate system, typified by ambient surface temperature, has been remarkably stable. Geological data suggest that temperatures have varied only between 275 and 305 K during the last 3.8 Ga (1 Ga = $10^9$ years BP). The first glaciation appears to have been initiated about 2.3 Ga (Figure 12.3). This was around the time that free oxygen began

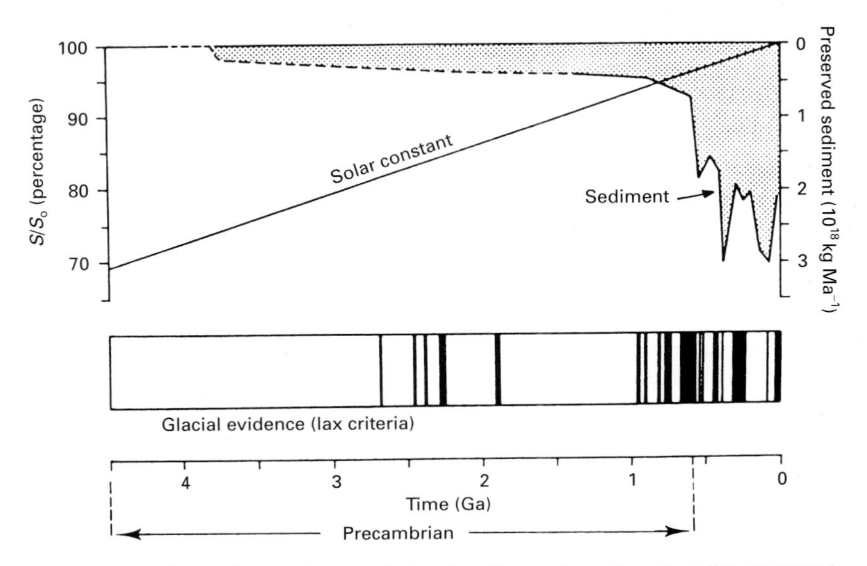


*Figure 12.3* The geological history of the glaciation and buildup of sedimentary material as a function of time. The increasing value of the solar constant (expressed as a percentage of present day) is also shown. The increasing mass of sedimentary material makes the retrieval of more recent geological information easier than retrievals of data pertaining to Precambrian conditions. In order to try to show as much information as possible fairly lax criteria for glacial activity have been used.

to evolve in the atmosphere, and since then there have been a series of alternations between glacial and interglacial periods suggesting relatively minor temperature excursions. Glacial periods seem to have been relatively common in these early times, only to be replaced by a long period without glaciers prior to the onset of a new cooler phase around 1 Ga. This evidence suggests that there is a very long-term fluctuation in climate, which is of relatively small magnitude. Certainly there has been a long-term fluctuation in the carbon dioxide content of the atmosphere, largely dependent on the level of geophysical activity. This has certainly played a role in creating the fluctuations. However, it has also been hypothesised that the general long-term stability is the result of the interaction between the climate and the biosphere. This **Gaia hypothesis** suggests that biologically induced surface and atmospheric changes generate negative feedbacks which stabilise the whole system.

Over geological time there have been significant changes both in the types of rocks produced and in the biospheric composition, as preserved in the fossil record. These changes are displayed systematically in the geological column (Table 12.2). Climate and climate change must be causal factors in these features. However, it is extremely difficult to isolate the climatic effect from others. Furthermore, even if the climate can be estimated, continental drift creates a problem of locating the geographical position of the rocks during their formative period. Nevertheless geological information permits the description of the long-term history of the climate.

For the **Precambrian**, as discussed above, the data are sparse and relate mainly to glacial events. The earliest of these seems to have occurred around 2.7 Ga, but evidence becomes widespread enough to be termed 'global' between 2.5 and 2.3 Ga. A second cooling seems to occur around 1.9 Ga (these data are shown in Figure 12.3). It must be noted that all these data are sufficiently uncertain that these two cold periods may have coincided. The **Late Precambrian** glacial period (600 Ma) (1 Ma = $10^6$ years BP) includes one or more ice ages beginning around 850 Ma. It is possible that a sharp, thermal gradient over a continental land mass that includes a pole of rotation may be required to explain the evidence.

On this broad geological time scale it appears that since the end of the Precambrian there have been two major warm periods, one in the early Palaeozoic, the other in the Mesozoic. These appear to be associated with high atmospheric carbon dioxide concentrations and thus with an enhanced greenhouse effect. However, the climatic information has been inferred partly from high rock weathering rates, which are indicative of high temperatures. Such weathering is itself an important release mechanism for atmospheric $CO_2$, so it is by no means clear what is driving the greenhouse effect. In any event, during the warm early Palaeozoic there was an **Ordovician** ice age from approximately to 450–430 Ma with a duration of up to 20–25 million years. The evidence for this ice age, which comes from the tropical regions, suggests that it affected most of the area of the supercontinent of Gondwanaland. In the longer lasting subsequent cool period, culminating in the **Permo-Carboniferous** glacial age (300 Ma), Earth's land was in a single continent, which was symmetric about the equator, with a mid-latitude continental concentration.

Much of the Mesozoic was warm, but the **Cretaceous** ice age began around 100 Ma, at a time when there was an essentially meridional configuration of the continents which, combined with shallow ocean ridges, prevented circumpolar ocean currents in either hemisphere. During the later part of the Mesozoic, the upper Cretaceous (100–65 Ma), there was a generally warmer global climate and the polar regions had no ice-caps. Around 55 Ma, the global climate began a long cooling trend known as the **Cenozoic climatic decline**, and around 35 Ma, Antarctic waters underwent a significant cooling.

The **Oligocene** epoch (approximately 35–25 Ma) was a period of global cooling. Around 30 Ma, Australia moved far enough northeast to allow an Antarctic circumpolar current to develop. By 25 Ma, ice reached the edge of the continent in the Ross Sea area. During the early **Miocene** (20–15 Ma) warm climates existed in low and mid-latitudes, but not in high southern latitudes. In the mid **Miocene** (approximately 15–10 Ma) there was widespread incidence of further cooling, leading to substantial growth of Antarctic ice and to the development of some ice at the North Pole.

Around 5 Ma, the already substantial ice sheets in Antarctica underwent rapid growth and attained their present volume. At about the same time uplift of 'alpine' mountains in both hemispheres led to the growth of mountain glaciers. Not until as recently as 2 Ma did ice sheets became re-established in the Northern Hemisphere, spreading from a centre in the extreme north of the Atlantic Ocean. However, during at least the last 1 million years ice cover in the Arctic Ocean has never been less than it is today.

During the **Quaternary** period, covering the last 2 million years, glacial and interglacial periods alternated. Some workers suggest that there were as many as 20 glacial events; certainly there were at least seven. Interglacials, each characterised by a warm

*Table 12.2* Scale of geological time and biospheric evolution

| Subdivisions of geological time: **Eras** / Periods | Epochs | Apparent ages (years before present) | Notable events in the evolution of organisms |
|---|---|---|---|
| **Cenozoic** | | | |
| Quaternary | Holocene (Recent) | | |
| | Pleistocene | | Man appears |
| | | $2 \times 10^6$ | |
| Tertiary | Pliocene | | Elephants, horses, large carnivores become dominant |
| | | $10 \times 10^6$ | |
| | Miocene | | Mammals diversify |
| | | $25 \times 10^6$ | |
| | Oligocene | | Grasses become abundant, grazing animals spread |
| | | $36 \times 10^6$ | |
| | Eocene | | Primitive horses appear |
| | | $58 \times 10^6$ | |
| | Palaeocene | | Mammals develop rapidly. Dinosaurs become extinct, flowering plants appear |
| | | $63 \times 10^6$ | |
| **Mesozoic** | | | |
| Cretaceous | | | |
| | | $135 \times 10^6$ | |
| Jurassic | | | Dinosaurs reach climax. Birds appear |
| | | $180 \times 10^6$ | |
| Triassic | | | Primitive mammals appear; conifers and cycads become abundant. Dinosaurs appear |
| | | $230 \times 10^6$ | |
| **Palaeozoic** | | | |
| Permian | | | Reptiles spread, conifers develop |
| | | $280 \times 10^6$ | |
| Carboniferous[a] | | | Primitive reptiles appear, insects become abundant. Coal-forming forests widespread. Fishes diversity |
| | | $340 \times 10^6$ | |
| Devonian | | | Amphibians, first known land vertebrates, appear. Forests appear |
| | | $400 \times 10^6$ | |
| Silurian | | | Land plants and animals first recorded |
| | | $440 \times 10^6$ | |
| Ordovician | | | Primitive fishes, first known vertebrates, appear |
| | | $500 \times 10^6$ | |
| Cambrian | | | Marine invertebrate faunas become abundant |
| | | $570 \times 10^6$ | |
| **Precambrian** | | | |
| Proterozoic | | | Life forms abundant |
| | | $2.5 \times 10^9$ | |
| Archaean | | | Primitive life forms (e.g. blue-green algae) |

[a] In North America the Carboniferous is divided into two periods (the Pennsylvanian and the Mississippian) at about $310 \times 10^6$ years.

period lasting 10 000 ± 2000 years, have occurred on average every 100 000 years during at least the past 0.5 million years. Around 125 ka (1 ka = $10^3$ years BP) the **Eemian** (**Sangamon**) interglacial commenced and the warmest part persisted for about 10 000 years. Between 115 ka and the **Pleistocene** glacial maximum at 18 ka, there were marked fluctuations superimposed on a generally declining temperature. Temperatures thereafter increased so that at present we are within the **Holocene** interglacial. To find conditions as warm and ice-free as at present it is necessary to go back to the Eemian, 125 ka.

Although we have emphasised glaciation and temperature conditions, the geological and geomorphological evidence also includes information about sea level and aridity. Consequently, except for the earliest periods, our inferences are based on evidence from several locations. Nevertheless, all data are highly specific for the locality from which they come, and thus they may easily represent local anomalies, rather than large-scale variations in the climate.

### 12.2.2 Climates of the last 1 million years

Data relating to conditions during the last 1 million years are relatively abundant and diverse compared with earlier periods. Hence we can say with some confidence, for example, that during the last 700 000

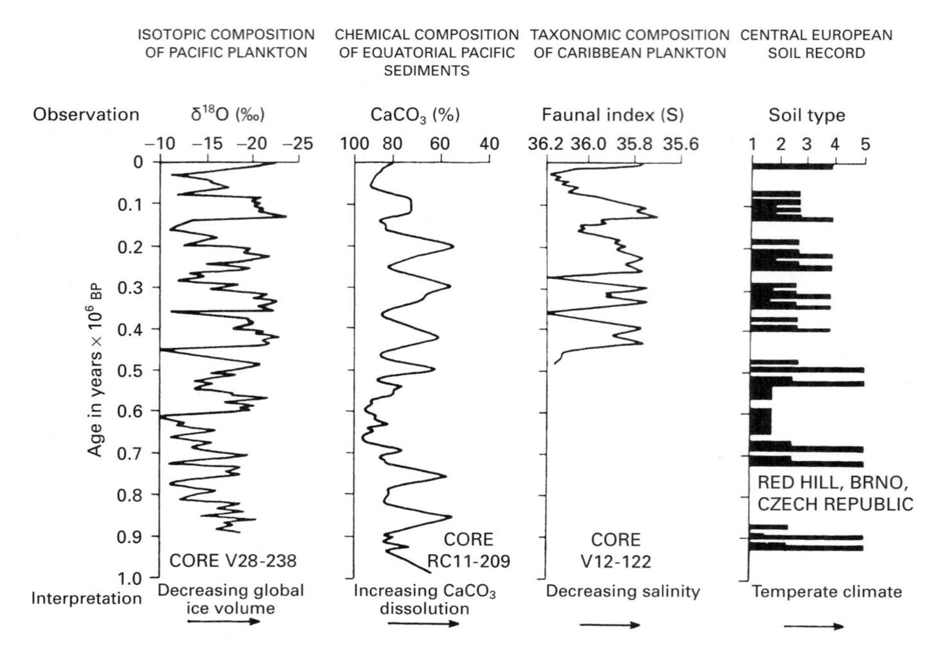


*Figure 12.4* Climate records of the last million years as derived from four sources: the oxygen isotope curve from Pacific deep-sea cores, reflecting global ice volume; calcium carbonate percentage in equatorial Pacific cores (low values are taken to indicate periods of rapid dissolution by bottom waters); faunal index, reflecting the changing composition of the Caribbean foraminiferal plankton and the associated salinity; and the sequence of soil types at Brno, Czech Republic, reflecting fluctuations in temperature and, possibly, precipitation.

years, the globe was as warm as or warmer than today for only 8% of the time. Much of this confidence arises because several independent methods of obtaining climate-related information are available. Figure 12.4 illustrates climatic records for the last $10^6$ years from four different observational techniques: oxygen isotope curves from deep-sea cores; calcium carbonate percentages in ocean sediments; variations in populations of foraminifera in tropical waters; and mid-European soil type sequences. Techniques for the interpretation and correlation of these types of record on these time scales are well established, and estimates of regional climates are emerging. The main events towards the end of this period, embracing the last great ice ages, are summarised below.

The last **Pleistocene** ice age reached a maximum 22 000–14 000 years ago (Figure 12.5). During this period two large ice sheets have been identified in the Northern Hemisphere: the Laurentide Ice Sheet covering parts of eastern North America, and the Scandinavian Ice Sheet covering parts of northern Europe. The Cordilleran Ice Sheet over western North America achieved a maximum somewhat later than these, about 14 000 years ago. During this last ice age, the maximum area of the Northern Hemisphere ice sheets was equal to approximately 90% of the maximum achieved during the last million years of the Pleistocene. At the glacial maximum the sea level dropped by approximately 85 m and sea-surface temperature fell by as much as 10 °C in mid-latitudes of the North Atlantic and 3 °C in the Caribbean.

Widespread **deglaciation** began rather abruptly approximately 14 000 years ago. The Cordilleran Ice Sheet melted rapidly and was gone by approximately 10 000 years ago. The Scandinavian Ice Sheet lasted only slightly longer. Deglaciation was later in North America and the melting of the Laurentide Ice Sheet lagged by approximately 2000 years. By 8500 years ago the conditions in Europe had reached their present state while this situation was achieved in North America about 7000 years ago. Within this overall warming period were times of widespread cooling and glacial advance, seeming to occur about every 2500 years (Figure 12.6). One of these is called the **Younger Dryas event**, which established itself within 100 years and lasted for 700 years (10.8–10.1 ka).

The **post-glacial climatic optimum** culminated between approximately 7000 and 5000 years ago.

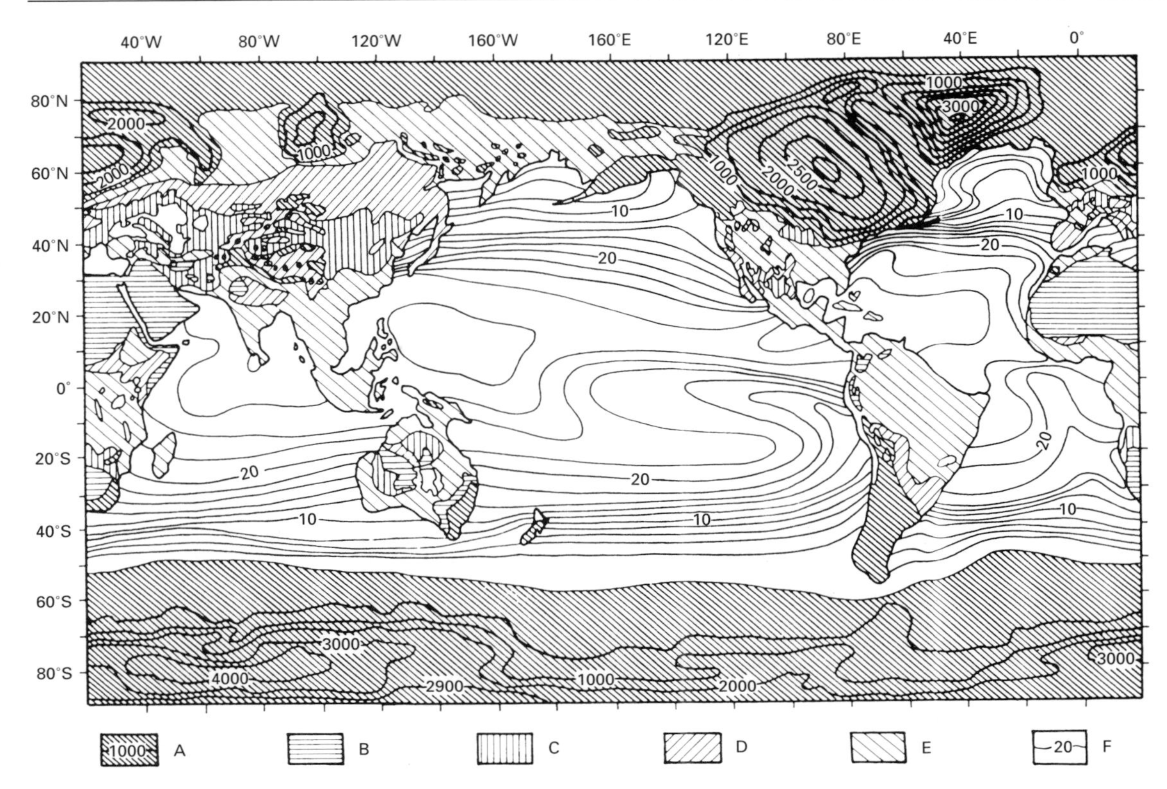


*Figure 12.5* Sea-surface temperatures, ice extent, ice elevaton and continental albedo for Northern Hemisphere summer (August) as reconstructed by the CLIMAP project members for 18 ka. Contour intervals are at 1 K for isotherms and 500 m for ice elevation. The continental outline represents a sea-level lowering of 85 m. Albedo values are given by the following key: A, >0.4 – snow and ice (isolines show elevation (m) of ice sheet above sea level); B, 0.30–0.39 – sandy deserts, patchy snow and snow-covered dense coniferous forest; C, 0.25–0.29 – loess, steppes and semi-desert; D, 0.20–0.24 – savannas and dry grasslands; E, 0.10–0.20 (mostly 0.15–0.18) – forested and thickly vegetated land; F, <0.10 – ice-free oceans and lakes (isolines indicate sea-surface temperature (°C)). (From McIntyre *et al.*, 1976 figure 1.)

Summer temperatures in both Antarctica and Europe were 2–3 °C higher than they are today. This led to a reduction of ice, mainly on land, but the sea ice in the Arctic Ocean was also reduced. Sea level rose rapidly, possibly reaching a peak of about 3 m above present levels approximately 4000 years ago, illustrating the lag between glacial melt and the sea-level rise.

Events of this type are sufficiently well established, and there is a sufficient world-wide network of data points, to allow attempts at reconstruction of the global climate at specific dates. Figure 12.5 shows an early, and perhaps the most successful, reconstruction: the 'CLIMAP' reconstruction for the time of the last glacial maximum at approximately 18 ka. The effects of the ice sheets on continental elevation and albedo and on the lowered ocean levels are enormous. This type of reconstruction is becoming increasingly important as a source of data with which to test climate models.

Data on this time scale are also sufficiently numerous to encourage attempts to investigate the level of organisation of the fluctuations of the type seen in the curves in Figure 12.4. For example, cyclic and quasi-cyclic activity seems to be suggested by some of the shorter period (<100 000 years) fluctuations. Such investigations are beginning to shed light on the possible mechanisms responsible for climatic change.

### 12.2.3 Historical climatic change

As we approach the present time the range and diversity of climatic information increases. During the last two or three thousand years much can be inferred from archaeological evidence. Within the last thousand years documentary evidence is available, while starting in the seventeenth century direct observations become ever more numerous. Consequently, for this

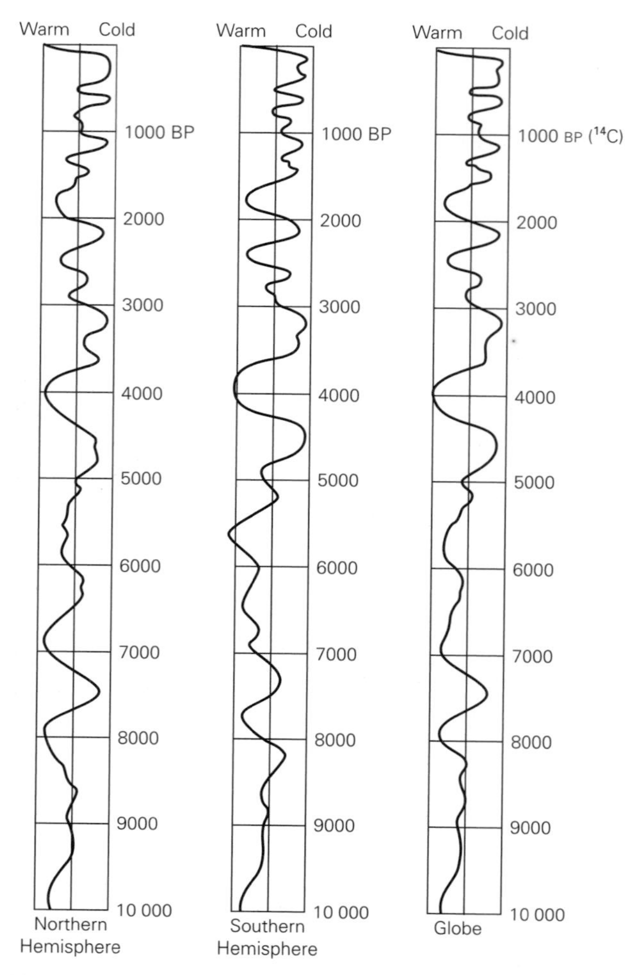


*Figure 12.6* Alpine (mountain) glacier advances and retreats in the Northern and Southern Hemispheres, and a composite curve. Note that although both hemispheres tend to respond together, this is by no means true of all of the fluctuations. (From Wigley, 1991.)

historical period we can have more, but by no means complete, confidence in evidence for regional-scale variations over relatively short time periods.

The **Iron Age**, occurring at 2.9–2.3 ka (900–300 BC), was generally wet and culminated in a cold epoch (probably around 500–300 BC). There is widespread evidence of regrowth of bogs throughout Europe.

The **Secondary Optimum** in climate (AD 1000–1200) was a warming to a lesser degree and of a shorter duration than the post-glacial optimum. This led to melting of the pack ice and drift ice of the Arctic margins and was coincident in time with the founding of Norse colonies throughout the North Atlantic land areas as far as America. Summer temperatures were probably 1 °C higher than today in western and central Europe, where the period lasted until AD 1300. Although Antarctica also seems to have been warmer during this Secondary Optimum, in China there may have been a cooling.

The **Little Ice Age** occurred between 1430 and 1850, with two intense periods around 1470 and in the late 1600s. It was most harsh in Britain during the second half of the seventeenth century. There was a preponderance over the entire period of harsher

*Figure 12.7* The severity of winters in northern Europe during the Little Ice Age was well documented by many artists. *A scene on the ice near a town* by Hendrick Avercamp (1585–1634) is a typical example. (Courtesy of the National Gallery © National Gallery.)

conditions than today (Figure 12.7), but there were also some periods of relatively equable climate. Glaciers advanced in Europe, Asia and North America, although there were also times when summer temperatures were at or above present values. During this whole time the Arctic pack-ice expanded, with important detrimental effects for Greenland and Iceland; by 1780–1820 (and probably by the late 1600s), the temperature across the North Atlantic north of 50°N was some 1–3 K less than today. The Southern Hemisphere seems to have partly escaped the cold period until 1830–1900. It may even have been warmer in Antarctica during the period 1760–1830.

The **warming trend** of the 1880s to 1940s was especially noted in the Atlantic sector of the Arctic and in northern Siberia. In the Southern Hemisphere, south of 30°S, there appears to have been a cooling.

A **cooling trend** was observed in the Northern Hemisphere following the high temperatures of the 1940s. Cooling at high latitudes (which seem to be predictive for other latitudes) stopped in the mid-1960s. Overall, sea-surface temperatures in the Northern Hemisphere cooled by 0.75 K but those in the Southern Hemisphere may have warmed during the 1935–1970 period by up to 1 K, especially at middle and high latitudes. This has led to concern about the West Antarctic Ice Sheet, which is grounded on islands on the sea floor and so may be influenced by sea-temperature or sea-level changes. Increasing temperatures may cause a more rapid disintegration of this ice sheet and lead to a rise in sea level.

### 12.2.4 Scale of temperature changes

The scale of temperature changes in the recent past can be summarised and put into perspective by considering temperatures over the last 150 ka (Figure 12.8). Features such as the thermal maximum of the 1940s and the lowered temperatures of the Little Ice Age are clearly seen. However, glacial/interglacial temperature variations are of the order of 6–7 K while the temperature range over the last century is approximately 0.5 K. The present conditions are relatively warm since the temporal extent of glacial times seems to be longer than that of interglacial periods. Again the temperature fluctuations appear to be somewhat cyclic and suggest lines of approach for understanding the causes of climatic changes.

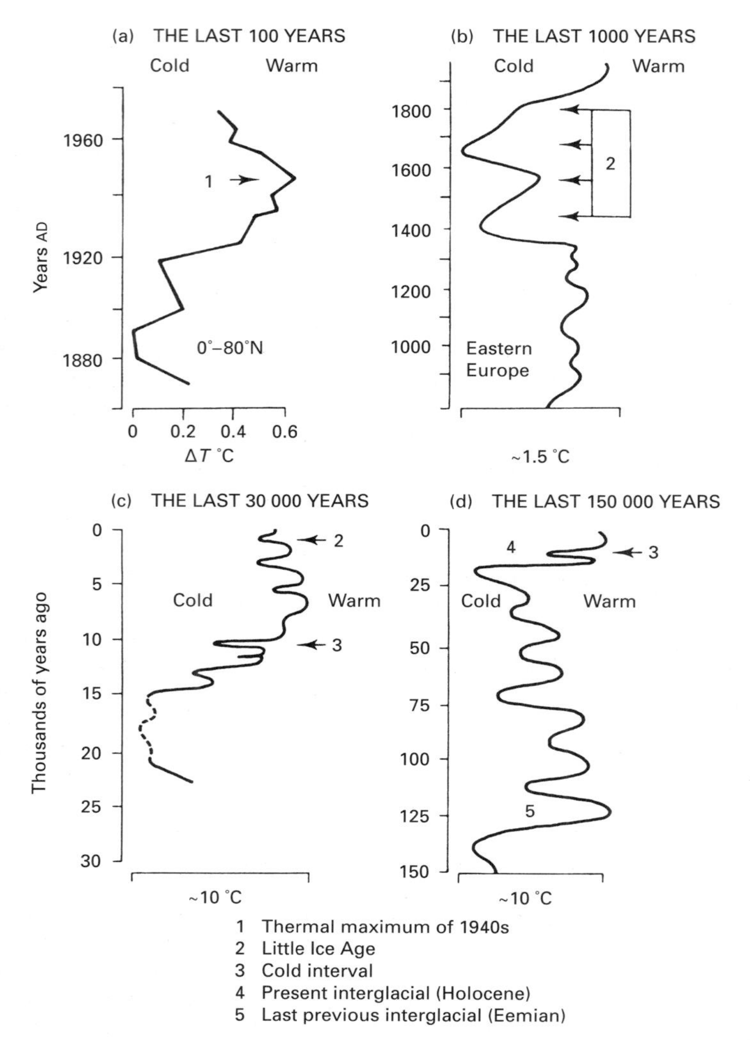


*Figure 12.8* Generalised trends in climate represented by approximate mid-latitude air temperature trends over a variety of time scales for the past 150 000 years.

## 12.3 Mechanisms of climate change

Climate changes can result from the action of any of the processes affecting the climate system. However, changes that affect the whole Earth, or major portions thereof, for at least several years, are likely to arise from a relatively small number of causes. These changes can be split into **external**, in which the agent of cause is outside the climate system, and **internal**, where the initial alteration is within the system itself (Figure 1.1). Isolating a particular cause for a particular change, however, is extremely difficult because the interlinked nature of the system ensures that there are feedbacks, so that a change in one component leads to a change in most, if not all, other components.

This feedback effect can be illustrated by considering climate changes on the longest possible time scale, that of the evolution of the Earth. It is generally agreed that solar luminosity has increased by between 20 and 40% during the $4.5 \times 10^9$ years of the Earth's existence. Low incoming solar radiation early in the history of the Earth suggests that surface temperatures should have been low, certainly below the freezing point of water. Geological evidence suggests that this was not the case, leading to the 'weak Sun–enhanced early surface temperature' paradox. The explanation appears to be intimately bound with the evolution of the atmosphere and the surface. The higher concentrations of carbon dioxide in the primeval atmosphere enhanced the greenhouse effect and created higher temperatures. With the advent of biological activity the amount of oxygen in the atmosphere increased while the $CO_2$ concentration decreased. This, combined with associated surface changes, ensured that temperatures remained stable as the Sun's output increased.

### 12.3.1 External causes of climatic change

Potential external causes of climatic change include changes in the luminosity of the Sun and in the astronomical relationship between the Earth and the Sun. In addition, changes in the character of the Earth's surface as a result of continental drift and mountain building forces are usually regarded as external. They, unlike changes in surface vegetation, for example, are not themselves influenced by the climate. In addition, changes in the polarity of the Earth's magnetic field may influence the upper atmosphere and thus the whole climate. Of these, possible changes resulting from astronomical changes and luminosity changes in sunspots have received the most attention and are considered here.

The astronomical relationships between the Earth and the Sun, expressed as a series of cyclical changes called the **Milankovitch cycles**, provide some measure of explanation of some climate changes. There are several different ways in which the orbital configuration of the Earth around the Sun can affect the received radiation and thus possibly the climate. They are as follows (Figure 12.9):

1. *Eccentricity variations* The Earth's orbit becomes more circular and then more elliptical in a pseudo-cyclic way, completing the cycle in about 110 000 years. The mean annual incident flux varies as:

$$\text{flux} \propto (1 - e^2)^{0.5} \qquad (12.1)$$

where the eccentricity of the orbit is $e$. For a larger value of $e$ there is a larger incident annual flux. The current value of $e$ is 0.018. In the last 5 million years it has varied from 0.004 83 to 0.060 791. These variations would result in changes in the incident flux of from −0.17 to +0.014%.

2. *Obliquity* The obliquity, the tilt of the Earth's axis, is the angle between the Earth's axis and the plane of the ecliptic (the plane in which the bodies of the solar system lie). This tilt varies from about 22 to 24.5°, with a period of about 40 000 years. The current value is 23.5°. Seasonal variations depend upon the obliquity (Figure 2.4). If the obliquity is large, so is the range of seasonality. However, the total received radiation is not altered, but a greater seasonal variation in received flux is accompanied by a smaller meridional gradient in the annual radiation.
3. *Precession of the location of perihelion* The orbit of the Earth is an ellipse around the Sun, which lies at one of the foci. The point in the orbit when the planet passes closest to the Sun is called the perihelion point. Due to gravitational interaction with the other planets, primarily Jupiter, this point moves in space so that the ellipse is moved around in space. This will cause a precession in the time of the equinoxes. These changes occur in such a way that two periodicities are apparent: 23 000 and 18 800 years. This change, like (2), does not alter the total received radiation but affects the distribution of the energy in both time and space.

For long-term temperature data, spectral analysis, which permits the identification of cycles, has shown the existence of cycles with periods of approximately $20 \times 10^3$, approximately $40 \times 10^3$ and approximately $100 \times 10^3$ years (Figure 12.10). These correspond rather closely with the Milankovitch cycles. However, the strongest signal in the observational data is in the longest time period, with a cycle of 100 000 years. This would be the result of eccentricity variations in the Earth's orbit, which produce the smallest insolation changes. Hence, the effect of the Milankovitch cycles is far from clear. While observations and models indicate that they play a role in climatic change, they by no means constitute an explanation, even for long-term, cyclic climatic changes. Almost certainly these external changes trigger large feedback effects in the climate system which are yet to be fully understood.

Variations in the climate during historical times have been linked with the sunspot cycle, the second possible external cause of solar-produced climatic change. This cycle occurs with a 22-year periodicity,

(a)

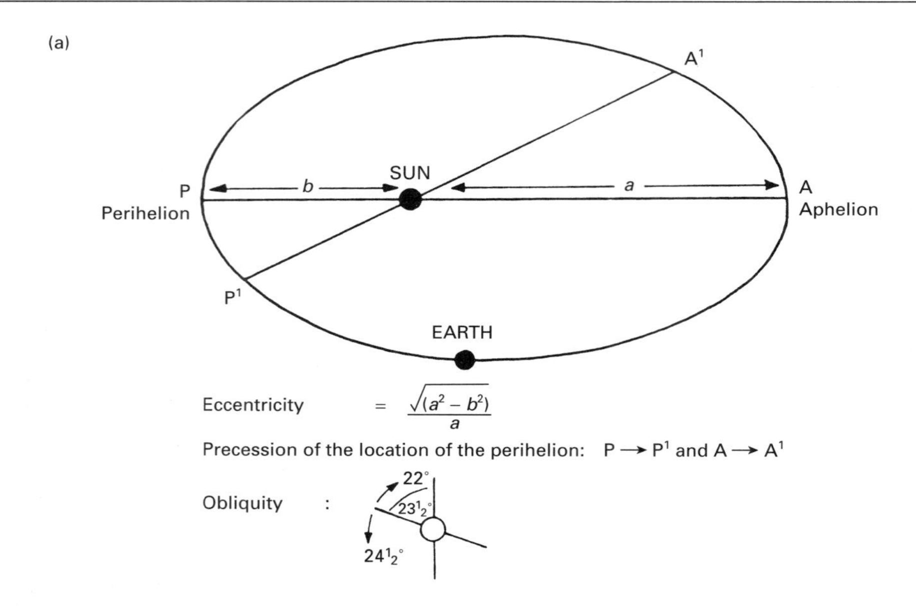


(b)

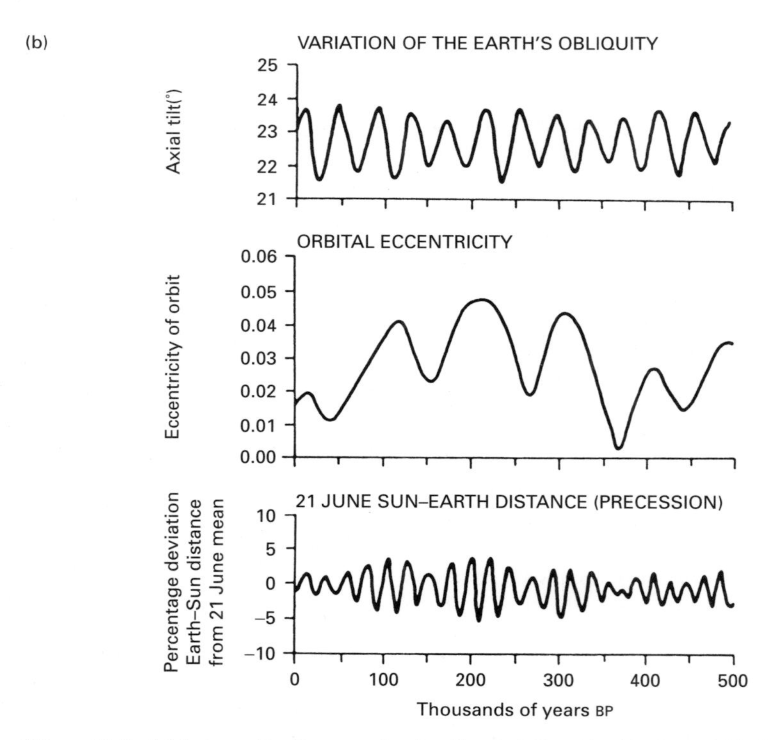


*Figure 12.9* (a) Schematic diagram showing the variations in the three orbital components: eccentricity, precession of the location of the perihelion and obliquity. (b) Variations in these three components as a function of time. (From Broecker and van Donk, 1970.)

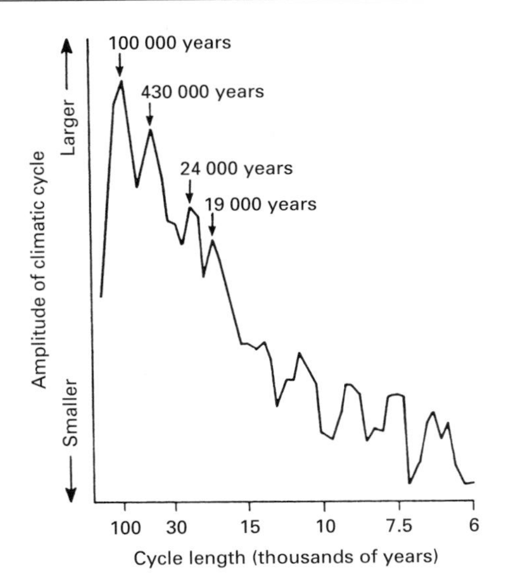


*Figure 12.10* Spectrum of climatic variations over the last 500 000 years. The graph shows the importance of the climatic cycles of 100 000 years (eccentricity); 43 000 years (obliquity) and 24 000 and 19 000 years (precession of the location of the perihelion). The curve is constructed from an isotopic record of two Indian Ocean cores. (From Imbrie and Imbrie, 1979.)

the **Hale double sunspot cycle** (Figure 12.11a). The overall amplitude of the cycles seems to increase slowly and then fall rapidly with a period of 80–100 years. There also appears to be a quasi-cyclic fluctuation of the order of 180 years.

There have been correlations of sunspots with very short-term changes in the solar constant and attempts have been made to link each of these apparent periodicities with suggested climate periodicities. In particular, the Little Ice Age has been linked to the **Maunder minimum** in sunspots (Figure 12.11b). Between 1650 and 1730 there was both a sudden and unaccountable reduction, almost to zero, in the sunspot activity of the Sun, and low surface temperatures on Earth. However, the actual period of the Little Ice Age itself seems to vary according to the geographical area from which the data are taken. Certainly as yet no mechanistic link has been shown to relate sunspot numbers directly to the mean surface conditions on Earth.

### 12.3.2 Internal causes of climatic change

Any change in any component internal to the climate system will lead to a climatic perturbation. Most current concern centres on anthropogenically created effects and their consequences, which could operate on the relatively short time scales necessary to create noticeable changes within the next century. The only natural effect likely to be important on this time scale is volcanic activity.

Volcanoes provide a huge local source of particulates, heat and water vapour for the atmosphere, while volcanic ejecta, in the form of lava flows, alter the surface character. Vulcanism can certainly produce measurable temperature anomalies of at least a few tenths of a degree (Figure 12.12a). However, an eruption and the resultant effect on the atmosphere is short-lived compared to the time needed to influence the heat storage of the oceans. Hence the temperature anomaly is unlikely to persist or lead, through feedback effects, to significant long-term climatic changes.

The energy and type of the volcanic eruption largely determine its climatic effect. Most eruptions inject aerosols into the troposphere at heights between 5 and 8 km. These particles are soon removed either directly by gravitational fallout or by rainout (Section 11.2) and their climatic effect is small. Violent eruptions can hurl debris into the upper troposphere or even into the lower stratosphere (15–25 km) (e.g. Mount Agung in 1963 and El Chichon in 1982). These are much less common, but can lead to more extensive climatic effects. At these heights there is a smaller chance of immediate removal by direct interaction with atmospheric water. Furthermore, the particles have a long residence time in the stratosphere, of the order of a year for particles of radii 2–5 μm, but as long as 12 years for smaller particles of radii 0.5–1.0 μm. Thus they can become widely distributed by the stratospheric circulation. For instance, after the Krakatoa eruption in 1883, red skies, blue moons and fantastically colourful sunsets were observed in many places for several months. One can only speculate on the optical effects of a massive event, or series of events, around 17.5 ka (Figure 12.13), which produced volcanic debris more than eight times more concentrated than anywhere else in the 50 000 year record.

Immediately following an eruption the stratosphere is dominated by particles which scatter radiation at wavelengths up to 10 μm very efficiently, being about 10 times better scatterers than the normal stratospheric particles. They also absorb visible radiation. The contribution of the particles to the 'clear sky' optical thickness, $\tau$, can rise to 0.1 (20 times the normal value) after large eruptions. After about 6 months there is an increase in sulphate production, which further increases the visible scattering and slightly increases the absorption in the infrared. These changes will affect the atmospheric heating rates. To attempt

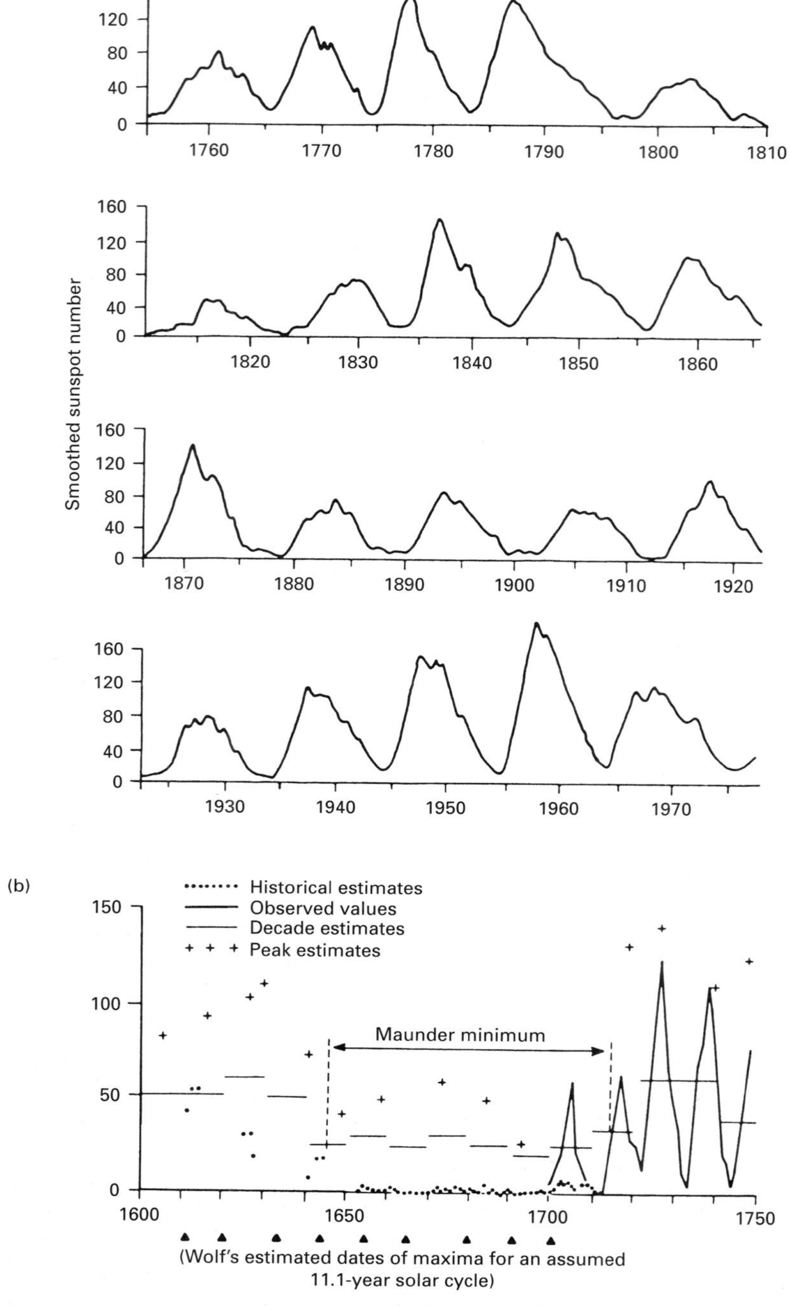


*Figure 12.11* (a) Variations in the annual mean sunspot number showing the 11-year sunspot number cycle, which, when the magnetic polarity of the sunspots is noted, is found to be half of the true 22-year cycle. (b) Estimated annual mean sunspot numbers in the period AD 1610 to 1750. The absence of sunspots during the Maunder minimum is clearly seen. (From White, 1977.)

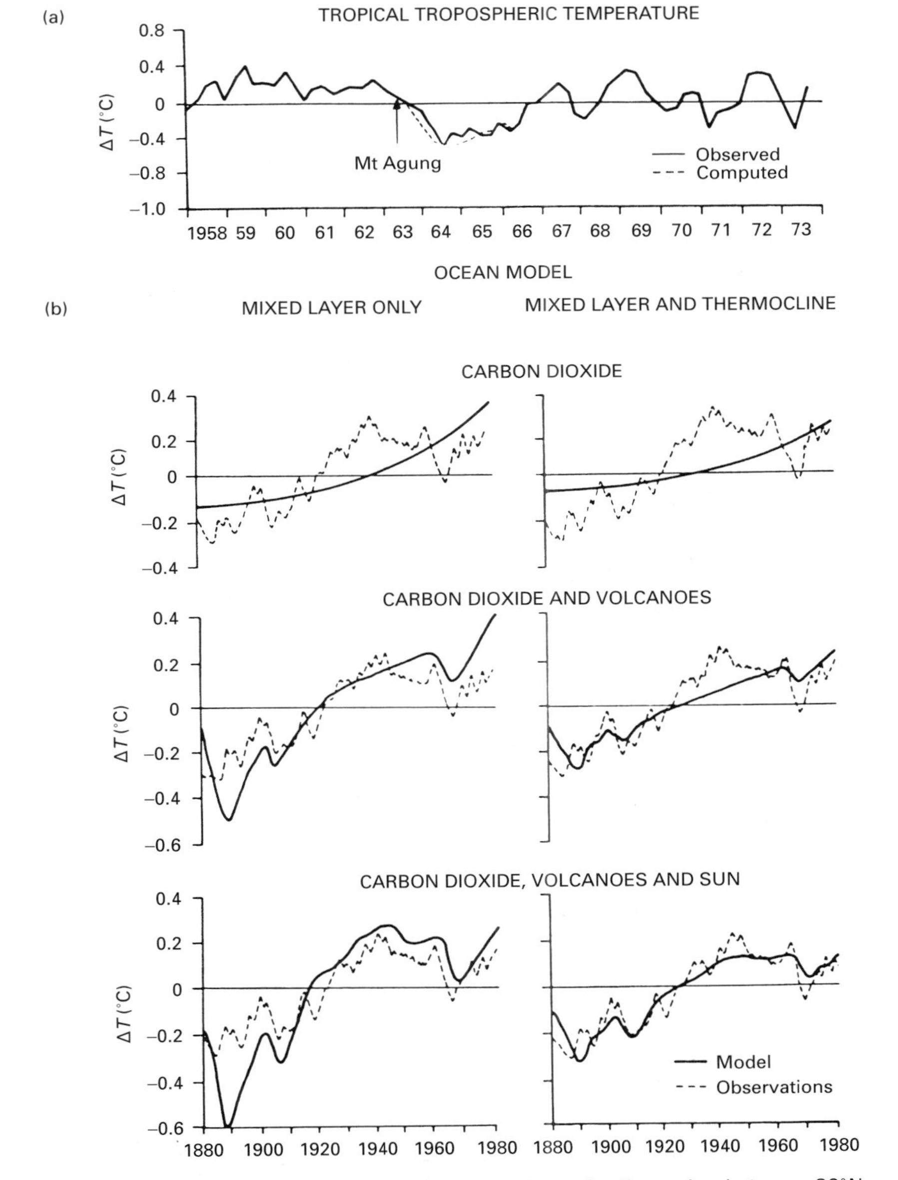


*Figure 12.12* (a) Observed tropospheric temperatures for the region between 30°N and 30°S and model-computed temperatures for the period after the eruption of Mount Agung in 1963. Both computed and observed temperatures exhibit a significant decrease as a result of the volcanic eruption. (b) Global temperature trends obtained from a one-dimensional radiative–convective (1-D RC) climate model taking changes in atmospheric $CO_2$, volcanic activity and solar luminosity into account. The results are in good agreement with the observed surface temperatures. (From (a) Hansen *et al.*, 1980; (b) Hansen *et al.*, 1981 – NASA – p.d.)

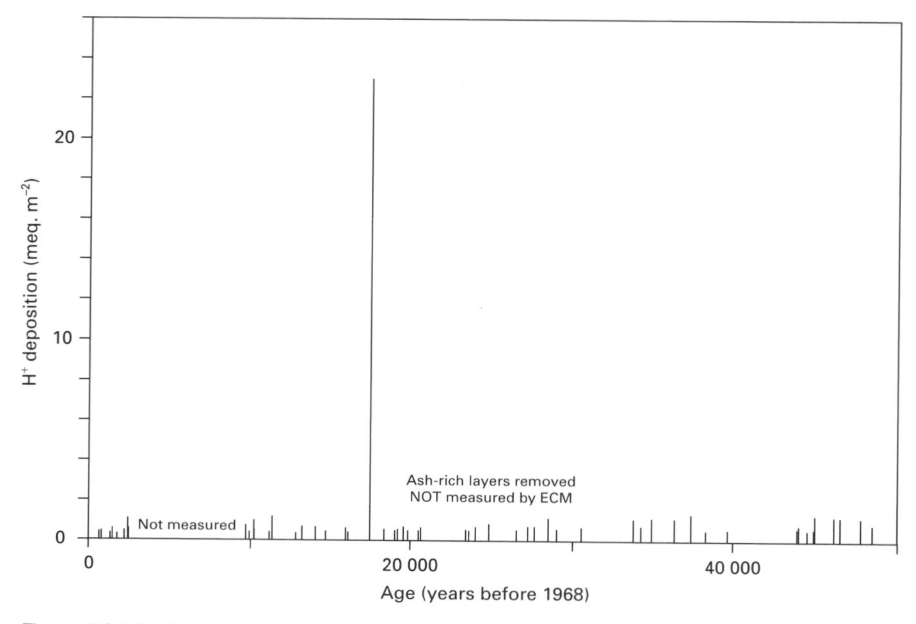


*Figure 12.13* Periods of volcanic activity over the last 50 ka deduced from an ice core extracted at Byrd Station, Antarctica. An electrical conductivity technique was used to determine ice acidity, which was then related to vulcanism. The major event at 17.5 ± 0.5 ka BP is clearly seen. (From Hammer *et al.*, 1997.)

to assess the resultant impact on the climate it is necessary to calculate how the ratio of absorption to backscatter for both short-wave and long-wave radiation alters relative to the normal situation. This seems to be partly dependent upon surface albedos. If the surface albedo is very high, warming can occur if the eruption causes approximate equality between absorption and backscatter, because the radiation that is now being absorbed was previously being lost to space. Over a surface with a lower albedo, however, most of the incident radiation was being absorbed anyway so that the only way to produce a net heating is if the absorption is altered to be greater than the backscatter. It is more likely in this case that backscatter from the volcanic particles will cause a local cooling.

Model simulations suggest that radiation effects will, overall, produce a global cooling when large-scale volcanic eruptions occur. Figure 12.12(b) shows a typical simulation attempting to reproduce recent temperature changes, taking into account changes in solar luminosity and atmospheric $CO_2$ in addition to volcanoes. The volcanic effect was established using a **dust veil index** developed from a catalogue of volcanic eruptions. The results clearly support the idea that volcanoes have a cooling effect. In all analyses, however, a complication arises because volcanic particles can serve as condensation nuclei, influencing the type and amount of cloudiness and thus possibly further altering the climate.

Aerosols, unlike volcanics, are continuously present in the atmosphere. In this regard they are akin to clouds and gases. The effects of these three on the radiative fluxes in the atmosphere are compared in Table 12.3. The particle effect is generally small except for the scattering of solar radiation. There is some indirect evidence of increasing particulate loading in the troposphere. The electrical conductivity of the Northern Hemisphere troposphere has been

*Table 12.3* Interaction between atmospheric constituents and electromagnetic radiation in short and long wavelength regions

| Agent | Effect | Magnitude of effect | |
|---|---|---|---|
| | | Short-wave | Long-wave |
| Gases ($CO_2$, $H_2O$) | Absorption | Moderate | Large |
| | Scattering | Small | Negligible |
| Aerosols (particles) | Absorption | Small | Small (large for human-made) |
| | Scattering | Moderate | Small |
| Clouds | Absorption | Small | Large (varies with thickness) |
| | Scattering | Large | Large, but dominated by absorption |

steadily increasing over the last 80 years. One reason proposed for this is an increase in the concentration of aerosols in the size range 0.02–0.2 μm, which, in turn, are likely to be attributable to human activities rather than natural sources. The change in particle amount and size distribution may be leading to variations in both radiative exchanges and cloud amount in the troposphere. In the stratosphere, however, the importance of sulphate aerosols in raising the planetary albedo is becoming increasingly apparent. Current theory suggests that aerosols are acting in opposition to the greenhouse role of carbon dioxide by decreasing the incident solar radiation at the surface. These aerosols are a by-product of industrial activity, and tend to be concentrated downwind of the world's major industrial areas and not yet completely dispersed. Hence they may have a regional impact, but the magnitude of their global effect is not clear.

One potentially important role for aerosols is to act as cloud condensation nuclei (Section 4.4). Any excess aerosols, either natural or anthropogenic, which are injected into the lower troposphere can become CCNs. However, they tend to be removed fairly quickly, and any effect on cloud or precipitation is likely to be local and short-lived. Similarly the input into the stratosphere is unlikely to affect cloudiness because the water vapour is generally unable to penetrate into the stratosphere. However, injections into the upper troposphere may have far-reaching effects. At heights of 10–18 km it is likely that some particles, whether volcanic debris or of human origin, could increase the formation of ice clouds. The amount of cirrus cloud may be critical as a climatic perturbing agent. An increase in cirrus cloud amount will increase absorption of the infrared radiation and therefore lead to a general climatic warming. However, variations in cloud type and amount can result from many processes, and aerosol concentration must be one factor among many.

A variety of other possible internal causes of climate change have been introduced in previous chapters. The potential role of changes in the surface character and in atmospheric composition can simply be noted here. Ways in which we can assess their potential effect using modelling techniques will be discussed in the next chapter, while consideration of some potential consequences are reserved for the final chapter.

### Box 12.A Climate and long-term ecological change

Many of the applications we have considered in previous chapters have been concerned with relatively short-term fluctuations of climate, primarily judged on the human time scale. The natural vegetation, however, frequently operates on a much longer scale, so that it responds to long-term climate variations of the order of thousands of years. Such variations are the focus of this analysis.

#### *12.A.1 Forest changes in eastern North America*

We are concerned here with the change in location of forest types in eastern North America over the last few thousand years. This particular example allows us to link directly with a palaeoclimatological technique discussed in Box 12.I. It also links observations and models, and thus looks forward to our study of models in the next chapter.

Any vegetation species has a range of environments in which it can grow. However, each species has a restricted range in which it does best, and the farther we go from this the more difficult it is for the species to survive. This may mean both poor health, because of unsuitable conditions, and the inability to compete with other species better adapted to that particular environment. Thus, in general, any environmental stress will decrease the survival ability of the individual and its species. This will occur on a short time scale, as when an ice storm or a hurricane removes a particular tree species. If the general climatic conditions remain unchanged, however, the same species, if not the same individual, will return. With long-term climatic changes, however, the species will not grow back after a particular stress, and the net effect should be a migration of the vegetation zones in response to climate change, in much the same way as the cultivation zones in Figure 11.4. In general, the long-term stress is provided either by temperature, which greatly influences an organism's ability to add new biomass, or by water, which is usually reflected in a stress associated with precipitation.

Changes in temperature regime have been the prime cause for the movement of vegetation zones throughout much of the northern mid-latitudes over the last few millennia. As the ice sheets retreated after the last great ice age 18 ka ago, the spruce, adapted to cold, moist conditions and associated with the margins of the Arctic climates, migrated northward (Figure 12.A.1). In this case, the migration of the ice sheet exposed new areas for spruce colonisation to the north, while species better adapted to warmer conditions were replacing spruce on its southern margin. As appears typical, the result was not a slow steady migration, but a series of expansions and contractions. This may reflect both the nature of the climatic fluctuations and the nature of the development of individual stands of spruce and their response to competition.

**Box 12.A (cont'd)**

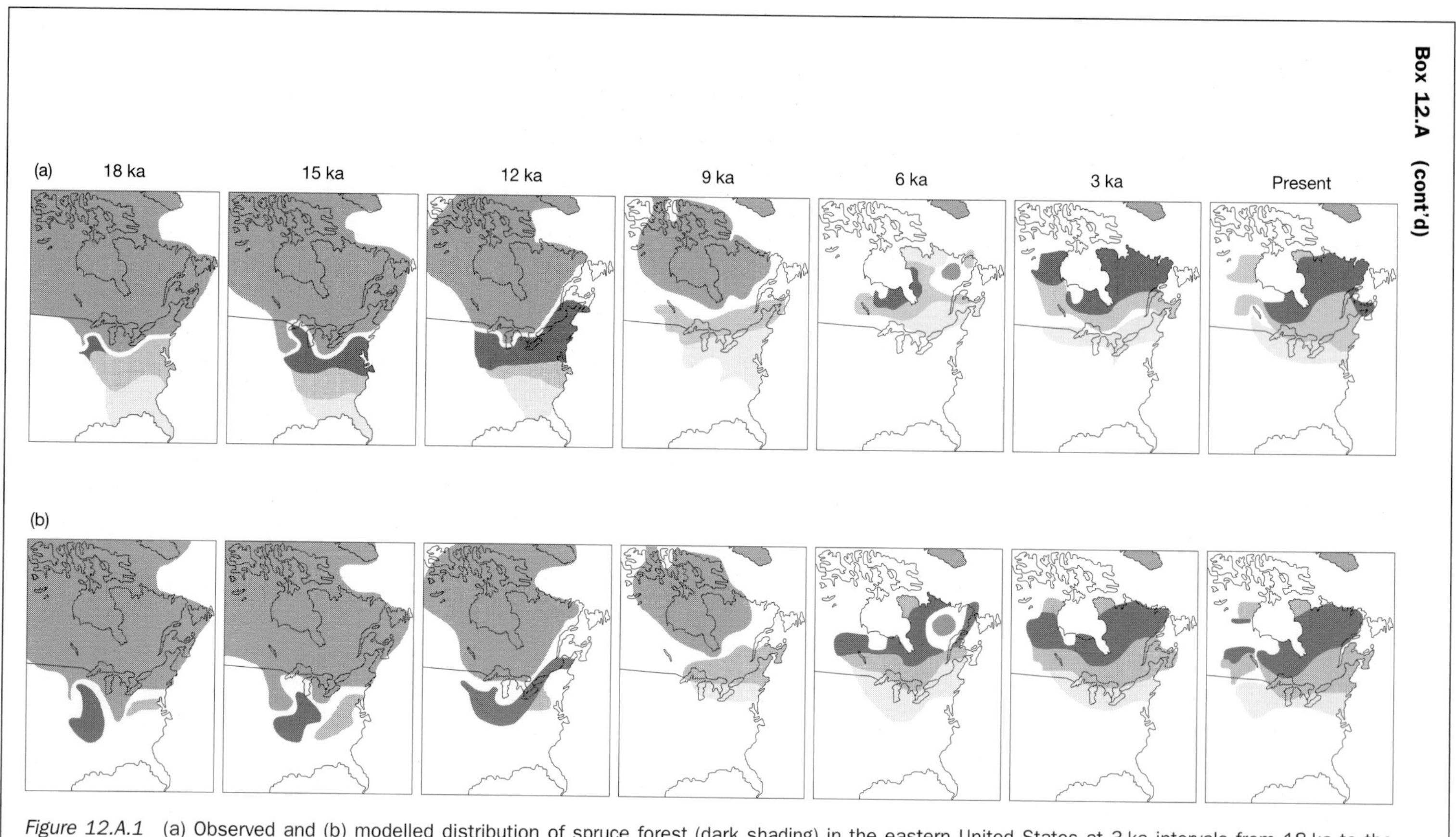


*Figure 12.A.1* (a) Observed and (b) modelled distribution of spruce forest (dark shading) in the eastern United States at 3 ka intervals from 18 ka to the present. The observations are based on pollen records, with the shading indicating approximate abundance. The lightly shaded area is the Laurentide Ice Sheet, which disappeared soon after 6 ka. The modelled information is based on a model of spruce growth driven by the temperature and precipitation outputs of a general circulation model. The good agreement between the two suggests that the model produces realistic values and supports the notion that the migration of spruce northward was largely a response to a changing climate. (From Webb *et al.*, 1987.)

### Box 12.I The use of proxy variables

There is a great deal that could be said about any and all of the techniques outlined in Table 12.1. However, we choose two as examples. The first, dendrochronology, continues our concern with trees. The second, use of written diary material, indicates the kinds of observations that can be used to infer a great deal about the climate.

*12.I.1 Dendrochronological techniques*

Every year every living tree grows by adding a ring of new wood. Early in the season the new cells tend to be large and well developed, but as the season draws to a close the cells tend to be smaller and less well developed. The result, however, is a clearly distinguishable ring. The amount of growth, and thus the width of the ring, depends on conditions during the growing season. In climatic regions close to the centre of a tree's range, conditions tend to be favourable in almost all years and the rings are of an approximately uniform size. Near the margins, however, where the tree is likely to be under stress, the ring width will depend on the ambient conditions. Thus there is a sequence of ring widths, which allow both dating of the tree and estimates of the climate once it has been dated.

The dating principle is simple. A core is extracted from a living tree and rings are counted towards the centre. For some long-lived species this can give a record that is hundreds of years long. In many cases, however, the record can be extended backwards by matching patterns with successively older wood (Figure 12.I.1). This older wood might be wood preserved in bogs and swamps, or timber cut centuries ago and used in building things ranging from boats to fence posts.

The relation between ring width and climate is not simple. Even when the tree is growing at its margin, and thus the ring width is most susceptible to climatic influences, there are many confounding factors. First, there are biological factors to be considered, including the variation in biomass productivity with tree age, or the influence of pests and diseases. Then there are climatic factors. Photosynthesis, and thus growth, one year may be impeded by fire, or an ice storm, removing most of the leaves. Further, if we are considering an area which is marginal for precipitation, it may be that a particular tree is missed by any convective storms, or alternately, gets a great amount of precipitation from several of them. In general, therefore, it can be misleading to attempt to use a single tree to get an annual climatic chronology. One tree may suggest decadal trends. A synthesis of results from a series of nearby trees, however, may suggest an annual pattern, as indicated in Figure 12.2.

*12.I.2 An observational diary*

Written records are an obvious and abundant source of proxy climate data for the last few centuries. Sometimes climatic snapshots from earlier periods are also possible. The failed invasion of Britain by Julius Caesar in 55 BC is one such snapshot. One great characteristic of Caesar is that he kept detailed written records of his activities. For this invasion, sea state, cloud type, wind direction and (qualitatively) speed, are known either from his direct observation or from reasonable inferences. Further, because the fleet was divided, we have two transects. Caesar's own ships travelled without climatic hindrance some time prior to those of his cavalry and transport fleet (Figure 12.I.2a). Caesar observed the changing, and deteriorating, conditions from the British mainland, while the cavalry fleet monitored them as it crossed the

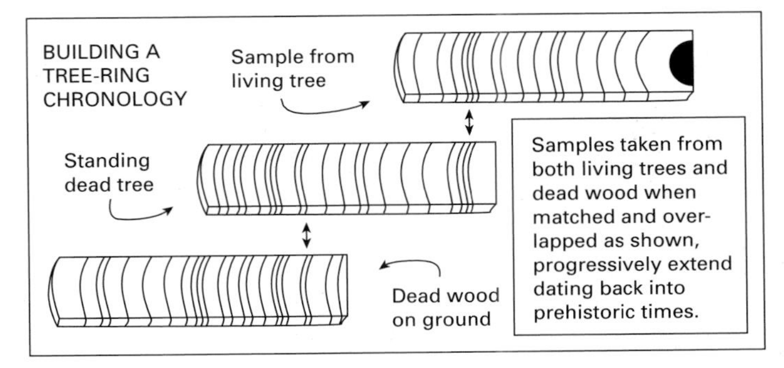


*Figure 12.I.1* Techniques for the construction of a tree-ring chronology. Although a simple three-stage process is shown, many stages, with several overlapping marker patterns, are common.

**Box 12.I (cont'd)**

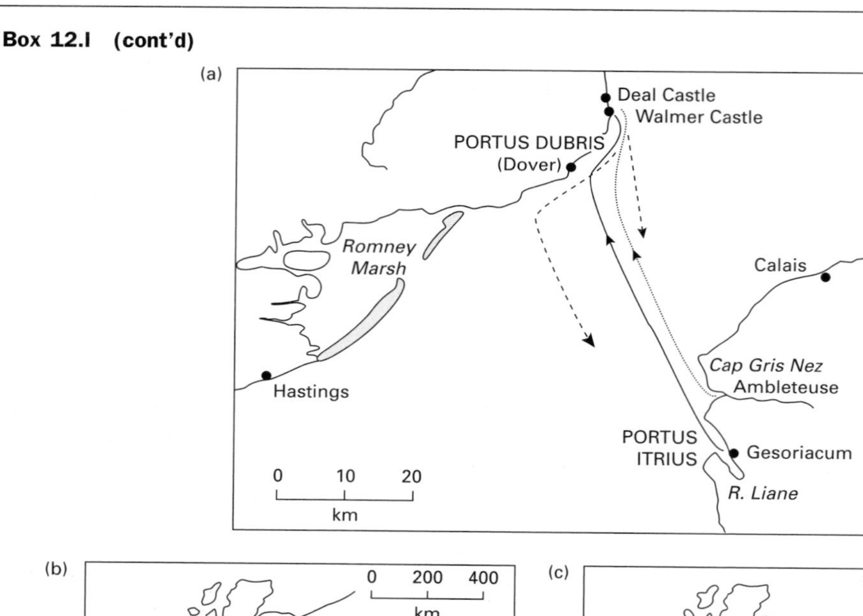


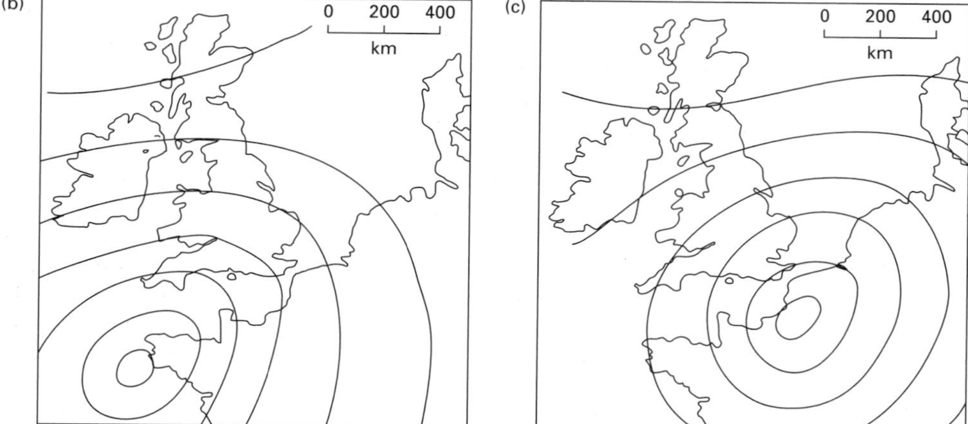


*Figure 12.I.2* (a) Route taken by Caesar's British invasion force in 55 BC, with the cavalry fleet crossing the English Channel on 30 August, and the postulated isobaric chart for (b) 0600 and (c) 1800 hours local time on 30 August. In (b) and (c) the actual pressure distribution is unknown, but gale force winds occurred in the English Channel. (After Meaden, 1976.)

English Channel. From the observations it is possible to infer that a severe, and probably deepening, depression moved up the Channel as the cavalry fleet crossed (Figure 12.I.2b, c). The fleet arrived off the coast in a gale and was forced to retreat. There are indications that this particular event was exceptionally severe, although it is impossible to compare it directly with more recent ones such as the great gale of 1987 in the same area. Nevertheless, records such as this provide direct indications that the current weather patterns are not confined to recent decades, but are likely to have been a characteristic of the climate for centuries, and probably millennia.

# CHAPTER 13 Modelling the climate

In the most general terms, a model is a means of simplifying reality in a way that yields useful information about that reality. A climate model is an attempt to apply the physical laws governing climate processes to simulate the climate and to understand and predict the effects of changes in the processes. The model itself is usually a series of equations expressing these laws. Thus to run a model, the initial climatic conditions are observed, the laws are applied, and the new conditions deduced. Simplification is needed because we neither understand all the laws perfectly nor observe the climate completely. Further, our current restricted computational ability limits the accuracy of our calculations, forcing additional simplifications.

The need for simplification means that no model can deal equally with all aspects of the climate. We need to identify the purpose of the model, which determines the climatic elements and time and space scales of major interest. Many of our previous applications boxes have used models for highly specific purposes. Here we shall pay most attention to models that deal primarily with climate processes themselves.

## 13.1 A simple global energy balance model

### 13.1.1 A zero-dimensional energy balance model

The concepts involved in the creation of a climate model can be illustrated by a discussion of a simple type of model: the **energy balance model (EBM)**. The simplest possible case is when single global annual averages are considered. This is a zero-dimensional model. Here the incoming and outgoing energy flows for the globe are balanced and the single parameter, the surface temperature, $T$, is calculated, i.e. $T$ is the dependent variable for which the 'climate equations' are solved. The rate of change of temperature, $\Delta T$, with time, $\Delta t$, is caused by a difference between the top-of-the-atmosphere (or planetary) net incoming $R\downarrow$ and net outgoing $R\uparrow$ radiant energies (per unit area):

$$\rho C \Delta T/\Delta t = R\downarrow - R\uparrow \qquad (13.1)$$

where $\rho C$ is the heat capacity of the system (per unit area). This is a very general equation with a variety of uses. Thus, for example, if the system we wish to model is an outdoor swimming pool we can calculate the rate of temperature change in time steps of one day from equation (13.1). Suppose the pool has surface dimensions $30 \times 10$ m, is well-mixed and is 2 m deep. Since 4200 J of energy are needed to raise the temperature of 1 kg of water 1 K, and 1 $m^3$ of water weighs 1000 kg, the pool has a total thermal capacity (i.e. the amount of energy (J) needed to raise the temperature by 1 K) equal to $2.52 \times 10^9$ J $K^{-1}$. If we assume that the difference between the absorbed radiation and the emitted radiation from the pool ($R\downarrow - R\uparrow$) is 20 W $m^{-2}$ for 24 hours, then the difference in energy content of the pool for each 24-hour time step is $20 \times 30 \times 10 \times 24 \times 60 \times 60$ J. Then, from equation (13.1):

$$2.52 \times 10^9 \, \Delta T = 20 \times 30 \times 10 \times 24 \times 60 \times 60, \text{ or}$$

$$\Delta T \text{ (in 1 day)} = (2 \times 3 \times 24 \times 36)/(2.52 \times 10^4) \cong 0.2 \text{ K} \qquad (13.2)$$

Thus at this rate it would take about a month to raise the temperature of the pool water by 6 K.

On the Earth the value of $\rho C$ is largely determined by the oceans. For instance, if we assume that the energy is absorbed in the first 70 m of the ocean (the average global depth of the top or mixed layer), and that approximately 70% of the Earth's surface is covered by oceans, then the value for $\rho C$ comes from:

$$\rho C = \rho_w c_w d 0.7 = 2.06 \times 10^8 \text{ J m}^{-2} \text{ K}^{-1} \quad (13.3)$$

where $\rho_w$ is the density of water, $c_w$ its specific heat at constant pressure, and $d$ is the depth of the mixed layer.

For our simple energy balance model of the Earth, the energy emitted, $R\uparrow$, can be estimated using the Stefan–Boltzmann law (equation (2.11)) and the surface temperature. This value must be corrected to take into account the infrared transmissivity of the atmosphere, $\tau_a$, since it is the planetary flux. Therefore we can write:

$$R\uparrow \approx \varepsilon\sigma T^4 \tau_a \quad (13.4)$$

The absorbed energy, $R\downarrow$, is a function of the solar flux, $S$, and the planetary albedo, such that $R\downarrow = S(1 - A)$.

Equation (13.1) therefore becomes:

$$\Delta T/\Delta t = [S(1 - A) - \varepsilon\sigma T^4 \tau_a]/\rho C \quad (13.5)$$

This equation can be used to ascertain the equilibrium climatic state by setting $\Delta T/\Delta t = 0$. Then:

$$[S(1 - A) = \varepsilon\sigma T^4 \tau_a] \quad (13.6)$$

Using values of $S = 342.5$, $A = 0.3$, $\varepsilon\tau_a = 0.62$ and $\sigma = 5.67 \times 10^{-8}$ gives rise to a surface temperature of 287 K, which is in good agreement with the globally averaged surface temperature today. Note that in this calculation the solar constant ($S_F$ = 1370 W m$^{-2}$) is divided by four to give $S$, since instantaneously the incoming radiation is incident on a smaller area than the area over which radiation is emitted. The ratio, 4 (Figure 13.1), makes $R\downarrow$ directly comparable with $R\uparrow$.

An alternative use of equation (13.5) is similar to the calculation of the swimming pool warming rate made above. Here a time-step calculation of the change in $T$ is made. This could be a response to an external forcing agent, such as a change in solar flux or in the heat capacity of the oceans resulting from changes in their depth or area. Alternatively, the response could be determined by an 'interactive' climate calculation when one of the internal variables (e.g. $A$) alters.

As an example of the change in an internal variable we can consider the variation in $A$ as a function of the mean global temperature. Above a certain temperature, $T_{\text{no ice}}$, the planet is ice-free and albedo is independent of temperature. As it becomes colder we expect the albedo to increase as a direct result of increases in ice and snow cover. Eventually the Earth becomes completely ice-covered, at temperature $T_{\text{ice}}$, and further cooling will produce no further albedo change. This could be expressed in the form:

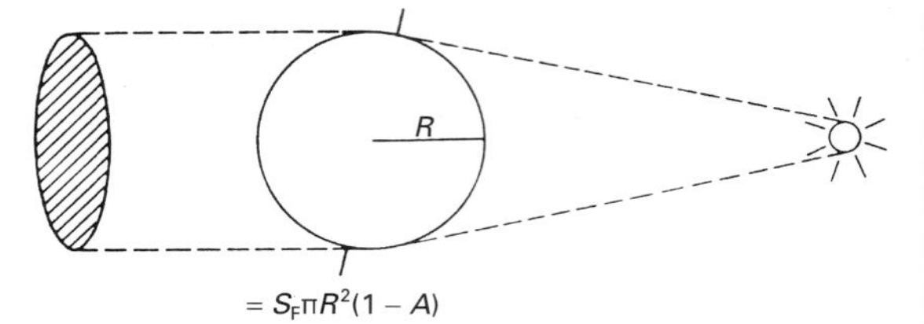


$= S_F \pi R^2(1 - A)$

EMITTED INFRARED RADIATION

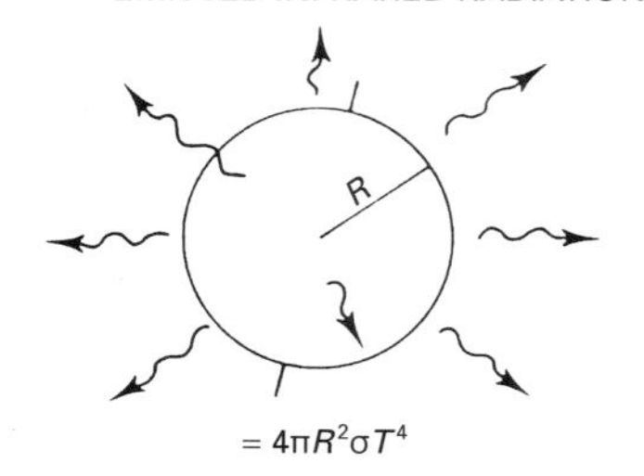


$= 4\pi R^2 \sigma T^4$

Radiation balance is $\sigma T^4 = \frac{S_F}{4}(1 - A) = S(1 - A)$

*Figure 13.1* Diagrammatic representation of the radiation balance of the Earth. Radiation is emitted over the whole area of the sphere ($4\pi R^2$), while an area $\pi R^2$ intercepts the solar radiation, $S_F$.

$$A(T) = A_{\text{ice}} \text{ for } T \leq T_{\text{ice}}$$

$$A(T) = A_{\text{no ice}} \text{ for } T \geq T_{\text{no ice}} \quad (13.7)$$

$$A(T) = \text{linear function of } T \text{ for } T_{\text{ice}} < T < T_{\text{no ice}}$$

$T_{\text{ice}}$ is usually assumed to be 273 K, but may range between 263 and 283 K. If we are concerned with equilibrium conditions, we can calculate $R\uparrow$ for a series of temperatures and $R\downarrow$ for the series of albedos and show the results graphic-ally. The point of intersection of the curves represents the equilibrium situation (Figure 13.2). Any slight imbalances between the absorbed radiation, $S[1 - A(T)]$, and the emitted long-wave flux, $\varepsilon\sigma T^4\tau_a$, lead to a change in the temperature of the system at the rate $\Delta T/\Delta t$, the changes serving to return the temperature to an equilibrium state. However, there are three equilibrium solutions

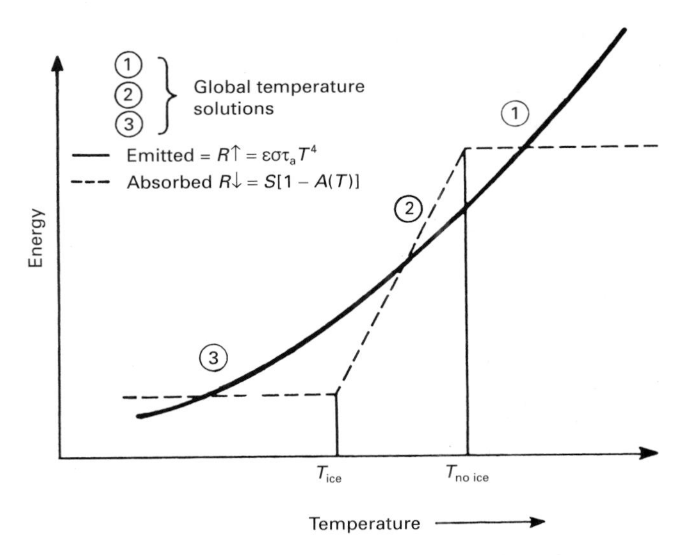


*Figure 13.2* The equilibrium temperature solutions for a zero-dimensional global climate model are shown at the intersection between the curves of emitted infrared radiation, $R\uparrow$, and absorbed solar radiation, $R\downarrow$.

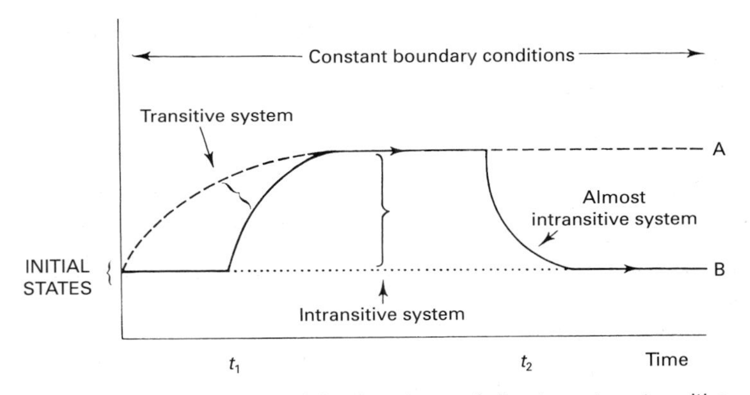


*Figure 13.3* The behaviour of the three types of climate system: transitive, intransitive and almost intransitive with respect to an initial state. In a transitive system two different initial states evolve to the same resultant state, A. An intransitive system exhibits the 'opposite' behaviour with two alternative resultant states. The characteristic of an almost intransitive system is that it mimics transitive behaviour for an indeterminate length of time and then 'flips' to an alternative resultant state.

of equation (13.5), as shown in Figure 13.2: a completely glaciated Earth (3), an ice-free Earth (1) and an Earth with some ice (2). All are possible.

This simple model has some very obvious limitations. Nevertheless, it not only shows typical lines of approach, but also indicates some of the more general problems associated with the solutions; in particular, the question of whether or not all three equilibrium states identified are 'stable' and capable of persisting for long periods of time. This is a concern with any non-linear system, even one that is far simpler than the climate system, which can have a characteristic behaviour termed **almost intransitivity** (Figure 13.3). If we assume that the boundary conditions do not change with time, we can postulate that any system with two possible starting conditions can evolve in two different ways. One is **transitive**, where the systems converge to the same resultant state; the other is **intransitive**, where the result is two stable, but separate, states. Difficulty arises when a system exhibits behaviour that mimics transitivity for some time, then flips to the alternative state for another (variable) length of time and then flips back again to the initial state and so on. In such an almost intransitive system it is impossible to determine which is the 'normal state', since either of two states can continue for a long period of time, to be followed by a quite rapid and unpredictable change to the other.

At present, geological and historical data are not detailed enough to determine which of these system types is typical of the Earth's climate. It is easy to see that should the climate turn out to be almost intransitive it will be extremely difficult to model.

### 13.1.2 One-dimensional models

A relatively simple extension of the above principles allows the incorporation of latitudinal variations. This produces a one-dimensional model. Vertical variations are ignored and the models are used with surface temperature as the dependent variable. Since the energy balance is allowed to vary from latitude to latitude, a horizontal energy transfer term must be introduced, so that the basic equation for the energy balance at each latitude, θ, is:

$$\rho C[\Delta T(\theta)/\Delta t] = R\downarrow(\theta) - R\uparrow(\theta) + \text{transport into zone } \theta \qquad (13.8)$$

where $\rho C$ is the heat capacity of the system and can be thought of as the system's 'thermal inertia'.

The radiation fluxes at the Earth's surface must be parameterised with care since conditions in the vertical are not considered. To a large extent the effects of vertical temperature changes are treated implicitly. In a clear atmosphere, convective effects tend to ensure that the lapse rate remains fairly constant. However, cloud amount depends only weakly on surface temperature, so that cloud albedo is only partially incorporated in the model. In particular, clouds in regions of high temperatures, such as the ITCZ, are ignored in the parameterisation of albedo.

The outgoing infrared radiation term in equation (13.8) is expressed in the form:

$$R\uparrow = B_1 + B_2 T \qquad (13.9)$$

where $B_1$ and $B_2$ are constants. These are obtained empirically by relating observed values of $R\uparrow$ to $T$. The relationship obtained combines the effects of surface emissivity and atmospheric transmissivity and thus provides a practical way of using equation (13.4).

When using the model for annual average calculations, the surface albedo can be regarded as constant for a given latitude. This type of model, however, can also be used for seasonal calculations. In this case it is usual to allow the albedo to vary with temperature, as in equation (13.7), to simulate the effects of changes in sea-ice extent.

The transport term must also be approximated since the dynamical events of the real world are not modelled in an EBM. It is assumed that a 'diffusion' approximation is adequate, relating energy flow directly to the latitudinal temperature gradient. This flow is usually expressed as being proportional to the deviation of the zonal temperature, $T$, from the global mean, $\bar{T}$. Thus the heat transport is set equal to $D(T - \bar{T})$ where $D$ is a constant of proportionality which is obtained from direct observations of latitudinal energy flux (e.g. Figure 7.2).

The final energy balance equation for the annually averaged case then has the form:

$$S(\theta)[1 - A(T,\theta)] - [B_1(\theta) - B_2(\theta)T(\theta)] = D(T(\theta) - \bar{T}) \qquad (13.10)$$

where θ is the latitude. This equation can be solved for the complete set of latitude zones. In simple climatic simulations such as this, it is generally assumed that the Earth is symmetric about the equator and the solution for only one hemisphere is found. The basic constraints or boundary conditions placed on the model to ensure that it remains realistic are that there should be no temperature gradient or heat transport at either the equator or the poles.

Early energy balance models were originally found to be stable only for small perturbations of the present-day conditions. For instance, they predicted the existence of an ice-covered Earth for only slight reductions in the present solar constant, suggesting that the climate system is almost intransitive (Figure 13.3). However, a number of stable solutions are possible, dependent upon the values chosen for the constants $B_1$, $B_2$ and $D$ in the parameterisations employed in the particular EBM.

In any climate model, great care must be taken in choosing the constants for any parameterisation scheme.

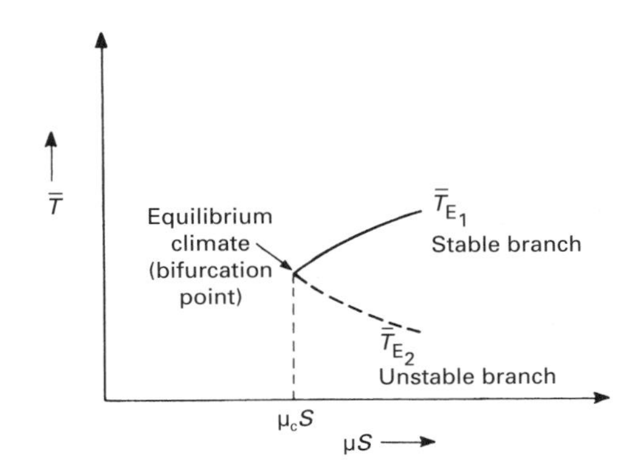


*Figure 13.4* The equilibrium climate bifurcation point. For values of the solar flux $\mu S > \mu_c S$ there are two solutions, whereas below this critical value no solutions exist. Changes in solar radiation lead to either a stable or unstable equilibrium climate illustrated here by the two equilibrium branches.

The more 'accurate' they are for the present day, the more closely the model is tied to predictions of the present-day situation, but the less likely it is to be able to respond realistically to perturbations.

The solution of equation (13.10) gives a climate prediction of the mean global temperature, $\overline{T}$, as a function of all the other parameters. As the equation is non-linear, there will be more than one solution. The inherent stability of these solutions to both internal and external perturbations must then be tested.

For the 'external stability' we can consider the variation of $\overline{T}$ as a function of solar variation since an alteration of the solar constant is a convenient method of exploring the characteristics of climate model structure. Figure 13.4 shows the way in which $\overline{T}$ changes as the total incident radiation, $\mu S$, changes. As the solar flux is reduced to some critical value ($\mu_c S$) the number of solutions is reduced from two to one at the critical value. Below $\mu_c S$, no solution is possible. This critical point is termed the bifurcation point. For values of incoming radiation less than $\mu_c S$, temperatures are so low that the albedo, $A(T,\theta)$, becomes very close to or equal to one and thus it is impossible to regain energy balance and the global temperature, $\overline{T}$, approaches minus infinity. If some limit is put on how high the albedo may get (e.g. $A \leq 0.75$), the solution ends up as an 'ice-age Earth'.

The 'internal stability' concerns the response of each branch in Figure 13.4 to perturbations from equilibrium created by internal factors. To determine if temperatures will return to equilibrium after the perturbation we can use the time-dependent formula in equation (13.5) and postulate a new value for $\overline{T}$, which is close to the equilibrium one already calculated at that level of $\mu S$. Rewriting equation (13.10) in the form of equation (13.5) allows computation of $\Delta\overline{T}/\Delta t$. This change can be computed iteratively until it is determined whether the values do regain the original $\overline{T}$ solution. If it is regained, then the solution is said to be 'internally stable'. In the case shown in Figure 13.4, only the top branch is stable. This method of evaluation of internal stability is the way in which the transitivity or intransitivity of the climate model is established. It would not, however, permit differentiation between an almost intransitive model and an intransitive one.

One important reason why the system type is difficult to establish concerns the differing heat capacities of the components of the climate system. Each component responds to a change in energy at a different rate, usually expressed by the thermal relaxation time, the time required for the component's temperature to change by a fraction $1/e$ of the imposed temperature change. This feature is easy to illustrate by returning to our swimming pool analogy. In the heating example described above it was discovered that the pool's temperature would increase by approximately 0.2 K each day. However, suppose that the pool is enclosed in a glass building with dimensions twice that of the pool (i.e. $60 \times 20 \times 4$ m). The thermal capacity of the air in the building is $C_{air} = 5.79 \times 10^6$ J K$^{-1}$. Using equation (13.1), it is easy to show for the building without the presence of the pool that $\Delta T_{air} \approx 360$ K. Thus, by the end of one day, the temperature in the building alone would be unbearable. If, however, the pool is filled with water then the two thermal capacities must be combined, leading to an overall temperature rise of $\Delta T_{total} \approx 0.8$ K. Of course we have cheated a lot here in holding the net energy input constant. In all cases, as the temperatures rose, the air and water would begin to re-emit increasingly larger amounts of infrared radiation, thus rapidly decreasing the assumed 20 W m$^{-2}$ net input energy. However, the message of the analogy is sound. The thermal capacity of water is so much greater than that of air that the thermal inertia of a combined atmosphere–ocean system is *much* larger than that of the atmosphere alone.

This analogy emphasises the importance of modelling the complete climate system. In particular, the vital role of the oceans is clear. Calculations suggest that a reasonable estimate of the thermal response time of the mixed layer (approximately 70 m) of the global ocean is around 6–7 years. This time scale is important for two reasons. First, it is most unlikely that any perturbation which operates on time scales significantly shorter than 6–7 years can cause a significant climatic response; these short period fluctuations will be damped and finally smoothed by the thermal inertia of the oceans. Second, model simulations must cover at least this time period so that the mixed layer of the oceans can come to equilibrium. In fact, oceanic transport and deep ocean mixing may cause the necessary time periods of simulation to be increased to the order of 10–20 years.

The advantage of energy balance models is their simplicity of formulation. Their applicability, like that of all climate models, is restricted by the limited way in which such straightforward sets of equations can capture physical reality. We can compensate for the alteration of the solar constant by varying any or all of the parameters in the model so that a new equilibrium climate can be established by changes in the albedo, infrared emissivity or the heat transport. However, the assumptions made about the nature of these changes can affect the equilibrium climate predicted. For example, a survey of recent literature concerning EBMs shows that the change in solar constant necessary to create a totally glaciated Earth varies from 4 to 21%, depending upon the albedo and long-wave parameterisations used in a particular model.

## 13.2 Global-scale climate models

It is theoretically possible to continue to refine the energy balance models considered above by adding consideration of longitudinal and altitudinal variations. However, such an approach precludes explicit consideration of other climatic parameters, such as winds, clouds or precipitation. Hence, while it is useful to retain the concept of continued refinement, it is advantageous to change the focus somewhat and consider the global-scale climate models which are designed to simulate the whole climate system (Figure 1.1).

### 13.2.1 General circulation climate models

The general circulation models (GCMs) treat the full three-dimensional nature of the atmosphere and the full range of climatic elements and processes. This requires the solution of a series of equations that describe the movement of energy and momentum and the conservation of mass and water vapour (Table 13.1). Physical processes, such as cloud formation and heat and moisture transports within the atmosphere and between the ground and air, must be included. Since our present understanding of climate dynamics dictates that temperature must be the element of prime concern, these equations are solved for the **wind field** as a function of **temperature**. The first step in obtaining a solution is to specify the atmospheric conditions at a number of grid points, obtained by dividing the Earth's surface into a regular grid. Conditions are specified at each grid point for the surface and several layers in the atmosphere. The equations (Table 13.1) are then solved at each grid point using numerical techniques. Various techniques are available, but all use a time-step approach and an interpolation scheme between grid points. Once the new conditions are established, the process is repeated, such repetition continuing until the final target time is reached. It is common to treat that target as a new equilibrium climate, as a response to a prescribed change in one of the climate forcing functions. Here we are primarily concerned with the simulation of the current climate. Some results associated with a changed climate are discussed in Chapter 14.

*Table 13.1* Fundamental equations solved in GCMs

*Conservation of momentum* (Newton's second law of motion) in three orthogonal directions:
Force = mass × acceleration
$F = ma$

*Conservation of mass* (continuity equation):
Sum of the gradient of ($\rho \mathbf{v}$) in three orthogonal directions is zero (i.e. matter may not be created or destroyed)

*Conservation of energy* (first law of thermodynamics):
Input energy = increase in internal energy + work done

*Ideal gas law* (approximate equation of state):
(pressure × volume)/absolute temperature = gas constant
$pV/T = R$

The processes usually incorporated into general circulation models are shown in Figure 13.5. The basic radiation streams, and resultant energetics, provide the driving force. This creates the pressure field which drives the wind field. Calculations are made for various layers, and vertical motions between layers are determined. Commonly there are several near-surface layers because of the need to specify as closely as possible the interaction between the surface and the overlying atmosphere. Similarly, the underlying

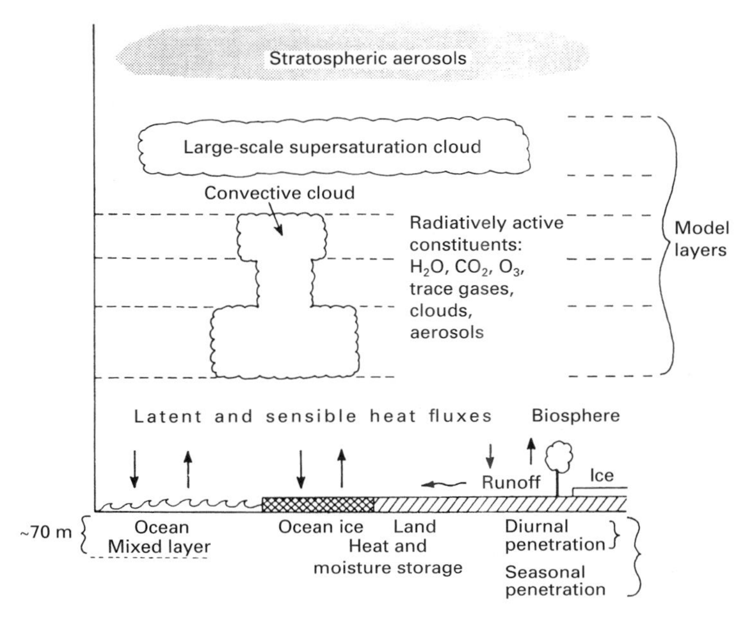


*Figure 13.5* Schematic representation of the subgrid-scale structure of a general circulation climate model. In this GCM two types of cloud are treated: grid-scale supersaturated clouds and subgrid-scale convective clouds. The surface interaction with the atmosphere is fairly complex and heat and moisture storage within the ground are treated by the use of a two-layer ground model. (From Hansen *et al.*, 1983.)

surface, whether land, ice or ocean, is layered to mimic the real situation. Two layers are shown here.

The models must simulate the cloud formation processes, and commonly incorporate two types of cloud into each grid cell. One, representing convective processes, is developed from energy exchange considerations. In essence, a temporal sequence of environmental lapse rates (Figure 4.10) is developed from a series of detailed calculations of radiative exchanges in the various atmospheric layers. The stability, and any resulting vertical motions and condensation, are determined at each time step. Any cloud formed, of course, interacts with the radiation streams to modify the stability, and eventually to re-establish stability and evaporate the cloud droplets. This sequence of steps represents a simple 'radiative–convective' climate model, now usually incorporated as part of a more complex GCM. However, when used in isolation, it clearly shows the number of uncertainties and assumptions involved in modelling clouds, and the resultant impact on temperature estimates (Table 13.2). This type of model was also

*Table 13.2* Equilibrium surface temperature increase due to doubled $CO_2$ (300–600 ppmv): results from a suite of 1-D RC model sensitivity experiments

| Model | Description | $\Delta T$ (K) | Feedback factor |
|---|---|---|---|
| 1 | Fixed absolute humidity, 6.5 K km$^{-1}$ Fixed cloud altitude | 1.22 | 1 |
| 2 | Fixed relative humidity, 6.5 K km$^{-1}$ Fixed cloud altitude | 1.94 | 1.6 |
| 3 | Same as 2, except moist adiabatic lapse rate replaces 6.5 K km$^{-1}$ | 1.37 | 0.7 |
| 4 | Same as 2, except fixed cloud temperature replaces fixed cloud altitude | 2.78 | 1.4 |

*Note*: Model 1 has no feedbacks affecting the atmosphere's radiative properties. The feedback factor specifies the effect of each added process on model sensitivity to doubled $CO_2$.

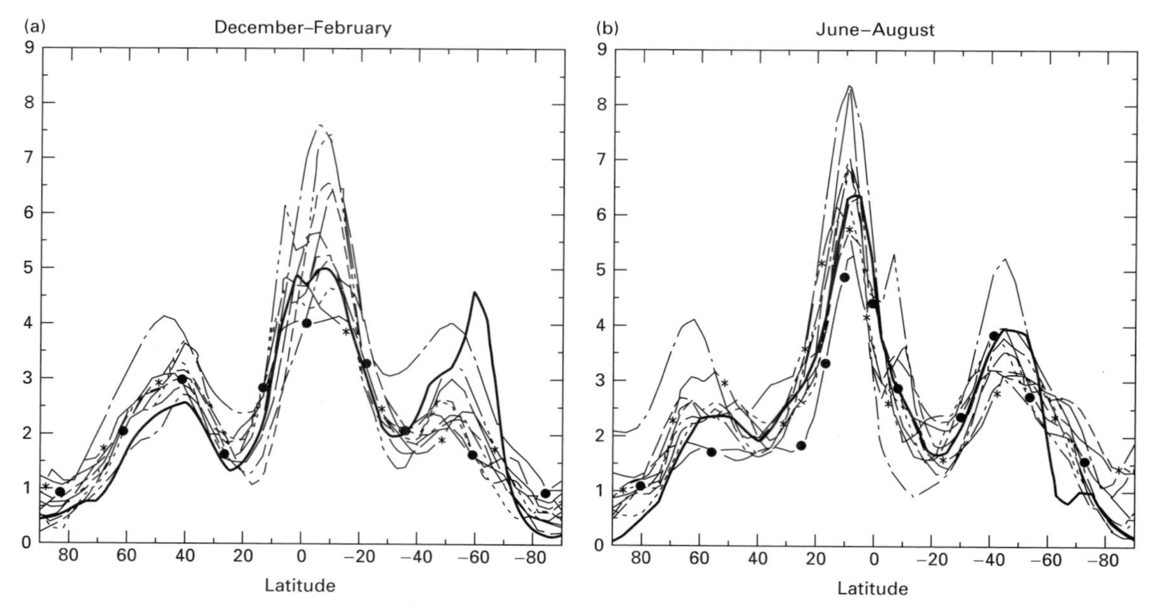


*Figure 13.6* The zonally average precipitation rate (mm $day^{-1}$) as observed (thick line) (Jaeger, 1976) and as determined by 11 GCMs (thin lines), for (a) December–February and (b) June–August. (From Gates *et al.*, 1996.)

used to investigate the role of volcanic activity on temperature (Figure 8.12). For precipitation estimation another step is needed. The creation of a process model, which translates cloud formation rates and moisture amounts to precipitation that arrives at the ground is proving to be very difficult. In most cases statistical relationships based on observations of the probability of a cloud giving precipitation, are used. Partly as a consequence, although most GCMs simulate the latitudinal pattern of precipitation rather well, estimates of actual values vary widely (Figure 13.6).

The other cloud modelled in GCMs (Figure 13.5) is horizontal in nature and is usually derived in height and location from the horizontal flow patterns which create convergence and vertical motions. Hence this cloud formation depends on accurate specification of the evolution of the wind field and any associated stability patterns. The wind field itself is calculated using the basic laws of motion. Indeed, in this area the climate models come close to operating in the same way as the models used to generate the daily weather forecasts. Although considerations of time and space scales (Section 13.3) prevent precise duplication, this similarity is important in at least one practical way. The daily models can be evaluated on a daily basis, rather than on the decadal scale, or use of palaeoclimatic information, common for climate models. Hence it is possible to run, test, refine, rerun and retest in a much more rapid cycle. Progress should be equally rapid. Indeed, this linkage has proved extremely beneficial for the continued development of GCMs.

General circulation models formulated in this way have the potential to approach the real atmospheric situation very closely. The IPCC evaluated the results of 11 models (as in Figure 13.6) and concluded that global temperature and precipitation averages were well simulated on a seasonal basis (Figure 13.7). In some cases, regional results were also realistic. With the great emphasis on atmospheric energetics, it is not surprising that this is particularly true for temperature (Plate 6). The models, on average, are too cold at high latitudes in the cold season and, more generally, over land areas (Plate 6a,b). The zonally averaged results for individual models (Plate 6c,d) indicate mid-latitude and tropical simulations within ±4 °C. Temperatures are rather poorly reproduced at high latitudes, but the small actual area, the sensitivity of results to assumptions about albedo, and the role of the polar cell in the general circulation, may create misleading impressions of the significance of the uncertainty.

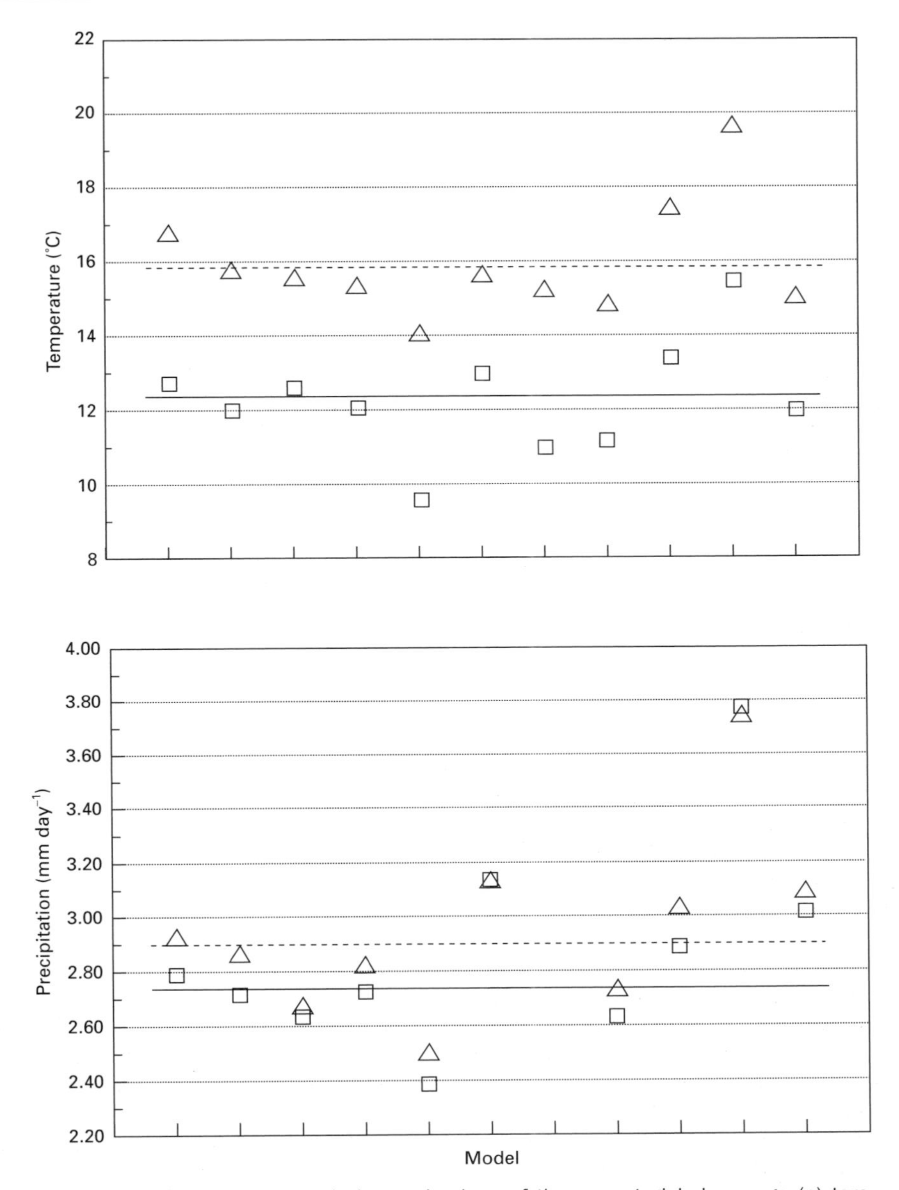


*Figure 13.7* Model estimates and observed values of the current global average (a) temperature (°C) and (b) precipitation rate (mm day$^{-1}$). Triangles indicate model estimates, and the dotted line the observations, for June–August. The squares and solid line represent December–February conditions. (After Gates *et al.*, 1996.)

There are a number of practical and theoretical limitations on GCM development. The prime practical consideration is the time needed for the calculations. To compute each of the basic atmospheric parameters at each grid point requires that roughly $10^5$ numbers be retrieved, calculated and re-stored at each time step. This places a strain on the resources of even the largest and fastest computers. Since the accuracy of the model partly depends on the spatial resolution of the grid points and the length of the time step, a compromise must be made between the resolution desired and the computation facilities available. At present, the grid points are usually spaced in the order of a few degrees of latitude and longitude apart and time steps of a few minutes are used. Vertical resolution is obtained by dividing the atmosphere into about ten levels, commonly with two layers below the surface. Each grid cell usually has a homogeneous surface type. There are, in addition, theoretical problems associated with increased resolution, and these will be considered in Section 13.3.

### 13.2.2 Climatic feedback effects

Up to this point we have effectively assumed that the perturbation of a single parameter, whether external or internal, causes a straightforward alteration in the element of interest. We have failed to consider interactions internal to the climate system, the **feedback** mechanisms. These can operate either in a positive sense so as to modify the monitored parameter further in the same direction as the perturbation, or in a negative sense acting to modify it in the sense opposite to the initial perturbation (Figure 13.8a). Feedbacks can occur anywhere in the climate system. Some selected examples follow.

If some external or internal perturbation acts to decrease the surface temperature then this is likely to lead to the formation of additional snow and ice masses. The surface, and probably the planetary, albedo increases and a greater amount of solar radiation is reflected away from the planet, causing temperatures to decrease further. This further decrease in temperature leads to more snow and ice and so on. This positive feedback mechanism is known as the **ice–albedo feedback mechanism**. Of course, the ice–albedo feedback mechanism is positive in the other direction as well. With higher temperatures, a portion of the permanent cryosphere may be removed, reducing the albedo and leading to further enhancement of temperatures.

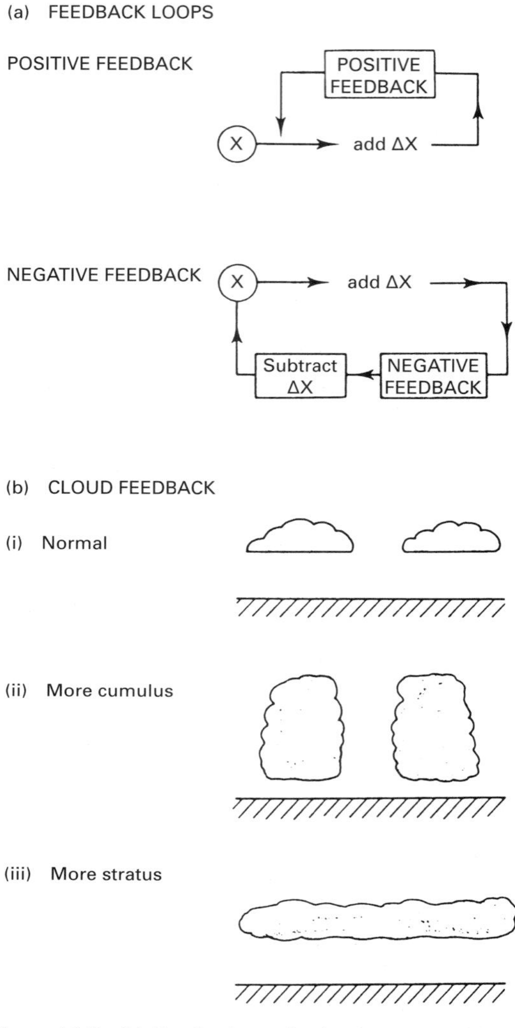


*Figure 13.8* (a) Feedback mechanisms operate either in a positive sense to enhance the perturbation further or in a negative sense to damp it. (b) Cloud feedbacks are complex. A particular difficulty is whether increased cloudiness will manifest itself as (ii) increased cumuliform clouds or (iii) increased stratiform clouds. The surface area covered and hence the planetary albedo differs significantly in these two cases.

A second, but less dramatic, positive feedback mechanism occurs with the increase of atmospheric water vapour from evaporation as temperatures rise. The additional greenhouse effect of this vapour further enhances the temperature increase. Again, if temperatures fall, this positive feedback operates in the opposite direction.

It is difficult to make general statements about the direction of the feedback associated with clouds since they are both highly reflective (thus contributing to the albedo) and composed of water and water vapour (thus contributing to the greenhouse effect). For low and middle level clouds the albedo effect will usually dominate over the greenhouse effect, providing a **negative feedback**, so that increased cloudiness will result in an overall cooling. On the other hand, cirrus clouds, which are optically thin, have a smaller impact upon the albedo so that their overall effect is to enhance the greenhouse mechanism. Thus cirrus tends to lead to warming.

Unfortunately cloud feedback is not this straightforward. There are two additional problems. First of all, it is uncertain whether increased temperatures will lead to greater or less total cloud cover. Increased temperatures will cause higher rates of evaporation and hence make more water vapour available for cloud formation, but it is unclear whether any additional clouds will be of cumuliform or stratiform type. Even for the same volume of new cloud, an increased dominance of cumuliform clouds would lower the percentage of the surface covered by clouds. Stratiform clouds would increase the area covered (Figure 13.8b). Thus increased temperatures, if they led to increased cumuliform cloudiness, could lead to a positive cloudiness feedback. The second uncertain factor about cloud changes as a feedback response to a climate perturbation concerns their level of formation, as demonstrated by Table 13.2.

The example of cloud feedbacks indicates that more than one feedback effect is likely to operate within the climate system in response to any perturbation. These feedback effects combine in a complex way. Consider a system in which a change of surface temperature of magnitude $\Delta T$ is effected. If there are no internal feedbacks then this temperature increment is equal to the change in the equilibrium temperature, $\Delta T_{eq}$. When feedbacks occur there will be an additional temperature change due, for example, to an increase in the atmospheric water vapour content, and the value of $\Delta T_{eq}$ will be:

$$\Delta T_{eq} = \Delta T + \Delta T_{feedbacks} \quad (13.11)$$

where $\Delta T_{feedbacks}$ can be either positive or negative. This equation can be rewritten, algebraically, as:

$$\Delta T_{eq} = f\Delta T \quad (13.12)$$

This **feedback factor**, $f$, can be related to the amplification or **gain**, $g$, of the system, which is defined, using the analogy of gain in an electronic system, by:

$$f = 1/(1 - g) \quad (13.13)$$

The most important result of this analysis occurs when more than one feedback is considered. It is easily shown that the gains are additive, i.e. the total gain, $g$, can be expressed as the sum of the individual gains, $g_i$, derived for each of $i$ feedback processes. Thus:

$$g = \Sigma g_i \quad (13.14)$$

The overall feedback factor, $f$, must be calculated using equation (13.13) and substituting for $g$ from equation (13.14). This gives, for two feedbacks:

$$f = (f_1 f_2)/(f_1 + f_2 - f_1 f_2) \quad (13.15)$$

In other words the feedback factors are neither simply additive nor multiplicative. Consider a feedback with $f_1 = 1.5$, operating alone. This would result in a 50% increase in the response (equation (13.12)). If a second feedback now operates with $f_2 = 2.0$, by substituting into equation (13.15) it is seen that the overall feedback factor is 6.0. Hence an additional feedback may cause a significant increase in the response.

The importance of the combination of many feedback factors can best be illustrated with an example. Suppose, as seems reasonable from published model results, that the change in surface temperature predicted by a climate model without feedbacks after a doubling of atmospheric $CO_2$ is $\Delta T = 1.2$ K. If a snow and ice feedback factor, $f_{ice} \approx 1.2$, and a water vapour feedback factor, $f_{water\ vapour} \approx 1.6$, are incorporated, then from equation (13.15), the joint value of $f \approx 2.18$. Hence:

$$\Delta T_{eq} = 2.18 \times 1.2 = 2.6\text{ K} \quad (13.16)$$

This result is in general agreement with the evaluations of the possible effect of doubling $CO_2$ when cloudiness changes are neglected. Suppose now that two cloudiness feedback factors, both positive, are discovered, where the effect of increased temperatures leads to increased cumuliform cloud and thus to (1) decreased cloud cover and (2) an increased cirrus greenhouse. Suppose, in the absence of any better information, both these cloudiness feedback factors have the same magnitude as $f_{ice}$. Then just one of them (i.e. snow/ice plus water vapour plus one cloud feedback) leads to a value of $\Delta T_{eq} = 4.1$ K. With the extra cloud feedback the overall feedback factor

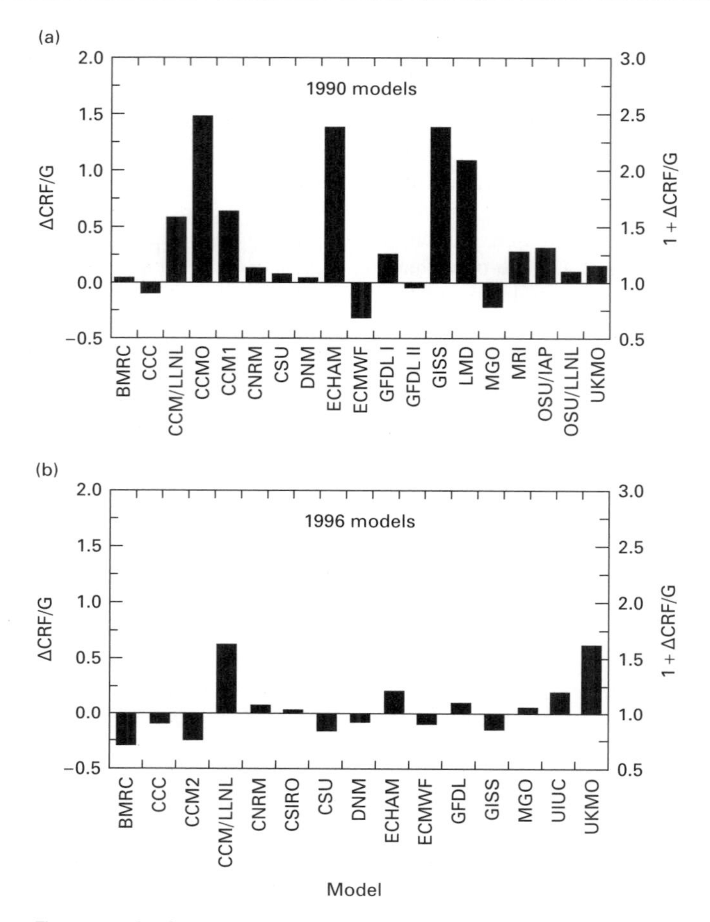


*Figure 13.9* Cloud feedback parameters (cloud radiative forcing/overall energy change) for 19 GCM models as assessed in (a) 1990 (Cess *et al.*, 1990) and (b) 1996 (Cess *et al.*, 1996). The results suggest that the feedback is much smaller than originally estimated. In particular, no models now suggest a large positive forcing. (After Dickinson *et al.*, 1996.)

is $f_{tot} \approx 8.00$ and thus there is a total temperature change of:

$$\Delta T_{eq} = 8.00 \times 1.2 = 9.6 \text{ K} \qquad (13.17)$$

Of course it is just as likely that some or all of the cloudiness feedback effects could act in a negative sense, but the power of including more and more facets of the climate system is seen to be immense.

This analysis indicates that it is important to establish the magnitude of the feedback factors of the climate system, as well as any changes that occur over time. Rapid advances are being made, but values are still highly uncertain (Figure 13.9), and the nature of the various feedback mechanisms remains one of the least understood features of the climate system.

## 13.3 Regional and local climate models

Atmospheric modelling is by no means confined to the global scale or to specifically climatic processes.

Active development of models on a variety of scales, from small-scale short-term mesoscale ones exploring tornado dynamics through synoptic models for daily weather forecasts to those simulating climate processes on a regional scale, is under way. It is increasingly clear that linking models, so that each model operates optimally on its own most appropriate time and space scales, is highly advantageous. At present for climate modelling, the major linkage is **nesting**, whereby a regional model is driven by a global model (Figure 13.10). Figure 8.12 shows the practical strength of this approach, since the result could not readily be obtained by a GCM operating alone. The two major reasons for this are considered in the following sections.

### 13.3.1 Time and space considerations

Use of any particular model resolution implies that the active processes are known on that scale, and that observations match the scale. In almost all models, conditions are calculated at the grid points, with conditions between them, and between time steps, being interpolated. Some 'smoothing' always occurs. With a coarse grid spacing, small-scale atmospheric motions, such as thundercloud formation, cannot be modelled, however important they may be for atmospheric dynamics. Similarly, with a coarse time step, individual day-to-day events cannot be considered.

Global-scale models are usually concerned with long-term average conditions. It is common, for example, to develop models which output monthly average values. These are similar to the monthly normals associated with the observational record, and should not be interpreted as estimates for any particular single month. It is always implied that there will be variability about the mean in the same way as there will be spatial variability about the grid-cell average. The nature of either variability is not specified by the model. In particular, there is no necessary reason why the variability in the 'new' climate should be the same as that in the old one. The model cannot resolve detail at that level. Similarly, there may be differences between the old and new climate in the day-to-day variability within the month. That cannot be resolved either. Note that the actual time step used in the calculation may be in the order of a day. In some cases it is possible to determine the distribution of daily values about the monthly mean for the old and new climates. This may be useful information, but it cannot be interpreted as a day-by-day time series of values, and must be used with much less confidence than the monthly values.

Many models need actual observed values for part of their input and all require observational data with which to compare the results. Some parameters, such as surface temperature, are readily available world-wide. Others, however, are sparse in either time or space. We only have a short record of sea-ice extent, for example, so that models which consider the ice–albedo feedback are constrained by lack of data. Similarly, global coverage for cloud observations is relatively recent (e.g. Plate 3), so that it is difficult to initialise models which require the present cloud distribution as an input. On the other hand, stations with long humidity records are sparse, so that grid cells would have to be large, or much spatial extrapolation used, if humidity observations are required by any model.

### 13.3.2 Incorporating interactions with the surface

More accurate representations of local areas demand more accurate specification of the local conditions and the processes acting therein. While this is true for all aspects of the climate, most pertinent here is accurate specification at the surface. This involves both the incorporation of the detail of the interface itself and the pertinent time scales of the processes on either side of that interface.

The interaction between the surface and the near-surface layer of the atmosphere must be parameterised in any model. For many GCMs it has been found that a 'two layer' land surface is adequate (Figure 13.5). The surface fluxes of momentum, sensible heat and moisture are taken to be proportional to the product of the surface wind speed and the gradient of the property away from the surface, and diurnal and seasonal heat and moisture storage and exchange can be simulated. This presupposes a rather simple surface, however, and greater refinement is needed for incorporation into regional models.

Several schemes have been proposed. One example (Figure 13.11) treats the moisture and sensible heat flows in the resistance framework mentioned in Section 10.1. The vegetation itself can have several layers, which may change configuration and radiative properties with the seasons. Thus the energy balance involves not a simple single horizontal surface, but several layers each interchanging energy.

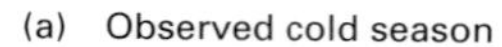

(a) Observed cold season

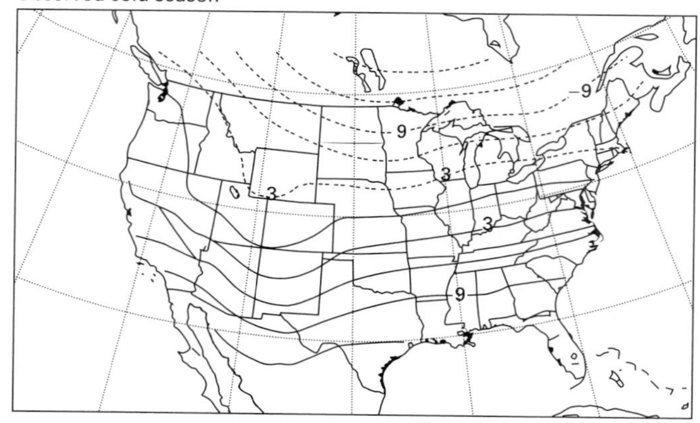


(b) Nested MM4 cold season

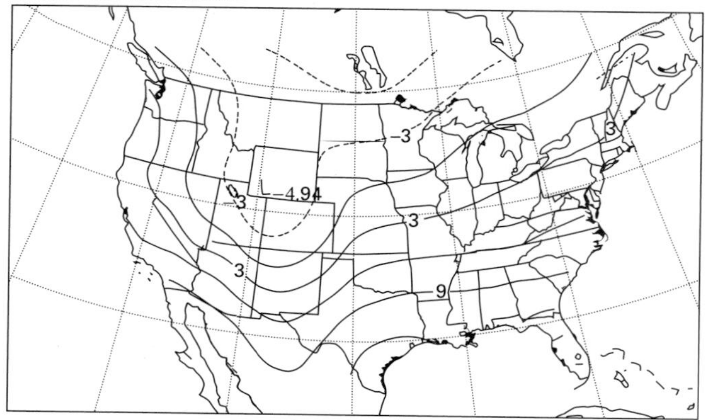


(c) Driving CCM cold season

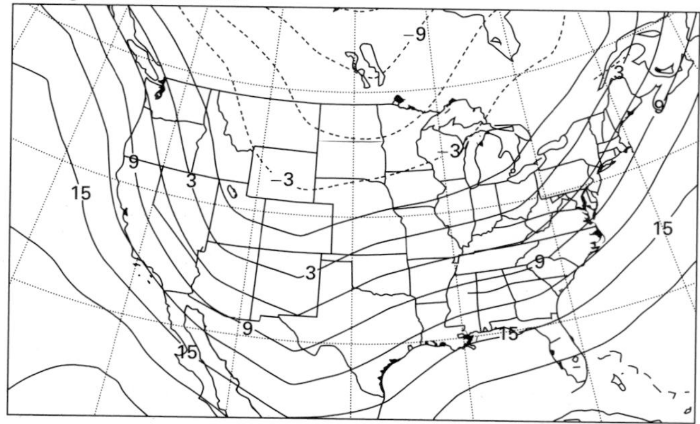


*Figure 13.10* Long-term average cold-season temperatures (°C) for the United States as determined (a) from observations, (b) using a general circulation model with a resolution of 7.5° latitude × 4.5° longitude (approximately 800 × 400 km), and (c) using a regional model with grid-point spacing 60 × 60 km nested within the GCM. Note the smoother, more detailed, and more realistic pattern of the nested model results. (From Giorgi *et al.*, 1994.)

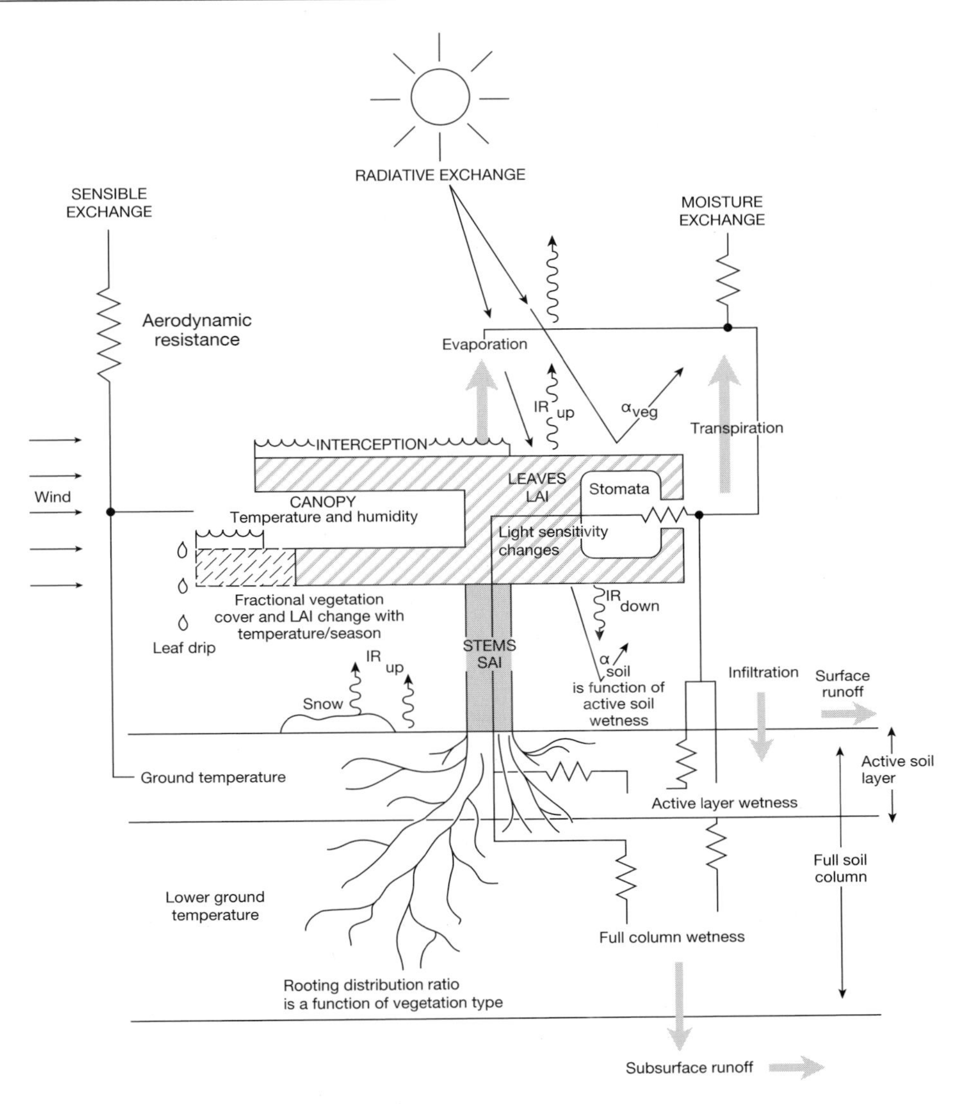


*Figure 13.11* The major features incorporated into a scheme for interaction between the land surface and the atmosphere suitable for incorporation into a regional climate model. LAI is the leaf area index, SAI the stem area index and IR stands for infrared. This scheme parameterises many of the processes discussed in Chapter 10. (From Wilson *et al.*, 1987.)

The low level leaves, for example, may be protected from high daytime or low night-time temperatures by the layer above. This in turn influences the surface energy, evaporation and water infiltration. The complex canopy also influences the wind flow. Below the surface the root structure is better modelled than in the two-layer model, so that the water movement and storage is more realistic.

The corresponding interaction between the ocean and the atmosphere can be considered in comparable ways. Rather simple is the 'swamp' model, where the ocean acts only as an unlimited source of moisture. The simplest models use fixed ocean temperatures based on observed average values. However, the oceans are, in effect, the thermostat of the climate, so it is very difficult to disturb the climate away

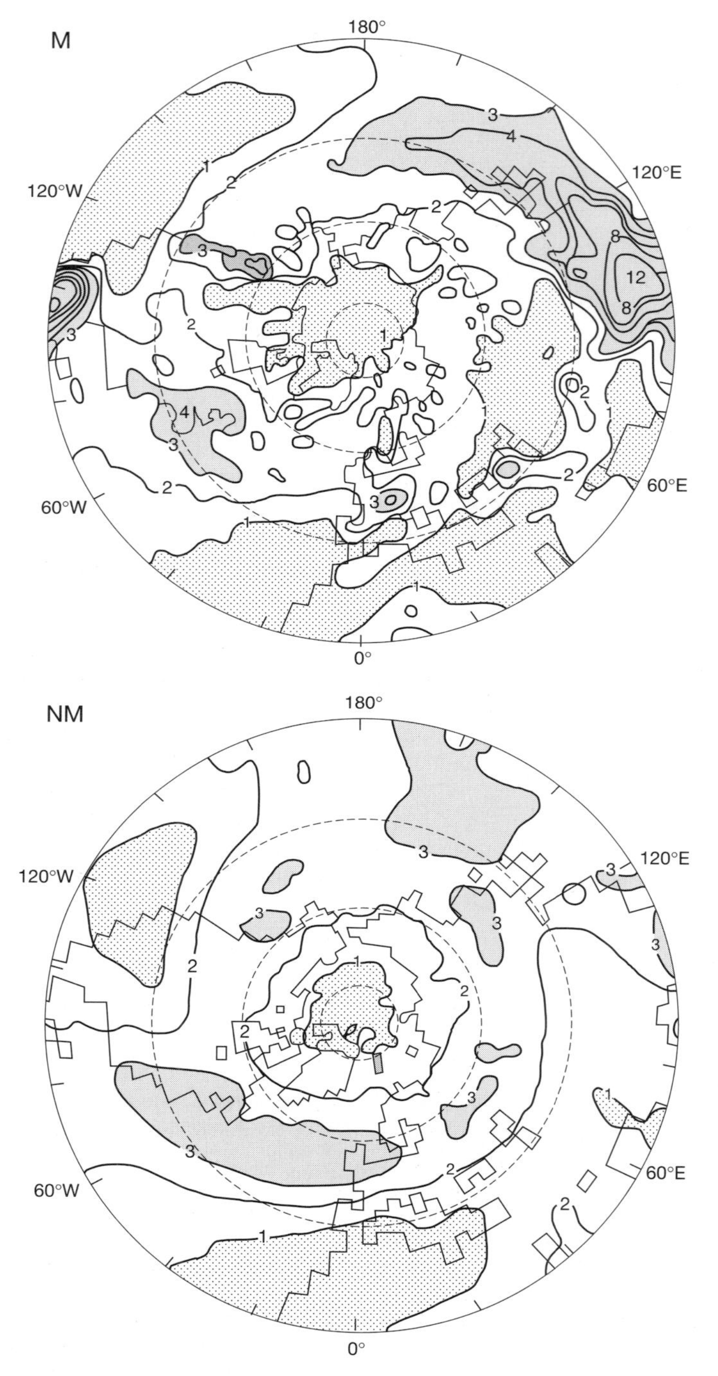


*Figure 13.12* Model simulation of the precipitation rate in the Northern Hemisphere using a general circulation model (a) with and (b) without the presence of major topographic barriers. (a) simulates the main features of the current climate, with (b) indicating much more restricted dry areas west of the Rockies in North America and north of the Alps–Himalayas in Eurasia. Low polar values occur in both simulations. (From Manabe and Broccoli, 1990.)

from the present-day configuration, while sea-surface temperatures are specified in an unchanging manner each month and year after year. A more sophisticated approach is to compute the heat storage of the mixed layer of the ocean and deduce from this the mixed layer depth and the sea-surface temperature. Often the maximum depth of the mixed layer is prescribed to be around 70–100 m, allowing the lower deep ocean layer to act only as an infinite source and sink for water. More recently developments which allow variation in mixed layer depth have been incorporated.

Oceans also perform the very important duty of advection of heat around the globe. Progress in simulations of the oceanic circulation has been closely related to the evolution of atmospheric models. Although the processes are similar, close coupling in a way similar to that for land has proved difficult. Momentum transfer, considered primarily as a feature of the wind profile over land surface, becomes significant over the oceans. Momentum can be transferred in either direction. Hence there is an extra direct link to consider. Further, the transports and temperature changes in the oceans take very much longer to accomplish than those in the atmosphere. Thus even if satisfactory oceanic circulation models are developed, there remains the problem of linking an atmospheric model with a time step (i.e. scale) of around 20–30 minutes (or less) to an oceanic model that needs between 20 and 30 years to come to equilibrium.

### 13.3.3 The use of climate models

The fundamental use of a climate model is to enhance our understanding of climatic processes. Once the model realistically simulates the current climate and its variability, or any pertinent facet thereof, it can be used in an experimental mode to explore the impact of selected imposed changes. We have incorporated the results of such experiments, whether explicitly as in our discussion of reforestation in the Sahel (Figure 8.12), or implicitly when considering the mechanisms of climate change in Chapter 12. One further example is given here (Figure 13.12). We stated in Chapter 9 that the presence of the Rocky Mountains plays a vital role in the creation of the North American climatic regions, and that the topographic barrier is a prime cause of differences within Eurasia. The model simulation supports this, but emphasises that the dry interior climates of North America are a response to the subsidence induced by the southward-moving Rossby wave flow, rather than direct descending motions off the mountains themselves.

Climate models are increasingly being used for practical applications. The model illustrated in Figure 13.11, for example, is being developed partly as a tool for the assessment of water supplies in the western USA. Model use for climate forecasting will be considered in the next chapter. The most visible role of climate models, quantification of the potential effects of greenhouse gas increases in global climate, will also be treated in the next chapter.

**Box 13.A Spatial models for applications and impacts**

Most of the applications we have considered so far have been of the type which required use of data from a single observing station, or where observations from stations in a network could be combined in a relatively simple way to provide results. This has in some cases meant that we have greatly simplified the real problem, and here we need to consider in more detail how we can develop models which allow us to use a network of observations to provide more realistic and useful spatial information. Commonly this requires us to use a combination of various types of model. So we will consider model types prior to giving a specific example involving precipitation distribution.

*13.A.1 Combining model types*

When we consider that a model is essentially a means of simplifying reality for some useful purpose, there is a large range of possible model types. Consequently there is also a large range of possible model classification schemes. Here we will just divide models into two classes: physical (mathematical) and statistical. The main part of this chapter has considered physical models. Indeed, the whole thrust of climate modelling is to use what we know of climate processes to simulate the climate. As far as possible, minimal information other than the physical laws, and the observations necessary to initialise the model, should be used. This could be considered as a 'pure' physical model. The opposite extreme would be a 'pure' statistical model, where prediction of a particular climatic feature or event is achieved statistically without any regard to atmospheric processes. In practice, neither extreme is likely to provide useful models. Almost all models we use contain elements of both.

**Box 13.A (cont'd)**

A physical climate model must contain some statistical information. The near-surface turbulent wind field, for example, is vital in maintaining sensible and latent heat fluxes. Micrometeorological investigations may examine the nature of this turbulence, but for climatic purposes we are usually content to accept the structure of the field as deduced by these investigations and use the resulting statistical summaries. Further, although nesting more and more sophisticated models may appear to be working towards a pure physical model, this is unlikely to be the case. It is commonly advantageous to use a physical model somewhat simpler than theoretically possible, using statistical approaches for processes outside our immediate interest, in order to emphasise the particular elements of interest. The radiative convective models of Figure 8.12 and Table 13.2 are examples. Indeed, this kind of approach is frequently used when developing application models.

Many of our application models emphasise statistical relationships. Examples include the regression model for energy demand calculations and the development of the physiological perception component of the comfort indices. Indeed, at times it may seem that such statistical approaches are the only way forward. This may be most acute when considering climate impacts, when the climate is only one of many factors. Even then, however, progress in understanding and prediction is most likely to be made if there is some physical connection between the element being considered and the impact. Thus it is clear, if perhaps trivial, to emphasise temperature rather than wind speed when considering the post-glacial migration of tree species.

We have tended to treat our application models in relatively simplistic terms, emphasising the application techniques rather than the modelling strategy. However, in most ways the applications models themselves can be as complex as the climate models. Certainly the system under consideration is often complex, so that there is a need to identify and parameterise the important active processes, incorporate feedback mechanisms where needed, calibrate the models for particular conditions and localities, and test their predictions using real data.

*13.A.2 Precipitation in mountainous areas*

The linking of statistical and physical approaches, and the potential complexity of the system being studied, can be exemplified by consideration of precipitation in mountainous terrain. We confine ourselves at the moment to techniques for the interpolation of surface-based observational (rain gauge) data.

In most mountainous areas the observation sites are concentrated in valleys, where observers live. From what we know of orographic precipitation and rain shadow effects (Section 4.4) these valley sites are not representative of the whole mountainous area. Indeed, they are rarely representative of an area more than a few metres from the gauge itself. Even if we assumed flat terrain and drew isohyetal maps, the patterns would be complex and unreliable (e.g. Figure 9.I.1). Thus the approach often used for temperature, assuming an underlying uniform trend with topographically forced deviations superimposed, is not easily applied. Nevertheless, the most common strategy is to use any available observational data to develop statistical interpolation procedures which, as far as possible, incorporate elevation and exposure. Various forms of **kriging** are often used (Figure 13.A.1a). A possible refinement is to incorporate the effect of wind direction on exposure and thus on the altitude relationship. However, determination of wind direction in these regions, where channelling effects are paramount, is itself very difficult.

Recent advances in computational techniques, along with digital elevation models and Geographic Information Systems, make it possible to combine a physically based rainfall distribution model and more statistically based interpolation techniques to interpolate between data points (Figure 13.A.1b). These methods suggest that it should be possible to create realistic grid point precipitation estimates for monthly totals, so that the potential exists to input the observed surface field directly into atmospheric models.

A major concern when considering climatic elements such as precipitation is the way in which the spatial distribution varies as a function of observation period. In general, the longer the period for which the precipitation is totalled, the smoother is the spatial field. It is much easier, and commonly much more reliable, to interpolate on a smooth field than on a disjunct one. It is common to assume that in mid-latitudes either annual totals or monthly normals are smooth fields, often in much the same way that daily temperatures offer generally smooth fields. For shorter periods, this is not the case, and it is not clear whether it holds for tropical conditions. Certainly this lack of spatial conservatism is one reason why it is difficult to make many meaningful statements about the small-scale, short-term distribution of rainfall.

**Box 13.A (cont'd)**

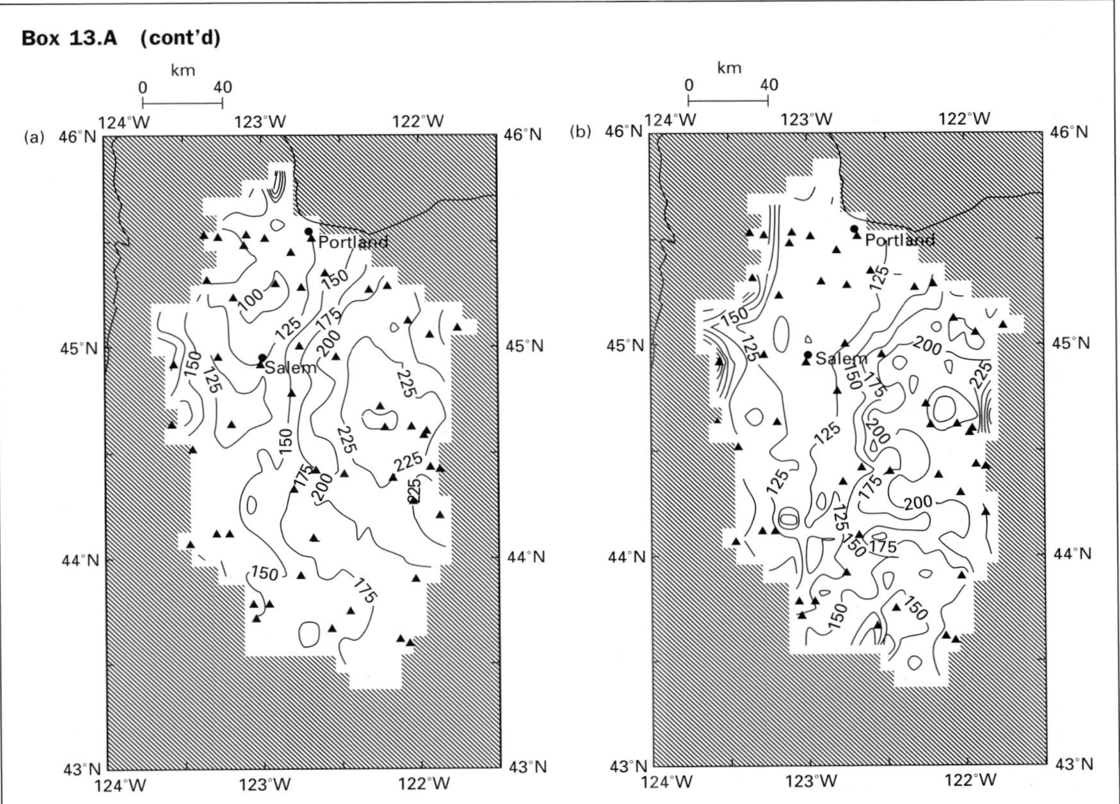


*Figure 13.A.1* Estimates of average annual precipitation in a domain with elevation ranging from below 250 m in the west to over 1250 m in the east, and having 52 precipitation observing stations, using (a) the standard geostatistical technique of detrended kriging and (b) a model using climatological and statistical components (PRISM). (From Daly *et al.*, 1994.)

**Box 13.I Linking models and observations**

We have noted in several places that there must be an intimate connection between models and observations. Almost all models require some observational information in their initialisation, and all require data to compare model projections with reality. Here we consider an example where a model development and pertinent observations were closely linked.

*13.I.1 Walking in the rain*

The applications models considered previously tended to require many high quality data, complex and refined models, and sophisticated analysis techniques. Often this is the end result of an investigation which started out in a much simpler way. We examine one such case here.

The fundamental question is whether it is better to run to shelter when it starts raining rather than to walk. This may not sound too climatological initially, but as the analysis proceeds, climatological implications appear. The initial analysis treats a person as a cuboid who gets wet by (a) water falling on its head, and (b) by sweeping out a volume of raindrops as it moves forward. Assuming that this cuboid person has a top surface area of $T$, a forward-facing cross-sectional area of $S$, travels a distance, $l$, at a speed, $v$, to shelter from a rain having a rate, $R$, composed of

**Box 13.I (cont'd)**

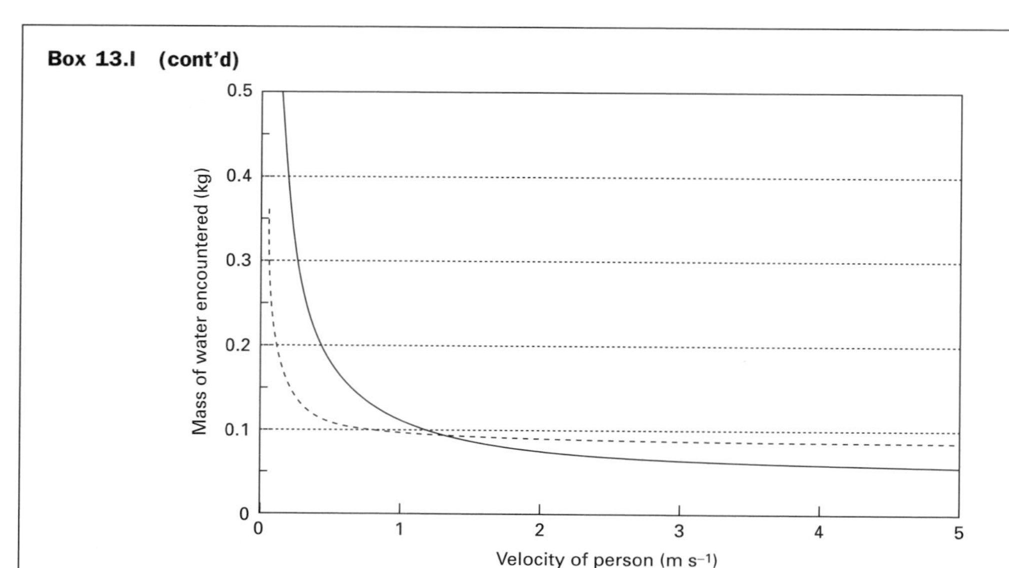


*Figure 13.I.1* The mass of water encountered as a function of travel velocity for a 100 m journey during (a) a heavy convective shower (rate 25 mm $h^{-1}$, drop fall velocity 10 m $s^{-1}$) (solid line) and (b) a slight drizzle (rate 5 mm $h^{-1}$, drop fall velocity 1 m $s^{-1}$) (dashed line) at various travel velocities. (After Peterson and Wallis, 1997.)

drops with fall velocity, $u$, we can use simple geometry to create a model. The mass of water, $M$, encountered on the journey to shelter can be expressed as:

$$M = R\rho(Tl/v) + R\rho(Sl/u)$$

where $\rho$ is the density of water (introduced to ensure that the water mass is related to the volume calculated from the various velocities). The first term on the right represents the amount hitting the head and is dependent on the speed with which the person moves. The second term, independent of forward velocity but dependent on the drop fall velocity, represents that swept out in front of the body. Both, of course, depend on the rain rate. Provided care is taken in ensuring consistency of units, relatively simple calculations provide the output for the simple model (Figure 13.I.1). Two situations are illustrated here, using data from Figure 5.4 and Table 5.1: a rather heavy, although not unusual, downpour, and a gentle drizzle. For a 'caught in the rain' situation, a brisk walk corresponds roughly to 1.5 m $s^{-1}$ and clothed, encumbered running to 4 m $s^{-1}$. For the drizzle, running avoids approximately 7% of the water, but this increases to 33% in the downpour. Note that increased running speed has little effect.

This model demands experimental verification. And indeed, it has been tested by at least one pair of travellers. Equipped with identical water-absorbent clothing, underlain by plastic to prevent interaction with the body, the pair traversed identical courses during a period with a known rain rate. The water encountered was deduced by the weight change in the clothing, and it was found that the runner collected about 40% less water, a somewhat greater difference than was predicted by the model for the ambient conditions they encountered. Naturally, refinement of both the model and the observations was suggested. For the model, incorporation of a runner's increased body tilt, and hence increased top but decreased front surface, inclusion of a formulation for the influence of wind, or the aerodynamic effects of a body moving through the air, are possible, although the complexity of the model, and its data requirement, increases rapidly. For the observations, inclusion of water collected by splash, and the influence of puddles or perspiration, need to be considered. Hence there is room for improvement in model and measurement. Climatologically, however, it does suggest that running ability is a more useful attribute in climates where convection is dominant than in those where stratus is the main precipitation-producing cloud.

# CHAPTER 14 Climate, climate change and the future

Our increasing ability to observe, model and understand the Earth's climate is allowing us to predict that climate for several months in the future. It is also suggesting that human activities are changing the climate on a world-wide scale and in ways which may have adverse impacts for the planet. This concern is encouraging us to explore ways of estimating climatic conditions decades, or even centuries, into the future.

## 14.1 Implications of climate variability and change

Throughout Earth's history, the climate has varied on many time and space scales. In that sense, climatic change is not new, and humanity has had to adapt to a constantly changing climate. Human action has itself had an impact on that climate. Until recently, the impact was local, but now there is concern that human activity, primarily through the emission of greenhouse gases into the atmosphere, is leading to a world-wide climatic change which is more rapid and more extreme than any encountered during human history. The ability of the whole planetary ecosystem, including the human race, to thrive in these changing conditions, is being questioned. In response to these major concerns, the United Nations has established the Intergovernmental Panel on Climate Change, to investigate the likelihood of changes, to assess their potential significance, and to suggest possible responses.

### 14.1.1 The Intergovernmental Panel on Climate Change

The Intergovernmental Panel on Climate Change (IPCC) was established jointly by the World Meteorological Organisation and the United Nations Environment Programme in 1988. The Panel was organised into Working Groups with three different purposes:

1. to assess the available scientific information on climate change;
2. to assess the environmental and socio-economic impacts of climate change; and
3. to formulate response strategies.

Each Working Group aimed to establish the known facts, uncertainties, possibilities and research needs in its area. Formal reports were published which have become major sources of information, some of which are included in this chapter. They are also documents intended to aid national governments in their deliberations about global warming.

Virtually all the material in this book is most closely associated with the activities of Working Group 1. Their aim was to provide the scientific background that would facilitate the creation of scenarios of climatic changes caused by the enhanced concentration of greenhouse gases in the atmosphere. This requires prior knowledge of those concentrations, which depends on many non-climatic factors (Table 14.1).

*Table 14.1* Non-climatic concerns influencing future climates and which must be considered in developing scenarios of future climate

| |
|---|
| Concerns: |
| Rate of global development: |
| greenhouse gas concentration |
| surface modification |
| Direct factors: |
| Level of economic activity |
| Agricultural technology development |
| Fuel types used |
| Government policies |
| Compounding factor: |
| Regional differences |

*Source*: summarised from Houghton *et al*. (1996).

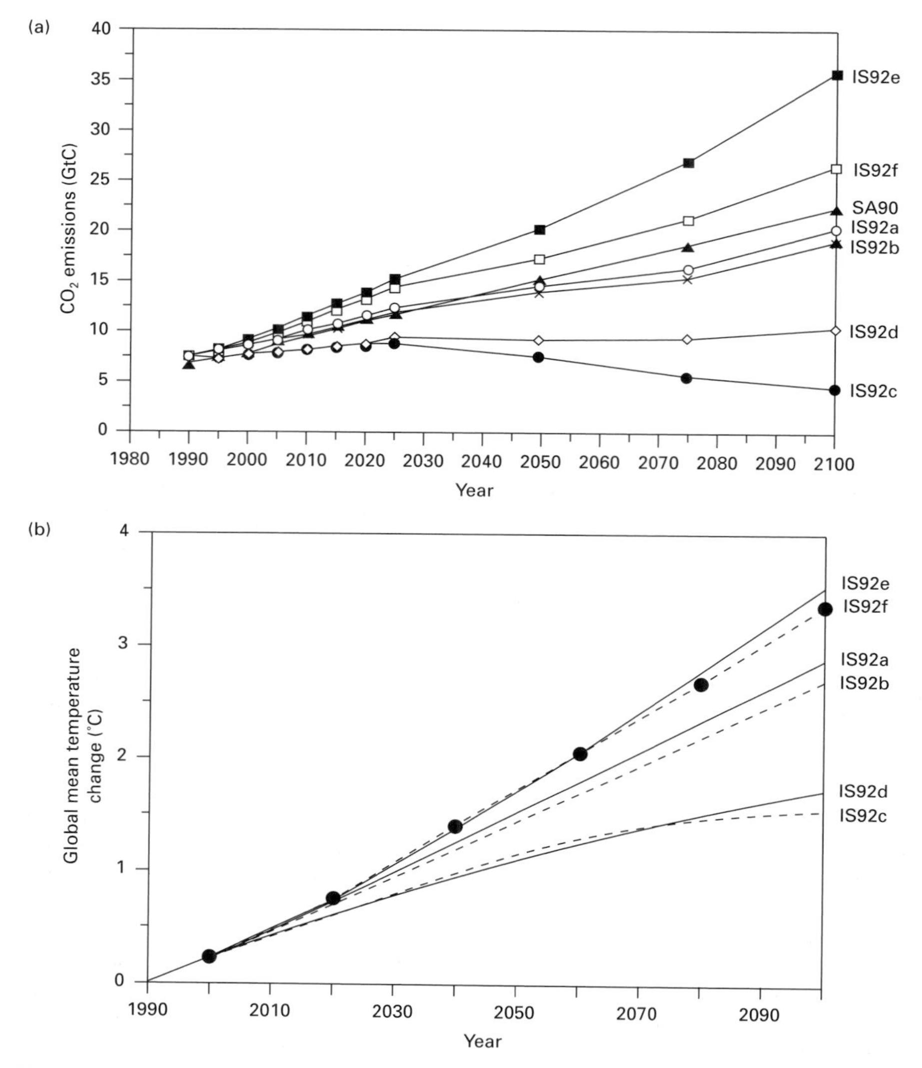


*Figure 14.1* continued on next page

The rate and magnitude of the impact of these factors is uncertain, particularly as one influence on policy may be the reports the Working Group itself produces. Climatically the factors may influence surface type as well as atmospheric $CO_2$ concentrations. Consequently the Working Group developed a suite of scenarios of potential greenhouse gas emissions which were used to postulate future global temperatures using a suite of forecasting methods, clearly indicating the uncertainty throughout the development process (Figure 14.1).

Our discussions of the impact of climate on human activity in Chapter 11 and in most of the applications boxes parallel many of the deliberations of Working Group 2, and we say nothing further here. The development of strategies to respond to the potential effects of global climate change, the focus of Working Group 3, may involve actions to limit greenhouse gas emissions or the adoption of strategies to adapt economic activities to the changed climate and environmental conditions arising from elevated $CO_2$ levels. Such decisions involve political, social, legal and national considerations, which take us far beyond the scope of this book.

The fundamental concept behind the concern for global warming, the greenhouse effect, was discussed

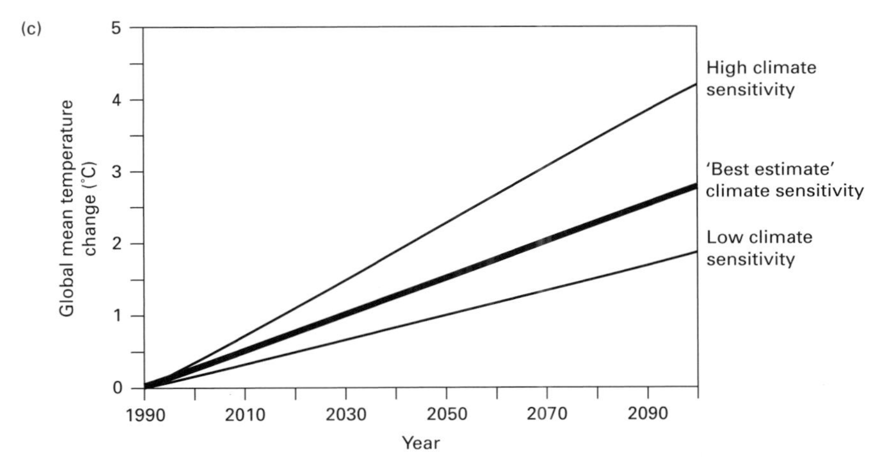


*Figure 14.1* (a) Estimates of annual $CO_2$ emissions for 1990–2100, based on various assumptions about global economic activity; (b) temperature projections for a single GCM based on these economic scenarios, indicating the range of 'non-climatic' uncertainties in future temperatures, and (c) temperature projections for that model and a single emission scenario, incorporating various modelling assumptions to indicate the 'climatic' uncertainty in any projection. (From Houghton *et al.*, 1992.)

in Chapter 2. Here we briefly consider how $CO_2$ and the associated greenhouse gases may get into the atmosphere and what their concentrations may be.

### 14.1.2 Cycling of greenhouse gases

Carbon dioxide is, from the standpoint of global climate change, the major greenhouse gas (Section 2.5). The others, in rough order of current importance, are methane, nitrous oxide, halocarbons (particularly chlorofluorocarbons and carbon tetrachloride), ozone (including precursor species such as carbon monoxide and various oxides of nitrogen) and sulphur-containing gases. Taken together, they are often called the **radiatively active trace gases**, although sometimes $CO_2$ is excluded from this designation since it is hardly a trace gas (Table 1.1). They all interact with radiation at specific wavelengths, so that there is no simple connection between, for example, doubling the concentration or adding a single molecule, and the energy absorbed. Nevertheless, it is common to calibrate the gases so that climate model calculations can be based on an 'effective' carbon dioxide concentration, even if the exact mix of gases is unknown. Note that water vapour is, in absolute terms, the most important greenhouse gas. However, its atmospheric concentration is not changing appreciably as a consequence of human action, only in response to climate changes induced by such actions. Hence it is not considered here.

The list of radiatively active trace gases includes naturally occurring gases which cycle through the atmosphere and commonly have sources and sinks within the land or the oceans. Human activities are increasing the source strengths, but as yet there has not been a natural compensating increase in the strength of the sinks. Some of the greenhouse gases are of human origin, with few or no natural sources, and consequently no existing sinks. The resulting imbalance leads to an increase in atmospheric concentrations. While the perpetual cycling and current imbalance is true of all the greenhouse gases, it is best known for carbon dioxide, which we can use as an example here.

Since the industrial revolution, the atmospheric concentration of $CO_2$ has been increasing as a direct consequence of fossil fuel combustion. Pre-industrial levels were probably around 275 ± 5 ppmv (Figure 14.2a). The record from the observatory atop Mauna Loa in Hawaii, USA, shows a nearly exponential rise in $CO_2$ since accurate recording began 40 years ago, with the mean yearly level increasing from 315 to 360 ppmv in that period (Figure 14.2b) This trend has been calculated to be in close agreement with that expected from the fossil fuel production figures. $CO_2$ is readily mixed throughout the troposphere, so despite the concentrated nature of the major sources in the industrial centres of the developed world, and despite short-term sensitivity to seasonal and other

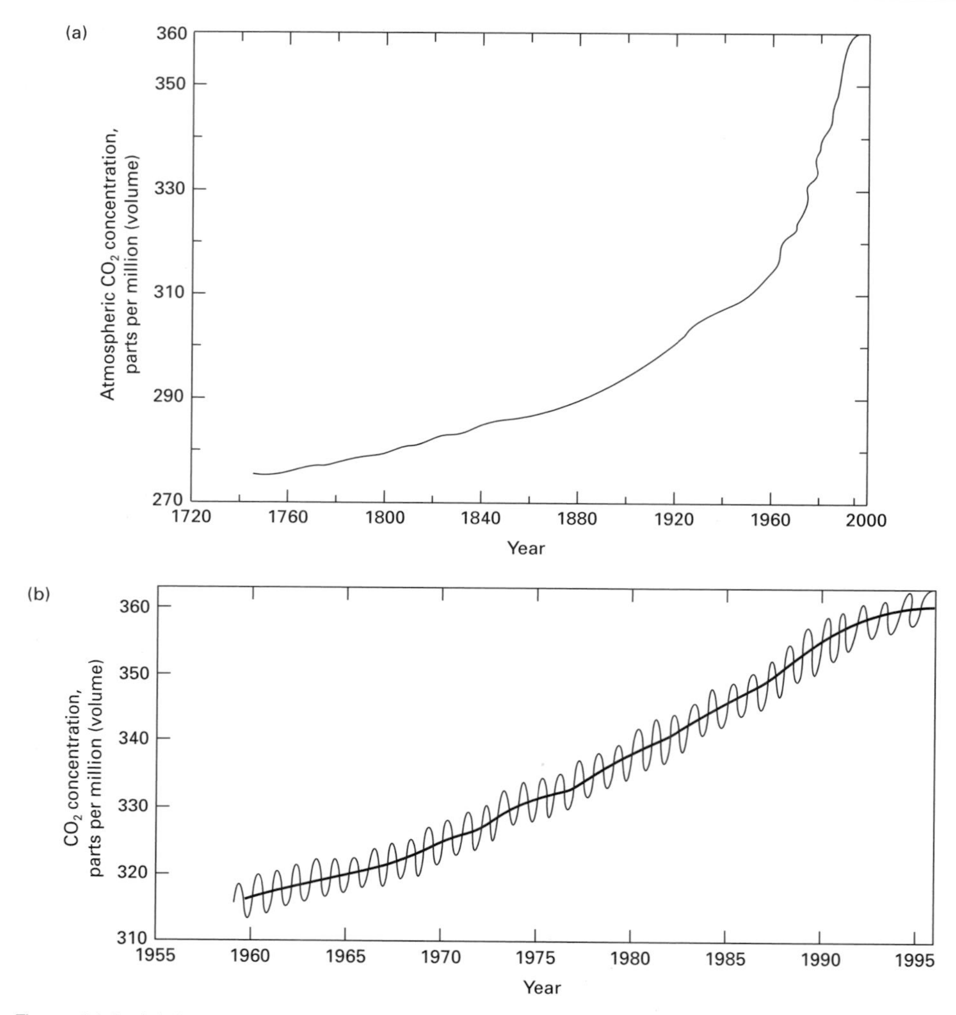


*Figure 14.2* (a) Atmospheric carbon dioxide concentrations for the last 300 years, based on ice-core data and atmospheric measurements. The values for the late eighteenth century are usually taken to represent pre-industrial $CO_2$ levels. (b) Monthly measurements of atmospheric carbon dioxide concentration at Mauna Loa Observatory, Hawaii, USA, 1958–1995. This is the longest continuous record of observations available. The seasonal cycle due to biospheric uptake and release is clearly seen. (From MacKenzie and Mackenzie, 1995.)

environmental changes, this rise is a world-wide phenomenon.

There is a constant cycling between the major reservoirs of carbon (Figure 14.3). $CO_2$ released by respiration and decay of vegetation is placed in the atmosphere, only to be reincorporated into vegetation as new growth develops, leading to the marked annual cycle in the atmospheric concentration (Figure 14.2). Exchange between the atmosphere and the oceans depends on sea-surface temperatures and the oceanic circulation as well as the atmospheric and oceanic concentrations. Slow changes in the oceanic circulation, notably the Southern Oscillation, may well be the cause of long-term variations in atmospheric concentration.

All of the exchanges shown in Figure 14.3 are attempts to create and maintain a dynamic equilibrium within the global carbon cycle. Human intervention, both through industrial production and agricultural activity, has increased the atmospheric component and

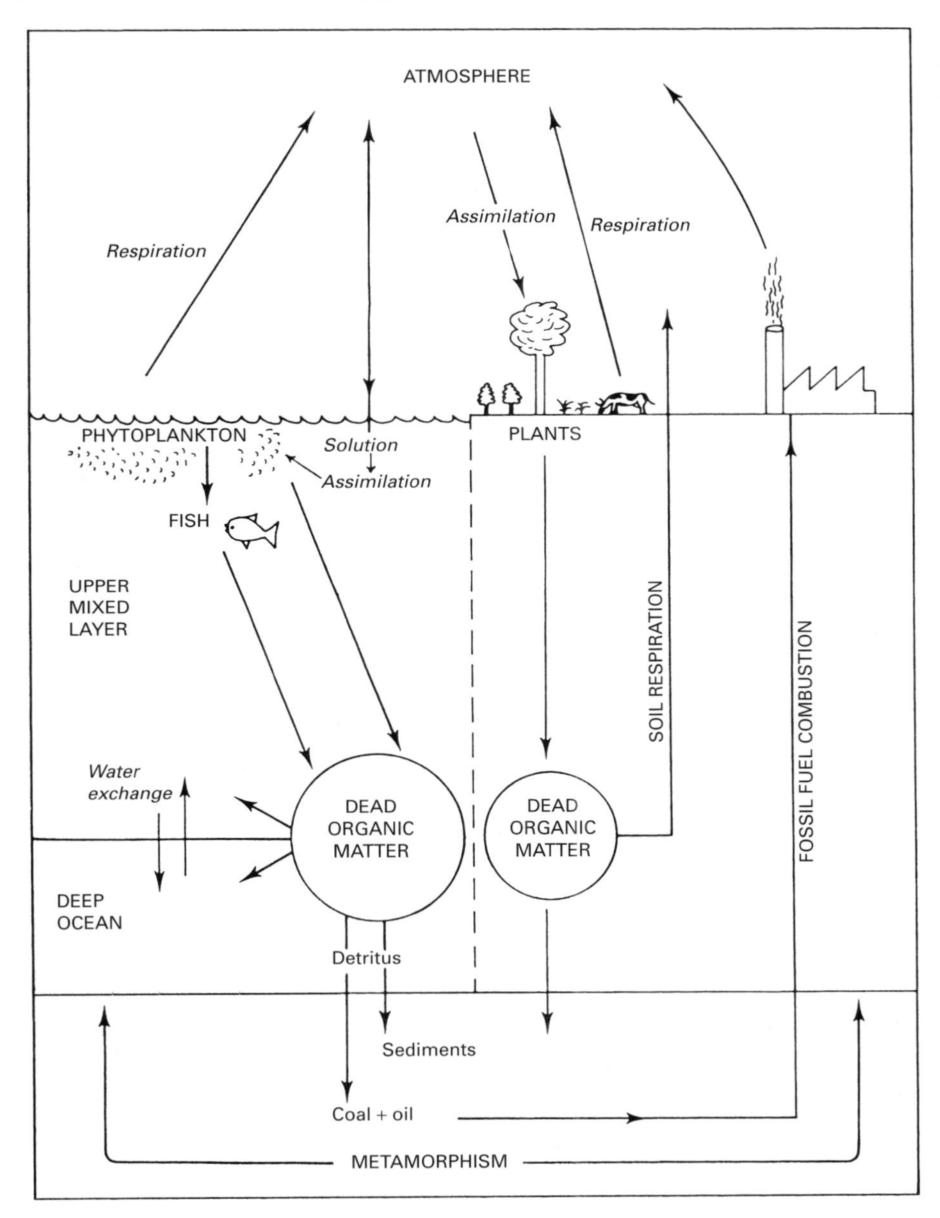


*Figure 14.3* The global carbon budget is primarily controlled by the biosphere, although there are sources and sinks for $CO_2$ other than those shown here (e.g. volcanoes). Note that the atmospheric $CO_2$ is the result of separate land and ocean surface interchanges. The carbon reserves represented by land, animals and fish are very much smaller than those represented by land plants and phytoplankton. The carbon reserve incorporated into the oceanic dead organic matter is believed to be about five times larger than the terrestrial reserve. The sediments are a massive carbon sink containing 30 000 tonnes more carbon than the atmosphere.

the system is currently out of balance. As a consequence the fluxes will naturally change in an effort to restore equilibrium. Predicting the rates of these changes is extremely difficult. The exchange mechanisms are not well known, although it is clear that they act on various time scales. For example, the biotic response is relatively rapid, but oceanic uptake is very slow. Although the biotic response is rapid, vegetation represents a relatively small sink, so that most of the excess, if it is to be removed, must go into the oceans. Their slow uptake means that a considerable time will be needed before the oceans and atmosphere can come into equilibrium. It is estimated, for example, that if anthropogenic production stopped now, it would take several thousand years to establish a new equilibrium, with about 85% of the current excess of atmospheric $CO_2$ going into the oceans. A reasonable estimate of the partitioning of this excess on the decadal or century time scale is that 57% remains in the atmosphere, 34% goes into the ocean, while vegetation receives 9%.

## 14.2 Climate forecasts

### 14.2.1 Types of forecasts

There are two types of atmospheric forecast: the **deterministic**, which gives a prediction of what will happen at a particular time and place; and the **probabilistic**, which gives an indication of the probability of a particular event within a particular time frame or geographic location. Almost any forecast of future atmospheric conditions will require elements of both. In general, the short-term weather forecasts, whether for a few minutes or a couple of days in advance, will emphasise the deterministic, while the longer-term ones will be increasingly probabilistic.

In general, the deterministic portion of any forecast will come primarily from model results, and the probabilistic portion will come from the historical record. One example of where overlap often occurs is in precipitation prediction. A model may indicate that unstable conditions and cloud formation are likely, along with values for water available and updraft velocities, but our knowledge of the precipitation formation mechanism is insufficient to state definitely whether precipitation will occur. It is necessary to review past conditions when the environment was in a state similar to that forecast by the model and then give a probability estimate of whether or not it will actually rain.

For many years the probabilistic approach was the sole means of climatological forecasting, and several examples have been used in previous applications boxes. The reliance on the historical record always implied that the future climate would be a direct continuation of that in the past, and such forecasts were often called 'anytime' forecasts since they are identical irrespective of how far into the future they project. The probability of frost (Table 3.A.1) is the same irrespective of which year is of concern. Although statistical techniques allowed us to incorporate 'climate change' in our estimates, for example by allowing the mean to change linearly with time, there was always an underlying assumption that the climate was stable.

The distinction between forecasts emphasising deterministic and probabilistic approaches was used to represent a distinction between weather and climate forecasts. Now, however, there are a group of forecasts which are clearly climatic, but use deterministic information. They are considered in the rest of this section. The following section treats the longer time scales, where forecasts must still be probabilistic.

### 14.2.2 Long-lead climate forecasts

The forecasts considered here are clearly climatic since they refer to conditions averaged over a three-month period, and expressed in terms relative to the long-term normal conditions. The term **long-lead** refers to a forecast for some time in the future with a gap between the time of issue and the period which is actually forecast. Thus, for example, rather than representing a forecast for 'the next three months', it is a forecast for the three months starting 'next 15 February'. This gap is the lead time. For practical applications of forecasts, this is frequently an important consideration. For an oil company, a winter that is colder than normal may increase the demand for heating oil at the expense of petrol. A forecast for a colder than normal winter issued the previous summer allows the company time to configure its refinery to produce more heating oil.

The US National Weather Service has been able to develop a long-lead climate forecasting system based on the slow and somewhat predictable evolution of conditions in the equatorial Pacific. The observational information (Figure 8.I.1) is used in process models to monitor the Southern Oscillation

Index (Figure 8.22) and provide an estimate of the future evolution of the ENSO. At present, the models produce circulation patterns rather than weather conditions. Therefore statistical techniques, based on past experience, are used to produce the links between the Pacific circulation and future weather in various parts of the globe. At present, forecast information is given with a lead up to one year in advance of the predicted conditions.

The modelling and the statistical analysis both contain major uncertainties, so that a precise forecast cannot be given. The forecasts that are issued (Figure 14.4) explicitly indicate uncertainty. The forecast itself is given in categories, not in absolute terms. For any area the temperature and precipitation forecasts indicate the likelihood of above, near and below normal conditions. These categories are derived from the historical observational records. With no forecast information, the probabilities of occurrence are 33.3% for each category. The forecast adjusts the percentages to indicate the most likely category and the confidence which can be placed in that statement. Thus for the central California coast of the USA in Figure 14.4 the 5% confidence indicates that the probabilities are 38.3, 33.3 and 28.3% for above, near and below normal temperatures, respectively. In general, the confidence will be highest when there are clear signals that an El Niño event is imminent. In some cases the signals may be ambiguous or absent, and no deterministic forecast is possible. Here the more traditional probabilistic forecast, heralded by the CL for 'climatology' in Figure 14.4, must be used.

### 14.2.3 Seasonal hurricane forecasts

Conditions in the eastern Atlantic and over west Africa appear to have an influence on the frequency of hurricanes in the western Atlantic a few months later. This suggests the presence of a phenomenon in the Atlantic similar to the ENSO in the Pacific, and there has been a search for similar physical and statistical links. The results have now been formalised into a seasonal hurricane forecasting system for the Atlantic basin. The hurricane season stretches from June to December, and the first forecast for the coming season is issued in late November of the previous year. Updates are made in early April, June and August. Certain facets of hurricane activity, such as the number of tropical storms of all intensities, the number of intense hurricanes, and the potential destruction from wind or storm surge, are forecast.

Currently a suite of potential predictors have been identified (Figure 14.5). Some, such as the high level (30 hPa) winds in the tropical Atlantic, the surface pressure and temperature anomalies in the North Atlantic generally and the Caribbean basin in particular, and the Sahel conditions, appear to have direct links. They influence the strength of the Hadley circulation, the location of the ITCZ, and the distribution of warm ocean surface waters in ways which have been considered in earlier chapters. The connections between hurricanes and ENSO, or with Pacific conditions in general, are perhaps less clear. Indeed, the Atlantic predictors are commonly employed in a fairly rigorous, statistically based analysis which utilises the historical record to estimate the strength of the relationships and the similarity of previous patterns to the present ones. The Pacific predictors are presently used more qualitatively, but allow the incorporation of some known relationships between various parts of the global atmosphere. More recently the influence of surface pressure and temperature anomalies in the North Atlantic and high level temperature differences above Indonesia have been found to have the potential to improve the forecast, although the physical processes acting to create the link are not completely clear.

The combination of predictors used for a particular forecast depends on the atmospheric situation and the aspect of hurricane activity of concern. As with the National Weather Service long-lead forecasts, there are some situations, or some events, which can be forecast with much greater confidence than others. Hence while the forecasts are issued in a regular routine way, they are likely to vary in quality and reliability. Improvements can be anticipated as our understanding increases.

### 14.2.4 Teleconnections and circulation indices

Both the long-lead and the hurricane forecasts depend in part on a statistical relationship between conditions in two or more locations which have no necessary or apparent direct connection. Such a relationship is a **teleconnection**. The recognition that phenomena in one region are connected with those in another, and that monitoring changes in one may allow forecasts of changes in the other, has stimulated research striving to provide physical support for the statistical links.

At present the teleconnection idea is frequently used in association with **circulation indices**, which are

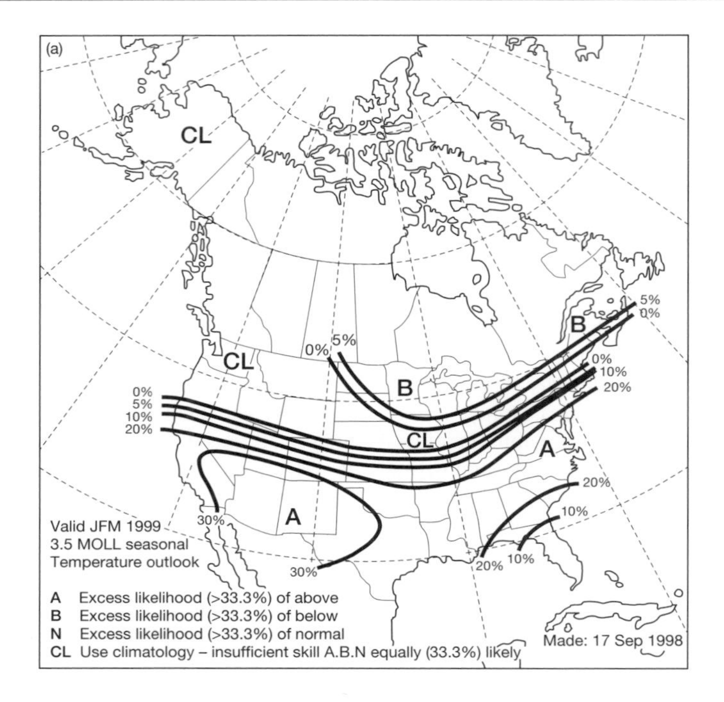
(a)
CL
CL
0%
5%
10%
20%
0%
5%
B
5%
0%
0%
10%
20%
B
CL
A
20%
10%
A
30%
30%
20%
10%
Valid JFM 1999
3.5 MOLL seasonal
Temperature outlook
A Excess likelihood (>33.3%) of above
B Excess likelihood (>33.3%) of below
N Excess likelihood (>33.3%) of normal
CL Use climatology – insufficient skill A.B.N equally (33.3%) likely
Made: 17 Sep 1998

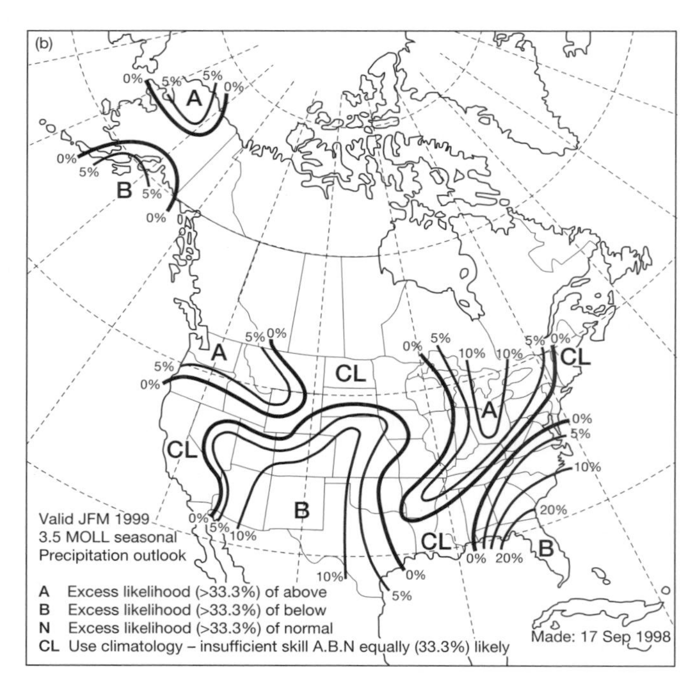
(b)
0%
5%
5%
0%
A
0%
5%
B
5%
0%
5%
0%
A
5%
0%
CL
0%
5%
10%
10%
5%
0%
CL
A
0%
5%
10%
CL
B
20%
Valid JFM 1999
3.5 MOLL seasonal
Precipitation outlook
0%
5%
10%
10%
0%
5%
CL
0%
20%
B
A Excess likelihood (>33.3%) of above
B Excess likelihood (>33.3%) of below
N Excess likelihood (>33.3%) of normal
CL Use climatology – insufficient skill A.B.N equally (33.3%) likely
Made: 17 Sep 1998

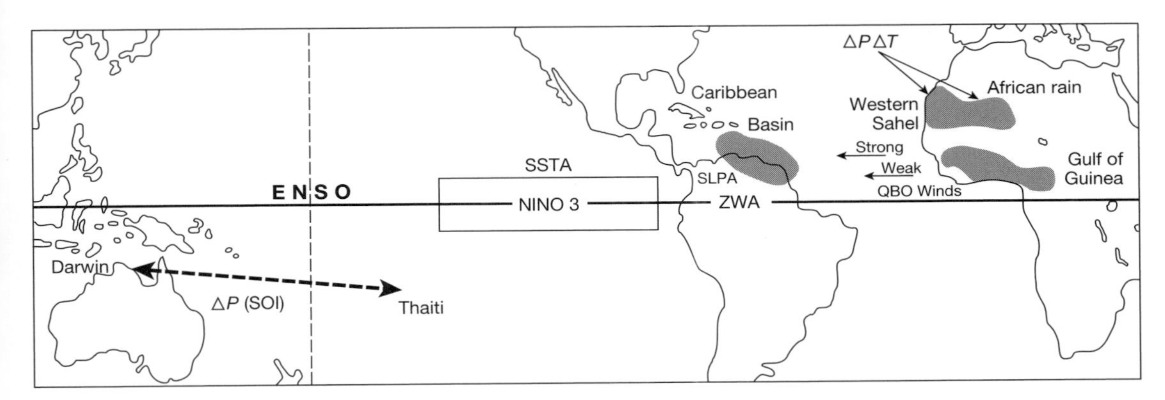


*Figure 14.5* Schematic diagram of some of the factors incorporated into current forecasts of seasonal hurricane activity. In the Atlantic basin the major factors are anomalies of precipitation and temperature ($\Delta P$ and $\Delta T$) in west Africa, sea-level pressure (SPLA) in the Caribbean, and the quasi-biennial wind oscillation (QBO) and the 200 mb zonal wind anomaly (ZWA). In the Pacific those factors affecting ENSO are included. (Courtesy of William Grey and Colorado State University.)

usually based on pressure differences between two or more locations. They provide a convenient method of summarising the often complex atmospheric motions, and hence processes, in an area. Indeed, the use of indices based on spatial pressure differences were explored earlier this century as aids in mid-latitude weather forecasting. Experience suggested that they were unsuitable for such short-term tasks, but did have potential for long-term forecasting. Now the best known index is the Southern Oscillation Index (SOI), used for both the long-lead and the hurricane forecasts. Probably equally important from the human perspective, but not yet as well developed, is use of the East Asian summer circulation index as a predictor of monsoonal rainfall (Figure 8.16b).

The predictive ability of mid-latitude indices is not well developed, but there appear to be some teleconnections akin to those in the tropics. First, directly analogous to SOI, but of particular relevance to Europe, is the **North Atlantic Oscillation** (NAO). This is loosely defined as the pressure difference between the Azores High and the Icelandic Low (Figure 6.1), with the stations at Lisbon, Portugal, and Stykkisholmor, Iceland, being most often used to provide an index. A high value implies a strong zonal (west–east) flow, and a low index a more meridional (north–south) one. It has recently been suggested, for example, that the relatively wet period of the early 1990s in Britain was a response to an unusually persistent high NAO. The other commonly used index is the **Pacific–North America Index** (PNA). This also has been variously defined, but has the form:

$$\mathrm{PNA} = (-z_a + z_b - z_c)/3 \qquad (14.1)$$

where $z$ is the normalised sea-level pressure and the subscripts refer to locations (a) off the Aleutian Islands in the Gulf of Alaska, (b) over the US–Canadian (Alberta–Montana) border, and (c) in the Florida panhandle in the USA, respectively. This is also an index for the amount of meridional or zonal flow in a portion of mid-latitudes and has been used to suggest, for example, a relationship between flow patterns and the frequency and intensity of rainstorms in the southeast United States (Figure 14.6). Like most indices, it changes relatively slowly, so that the establishment of teleconnections enhances the possibility of other long-range forecasting systems akin to those already noted.

*Figure 14.4* The US National Weather Service Climate Prediction Center long-lead forecast for January–March 1999, issued 17 September 1998. (a) The temperature panel indicates there is a high degree of confidence that the three-month period will be warmer than normal in the southwest and, with lesser confidence, for much of the rest of the south. There are also some indications of a cooler than normal period around the Great Lakes. (b) The precipitation panel suggests above normal amounts for the northwest, the lower Mississippi Valley and the Great Lakes, and below normal rainfall for the southwest and in Florida. Subsidiary information is available to determine the normal precipitation for any region. (Courtesy of NOAA.)

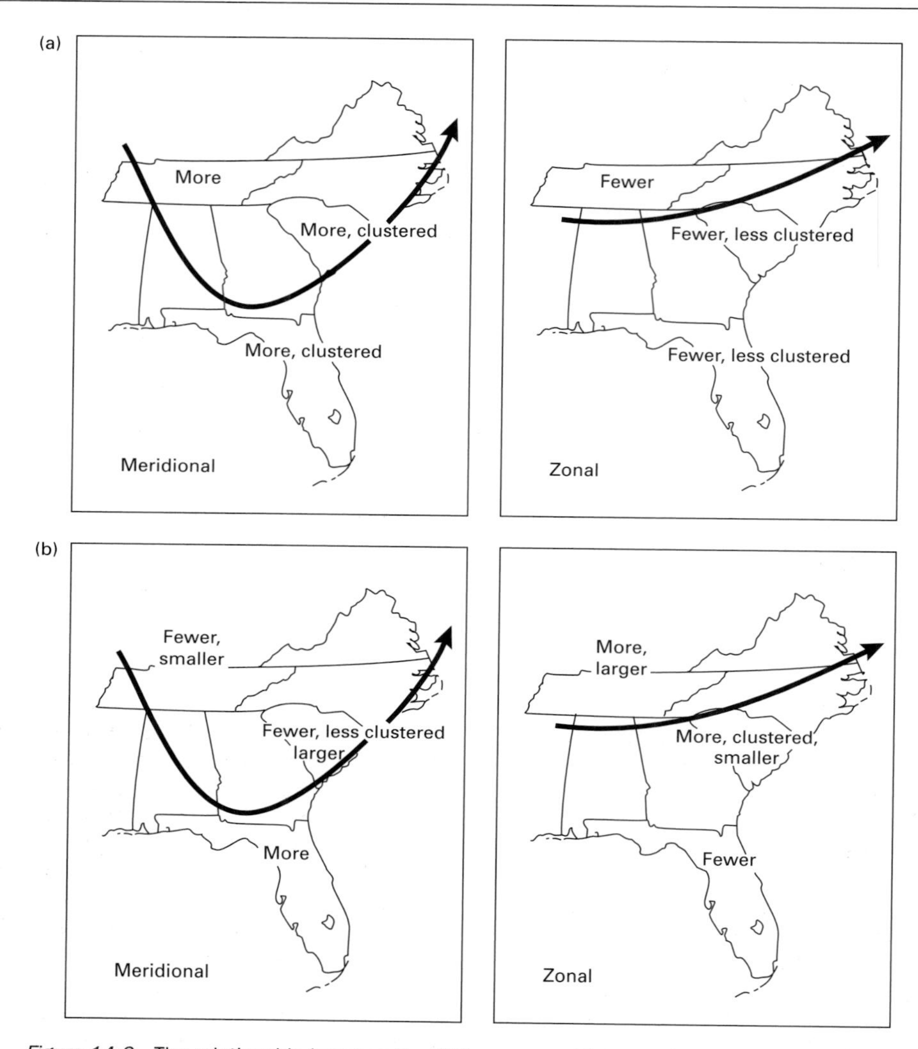


*Figure 14.6* The relationship between the PNA, expressed here as the generalised flow pattern represented by the heavy arrow, and the number, time sequence and size of rainstorms in the southeastern United States for (a) summer and (b) winter. The relationship is much more clearly established for winter than for summer. (From Henderson and Robinson, 1994.)

## 14.3 Scenarios of future climates

Projections of future climate conditions associated with greenhouse gas emissions and 'global warming' require a forecast time scale of decades. Here the models which we associated in the previous section with deterministic forecasts are themselves highly uncertain. Further, it is often necessary to incorporate historical observational data, even if they refer to a different atmospheric composition, if we are to produce the detailed information commonly needed for impact assessment. Rather than attempt to produce a single forecast, therefore, we create **climate scenarios**, sets of possible future climates, each member of the set having some characteristics of a deterministic forecast, while the ensemble has more probabilistic properties. All members must be internally consistent in maintaining the physical constraints imposed by the nature of the atmosphere while providing useful

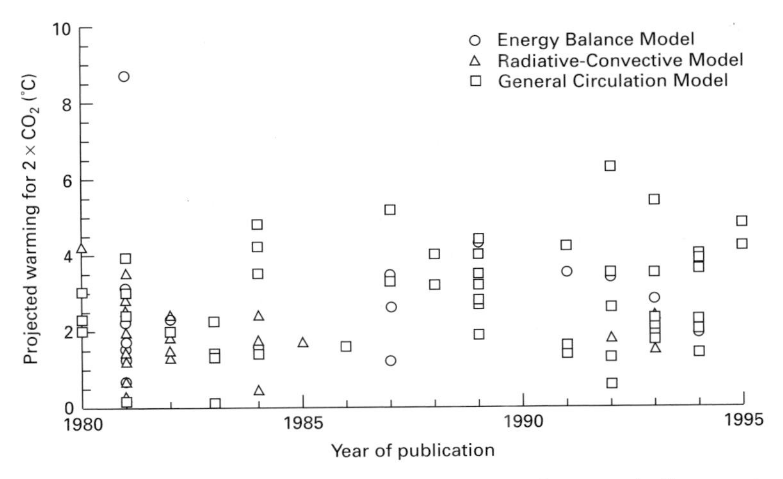


*Figure 14.7* The results of 108 global temperature change projections made between 1980 and 1995 from a variety of model sources. Although not all models are strictly comparable, it is clear that an atmospheric response centring on a warming around 2–3 °C was quickly established and remains the most common projection now. (From Kacholia and Reck, 1997.)

impact information. Here we discuss future climates on this time scale by considering models, observations and scenarios in turn.

### 14.3.1 The use of climate models

Until recently, most models addressing problems of the climate response to greenhouse gases have concentrated their efforts on situations where the gas amounts have reached a specified level, usually double the present value. This avoids the problem of attempting to predict the actual rates of increase in $CO_2$. Attention is thereby focused on the actual temperature increase, model comparisons are possible, and the role of the various feedbacks can be assessed (Figure 14.7). These models in effect simulate the equilibrium climate in the current conditions, assume a step change in carbon dioxide concentration, and then simulate the new equilibrium climate. The difference gives the 'climatic change'. These results can give a misleading impression. There is nothing inherently special about a doubling of $CO_2$ beyond computational convenience, and it is unlikely that concentrations will stabilise at that level (Figure 14.1). Certainly, while the temperature change identified by these models may provide a useful benchmark for comparisons, it cannot be interpreted as a 'before and after' situation.

To overcome the step-change drawback, more realistic models have been developed. These are **transient** in that they explicitly incorporate the slowly changing carbon dioxide concentration as they themselves allow the climate to change. This approach allows direct comparison of the models with the historical record (Figure 14.8). A major problem introduced by the use of transient models, however, involves the establishment of links between the atmosphere and the ocean. These two operate on very different time scales, with the ocean warming much more slowly than the land (Figure 1.2). Since it is a common practical strategy to initialise GCMs by running the atmospheric and oceanic components separately, there must be compensation for this difference before the models are joined. Without compensation the fluxes of heat, moisture and momentum between atmosphere and ocean would be out of balance and a slow, model-induced climate drift could occur. One approach is to make an empirical **flux adjustment** to compensate, the other is to minimise the time when the two models are uncoupled and manage the small residual drift. Indeed, suitable tuning of the managed parameters can alleviate the problem. The case of flux adjustment is just one of many where assumptions or compromises must be made in running transient models. Nevertheless, most indicate similar patterns of temperature change (Figure 14.9), with

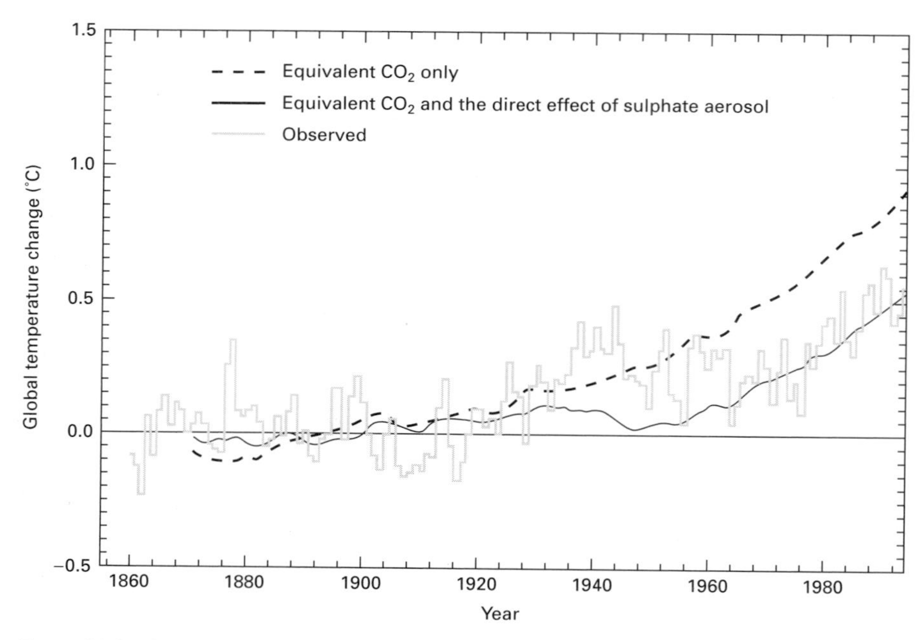


*Figure 14.8* Comparison of the global mean annual temperature of a transient general circulation model with observations, expressed as a departure from the 1880–1920 period. The model incorporating only the greenhouse effect (dashed line) fails to reproduce the cooling of the 1940–1960 period. When the effects of sulphate aerosols (increasing the albedo) are included (thin solid line), the historical record (dotted line) is more faithfully reproduced. (From Kattenberg *et al.*, 1996.)

maximum warming in polar latitudes, especially in the north, and relatively small changes over the tropical oceans. There are numerous regional differences between the two model results illustrated here, and between them and other models. Consequently it is possible to suggest that global average temperature changes are reasonably well established (Figure 14.7), latitudinal changes can be cautiously estimated, and regional changes are highly uncertain.

For precipitation, model projections are much less certain than those for temperature. This is to be expected since this holds even for the simulations of current conditions (Figure 13.6 and Plate 6). Hence, although regional results are possible (Figure 14.10), they must be treated with extreme caution. Our present level of understanding and modelling of the atmosphere makes it unwise to place much faith in any estimates of future precipitation. This remains a major challenge for climatology.

It is clear that the climate is changing. The changes are well documented. Whether there is an anthropogenic signal superimposed on the natural variability, or whether the changes are purely natural, is a source of great controversy, as well as of great significance in developing policies to respond to future climate changes. Although the causes are being investigated, it is not clear exactly how to proceed. The physical models alone cannot provide a definitive answer, since they frequently have recourse to observational data for calibration, which obscures any simple cause and effect analysis. Nevertheless, the close agreement between observations and the models incorporating sulphates with the greenhouse gases does suggest that there may be an anthropogenic link. A purely statistical approach to the isolation of causes is equally difficult, since that requires an estimate of the probability that the rate and magnitude of recent climatic change is entirely natural. The detailed observational record is too short, and the long-term palaeoclimatic record is insufficiently detailed to allow statistically satisfying answers.

Recent interest has centred on identifying potential **fingerprints** of climate change. One means, using the long record of surface observations, is suggested in Box 14.I. Another uses pattern analysis. Physical principles indicate, for example, that a greenhouse

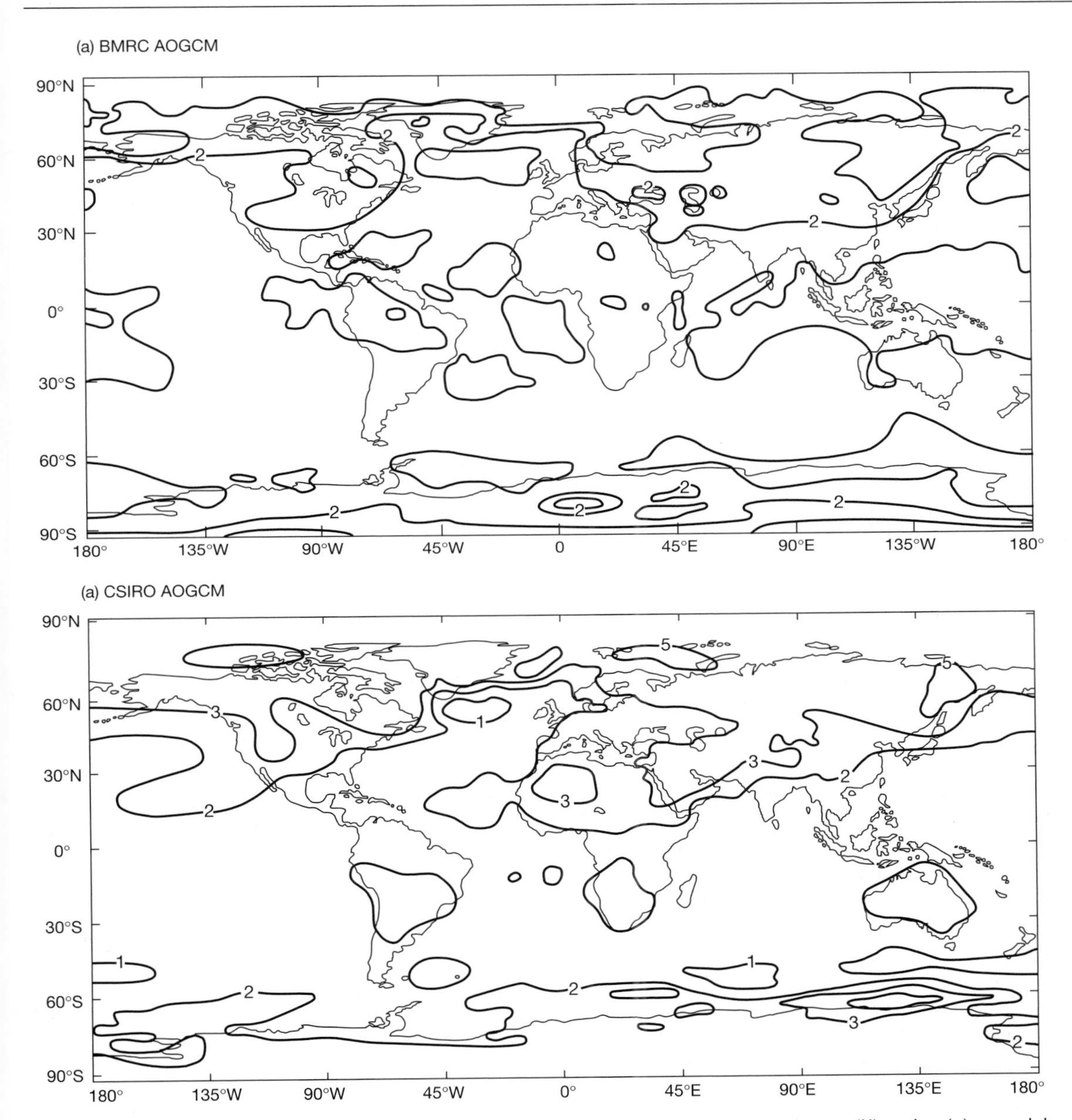


*Figure 14.9* Examples of transient model projections of mean annual temperature change (K), using (a) a model with no adjustment for the different response times of the ocean and the atmosphere (Australian Bureau of Meteorology Research Centre), and (b) a model with such an adjustment (Commonwealth Scientific and Industrial Research Organization, Australia). Both models had a 1% year$^{-1}$ increase in $CO_2$ and show conditions after an approximate doubling of the present concentration. (From Kattenberg *et al.*, 1996.)

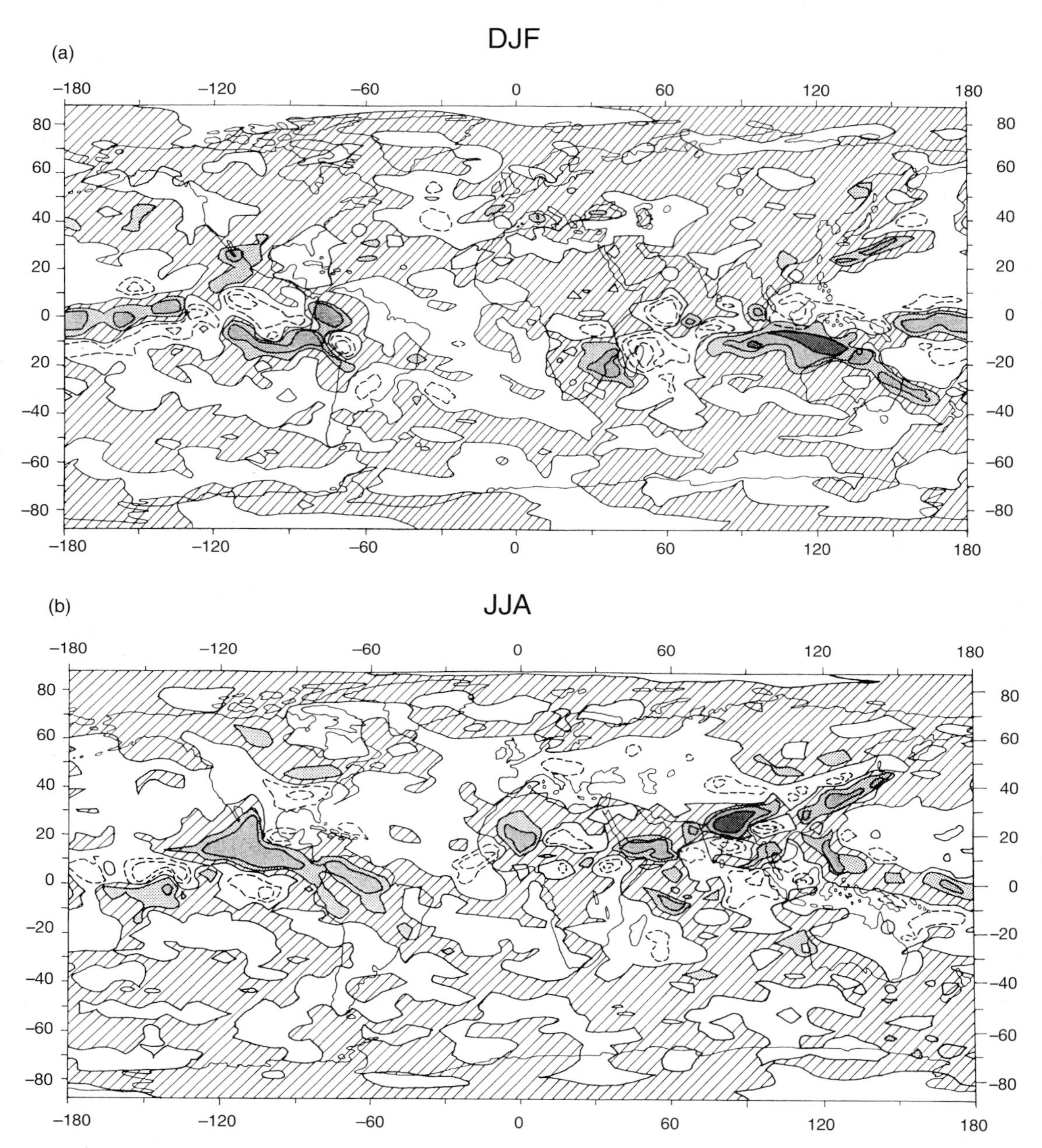


*Figure 14.10* Changes in precipitation (mm day$^{-1}$) following a doubling of $CO_2$ concentration at a rate of 1% year$^{-1}$ as projected by the same model as in Figure 14.9(a) for (a) December–February and (b) June–August. Areas of increase are shaded in the sequence 0–0.5, 0.5–1, 1–2, 2–4 and >4. Negative contours are dashed. The complexity of the change, especially in the topics, is readily seen. (From Kattenberg *et al.*, 1996.)

**Box 14.1 Monitoring for climate change**

In previous 'instrument' boxes we have looked at various instruments, networks and analytic techniques. In this final box we consider ways of using our observational information to address two aspects of climate change: first, the monitoring of the actual climatic conditions in a way that aids in the detection of significant climate change; and second, a suggestion of a way of monitoring so that the impact of climate change can be assessed.

*14.1.1 Monitoring the natural attributes of climate*

From the perspective both of the physical laws governing climate and of suggestions of future climates obtained through models and analogues, an enhanced greenhouse effect should have several distinct influences on various elements of the surface climate. Some relationships, such as increased temperatures, particularly an increase in night-time minima, are reasonably clearly established. Others, such as the consequence of an enhanced hydrological cycle, are less clear. The response could be more severe floods, more severe droughts, more frequent heavy rains, or a combination of these. Other effects, such as enhanced wind speeds, can be postulated. An observational analysis may indicate specific trends, but each of these effects, taken individually, does not provide unequivocal information about the causes of climate change. It can be suggested that an amalgamation of these effects could be used as a fingerprint, helping to identify whether all the trends were consistent with the greenhouse effect.

It is not immediately apparent how to combine the various elements into a useful and pertinent index. Two have been suggested, using percentage area covered by anomalous conditions as the major variable (Table 14.I.1). This avoids the problems associated with the potential impact of local conditions on individual stations, and has the potential to provide a global index comparable with the global model outputs. So far they have been calculated for the conterminous United States (Figure 14.I.1). Each uses five components expressed as the percentage of the conterminous United States under the indicated conditions (Table 14.I.1). An unusual or extreme event is expressed as one falling in the upper/lower tenth percentile of the historical record. The first index is the US Climate Extremes Index, designed to provide a measure of the status of the country with respect to extremes. The other, the US Greenhouse Climate Response Index, is more directly related to factors which might provide a global warming fingerprint. Both indicate an increase over time. While to some extent this is expected, given the observed global temperature increase, the choice of a relatively restricted set of components for the index, clearly related to the greenhouse effect, lends support to the contention that there is an anthropogenic warming under way, at least in the United States. There seem to have been major changes in both indices in the early 1920s and in the mid-1970s. For the latter there are a suite of surface and upper air observations available, which suggest that this was a time of a major change in the flow regime of the Northern Hemisphere, linked, at least to some extent, to the increased frequency and intensity of ENSO events. This clearly indicates that all aspects of the climate, not just temperature, are changing.

*Table 14.I.1* Elements incorporated into climatic indices

*Climate Extreme Index*

Annual arithmetic average of:

1. area with maximum temperatures much below normal + area with maximum temperatures much above normal
2. area with minimum temperatures much below normal + area with minimum temperatures much above normal
3. area in severe drought + area with severe moisture surplus
4. 2 × area with proportion of precipitation derived from extreme one-day precipitation events much above normal
5. area with number of days with precipitation much above normal + area with number of days with no precipitation much above normal

*Greenhouse Climate Response Index*

Annual arithmetic average of:

1. area with mean temperatures much above normal (daily mean = 0.525 (minimum) + 0.475 (maximum) daily temperature)
2. area with cold season (October to April) precipitation much below normal
3. area with warm season (May to September) severe drought
4. area with proportion of precipitation derived from extreme one-day precipitation events much above normal
5. area with day-to-day temperature differences much below normal

N.B. All areas are percentages of conterminous United States.
*Source*: after Karl *et al.* (1996).

**Box 14.I (cont'd)**

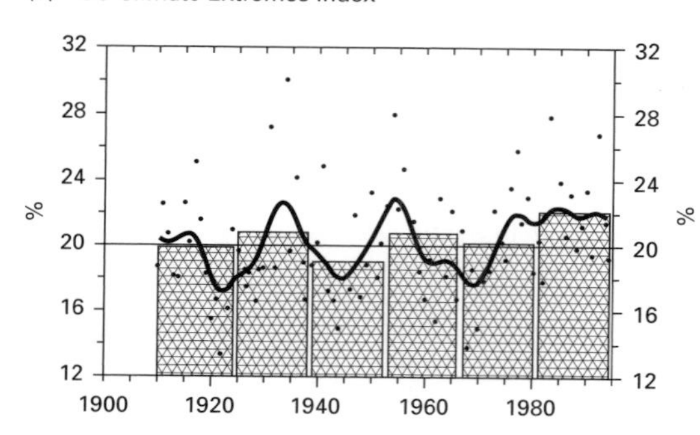


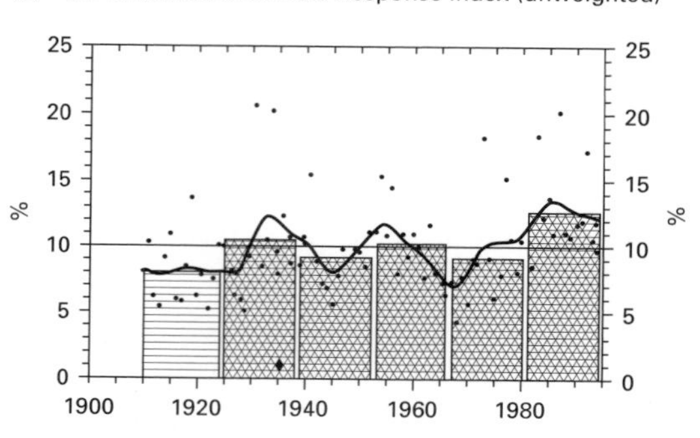


*Figure 14.I.1* Annual values for the conterminous United States of (a) the US Climate Extremes Index and (b) the US Greenhouse Climate Response Index. Dots represent annual values, the curve a 21-year smoothing and the bars 14-year averages. For (a) 20% represents the average (expected) value; for (b) the corresponding value is 10%. (From Karl *et al.*, 1996.)

*14.I.2 Monitoring the human role in climate*

The indices of Figure 14.I.1 are strictly climatic, designed to identify climatic trends. It has been suggested that similar indices more closely related to the needs of impact assessment could be developed. They would be akin to, and indeed serve a similar function to, national economic indicators, providing a measure of the climatic health of the nation. Although similar in form to those of Table 14.I.1, the pertinent climatic elements or events, and the appropriate weighting, whether by area or population, would have to be devised in consultation with those most likely to benefit from them. A suggested suite of indices is given in Table 14.I.2.

These indices have not yet been implemented. They do, however, provide an opportunity for the climatologist, working in co-operation with impact assessors, to provide pertinent information to a broad community in a variety of fields. The original suggestion was to use the past data to analyse the nature and causes of impacts. The potential to use forecast information to produce a predictive index, again akin to the familiar economic forecasts, should be investigated.

*Table 14.I.2* Examples of possible climate indicators for impact assessment

*Climatic hazards*

Annual average of the sum (relative to climatic normals) of:

1. 2 × area of Niño-3 with much above normal seasonal SST
2. national area in severe drought
3. national area with severe moisture surplus
4. 2 × national area with severely depleted groundwater levels
5. national area with much above normal snow-pack
6. national area with much below normal snow-pack

*Ecosystem health*

Annual average of the sum (relative to climatic normals) of:

1. % of marginal areas with heat stress much above normal
2. % of marginal areas with heat stress much below normal
3. % of marginal areas with ratio of potential evapotranspiration to precipitation much above normal
4. % of marginal areas with ratio of potential evapotranspiration to precipitation much below normal

*Energy demand (heating and cooling)*

Annual average of the sum (relative to (a) heating/cooling season-to-date normals; (b) optimal predictive averaging period normals, and (c) long-term climatic normals) of:

1. 2 × population weighted national area with HDD much above normal
2. 2 × population weighted national area with CDD much above normal

N.B. All areas expressed as percentages. For the location of Niño-3, see Figure 14.5. SST = sea surface temperature, HDD = heating degree days, CDD = cooling degree days.
*Source*: after Easterling and Kates (1996).

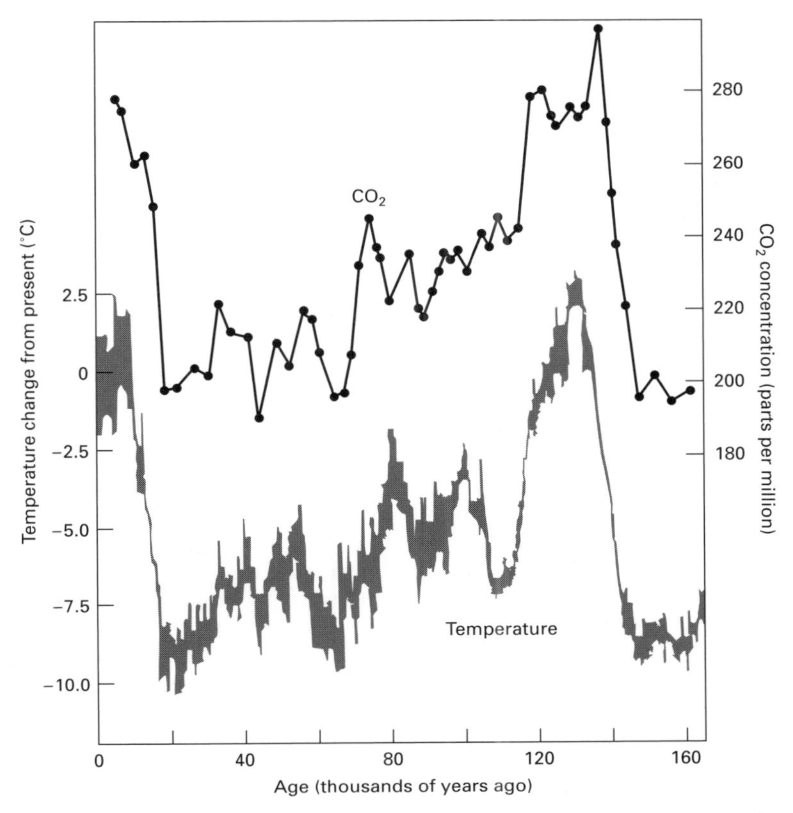


*Figure 14.11* Relationship between temperature and $CO_2$ concentrations for the last 160 ka. The dating for the values is not exact, and it is not clear whether the $CO_2$ increase around 140 ka, for example, preceded or followed the temperature increase. (From Houghton *et al.*, 1996.)

effect dominated by $CO_2$ should have warming in the lower and mid-troposphere and cooling in the stratosphere. Surface albedo changes should lead to changes most marked in the near-surface layers and in polar regions. Consequently there have been a series of experiments to see if it was possible to detect such changes (Plate 7). The results also give hints of an anthropogenic role, but as yet the signal is not completely convincing.

### 14.3.2 Climatic analogues

A second approach for investigating possible $CO_2$ induced climate changes is to use relevant historical and geological **analogues**. The role of palaeoclimatology in understanding and predicting climate was explored in Chapter 12. Some of the material discussed there could be treated as the basis for analogues, with Figure 12.5 serving as a complete example for a postulated cooler future. Here we can emphasise that there is a clearly established link between $CO_2$ levels and temperature (Figure 14.11), and that there have been several periods in the past when conditions similar to those likely to be caused by increased $CO_2$ levels may have occurred (Table 14.2).

One relatively recent period of elevated $CO_2$ levels was the Altithermal some 6 ka ago. A reconstruction of possible conditions then (Figure 14.12) provides information about the spatial distribution of precipitation, which is relatively poorly reproduced by GCMs. The distribution does bear some similarities with the GCM output (Figure 14.10), particularly the increasing moisture in the tropics. For the mid-latitude continents the models tend to suggest slightly more precipitation in winter, somewhat less in summer. This, combined with the higher temperatures and enhanced potential evapotranspiration, should lead

*Table 14.2* Comparison between warm climate phases of the past and the $CO_2$ levels currently believed to be able to produce similar changes in the average global surface temperature ($\Delta T$)

| Period | Date | Increase $\Delta T$ (K) | Estimated $CO_2$ concentration (ppmv) |
|---|---|---|---|
| Medieval | AD 1000 | +1.0 | 385–430 |
| Holocene (Altithermal) | 6000 BP | +1.5 | 420–490 |
| Eemian Interglacial | 120 000 BP | +2.0/+2.5 | 460–555/500–610 |
| ?(Ice-free Arctic but glaciated Antarctic) | 3–15 Ma | +4.0 | 630–880 |

to some drying. This appears to have occurred in North America, but not in Europe.

A second analogue 'reconstruction' uses data relating to the five coldest and five warmest years in a recent 50-year period (Figure 14.13). For both temperature and precipitation there are decreases and increases of similar magnitude. For temperature there is a rather regular decrease in temperature difference away from the Pole, while the precipitation distribution is more complex. Although this may suggest possible responses to the effects of doubled $CO_2$, the mean temperature increase in this analogy is only 0.6 K as compared with the anticipated 2–3 K as a result of doubling $CO_2$.

There are some similarities between Figures 14.9, 14.10, 14.12 and 14.13. The temperature sensitivity of the polar regions, and the potential mid-continental drying, especially in summer, appear to be well established. However, for many areas, there are significant differences and no single prediction can be made.

### 14.3.3 Downscaling and scenario development

The two previous sections have identified general methods of estimating future conditions, emphasising the scientific issues in establishing future climates. Here we emphasise the practical needs for scenarios which allow impact assessments. This usually entails the provision of rather more detailed information than that considered above.

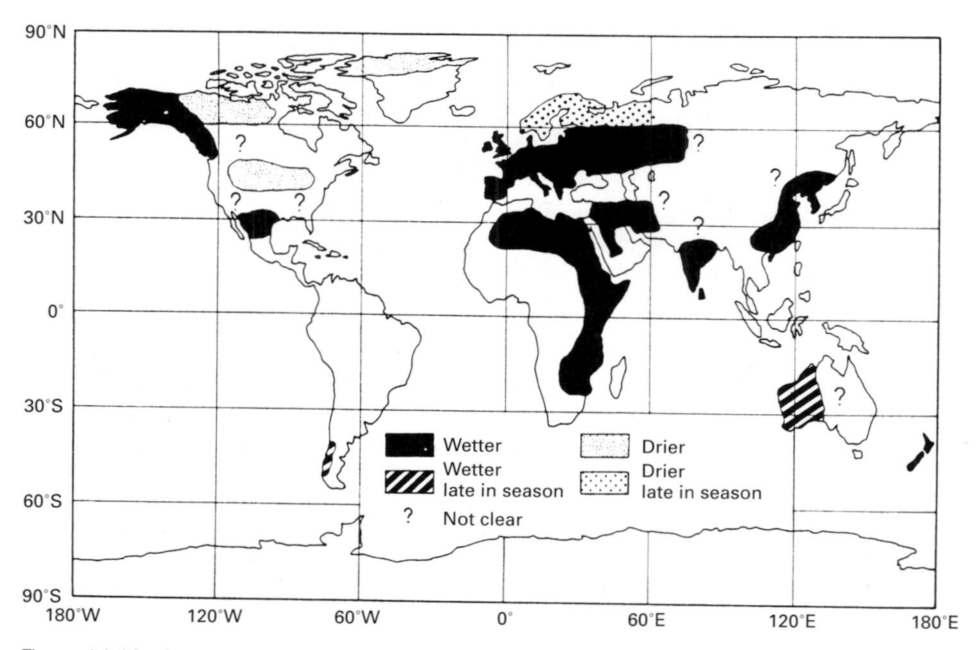


*Figure 14.12* A schematic attempt to reconstruct the distribution of rainfall (predominantly during the summer) during the Altithermal period of 4–8 ka when the world was a few degrees warmer than now. The terms 'wetter' and 'drier' are relative to the present. Blank areas are not necessarily regions of no rainfall change, just of no information. (From Kellogg, 1978.)

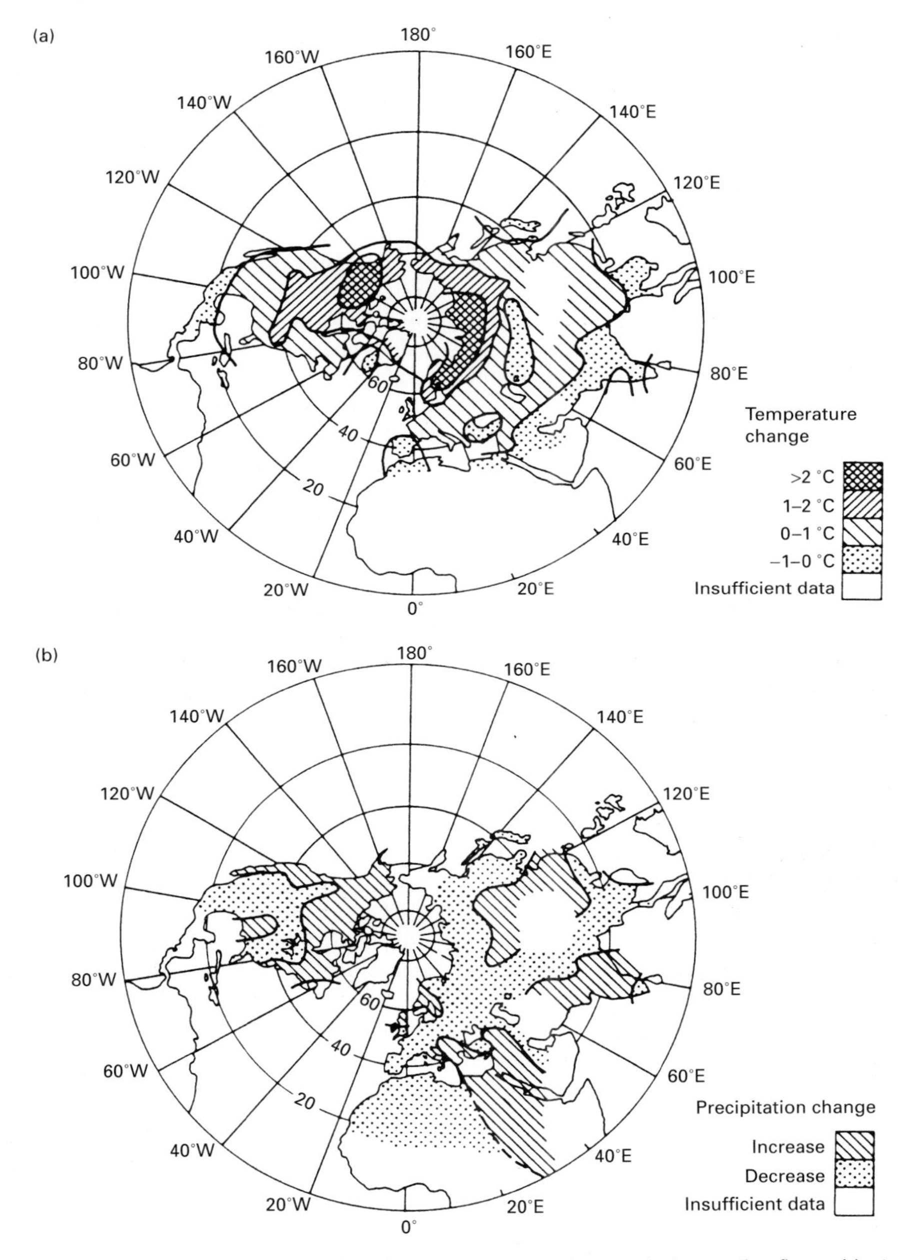


*Figure 14.13* (a) Mean annual surface temperature changes between the five coldest years and the five warmest years in the period 1925–1974 in the Northern Hemisphere. The corresponding change in hemispheric mean temperature was 0.6 K. (b) Mean annual precipitation changes between the same five coldest and five warmest years. (From Wigley *et al.*, 1980.)

There are two possible methods of providing this detailed information. The first is to adopt a complete modelling approach and use regional models embedded within GCMs. This was treated in Chapter 13. Although it is likely to be the long-term solution, our current ability to use it is severely limited. There are problems with linking the models of different scales, particularly in allowing a two-way interaction between them. At present, any regional development tends to be for a fixed location, so that the need for an analysis in a new area requires a new model development effort. Finally, from a practical standpoint, they are expensive to run, and cannot be used routinely to provide a large suite of scenarios.

The alternative approach is to use observational information to refine the GCM output. The major barrier here is that models give values at grid points or averaged over a grid cell, while the observations are for specific points. Linking the two, specifically taking the broad-scale model outputs and transforming them to the small-scale information needed for impacts, has become known as **downscaling**. The general problem of downscaling is one of needing to take the areal average produced by a model and distribute it realistically over the grid cell, using the current observational evidence as a guide. For scenario development, there is usually the need to distribute some change between current and future modelled conditions. For the development of downscaling techniques it is usually advantageous to use only the current climate. Even then, the method is not straightforward, because the observational information itself only samples the distribution, a problem already indicated in several of the boxes discussing instrumentation (see, for example, Box 9.I). The simplest technique, which might be appropriate for long-term average temperatures in regions of flat terrain, is a uniform distribution. A more complex technique, for precipitation in mountainous areas, might be based on a modification of the method discussed in Box 13.A. As a more pertinent example, we can consider the downscaling of precipitation over the Susquehanna River basin in the eastern USA (Figure 14.14). In this case the distributions were established using relationships between pressure patterns, near-surface humidity and precipitation, rather than a statistical relationship relying on precipitation alone. In an area such as this, with a complex mix of mountains, lowland and water, the GCM modelled precipitation field may be much less reliable than the spatially more conservative pressure and humidity fields. In this case, the GCM precipitation field indicated a maximum over the ocean, whereas the downscaled data and the observations indicated a minimum.

As an example of a scenario development for a practical problem, we can consider estimates of the impact of global climate change on the amount and distribution of permafrost (Table 14.3). Permafrost (see Section 9.4) has two roles, one as a major influence on the energy balance, the other as a major storage area for carbon. Any thawing releases this carbon as $CO_2$, which, if the greenhouse effect is responsible for the warming, creates a positive feedback. The maintenance of permafrost depends on the annual temperature cycle and on the amount of winter snow cover. A **frost index** has been developed linking thawing degree days, freezing degree days (both defined in the same way as for heating/cooling degree days, but using 0 °C as a base) and snow cover. The link between this frost index and permafrost has been established in the current climate, and the current distribution of permafrost established using observational data interpolated to a regular $0.5 \times 0.5°$ latitude/longitude grid. For the scenario of future conditions three transient models were run and the temperature and precipitation differences between the current and the (approximately) $2 \times CO_2$ situations determined for each model grid cell. These changes were then superimposed directly onto the observational grid, and new freeze/thaw degree days and snow cover determined. This led to a new frost index at each point and a consequent change in permafrost distribution (Table 14.3). The results from the three models can be taken as the scenario. They do not represent the bounding extremes, which are undefined, but clearly indicate that there is uncertainty in the estimates. In this case, two of the models gave similar results, but the other suggested a much greater reduction in the zone of continuous permafrost. The major difference was in Eurasia. This suggests a need to assess both the potential practical consequences and the performance of the models in this region. A comparison with earlier experiments also indicated that the transient runs gave smaller changes than those using equilibrium models with doubled $CO_2$.

The downscaling approach adopted for our final example is similar to that for the permafrost analysis. A single equilibrium model was used to determine the temperature and precipitation changes likely to occur as a result of $CO_2$ doubling for the grid cells covering Ontario, Canada. These changes were then imposed on the observational data for the current climate to produce the future scenario. The approach, giving the future climate with the same precision as

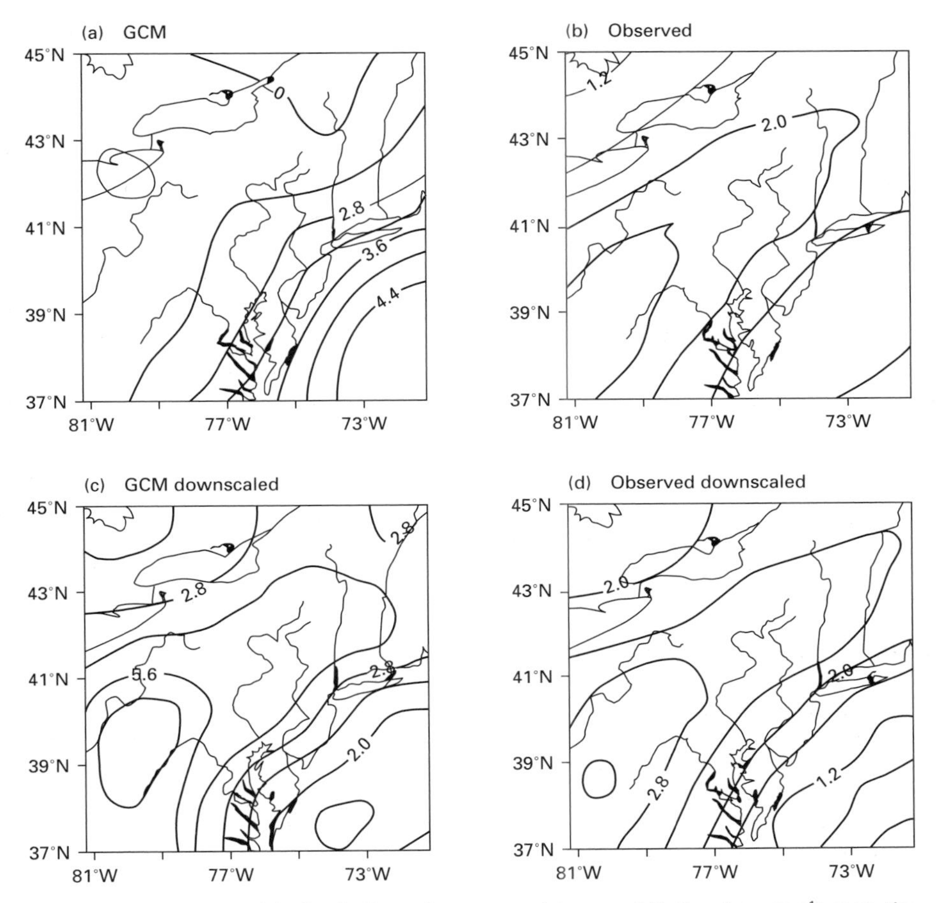


*Figure 14.14* The spatial distribution of average winter precipitation (mm day$^{-1}$) over the Susquehanna River basin and surrounding areas as determined from (a) a GCM run; (b) observational data with the same spatial resolution as the GCM; (c) the GCM after the imposition of downscaling; and (d) the observational data with the same spatial resolution as the downscaled GCM. Note especially the reversal in gradient across the coast in the south-east of the domain between (a) and the other panels. (From Crane and Hewitson, 1998.)

*Table 14.3* Area (km$^2$ × 10$^6$) of terrestrial Northern Hemisphere in each permafrost zone in current conditions and as estimated by three transient GCM models (percentage of current conditions given in parentheses) (Anisimov and Nelson, 1997)

| Model[a] | All zones | Continuous | Extensive | Sporadic |
|---|---|---|---|---|
| Current climate | 25.5 (100) | 11.7 (100) | 5.6 (100) | 8.1 (100) |
| GFDL | 22.4 (88) | 10.3 (88) | 4.9 (87) | 7.2 (88) |
| Max Planck | 22.4 (88) | 10.3 (88) | 4.8 (86) | 7.3 (90) |
| UKMO | 19.8 (78) | 7.8 (66) | 4.7 (85) | 7.3 (90) |

[a] Refers to organisations developing/running models.

*Table 14.4* Selected possible impacts of a $CO_2$-induced climate change on economic activity in Ontario, Canada

| Resource | Impact | Approximate annual gain/loss (Can$ × $10^6$) (1979) |
|---|---|---|
| Electricity demand | Reduced winter heating need (warmer winter); increased summer cooling need (hotter summer) | +(100–120) |
| Hydropower production | Reduced water availability (lower lake levels/flows) | –(35–65) |
| Shipping | Longer season (less ice cover); less cargo/vessel (lower channel depth) | –(10–30) |
| Agriculture | Extended growing season, northward expansion (warmer summers); reduced yield (increased heat and moisture stress) | –100 |
| Tourism | Greater outdoor summer activity (warmer summer); ski season reduced/eliminated (warmer winter) | +4 |
| Water use | Increased demand in summer (warmer summer) | –? |

*Source*: summarised and generalised from Cohen and Allsopp (1988).

the current climate, allowed a direct link to a set of economic impact models (Table 14.4). Although the precision of the new climate was the same, the accuracy of the scenario was highly uncertain. Hence the actual values in the table represent only gross estimates. They clearly demonstrate, as would be anticipated from the discussion of Section 11.1, that some economic sectors are likely to gain, some to lose, from climatic change. All, however, demonstrated the influence of climate on human activities.

In conclusion, it is clear that over the last few decades our understanding of the links between climate and human actions has increased tremendously. We are increasingly able to specify the impact of climate on human activities, and to develop means of both minimising adverse climatic impacts and maximising the use of climate as a resource. At the same time, we are becoming more aware of the potential impact of human activities on climate, particularly the possibility of global warming. The other great advance in climatology over the last few decades has been in our understanding of the processes creating climate. There are still many gaps in our knowledge, and much research is needed. Nevertheless, better observations and more refined theories are working together to allow the creation of more sophisticated and useful climate models. The practical result has been that, for the first time, realistic predictions for future climates are possible. These forecasts, experimental at present but clearly capable of much improvement, link the twin advances in understanding process and impacts, and point to an exciting future for climatology.

**Box 14.A Estimating future climates for decision-making**

*14.A.1 The climate for the next 10 000 years at Yucca Mountain*

Most climate scenarios involve impacts stretching over a few decades. Some may involve longer periods. One such example is the need to specify future climates for several thousand years at nuclear waste disposal sites in order to minimise the risk of an accidental radiation release stemming from a climatic hazard. One proposed site in the United States is in the Nevada Desert, and excessive rain could be a potential hazard. The area, at about 37°N, is an interior desert (see Section 9.3) with a climatic normal annual precipitation of somewhat less than 150 mm. As with most desert or semi-desert areas, there is a great deal of interannual and inter-decadal variability. Moisture sources are the depressions which get over the western mountains, mainly in fall and winter, and some monsoonal inflow in summer (Figure 9.19). The practical need is to specify the precipitation regime over the next 10 000 years. For our illustrative purposes, we can refine this and simply define the target as an estimate of the likely short-term (i.e. approximately one week total) maximum precipitation for a time 7500 years in the future.

There is no one preferred approach to this target. Indeed, this was chosen as the final example partly

**Box 14.A (cont'd)**

because virtually everything in this book could be brought to bear on the problem. This particular scenario development exercise involved five climatologists working independently and then seeking consensus. While recognising all of the non-climatic uncertainties associated with Table 14.1, it was agreed that the nature of the planet's surface be assumed to remain broadly similar to the current situation, and that the atmospheric composition would be close to that of the present. Even with these assumptions established, the approach to adopt was by no means clear. Some people elected to use a GCM suitably modified by the Milankovitch forcing; others used palaeoclimatological analogues to deduce potential flow patterns suggested by the Milankovitch radiation distribution (there being some pollen, tree ring and lake sediment information available for the region); others looked at historical data to reconstruct possible flow patterns for the Rossby waves and the summer monsoon; and others took a purely statistical approach using the existing precipitation records for the area and an extreme value analysis. Most used or combined several methods. The results (Figure 14.A.1) for these five people showed surprising agreement, although that agreement was by no means complete. There is, of course, no correct, or even testable, answer. The estimation of precipitation for this area this far in advance, leaving aside human-induced changes, presupposes a profound knowledge of the climate system and ways in which to analyse it. These, it is left to the reader to consider and, for the bold-hearted, to attempt.

*14.A.2 Creating your own scenarios*

One of the most exciting advances in climatology in recent years has been the increasing ease of access to climatological data and information. Although that contained on various Home Pages of the World Wide Web may be the most visible and obvious, it is increasingly common for national data centres to maintain climatological archives which can be accessed directly through the Internet (see Appendix B). Downloading vast amounts of data is now possible, frequently limited only by the storage capacity of the local computer or, in nations which charge for data, the pocketbook of the recipient. In some cases it is also possible to download the code for climate models, or access such models directly from your computer.

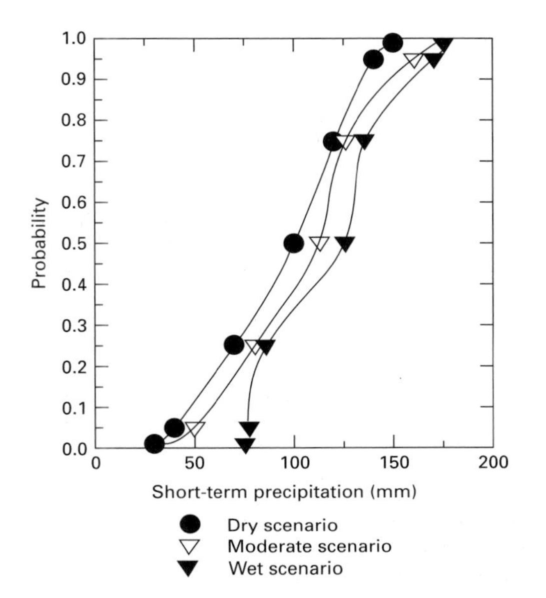


*Figure 14.A.1* Probability distribution of short-term precipitation (mm) at the proposed nuclear waste disposal site, Yucca Mountain, USA, 7500 years in the future. The curves represent the attempted consensus of five climatologists, who agreed to suggest that there are three possible general scenarios for the moisture conditions of the whole region. (After DeWispelare *et al.*, 1993.)

It is now possible, therefore, to undertake many kinds of climatological analysis without too much trouble. In this text we have made no attempt to provide step by step instructions for any analyses, but have suggested numerous possible lines of approach. A personal computer with a statistical analysis package opens up many possibilities. Indeed, you are encouraged to explore for yourself, with your own data and interests, using the text as a starting point.

Throughout the book we have also indicated uncertainty. This runs all the way from the original observation and archiving steps, through any necessary interpolation schemes to the use of process and statistical models to create the scenarios. So you are encouraged to use the freedom that the Internet gives, but beware of placing too much faith in your results. A good climatologist treats all observations and models, and the forecasts derived therefrom, with healthy scepticism.

APPENDIX A

# SI units

The Système Internationale (SI) is based on seven independent fundamental units. These are the units of length, the **metre**; mass, the **kilogram**; time, the **second**; temperature, the **kelvin**; electric current, the **ampere**; amount of substance, the **mole**; and luminous intensity, the **candela**. The last three units have not been needed in material presented in this book. SI units of other quantities, such as force (the **newton**), pressure (the **pascal**), power (the **watt**) and energy (the **joule**), are defined in terms of the fundamental units. Multiples or fractions of units, in powers of ten, are designated by the following prefixes:

| | | | | |
|---|---|---|---|---|
| p | pico- | one trillionth (US) | $10^{-12}$ | 0.000 000 000 001 |
| n | nano- | one billionth (US) | $10^{-9}$ | 0.000 000 001 |
| μ | micro- | one millionth | $10^{-6}$ | 0.000 001 |
| m | milli- | one thousandth | $10^{-3}$ | 0.001 |
| c | centi- | one hundredth | $10^{-2}$ | 0.01 |
| d | deci- | one tenth | $10^{-1}$ | 0.1 |
| da | deca- | ten | 10 | 10 |
| h | hecto- | one hundred | $10^{2}$ | 100 |
| k | kilo- | one thousand | $10^{3}$ | 1 000 |
| M | mega- | one million | $10^{6}$ | 1 000 000 |
| G | giga- | one billion (US) | $10^{9}$ | 1 000 000 000 |
| T | tera- | one trillion (US) | $10^{12}$ | 1 000 000 000 000 |

## Table of SI units and equivalents

| SI units | Conversions |
|---|---|
| | Length |
| metre (m) | 1 m = 3.28 ft; 1 ft = 0.304 m |
| centimetre (cm) | 1 cm = 0.3937 in; 1 in = 2.54 cm |
| kilometre (km) | 1 km = 0.621 mi; 1 mi = 1.61 km |
| micrometre (μm) | 1 μm = $10^{-6}$ m = $10^{-4}$ cm = $3.94 \times 10^{-5}$ in |
| | 1 Ångstrom unit (Å) = $10^{-10}$ m = $10^{-4}$ μm |
| | one degree of latitude = 111.1 km |
| | = 69.1 mi |
| | = 60 nautical miles |
| | one nautical mile = 1.15 statute miles |
| | = 1.85 km |

| SI units | Conversions |
| --- | --- |
| | Mass |
| kilogram (kg)<br>gram (g)<br>metric tonne (t) | 1 kg = 2.20 lb; 1 lb = 0.454 kg<br>1 g = 0.0353 oz; 1 oz = 28.35 g<br>1 tonne = $10^3$ kg = 1.10 short ton<br>1 short ton = 0.907 tonne |
| | Temperature |
| kelvin (K) | 1 K = 1 °C = 1.8 °F;<br>1 °F = 5/9 °C = 5/9 K<br>$T$ °C = ($T$ + 273.16) K |
| | Area |
| square metre ($m^2$) | 1 $m^2$ = 10.76 sq ft<br>1 sq ft = 0.093 $m^2$ |
| square centimetre ($cm^2$) | 1 $cm^2$ = 0.155 sq in<br>1 sq in = 6.45 $cm^2$ |
| square kilometre ($km^2$) | 1 $km^2$ = 0.386 sq mi<br>1 sq mi = 2.59 $km^2$ |
| hectare (ha) | 1 ha = 10 000 $m^2$<br>= 2.47 acres<br>1 acre = 43 560 sq ft<br>= 0.4047 ha |
| | Volume |
| cubic metre ($m^3$) | 1 $m^3$ = 35.3 cu ft<br>1 cu ft = 0.028 $m^3$ |
| cubic centimetre ($cm^3$) | 1 $cm^3$ = 0.061 cu in<br>1 cu in = 16.39 $cm^3$<br>1 litre (1) = $10^{-3}$ $m^3$ = $10^3$ $cm^3$<br>= 0.264 gal (US)<br>= 0.220 gal (UK)<br>1 gal (US) = 231 cu in<br>= 0.003 785 3 $m^3$<br>1 gal (UK) = 277 cu in<br>= 0.004 546 0 $m^3$ |
| | Speed |
| metre per second (m $s^{-1}$) | 1 m $s^{-1}$ = 2.24 mi $h^{-1}$ = 1.94 kt<br>1 mi $h^{-1}$ = 0.447 m $s^{-1}$ |
| centimetre per second (cm $s^{-1}$) | 1 cm $s^{-1}$ = 0.0328 ft $s^{-1}$<br>= 1.97 ft $min^{-1}$<br>1 ft $s^{-1}$ = 30.48 cm $s^{-1}$<br>= 0.592 kt |
| knot (kt) | 1 knot = 1 nautical mile per hour<br>= 0.515 m $s^{-1}$<br>= 1.15 mi $h^{-1}$ |

| SI units | Conversions |
|---|---|
| | Work and energy |
| joule (J) | 1 J = $10^7$ ergs<br>1 calorie (cal) = 4.186 × $10^7$ ergs<br>= 4.186 J<br>1 ft lb = 1.36 J<br>1 J = 0.738 ft lb<br>1 BTU = 1055 J |
| | Power |
| watt (W) | 1 W = 1 J $s^{-1}$ = 1.34 × $10^{-3}$ hp<br>1 horsepower (hp) = 33 000 ft lb $min^{-1}$<br>= 746 W<br>1 ly $min^{-1}$ (= 1 cal $cm^{-2}$ $min^{-1}$)<br>= 6.98 × $10^2$ W $m^{-2}$ |
| | Pressure |
| pascal (Pa) | 1 Pa = 1 N $m^{-2}$ = 10 dynes $cm^{-2}$<br>= 2.45 × $10^{-4}$ lb $in^{-2}$<br>1 mbar = $10^2$ Pa = 1 hPa<br>= $10^3$ dynes $cm^{-2}$<br>= 0.750 mm Hg<br>= 2.95 × $10^{-2}$ in Hg<br>1 mm Hg = 133.3 Pa |
| one standard atmosphere (atm) | = 101 325 Pa = 1013.25 hPa = 1013.25 mbar<br>= 760 mm Hg = 29.92 in Hg = 14.7 lb $in^{-2}$ |

APPENDIX B

# Selected World Wide Web sites

This is only a small selection of the many sites available. They have been selected primarily as gateways to the vast range of information, images, data, comment and opinion available on the Web. Most are for organizations closely associated with national and international weather data collection, processing and distribution, or organizations associated with professional climatology. In most cases the home page address only is given, some have a more detailed pointer (*in italics*) to specific items of interest.

http://cdiac.esd.ornl.gov
Carbon Dioxide Information Analysis Center (USA)
Worldwide carbon dioxide, trace gas and global change related data (including climatology)

http://gcmd.gsfc.nasa.gov
The Global Change Master Directory (USA)
Access to a broad range of environmental data bases worldwide

http://itu.rdg.ac.uk/rms/rms.html
Royal Meteorological Society (U.K.)
Information about professional society activities

http://rredc.nrel.gov
Renewable Resources Data Center (USA)
Solar radiation and wind resources (mainly for USA)

http://www.ametsoc.org/AMS
American Meteorological Society (USA)
Information about professional society activities, including abstracts from journal papers and meeting presentations

http://www.athena.auth.gr:80/ozonemaps/
Aristotle University of Thessaloniki (Greece) daily Northern Hemisphere ozone maps

http://www.bom.gov.au/*climate*
Bureau of Meteorology (Australia)
climate maps, data and information for Australia

http://www.cdc.noaa.gov
National Oceanic and Atmospheric Administration (NOAA) – Climate Diagnostics Center (USA)
much interactive data available mainly concerning recent climate worldwide (including oceanic and upper air)

http://www.ciesin.org
Center for International Earth Science Information Networks (USA-based)
Data to help analyse human interaction with climate worldwide

http://www.cmc.ec.gc.ca
Canadian Meteorological Centre Climate and Water Information (Canada)
climate and environmental data for Canada

http://www.dir.ucar.edu/esig
National Center for Atmospheric Research, Environmental and Societal Impacts Group (USA)
data and monitoring for climate impacts worldwide

http://www.ecmwf.int/
European Centre for Medium-Range Weather Forecasts (EC)
(includes direct links to other European Meteorological Services)
some long-lead forecasts, commonly in chart form

http://www.epa.gov
Environmental Protection Agency (USA)
Much policy related material concerning air pollution, global warming and similar problems

http://www.erin.gov.au/
Environment Australia:
Discussion of environmental issues in Australia

http://www.ipcc.ch/
Intergovermental Panel on Climate Change
Contains details of current activities

http://www.meto.gov.uk/
Meteorological Office (UK)
Some climate information, mainly for United Kingdom in 'Research' section

http://www.ncdc.noaa.gov
National Climatic Data Center (USA):
Major interactive source of US (and much worldwide) climate data

http://www.nnic.noaa.gov/cpc/
NOAA Climate Prediction Center (USA)
long-lead predictions and recent climate statistics, mainly for USA

http://www.niwa.cri.nz
National Institute of Water and Atmospheric Research Ltd. (NZ)

Climate information for New Zealand and some nearby South Pacific areas

http://www.sawb.gov.za/
South African Weather Bureau (South Africa)
Climate statistics mainly for South Africa

http://www.wmo.ch/
World Meteorological Organization (United Nations)
Mainly policy oriented, but much information, including recent research results, in World Climate Research Programme section

# Glossary

Terms are included in this glossary if they appear in textual locations separate from their initial definition or if a complete definition *in situ* would detract from textual continuity.

***Absorption***: The process by which incident radiation is taken into a body and retained without reflection or transmission. Absorption increases either the internal or the kinetic energy of the molecules or atoms composing the absorbing medium.
***Absorption band***: A wavelength region where the absorption of radiation by a particular substance, usually a gas, is large.
***Adiabatic process***: A process in which a system does not interact thermally with its surroundings (i.e. no heat is exchanged).
***Advection***: The process by which the property of a mass of air is transferred by movement, usually in the horizontal direction.
***Advection fog***: Fog resulting from warm air advection, i.e. the movement of warm air horizontally over a cooler surface, in which the air mass is sufficiently cooled that condensation occurs.
***Aerosol***: A collection of small liquid or solid particles suspended in the atmosphere.
***Albedo***: The ratio of reflected solar radiation to the total incoming solar radiation where both streams are measured across the complete wavelength range of solar radiation (approximately 0.3–4.0 μm). At a single wavelength or for a narrow waveband the ratio is termed spectral reflectance.
***Anabatic wind***: A valley wind system developed in daytime in which airflow is directed upslope. The upper valley slopes are preferentially heated compared to the valley bottom, causing the warmer air to rise and an upslope flow to develop. Anabatic flow occurs only when regional pressure gradients are weak.
***Analogue model***: Method of predicting a future climate by considering a known historical situation which had similar features to those anticipated in the future.
***Angular momentum***: The angular momentum of a particle rotating about a fixed axis is the product of the particle's linear momentum (its mass times the linear velocity) and its perpendicular distance from the axis of rotation.
***Annulus experiment***: A simulation of some features of the Earth's atmospheric circulation undertaken by differentially heating a liquid contained in a hollow cylindrical dish which is rotated on a turntable.

***Atmospheric pressure***: The pressure created by the constant motion of atmospheric gas molecules. They exert a force whenever they impact upon a surface. The total force per unit area is the pressure.
***Backing***: The wind direction changing in a counter-clockwise sense.
***Baroclinic***: The atmosphere has isotherms, which are not parallel to the isobars, i.e. there is a temperature gradient along the isobars.
***Barotropic***: The atmosphere has horizontally uniform temperatures at all heights. Thus pressure gradients can exist but temperature gradients cannot.
***Bergeron–Findeisen process***: One possible initial stage of precipitation formation in which ice crystals grow in preference to water droplets in a mixed (i.e. water and ice) cloud.
***Black body***: A hypothetical body which is both a perfect absorber and a perfect emitter of radiation.
***Boundary layer***: The atmospheric boundary layer is usually considered as the lowest approximately 1 km of the atmosphere where motion is strongly influenced by surface characteristics, predominantly frictional drag.
***Cloud***: A visible suspension of water droplets or ice crystals, collected together in an identifiable unit and with its base above the ground.
***Cloud seeding***: A method by which the precipitation process is artificially induced in a cloud by, for example, the injection of artificial condensation nuclei.
***$CO_2$ fertilisation***: The theory that increased $CO_2$ in the atmosphere may lead to an increase in the efficiency of photosynthesis possibly producing lush plant growth where the water supply is adequate.
***Conditional instability***: Consider a 'parcel' of air rising in the atmosphere. The range of temperature profiles for which a wet parcel is unstable and a dry parcel stable defines the region of conditional instability.
***Conduction***: Collisions between fast moving molecules (high temperature regions) and slow molecules (low temperature regions) cause the slower moving molecules to speed up. This mode of heat transfer through matter is known as conduction.
***Conductive capacity***: A measure of the ability of a substance to conduct heat, given as the product of the heat capacity and the square root of the thermal diffusivity. The higher the conductive capacity, the better the conduction.
***Continentality***: A measure of the extent to which a location is outside the moderating influence of the oceans.
***Convection***: A type of heat transfer which occurs in a fluid by the vertical movement of large volumes of the heated material by differential heating (at the bottom for the atmosphere) thus creating, locally, a less dense, more buoyant fluid.
***Convective instability***: When warm, dry air overlies cooler, moist air, separated by an inversion layer, lifting the air *en masse* causes the lower air to reach saturation rapidly and become unstable, while the upper air remains stable for a longer period of lifting. Thus lifting creates instability although the original air mass was absolutely stable.
***Convergence***: If a constant volume of fluid has its horizontal dimensions decreased it experiences convergence and, by conservation of mass, its vertical dimension must increase.
***Coriolis force***: A force experienced by any object moving over the surface of a rotating body such as the Earth. It serves to modify the direction

of travel, causing a turning to the right (left) in the Northern (Southern) Hemisphere.
***Cryosphere***: The Earth's snow and (sea and land) ice masses.
***Detritus***: A collection of debris from the erosion of rocks, the remains of animals or of plants.
***Dew-point temperature***: The temperature at which an air parcel would become saturated if it were cooled without a change in pressure or moisture content.
***Divergence***: If a constant volume of fluid has its horizontal dimensions increased it experiences divergence and, by conservation of mass, its vertical dimension must decrease.
***Downdraught***: A downward movement of air in the lee of an obstruction. Also, in thunderstorms a downdraught occurs when cold air descends from a thundercloud, 'hits' the ground and spreads out as a 'wedge' ahead of the storm.
***Downscaling***: The development of climate data for a point or small area from regional conditions.
***Downwash***: A downward movement of air caused by the negative pressure region behind a narrow obstacle.
***Dust veil index***: A scale which ranks volcanic dust veils in terms of mass of ejected material, duration and maximum extent of spread of the veil. The Krakatoa eruption (1883) is ranked as 1000, while the eruption of Mount Agung (1963) is 800.
***Easterly wave***: Troughs of low pressure sloping away eastward with height, which move slowly westward. These occur in the region of the ITCZ.
***Eccentricity***: A measure of the deviation of an ellipse from a circle. It is the distance between the two foci divided by the length of the major axis. The eccentricity of the Earth's orbit is 0.018 at present.
***Ecliptic***: The great circle the Sun appears to describe on the celestial sphere. It is inclined at 23.5° to the equator at present.
***Eddy***: A disturbance in a flow of fluid, which looks like a 'whirlpool'. Transient eddies move in space (e.g. cyclones), while stationary eddies remain in fairly fixed locations (e.g. anticyclones).
***Ekman spiral***: The change in the horizontal direction of fluid flow with the vertical coordinate. For the case of winds, there is a slow clockwise turning on ascent (anticlockwise in the Southern Hemisphere).
***Electrical conductivity***: The ability of a substance to conduct electricity.
***El Niño***: The reversal in direction of the Walker circulation, associated with replacement of the cool upwelling Peruvian coastal current by warm equatorial water, which, in turn, leads to heavy rainfall in the normally arid desert. Fisheries off the Peruvian coast are adversely affected during these periods of suppression of cold coastal upwelling.
***Emissivity***: The degree to which a real body approaches a black body radiator.
***Emittance***: The rate at which radiation is emitted from unit area.
***Equivalent barotropic***: The atmosphere has temperature gradients such that the isotherms are parallel to the isobars.
***Evapotranspiration***: The combined process of evaporation and transpiration.
***Fingerprint***: A sensor sequence of valves, often for elements combined, which characterises the climate signal.
***Foraminifera***: A family of small unicellular marine animals which secrete calcareous shells or 'tests', which are frequently preserved as fossils.

***Front***: A region of steep horizontal temperature gradients along the plane where two air masses having different origins and characteristics meet.
***Gaia hypothesis***: The hypothesis that the Earth's physical and biological systems are considered to be a complex and self-equilibrating entity.
***Gaussian distribution***: This, also called the normal distribution, is characterised by a bell-shaped, symmetrical curve, having its mean, mode and median at the point of symmetry.
***GCM***: A General Circulation Model is one in which the three-dimensional general circulation of the Earth's atmosphere, and sometimes oceans, is modelled.
***Geopotential heights***: The acceleration due to gravity, $g$, is approximately constant over the surface of the Earth since its magnitude is determined by the distance from the centre of the Earth. However, higher in the atmosphere this means that the value of $g$ decreases. Hence the use of formulae, such as the hydrostatic equation, at different heights in the atmosphere introduces a new variable: $g'(z)$. To obviate this variability with height a new height scale is introduced (the geopotential height), which is defined as the work done when lifting a body of unit mass against gravity (i.e. acceleration due to gravity multiplied by distance) divided by the value of $g$ at the Earth's surface.
***Geostrophic wind***: In the free atmosphere, when only the pressure gradient and Coriolis forces act on a parcel of air, they rapidly come into equilibrium to give the balanced or geostrophic wind, which blows parallel to the isobars.
***Greenhouse effect***: The effect whereby the Earth is warmed more than expected due to the atmospheric gases being transparent to incoming solar radiation, but opaque to outgoing terrestrial infrared radiation. The infrared radiation emitted from the surface is absorbed and re-emitted, some downwards, warming the Earth.
***Hadley cell***: A direct, thermally driven circulation which comprises an upward motion of air at the ITCZ and downward motion in the subtropics, poleward movement of air at high levels and an equatorward movement at low levels. The ascending motion is the combined result of convergence, of radiative imbalances and of latent energy released during cloud condensation.
***Harmonic analysis***: Periodic features are represented by summation of simple sine and cosine waves.
***Heat capacity***: The heat required to change the temperature of unit volume of a body by 1 K.
***Heating (cooling) degree days***: Defined as the number of degrees by which the average daily temperature falls below (exceeds) a threshold or base temperature. The number of heating (cooling) degree days in a season is the summation of the heating (cooling) degree days for all days.
***Hurricane***: Storms with a mean surface wind speed exceeding 34 m $s^{-1}$, in the shape of an intense circular vortex. Hurricane is the regional name in the Caribbean and Gulf of Mexico, and typhoon in the western north Pacific and tropical cyclone in the Indian Ocean.
***Hydrostatic stability***: The atmosphere is hydrostatically stable if vertical accelerations are negligible and hence there exists a hydrostatic balance between the vertical pressure force and the gravitational force.
***Hygrometer***: Instrument for measuring humidity.

***Hythergraphs***: Plots of precipitation and temperature or humidity and temperature, usually by month.
***Ice age***: A period of time when the Earth's cryosphere was approximately double its present area.
***Index cycle***: The term used to describe alternation between periods of zonal and meridional flow. The 'index' is the pressure difference between two latitude zones.
***Interglacial***: The period between two ice ages.
***Intertropical convergence zone***: The ITCZ is a region close to the equator where the trade winds converge. Ascent of air causes low atmospheric pressure, deep convective clouds and heavy precipitation. The ITCZ changes position during the year following the seasonal insolation cycle.
***Inversion***: The situation when the environmental temperature increases with height. It is a highly stable condition in which convection and vertical dispersion cannot occur and pollution is trapped, often near the surface.
***Irradiance***: The rate at which radiation is incident upon a unit area.
***Isotopic fractionation***: The separation of isotopes. Isotopes are atoms with the same atomic number (i.e. the same element) but a different number of neutrons (i.e. having a different weight from other isotopes of that element).
***Jet stream***: A 'ribbon-like' belt of rapidly moving air found at or just below the tropopause. In mid-latitudes it is the 'core' of the Rossby or planetary waves.
***Katabatic wind***: A valley wind system developed on clear nights when regional pressure gradients are weak. Long-wave radiant cooling of the upper valley slopes leads to downslope flow.
***Kinetic energy***: The energy of movement of a body. It is given by half the product of the mass and the square of the linear velocity of the body.
***Lapse rate***: The rate of decrease of temperature with height at a given time and place, i.e. a negative temperature gradient is a positive lapse rate.
***Lysimeter***: An instrument for measuring the amount of water lost by evapotranspiration.
***Milankovitch periodicities***: Periodic changes in the Earth's orbital parameters, which are believed to control to some degree the onset of ice ages.
***Mixing ratio***: More strictly the 'water vapour mixing ratio', which is the ratio of the mass of water vapour to the mass of dry air occupying the same volume.
***Momentum***: The product of mass and velocity of a body.
***Monsoon***: A seasonal reversal of wind, which in the summer season blows onshore, bringing with it heavy rains, and in the winter season blows offshore. The name is derived from the Arabic word 'mausin', meaning a season.
***Nephanalysis***: A cartographical representation of cloud amount and, often, cloud type. From the Greek 'nephos', meaning cloud.
***Noise***: Random fluctuations in a parameter caused by effects other than the one being studied.
***Oasis effect***: Hot dry air in equilibrium with the desert, flowing across an oasis edge, experiences rapid evaporation using sensible heat from the air as well as radiant energy. The air is cooled by this process until it reaches equilibrium with the new surface.

***Occlusion***: As a depression develops, the trailing cold front moves more rapidly than the leading warm front. The warm sector between them is narrowed until the cold front catches up with the warm front when occlusion takes place and the warm sector is effectively lifted off the ground.
***Optical thickness***: A measure of the attenuation of the solar radiation by the atmosphere due to scattering and absorption processes. The greater the optical thickness the greater the attenuation.
***Orographic***: Pertaining to the relief of mountains and hills.
***Photochemistry***: The study of chemical reactions which take place when substances are exposed to electromagnetic energy, especially ultraviolet and short wavelength visible radiation.
***Photosynthesis***: The process by which plants use light, absorbed through chlorophyll, to produce organic compounds from carbon dioxide and water, leaving oxygen as a by-product.
***Planck curve***: The curve describing the amount of energy being radiated as a function of wavelength by a black body at a fixed temperature.
***Potential energy***: The energy a body has by virtue of the work done against a restoring force in attaining its position, i.e. being raised above the ground against the force of gravity. It is the product of the mass of the body, its height above zero (ground level) and the acceleration due to gravity.
***Potential temperature***: The temperature an air parcel would have if brought adiabatically to a standard pressure of $10^3$ hPa.
***Psychrometer***: Dry and wet bulb thermometers in a frame or sling, which are ventilated so that a steady stream of air passes over the bulbs. The humidity parameters can be calculated from the two temperature readings.
***Radiation fog***: When net radiation is negative the air in contact with the ground is cooled. If there is sufficient moisture in the air or the cooling is sufficient, condensation will occur.
***Radiosonde***: A package, suspended below a balloon, consisting of instruments to sense and relay temperature, humidity and pressure as it ascends.
***Rain day***: A 24-hour period, usually starting at 0900Z during which 0.2 mm or more of precipitation falls.
***Rainout***: Pollution particles which act as condensation nuclei in the precipitation formation process are said to reach the ground by rainout.
***Rawinsonde***: A more sophisticated version of the radiosonde, which also measures wind speed and direction.
***Regression***: Statistical derivation of a relationship between a dependent variable and one or more independent variables.
***Return period***: The probable time period between the repetition of two extreme events.
***Rossby wave***: When the temperature gradient across a rotating fluid (such as in an annulus experiment) reaches a critical value the previously symmetric flow breaks down into a wave-like flow; the waves being termed Rossby waves after the Swedish meteorologist who first recognised the importance of transient mid-latitude disturbances to the general circulation of the atmosphere.
***Sahel***: A region in Africa south of the Sahara Desert with a marginal climate for agriculture.
***Satellite (geostationary)***: A satellite whose high altitude orbit (approximately 35 000 km) is in the equatorial plane and orbital velocity matches

that of the Earth so that its position remains constant with respect to the Earth.

***Satellite (polar orbiting)***: A satellite whose orbit is approximately Sun synchronous, low altitude (approximately 1000 km) and intersects the equator at approximately $\pi/2$ thus passing close to the poles on each orbit.

***Scanning radiometer***: An instrument, generally on board a satellite or aircraft, which scans along a line perpendicular to the flight path detecting the radiant energy.

***Scattering***: The process by which some of a stream of radiation is dispersed to travel in all directions by particles suspended in the medium through which the radiation passes. Usually refers to solar photons, e.g. Rayleigh or Mie scattering.

***Shear instability***: In a stably stratified atmosphere a wave-like motion can develop when the vertical wind shear exceeds a critical value defining the onset of shear instability.

***Shelter belt***: An increase in surface roughness, such as that caused by a belt of trees, causes a reduction in the wind speed and thus provides protection to areas downwind.

***Solar constant***: The amount of energy passing in unit time through a unit surface perpendicular to the Sun's rays at the outer edge of the atmosphere at the mean distance between the Earth and the Sun. Currently believed to be 1370 W $m^{-2}$.

***Solar radiation***: The radiation the Earth receives from the Sun at wavelengths between 0.3 and 4.0 μm.

***Southern Oscillation***: A fluctuation in the intertropical atmospheric and hydrodynamical circulations which manifests itself as a quasi-periodic (2–4 year) variation in sea-level pressure, surface wind, sea-surface temperature and rainfall over a wide area of the Pacific Ocean. It is dominated by an exchange of air between the southeast Pacific subtropical high and the Indonesian equatorial low.

***Spectral analysis***: A type of Fourier (harmonic) analysis which identifies cycles in atmospheric features.

***Squall line***: A series of thundercells (cumulonimbus towers) aligned at right angles to the direction of motion. A leading cell is often 'fed' by adjacent cells so that the storm can persist for longer than individual cells.

***Standard deviation***: A measure of the scatter of observations (points) about the mean.

***Synoptic scale***: Literally 'at the same time'; usually pertaining to waves or eddies with horizontal scales of hundreds to thousands of kilometres. Often used to describe features (such as depression systems and anticyclones) which control the day-to-day variations in the weather.

***Teleconnection:*** A statistical connection between climate fluctuations in two or more physically separate regions. Physical and meteorological linking mechanisms have not yet been fully established.

***Tephigram (T$\phi$-gram)***: A form of thermodynamic diagram in which the axes are logarithms of potential temperature (or entropy) and temperature, such that the dry adiabats (which are lines of constant potential temperature) are straight. The graph is usually rotated through 45° so that lines of constant pressure are approximately horizontal.

***Terrestrial radiation***: The radiation emitted by the Earth, which is also known as long-wave radiation.

***Thermal conductivity***: The rate at which a substance can conduct heat.

***Thermal diffusivity***: A quantity defining the ease with which a substance propagates temperature differences, being the thermal conductivity divided by the product of the specific heat capacity and the density of the substance.
***Thermal relaxation (or response) time***: A measure of the time taken by a system to achieve a new equilibrium temperature following an imposed perturbation.
***Thermal wind***: A thermally induced gradient in wind speed.
***Thermodynamic diagram***: A diagram in which the thermodynamic variables may be plotted as functions of two other variables; for example, pressure as a function of temperature and potential temperature.
***Thermodynamics***: The science of the relationships between different forms of energy; particularly the ways in which energy can be converted from one form to another and work be done.
***Time (of observations)***: By international agreement many meteorological and climatological observations are made at fixed times in a 24-hour system identical to Greenwich Mean Time and denoted by the hour number followed by an upper case Z (e.g. 0900Z)
***Trade winds***: The quasi-geostrophic northeasterly (southeasterly) flow towards the ITCZ in the Northern (Sourthern) Hemisphere, which is the surface 'return flow' of the Hadley cell circulations.
***Transmittance***: The ability of a substance to allow radiant energy to pass through it.
***Transpiration***: The removal of water from the interior of a plant through pores (stomata) located predominantly in the leaves.
***Urban heat island***: The effect of an urban area on the regional temperatures, such that the city is a few degrees warmer than its surroundings.
***Vapour pressure***: The force per unit area created by the motions of the vapour molecules treated in isolation from all other gases in the atmosphere.
***Veering***: The wind direction changing in a clockwise sense.
***Venturi effect***: The increased velocity of a fluid flowing through a constriction, such as the funnelling of air between two converging buildings. A consequence of the requirement to conserve mass.
***Vorticity***: Twice the angular velocity of a fluid particle about a local axis through the particle. It is thus a measure of rotation of an air mass.
***Walker circulation***: A thermally driven longitudinal cellular circulation extending across the Pacific Ocean from Indonesia to close to the Peruvian coast and forming a component of the Southern Oscillation.
***Washout***: Pollution particles removed from the atmosphere by being incorporated into falling raindrops by collision and coalescence.
***Whiteout***: The multiple scattering of sunlight between the base of a low cloud layer and a snow-covered surface, making it difficult to see surface features and to judge where the horizon is located.
***Wind chill***: The cooling of a body caused by the wind removing sensible and latent heat. Thus the 'perceived' temperature will be lower than the actual thermometer reading.
***Wind shear***: A condition where the wind speed (and usually direction) changes with height.
***Zenith angle***: The angular distance between the Sun's position in the sky and the local vertical (or zenith).

# Bibliography


Alcamo J (ed.) 1994 *Image 2.0: integrated modelling of global climate change*. Kluwer Academic, Dordrecht, 328pp

Anisimov, O A and F E Nelson 1997 Permafrost zonation and climate change in the Northern Hemisphere: results from transient general circulation models. *Climate Change*, **35**: 241–58

Barrett E C and D W Martin 1981 *The use of satellite data in rainfall monitoring*. Academic Press, London, 340pp

Berger A, C Tricot, H Gallee and M F Loutre 1993 Water vapour, $CO_2$ and insolation over the last glacial interglacial cycles. *Phil. Trans. Roy. Soc. Ser. B*, **341**: 253–61

Bossert J E 1997 An investigation of flow regimes affecting the Mexico City region. *J. Appl. Meteorol.*, **36**: 119–40

Bradley R S (ed.) 1991 *Global changes of the past*. University Corporation for Atmospheric Research, Office of Interdisciplinary Earth Studies, Boulder, CO, 514pp

Briffa K R, P D Jones and M Hulme 1994 Summer moisture variability across Europe based on the Palmer drought severity index. *Int. J. Climatol.*, **14**: 475–506

Broecker W S and J van Donk 1970 Insolation changes, ice volumes, and the $^{18}O$ record in deep-sea cores. *Rev. Geophys.*, **8**: 169–98

Brutsaert W 1982 *Evaporation into the atmosphere: theory, history and application*. D. Reidel, Dordrecht, 299pp

Bryan K 1969 A numerical method for the study of the world ocean. *J. Comput. Phys.*, **4**: 347–76

Bryson R A 1978 Cultural, economic and climatic records. In A B Pittock, L A Frakes, D Jenssen, J A Peterson and J W Zillman (eds) *Climatic change and variability: a southern perspective*. Cambridge University Press, Cambridge, 316–27

Bryson R A 1997 The paradigm of climatology: an essay. *Bull. Amer. Meteorol. Soc.*, **78**: 449–55

Budyko M I 1974 *Climate and life*. (English translation by D H Miller) Academic Press, New York, 508pp

Cess R D 1985 Nuclear war: illustrative effects of atmospheric smoke and dust upon solar radiation. *Clim. Change*, **7**: 237–51

Changnon D and S A Changnon 1998 Climatological relevance of major USA weather losses during 1991–1994. *Int. J. Climatol.*, **18**: 37–48

Clay J W, D M Orr and A W Stewart (eds) 1975 *North Carolina atlas*. University of North Carolina Press, Chapel Hill

Cohen, S J and T R Allsopp 1988 The potential impacts of a scenario of $CO_2$ induced climatic change on Ontario, Canada. *Journal of Climate*, **1**: 669–81

Crane R G and B C Hewitson 1998 Doubled $CO_2$ precipitation changes for the Susquehanna Basin: down-scaling from the GENESIS general circulation model. *Int. J. Climatol.*, **18**: 65–78

Critchfield H J 1983 *General climatology* (4th edition). Prentice Hall, Englewood Cliffs, 453pp

Daly C, R P Neilson and D L Phillips 1994 A statistical–topographic model for mapping climatological precipitation over mountainous terrain. *J. Appl. Meteorol.*, **33**: 140–58

Davies J A 1965 Estimation of insolation for West Africa. *Q. J. R. Meteorol. Soc.*, **91**: 359–63

De Haan B J, M Jonas, O Klepper, J Krabec, M S Krol and K Olendrzynski 1994 An atmosphere–ocean model for integrated assessment of global change. *Water, Air and Soil Pollution*, **76**: 283–318

Dettwiller J 1970 Deep soil temperature trends and urban effects at Paris. *J. Appl. Meteorol.*, **9**: 178–80

DeWispelare A R, L T Herren, M P Miklas and R T Clemen 1993 *Expert elicitation of future climate in the Yucca Mountain vicinity*. Center for Nuclear Waste Regulatory Analyses, San Antonio

Diaz H F and V Markgraf (eds) 1992 *El Niño: historical and paleoclimatic aspects of the Southern Oscillation*. Cambridge University Press, Cambridge, 480pp

Dickinson R E, V Meleshko, D Randall, E Sarachik, P Silva-Diaz and A Slingo 1996 Climate processes. In J T Houghton, L G Meira Filho, B A Callander, N Harris, A Kattenberg and K Maskell (eds) *Climate change 1995: the science of climate change*. Cambridge University Press, Cambridge, 193–227

Douglass J E and W T Swank 1975 *Effects of management practices on water quality and quantity*. Coweeta Hydrologic Laboratory, North Carolina. USDA, 13pp

Dowladabati H 1995 Integrated assessment models of climate change: an incomplete overview. *Energy Policy*, **23**(4): 1–8

Easterling, W E and R W Kates 1996 Indexes of leading climate indicators for impact assessment. In T R Karl (ed.) *Long-term climate monitoring by the global climate observing system*. Kluwer Academic Publishers, 623–48

Flint R F 1971 *Glacial and Quarternary geology*. John Wiley, New York, 892pp

Gates D M 1965 Heat, radiant and sensible. *Meteorological monographs* **6**(28)

Gates W L 1992 AMIP: the atmospheric model intercomparison project. *Bull. Amer. Meteor. Soc.*, **73**: 1962–70

Gates W L, A Henderson-Sellers, G J Boer, C K Folland, A Kitoh, B J McAvaney, F Semazzi, N Smith, A J Weaver and Q-C Zeng 1996 Climate models – evaluation. In J T Houghton, L G Meira Filho, B A Callander, N Harris, A Kattenberg and K Maskell (eds) *Climate change 1995: the science of climate change*. Cambridge University Press, Cambridge, 229–84

Geiger R 1965 *The climate near the ground* (translation of the 4th German edition). Harvard University Press, Cambridge, 611pp

Giorgi F, C S Brodeur and G T Bates 1994 Regional climate change scenarios over the United States produced with a nested regional climate model. *J. Climate*, **7**: 375–99

Glantz, M, R Katz and M Krenz (eds) 1987 *The societal impacts associated with the 1982–83 worldwide climate anomalies*. National Center for Atmospheric Research, Boulder, CO 105pp

Glantz M H, R W Katz and N Nicholls (eds) 1991 *Teleconnections linking worldwide climate anomalies*. Cambridge University Press, Cambridge, 535pp

Gleick J 1987 *Chaos: the making of a new science*. Viking Penguin, New York, 352pp

Goodman S J and H J Christian 1993 Global observations of lightning. In R J Gurney, J L Foster and C L Parkinson (eds) *Atlas of satellite observations related to global change*. Cambridge University Press, Cambridge, 191–222

Goody R M and J C G Walker 1972 *Atmospheres*. Prentice Hall, Englewood Cliffs, 150pp

Hammer C U, H B Clausen and C C Langway 1997 50,000 years of recorded global volcanism. *Climatic Change*, **35**: 1–15

Hansen J 1980 Climate impact of increasing atmospheric carbon dioxide. *Annals of the New York Academy of Sciences*, **338**: 575–86

Hansen J, D Johnson, A Lacis, S Lebedeff, P Lee, D Rind and G Russell 1981 Climate impact of increasing atmospheric carbon dioxide. *Science*, **213**: 957–66

Hansen J, G Russell, D Rind, P Stone, A Lacis, S Lebedeff, R Ruedy and L Travis 1983 Efficient three dimensional global models for climate studies: Models I and II. *Mon. Wea. Rev.*, **111**: 609–62

Hansen J E, A Lacis, R Ruedy, M Sato 1992 Potential climatic impact of Mount Pinatubo eruption. *Geophys. Res. Lett.*, **19**: 215–18

Hare F K 1985 *Climate variations, drought and desertification*. Report WMO-653, World Meteorological Organisation, Geneva, 35pp

Harrison E F, P Minnis, B R Barkstrom and G G Gibson 1993 Radiation budget at the top of the atmosphere. In R J Gurney, J L Foster and C L Parkinson (eds) *Atlas of satellite observations related to global change*. Cambridge University Press, Cambridge, 19–40

Henderson K G and P J Robinson 1994 Relationships between the Pacific/North American teleconnection patterns and precipitation events in the south-eastern USA. *Int. J. Climatol.*, **14**: 307–23

Henderson-Sellers A (ed.) 1995 *Future climates of the world*. World Survey of Climatology, Vol. 16, Elsevier, New York, 568pp

Hesketh J D and D N Moss 1963 Variations in the response of photosynthesis to light. *Crop Science*, **3**: 107–10

Hess W N (ed.) 1974 *Weather and climate modification*. John Wiley, New York, 842pp

Ho C-R, X-H Yan and Q Zheng 1995 Satellite observations of upper-layer variabilities in the western Pacific warm pool. *Bull. Am. Meteorol. Soc.*, **76**: 669–79

Hobbs J E 1980 *Applied climatology*. Wm Dawson, Folkestone, 218pp

Houghton J T, B A Callendar and S K Varney (eds) 1992 *Climate change 1992: the supplementary report to the IPCC scientific assessment*. Cambridge University Press, Cambridge, 200pp

Houghton J T, L G Meira Filho, B A Callander, N Harris, A Kattenberg and K Maskell (eds) 1996 *Climate change 1995: The science of climate change*. Cambridge University Press, Cambridge, 572pp

Hughes M 1991 The tree-ring record. In R S Bradley (ed.) *Global changes of the past*. University Corporation for Atmospheric Research, Office of Interdisciplinary Earth Studies, Boulder, CO 117–37

Imbrie J and K P Imbrie 1979 *Ice ages: solving the mystery*. Macmillan, London, 224pp

Jaeger, L 1976 *Monatskarten des Niederschlags für die ganze Erde*, Ber. Deutschen Wetterdienstes, Nr. 139, 38pp

Jenne, R L 1975 Data sets for meteorological research. *NCAR Technical Note, NC-TN/1A-111*, NCAR, Boulder, CO, 194pp

Kacholia K and R A Reck 1997 Comparison of global climate change simulations for 2 × $CO_2$ – induced warming. *Climatic Change*, **35**: 53–69

Karl T R, R W Knight, D R Easterling and R G Quayle 1996 Indices of climate change for the United States. *Bull. Am. Meteorol. Soc.*, **77**: 279–92

Kattenberg A, F Giorgi, H Grassl, J F B Mitchell, R J Stouffer, T Tokioka, A J Weaver and T M L Wigley 1996 Climate models – projections of future climate. In J T Houghton, L G Meira Filho, B A Callander, N Harris, A Kattenberg and K Maskell (eds) *Climate change 1995: the science of climate change*. Cambridge University Press, Cambridge, 285–357

Kellogg W W 1978 Global influences of mankind on the climate. In J Gribbon (ed.) *Climatic change*. Cambridge University Press, Cambridge, 205–27

Kopec R J 1970 Further observations of the urban heat island in a small city. *Bull. Am. Meteorol. Soc.*, **51**: 602–6

Kraus H and A Alkhalaf 1995 Characteristic surface energy balances for different climate types. *Int. J. Climatol.*, **15**: 275–84

Kripalani R H, S V Singh and N Panchawagh 1995 Variability of the summer monsoon rainfall over Thailand – comparison with features over India. *Int. J. Climatol.*, **15**: 657–72

Kunkel K E, S A Changnon, B C Reinke and R W Arritt 1996 The July 1995 heat wave in the Midwest: a climate perspective and critical weather factors. *Bull. Am. Meteorol. Soc.*, **77**: 1507–18

Lamb P J and S A Changnon 1981 On the best temperature and precipitation normals: the Illinois situation. *J. Appl. Meteorol.*, **20**: 1383–90

Laws E A 1993 *Aquatic pollution*. John Wiley, New York, 611pp

Lockwood J G 1979 *Causes of climate*. Edward Arnold, London, 260pp

Lovelock J E 1991 *Gaia: the practical science of planetary medicine*. GAIA Books, London

MacKay R M and M A K Khalil 1991 Theory and development of a one dimensional time dependent radiative convective model. *Chemosphere*, **22**: 383–417

MacKenzie F T and J A MacKenzie 1995 *Our changing planet*. Prentice Hall, Upper Saddle River, NJ, 387pp

Manabe S and A J Broccoli 1990 Mountains and arid climates of middle latitudes. *Science*, **247**: 192–5

Manabe S and R F Strickler 1964 Thermal equilibrium of the atmosphere with a convective adjustment. *J. Atmos. Sci.*, **21**: 361–85

Mather J R and G A Yoshioka 1968 The role of climate in the distribution of vegetation. *Annals, Assoc. Am. Geogr.*, **58**: 29–41

McIntyre T C and CLIMAP 1976 The surface of the ice age Earth. *Science*, **191**: 1132–4

Meaden G T 1976 Late summer weather in Kent, 55 BC. *Weather*, **31**: 264–70

Meehl G A 1990 Development of global coupled ocean–atmosphere general circulation models. *Clim. Dyn.*, **5**: 19–33

Meteorological Office (UK) 1991 *Meteorological glossary*, 6th edition. HMSO, London

Monteith J L 1981 Climate variation and the growth of crops. *Q. J. R. Meteorol. Soc.*, **107**: 749–74

Monteith J L and M H Unsworth 1990 *Principles of environmental physics*. Edward Arnold, London, 291pp

Moron V 1995 Variability of the African convection centre as viewed by outgoing long-wave radiation records and relationships with sea-surface temperature patterns. *Int. J. Climatol.*, **15**: 25–34

Munn R E 1966 *Descriptive micrometeorology*. Academic Press, New York, 245pp

Neiburger M, J G Edinger and W D Bonner 1982 *Understanding our atmosphere*. W H Freeman, San Francisco, 293pp

Nicholls N, G V Gruza, J Jouzel, T R Kar, L A Ogallo and D E Parker 1996 Observed climate variability and change. In J T Houghton, L G Meira Filho and B A Callander (eds) *Climate change 1995: the science of climate change*. Cambridge University Press, pp. 133–92

Nicholson S E, M B Ba and J Y Kim 1996 Rainfall in the Sahel during 1994. *J. Climate*, **9**: 1673–6

Nordhaus W D 1991 To slow or not to slow: the economics of the greenhouse effect. *The Economic Journal*, **101**: 920–37

Oort A H 1983 Global atmospheric circulation statistics, 1958–1973. *NOAA Professional Paper 14*, National Oceanic and Atmospheric Administration, Washington, DC, 123pp

Oort A H and Piexoto J P 1983 *Theory of climate*. Academic Press, New York

Oort A H and T H Vonder Haar 1976 On the observed annual cycle in the ocean–atmosphere heat balance over the Northern Hemisphere. *J. Phys. Oceanogr.*, **6**: 721–800

Palutikof J, T Holt and A Skellern 1997 Wind: resource and hazard. In M Hulme and E Barrow (eds) *Climates of the British Isles*. Routledge, London, 220–42

Parry M L 1978 *Climatic change, agriculture and settlements*. Wm Dawson, Folkstone, 214pp

Parthasarathy B, K Rupa Kumar and V R Deshparde 1991 Indian summer monsoon rainfall and 200-mbar meridional wind index: application for long-range prediction. *Int. J. Climatol.*, **11**: 165–76

Peixoto J P and A H Oort 1992 *The physics of climate*. American Institute of Physics, New York

Penman H L 1948 Natural evaporation from open water, bare soil and grass. *Proc. Roy. Soc. (London), Part A*, **193**: 120–45

Penner, J E, R J Charlson, J M Hales, N S Laulainen, R Leifer, T Novakov, J Ogren, L F Radke, S E Schwartz and L Travis 1994 Quantifying and minimizing uncertainty of climate forcing by anthropogenic aerosols. *Bull. Am. Meteorol. Soc.*, **75**: 375–400

Peterson T C and R S Vose 1997 An overview of the Global Historical Climatology Network temperature database. *Bull. Am. Meteorol. Soc.*, **78**: 2837–49

Peterson T C and T W R Wallis 1997 Running in the rain. *Weather*, **52**: 93–6

Petterssen S 1969 *Introduction to meteorology* (3rd edition). McGraw-Hill, New York, 333pp

Powell M D 1993 Wind forecasting for yacht racing in the 1991 Pan American mes. *Bulletin of the American Meteorological Society*, **74**: 5–16

Reiter E R 1996 Jet streams. In S H Schneider (ed.) *Encyclopedia of climate and weather*. Oxford University Press, Oxford, 455–9

Riehl H 1965 *Introduction to the atmosphere*. McGraw-Hill, New York, 410pp

Robinson P J 1989 The influence of weather events on aircraft operations at the Atlanta Hartsfield International Airport. *Weather and Forecasting*, **4**: 461–8

Robinson P J 1996 A view of western North Carolina's climate. *The North Carolina Geographer*, **5**: 11–20

Robinson P J 1997 Modeling utility load and temperature relationships for use with long-lead forecasts. *J. Appl. Meteorol.*, **36**(5): 591–8

Robinson P J and W E Easterling 1982 Solar energy climatology of North Carolina. *J. Appl. Meteorol.*, **21**: 1730–8

Ropelewski C F and M S Halpert 1996 Quantifying Southern Oscillation–precipitation relationships. *J. Climate*, **9**: 1043–59

Rose C W 1966 *Agricultural physics*. Pergamon Press, Oxford, 326pp

Rossow W B 1993 Clouds. In R J Gurney, J L Foster and C L Parkinson (eds) *Atlas of satellite observations related to global change*. Cambridge University Press, Cambridge, 141–64

Roy M G, M N Hough and J R Starr 1978 Some agricultural effects of the drought of 1975–76 in the United Kingdom. *Weather*, **33**: 64–74

Santer B D, T M L Wigley, T P Barnett and E Anyamba 1996 Detection of climate change and attribution of causes. In J T Houghton, L G Meira Filho, B A Callander, N Harris, A Kattenberg and K Maskell (eds) *Climate change 1995: the science of climate change*. Cambridge University Press, Cambridge, 407–43

Schemenauer R S, H Fuenzalida and P Cereceda 1988 A neglected water resource: the Camanchaca of South America. *Bull. Am. Meteorol. Soc.*, **69**: 138–47

Sellers W D 1965 *Physical climatology*. University of Chicago Press, Chicago, 272pp

Semtner A J 1995 Modelling ocean circulation. *Science*, **269**: 1379–85

Sharon D and A Arazi 1997 The distribution of wind drived rainfall in a small valley: an empirical basis for numerical model verification. *J. Hydrol.*, **201**: 21–48

Shi N and Q Zhu 1996 An abrupt change in the intensity of the East Asian Summer Monsoon Index and its relationship with temperature and precipitation over east China. *Int. J. Climatol.*, **16**: 757–64

Shuttleworth W J 1988 Macrohydrology: the new challenge for process hydrology. *J. Hydrol.*, **100**: 31–56

Skiles J W 1995 Modelling climate change in the absence of climate change data. *Climatic Change*, **30**: 1–6

Smagorinsky J 1963 General circulation experiments with the primitive equations. I. The basic experiment. *Mon. Wea. Rev.*, **91**: 99–164

Strahler A N 1965 *Introduction to physical geography*. John Wiley, New York, 455pp

Suskind J 1993 Water vapor and temperature. In R J Gurney, J L Foster and C L Parkinson (eds) *Atlas of satellite observations related to global change*. Cambridge University Press, Cambridge, 89–128

Tanner G 1971 *A collection of selected climographs*. Eau Claire Cartographic Institute, Eau Claire, 45pp

Thornthwaite W C and F K Hare 1965 The loss of water to the air. *Meteorological Monographs*, **6**: 163–80

Thurow C 1983 *Improving street climate through urban design*. Planning Advisory Service #376, American Planning Association, Chicago, 34pp

Tinker R J (ed.) 1997 *The climate bulletin*. NOAA, 28 February, 46pp

Todhunter P E 1989 An approach to the variability of urban surface energy budgets under stratified synoptic weather types. *Int. J. Climatol.*, **9**: 191–201

Troen I and E L Petersen 1989 *European Wind Atlas* Roskilde, Riso National Laboratory, 656pp

Tucker C J, H E Dregne and W Newcomb 1991 Expansion and contraction of the Sahara Desert from 1980 to 1990. *Science*, **253**: 299–301

University Corporation for Atmospheric Research 1994 *Reports to the nation: El Niño and climate prediction*. University Corporation for Atmospheric Research, Boulder, CO, 25pp

Wallace J M and P V Hobbs 1977 *Atmospheric science*. Academic Press, New York, 467pp

Warrick R and M Bowden 1981 *Changing impacts of drought in the Great Plains*. Center for Great Plains Studies, Lincoln

Watson R T, M C Zinyowera and R H Moss 1996 *Climate change 1995. Impacts adaptation and mitigation of climate change: scientific–technical analysis*. Contributions of Working Group II to the Second Assessment Report of the Intergovernmental Panel on Climate Change, Cambridge University Press, Cambridge, 880pp

Webb T I, P J Bartlein and J E Kutzbach 1987 Climatic change in eastern North America during the past 18,000 years; comparisons of pollen data with model results. In W F Ruddiman and H E Wright (eds) *North America and Adjacent Oceans during the Last Glaciation*. Geological Society of America, Boulder, CO, 447–62

White O R (ed.) 1977 *Solar output and its variations*. Colorado Associated University Press, Boulder, CO, 526pp

Wigley T M L 1991 Climate variability on the 10–100 year time scale: observations and possible causes. In R S Bradley (ed.) *Global changes of the past*. University Corporation for Atmospheric Research, Office of Interdisciplinary Earth Studies, Boulder, CO, 83–101

Wigley T M L, P D Jones and M Kelly 1980 Scenario for a warm, high-$CO_2$ world. *Nature*, **283**: 17–21

Wigley T M L and S Raper 1992 Implications for climate and sea level of revised IPCC emissions scenarios. *Nature*, **357**: 293–300

Willett H C and F Sanders 1959 *Descriptive meteorology*. Acdemic Press, New York, 355pp

Wilson M F, A Henderson-Sellers, R E Dickinson and P J Kennedy 1987 Investigation of the sensitivity of the land-surface parameterization of the NCAR community climate model in regions of tundra vegetation. *J. Climate*, **7**: 319–44

World Meteorological Organization 1953 *World distribution of thunderstorm days*. World Meteorological Organization, Geneva, 56pp

World Meteorological Organization 1975 The physical basis of climate and climate modelling. *Global Atmospheric Research Programme (GARP) Publication Series*, **16**, World Meteorological Organization, Geneva, 236pp

Xue Y and J Shukla 1996 The influence of land surface properties on Sahel climate. Part II: Afforestation. *J. Climate*, **9**: 3260–75

Zeng N, R E Dickinson and X Zeng 1996 Climatic impact of Amazon deforestation – a mechanistic model study. *J. Climate*, **9**: 859–83

# Index